Women and smallholder farming

Addressing global inequities in agriculture

It is widely recognised that agriculture is a significant contributor to global warming and climate change. Agriculture needs to reduce its environmental impact and adapt to current climate change whilst still feeding a growing population, i.e. become more 'climate-smart'. Burleigh Dodds Science Publishing is playing its part in achieving this by bringing together key research on making the production of the world's most important crops and livestock products more sustainable.

Based on extensive research, our publications specifically target the challenge of climate-smart agriculture. In this way we are using 'smart publishing' to help achieve climate-smart agriculture.

Burleigh Dodds Science Publishing is an independent and innovative publisher delivering high quality customer-focused agricultural science content in both print and online formats for the academic and research communities. Our aim is to build a foundation of knowledge on which researchers can build to meet the challenge of climate-smart agriculture.

For more information about Burleigh Dodds Science Publishing simply call us on +44 (0) 1223 839365, email info@bdspublishing.com or alternatively please visit our website at www.bdspublishing.com.

Related titles:

Climate change and agriculture
Print (ISBN 978-1-78676-320-4); Online (ISBN 978-1-78676-322-8, 978-1-78676-323-5)

Key issues in agricultural ethics
Print (ISBN 978-1-80146-313-3); Online (ISBN 978-1-80146-314-0, 978-1-80146-315-7)

The sustainable intensification of smallholder farming systems
Print (ISBN 978-1-78676-430-0); Online (ISBN 978-1-78676-432-4, 978-1-78676-433-1)

Chapters are available individually from our online bookshop:
https://shop.bdspublishing.com

climate•SMART•publishing
IN AGRICULTURAL SCIENCE

BURLEIGH DODDS SERIES IN AGRICULTURAL SCIENCE
NUMBER 163

Women and smallholder farming

Addressing global inequities in agriculture

Edited by Professor Emerita Carolyn Sachs, Pennsylvania State University, USA and Dr Paige Castellanos, Oxfam America (OUS), USA

burleigh dodds
SCIENCE PUBLISHING

Published by Burleigh Dodds Science Publishing Limited
82 High Street, Sawston, Cambridge CB22 3HJ, UK
www.bdspublishing.com

Burleigh Dodds Science Publishing, 1518 Walnut Street, Suite 900, Philadelphia, PA 19102-3406, USA

First published 2025 by Burleigh Dodds Science Publishing Limited

Library of Congress Control Number: 2025939881

British Library Cataloguing in Publication Data
A catalogue record for this book is available from the British Library

ISBN 978-1-80146-805-3 (print)
ISBN 978-1-80146-806-0 (ePub)
ISBN 978-1-80146-807-7 (PDF)
ISSN 2059-6936 (print)
ISSN 2059-6944 (online)

DOI: 10.19103/AS.2024.0148

Typeset by Deanta Global Publishing Services, Dublin, Ireland

Contents

Series list

Title	Series number
Achieving sustainable cultivation of maize - Vol 1 From improved varieties to local applications Edited by: Dr Dave Watson, CGIAR Maize Research Program Manager, CIMMYT, Mexico	001
Achieving sustainable cultivation of maize - Vol 2 Cultivation techniques, pest and disease control Edited by: Dr Dave Watson, CGIAR Maize Research Program Manager, CIMMYT, Mexico	002
Achieving sustainable cultivation of rice - Vol 1 Breeding for higher yield and quality Edited by: Prof. Takuji Sasaki, Tokyo University of Agriculture, Japan	003
Achieving sustainable cultivation of rice - Vol 2 Cultivation, pest and disease management Edited by: Prof. Takuji Sasaki, Tokyo University of Agriculture, Japan	004
Achieving sustainable cultivation of wheat - Vol 1 Breeding, quality traits, pests and diseases Edited by: Prof. Peter Langridge, The University of Adelaide, Australia	005
Achieving sustainable cultivation of wheat - Vol 2 Cultivation techniques Edited by: Prof. Peter Langridge, The University of Adelaide, Australia	006
Achieving sustainable cultivation of tomatoes Edited by: Dr Autar Mattoo, USDA-ARS, USA and Prof. Avtar Handa, Purdue University, USA	007
Achieving sustainable production of milk - Vol 1 Milk composition, genetics and breeding Edited by: Dr Nico van Belzen, International Dairy Federation (IDF), Belgium	008
Achieving sustainable production of milk - Vol 2 Safety, quality and sustainability Edited by: Dr Nico van Belzen, International Dairy Federation (IDF), Belgium	009
Achieving sustainable production of milk - Vol 3 Dairy herd management and welfare Edited by: Prof. John Webster, University of Bristol, UK	010
Ensuring safety and quality in the production of beef - Vol 1 Safety Edited by: Prof. Gary Acuff, Texas A&M University, USA and Prof. James Dickson, Iowa State University, USA	011
Ensuring safety and quality in the production of beef - Vol 2 Quality Edited by: Prof. Michael Dikeman, Kansas State University, USA	012
Achieving sustainable production of poultry meat - Vol 1 Safety, quality and sustainability Edited by: Prof. Steven C. Ricke, University of Arkansas, USA	013
Achieving sustainable production of poultry meat - Vol 2 Breeding and nutrition Edited by: Prof. Todd Applegate, University of Georgia, USA	014

Introduction

Of the estimated 500 million smallholder farmers across the world, h an estimated 60-80% are women. Despite their contribution, many women smallholders remain unsupported and their contribution to agriculture is often devalued because of their gender. Women smallholders face immense obstacles due to restricted access to markets and education, as well as a lack of control over assets and property rights, including land. This collection highlights the day-to-day challenges faced by women smallholders across regions in Africa, Asia and South America, whilst also considering how key stakeholders across the agri-food supply chain can support, empower and sufficiently compensate women smallholder farmers for their contribution to agriculture.

The book is split into five parts. Part 1 discusses the different roles women smallholder farmers play in achieving global food security and the United Nations (UN) Sustainable Development Goals (SDGs). It also highlights the adoption of climate change strategies by women smallholder farmers. Part 2 provides insight on the impact of sociocultural factors on women smallholder farmers. It presents case studies from Africa, Asia, Latin America and the Middle East. Part 3 explores the role of gender in smallholder farming e.g. the challenges of patriarchal gender norms, racial and sexual stereotypes of women smallholder farmers, gender-based violence and the impact of migration on women smallholder famers. Part 4 addresses the role public and private sectors play in supporting women in smallholder farming, including access to resources, technology and training, It also discusses gender-transformative approaches and financial support initiatives to support women smallholder farmers and highlights the challenges faced by women small holder farmers in terms of land tenure and rights. Part 5 completes the book with a number of case studies. These include the role of women smallholder farmers in water resource management and agroforestry systems, as well as in growing horticultural and beverage crops, and in pig and poultry farming.

Part 1 Women and smallholders and global agricultural production

This group of chapters evaluates the role of women smallholder farmers in the global agricultural production sector. As Chapter 1 states, across the globe women smallholder farmers play an essential role in improving food security through their contribution to food production. However, gender inequalities undermine the full potential women smallholder farmers in contributing to sustainable food security. Addressing these barriers of discrimination requires gender equality and women empowerment strategies to improve food security at household, community and global levels. The chapter provides a synthesis of the critical role of women smallholder farmers in achieving global food security. It discusses ways to implement strategies to overcome barriers to achieve gender equality and harness the full potential of women in the agricultural sector.

Chapter 2 reviews how the UN Sustainable Development Goals (UN SDGs) do or do not reflect the needs of women smallholder farmers The chapter explores how visible women smallholder farmers are in the SDGs as well the key contributions made by women smallholder farmers in achieving the SDGs. The chapter concludes by proposing how the SDGs can ensure better engagement of women smallholder farmers as agents and beneficiaries in agrifood systems.

Chapter 3 explores the adoption of climate change mitigation strategies by women smallholder farmers. The chapter focusses on Australian agriculture which has faced particularly severe droughts and other extreme weather events related to climate change. It reviews the experience of coping with these challenges of a number of women who co-own and operate farms and run small-scale enterprises in the Australian farming sector, including inspiring rural women identified through national agricultural award schemes as leaders in their industries. The chapter provides an in-depth exploration of climate change mitigation strategies adopted by enterprising women smallholder farmers.

Part 2 Sociocultural factors and their effects on women smallholder farmers

This group of chapters provides insights on the impact of sociocultural factors on the agency of women smallholder farmers in Africa, Asia, Latin America and Middle East. Chapter 4 emphasises the importance of promoting women's empowerment and gender equality in enhancing sustainable agricultural systems in Africa. Current challenges include limited participation in overall decision making, as well as how cash crops with a higher market value (like cotton, cocoa, coffee and tree fruit) are reserved for men to produce. Studies also suggest that men are more likely to engage in market-oriented work (agribusiness) when compared to their female counter parts. Societal and gender norms are rooted in capitalist-patriarchal systems, limiting women's agency and agricultural opportunities. A gender-responsive agricultural policy needs to be implemented to address the gender inequalities that persist in the African agricultural sector.

Women smallholder farmers in Asia are rich in cultural diversity, religion, languages and geographical settings. Chapter 5 focuses on ethnic Thai young couples in Northern Vietnam in the Southeast Asian uplands where women's agricultural work is unpaid. The chapter highlights the way this unpaid work both reduces the visibility of women and limits their potential contribution. Understanding their experiences provides the first step into supporting marginalized women in the uplands in transformative ways. The chapter stresses the need to position ethnic minority women as key actors in agricultural development in Asia.

Chapter 6 discusses the present status of women smallholder farmers in Mexico, South and Central America and the Caribbean. The role of smallholder farmers, and particularly gender, is often ignored in official statistics. The first section of the chapter focuses on Mexico and Costa Rico. Then it is followed by women smallholder farmers in Brazil, Argentina and Colombia. The final section of the chapter involves the Caribbean region which includes the British, French, Spanish and Dutch Antilles. Barbados has been given special emphasis in the chapter as it has a high proportion of women smallholder farmers. The chapter highlights key trends in feminization of agriculture, including the role of male migration in leaving women back home to cultivate land to feed their families. Research suggests that women smallholders can introduce greater agrobiodiversity as well as benefit from tourism as an additional source of income. As the chapter points out, new ideas on sustainable development and gender introduced by international agencies have also produced positive changes for women smallholder farmers.

Chapter 7 focuses on the impact of sociocultural factors on women smallholder farmers in the Middle East and North Africa (MENA). In the MENA region, gender inequality persists due to a deeply rooted patriarchal system within households and communities and often reinforced by national institutions. Some studies suggest that about 50% of women engaged in agrarian labour are either not counted at all in national surveys or simply classified as economically inactive. The chapter highlights sociocultural factors affecting women smallholder farmers in the MENA region. These factors include women's marginalization from land rights, being labelled as helpers and housewives instead of as farmers (including in official documents), gender wage gaps and sexual harassment. Properly recognizing women as workers through measures such as framing and enforcing equal pay legislation for men and women in agriculture, as well as strengthening social protection laws, will benefit women in MENA countries working in the agricultural sector.

Part 3 Gender negotiations for women smallholders

Part 3 reviews different aspects of gender in agriculture particularly affecting women smallholder farmers. Chapter 8 explores the impact of a colonial legacy on much agriculture in the developing world in enforcing a gender and sexual hierarchy while erasing indigenous cultural knowledge . This has been compounded by capitalist models of agribusiness to create a hegemonic masculinity that commodifies land and bodies whilst emphasizing dominance and control of property for profit so it can normalize a certain type of land use to maintain global agricultural expansion. The chapter outlines the interventions to both research and policy to achieve a transformative change through reimagining agricultural masculinities.

As Chapter 9 points out, research and policy relating to sexuality has often viewed the topic too narrowly e.g. within the context of sexual violence. This risks perpetuating broader heterosexual and cis-gender norms. The chapter explores the relationship between agriculture, gender and sexuality. A full engagement with the field is necessary to enable the profoundchanges needed which can then more fully benefit women smallholder farmers, food security and environmental sustainability on a global scale.

Chapter 10 identifies a noticeable lack of empirical research examining the intersection of gender and race in an agricultural context. Race and ethnicity continue to have a significant impact on women smallholders. The chapter explores this impact through a case study of women agriculturalists from Pittsburgh (in the State of Pennsylvania in the United States). It identifies the additional barriers that race can add on top of gender and ways to support female smallholders overcome these barriers.

Focussing on Africa, Chapter 11 highlights how male dominated the agricultural sector remains. Women often lack power and control in key decision making processes. The chapter examines the ways innate gender bias within smallholder farming operates, in areas such as inadequate access to land, finance, markets, agricultural training and education. As the chapter argues, this both devalues and reduces the potential contribution women smallholder farmers can make towards sustainable development of agriculture and food production.

Chapter 12 discusses the particular effects of migration on women farmers, focussing on Ghana. Climate change is reinforcing existing economic pressures driving increasing levels of migration across national and regional borders. The chapter focusses on the experiences of those migrating, those returning and those staying at home. Women are both participants particularly in rural-rural migration as well as those who most often shoulder the responsibility for establishing and running smallholder farming operations whilst male members of the family are away. The findings from this chapter show that migration can reinforce the gendered, economic and sociocultural marginalization women smallholder farmers in Ghana continue to face.

Part 4 Facilitating access to resources and technology

The next group of chapters highlights the need to improve access to technical training and financial support as well as improve land tenure and rights.

Chapter 13 points out how historically, extension and advisory services have been disproportionately focussed on men with women benefitting less from the system. The Gender Integrated Continuum approach helps to map gender integration from those that discriminate or ignore gender (gender regressive or gender blind) to fully integrated (gender transformative) systems. The chapter utilises this approach to assess examples of extension programs from Africa, Asia and Latin America with integration of programs utilizing innovative techniques to fully meet the needs of male and female producers.

Chapter 14 discusses women smallholders and technology. A number of studies show that women can be particularly receptive to adopting new technology and practices to improve productivity. However, many studies focus only on the technological adoption strategies of the head of the household head. This underplays both the role of females in male-headed households and males in female-headed households. It is important to take account of differing gender power relationships within households in agricultural decision making, making gender-related technology adoption context specific.

Chapter 15 discusses the application of Gender Transformative Approaches (GTAs) to understanding and improving women smallholders' engagement in agrifood systems. The chapters shows both limitations in use of GTAs in current agricultural research, as well as their potential in supporting women smallholders and social justice more broadly through a more critical examination of power relationships.

Chapter 16 discusses the financial support available to women smallholder farmers in Bangladesh. The chapter utilizes data collected from the 2010 Household Income and Expenditure Survey (HIES) to show that just over 17% of women smallholder farmers have access to credit with an average loan of approximately Taka 58,000 (£380). Some of the key factors that influence access to credit are marital status of the household head, the number of young household members, minority status, business and asset ownership and district labour participation rate. Current financial policies are not supportive of women smallholders. Due to the significant contribution of the agricultural sector to economic growth, employment and food security, there needs to be an increase in the access to credit for women smallholders.

Chapter 17 reviews integrated approaches to help gender and agriculture practitioners to improve land rights and security of tenure for women smallholder farmers. Some of these approaches entail awareness raising and training in legal literacy, expansion of women's land access and governance roles in customary tenure systems, supporting gender-responsive, participatory certification of land rights, combining land tenure and agricultural interventions, and social norms and behaviour change (SNBC) interventions. The chapter goes on to list examples from the private sector which suggests that agri-food companies regardless of size can take action to strengthen women's land tenure and security.

Picking up on themes in Chapter 3, Chapter 18 highlights that, despite the growing importance of women in carrying out the essential daily work of farms, and often earning additional income both on and off farms that is vital to farm viability, their role remains invisible. This lack of recognition extends from governments down to the communities in which they live where farming is still viewed as an essentially male activity. As the chapter states, there is a need to focus attention not only to the attitudes of families in relation to inheritance practices, but also to address the imagery of farming that reinforces the dominant image of a male farmer. This will then lead to improving practices of institutions such as banks, farming organisations and industry bodies and governments. An example is to ensure that women smallholders have access to input subsidies and other government support schemes available to farmers. Governments should also examine very basic factors such as the provision of adequate services including child care support.

Part 5 Case studies

The last group of chapters in the book consist of case studies of women smallholder farmers and particular management practices and systems. Chapter 19 looks at women smallholders and water resource management, focussing on high, middle and low elevation gradients within the south, centre and north of Colombia's Magdalena River watershed. It utilizes a water, energy, food (WEF) nexus framework with a gender lens to analyze the experience of women smallholders in the fields and in the home in three rural communities. As the chapter shows, women in this area frequently lack ownership rights and even access to their own land, as well as reliable, affordable and safe sources of water for both domestic and agricultural use. They also lack safe sewage systems and sufficient energy for pumping water for both farming and domestic use. Solving these problems requires comprehensive policy interventions at different scales.

Chapter 20 focuses on the impact of agricultural feminization (as women play an increasing (if frequently undervalued) role in farming) through the case of an indigenous women's organization in Intibucá, a department in Western Honduras. The Association of Renewed Intibucán Women (AMIR, by its Spanish acronym) seeks to empower women smallholders by giving them a collective voice. The chapter discusses the experience of AMIR through a "new materialist sociology" framework which helps to better address gender issues in agriculture.

Chapter 21 focuses on the experience of women in coffee and tea growing. It discusses two primary factors affecting women: the feminization of agriculture and the rise of the fairtrade movement . The fair trade movement has resulted in greater efforts to create more equitable inclusion in formal institutions that supports women smallholder farmers in the beverage industry. It explores the rise of cooperatives and integration of women into these organization as formal members. It further highlights the support for addressing global inequities in agriculture for women smallholders in global supply chain in the tea and coffee industry.

Chapter 22 explores how women smallholders interact with agroforestry system. It employs the Gender at Work Framework and the Reach-Benefit-Empower-Transform Framework to help understand gender transformative change in and through agroforestry. The chapter explores examples of agroforestry projects that intentionally highlight and support greater gender equality and examine how they have performed in practice.

Chapter 23 looks at livestock research and development programs supported by the International Livestock Research Institute (ILRI) and the University of Illinois at Urbana-Champaign (USA) to support women smallholder in the Masaka district of Central Uganda involved in pig farming. The chapter shows that women smallholder farmers involved in pig farming require greater support for their role, initiatives to de-risk pig production as well as investment along pig value chains.

Chapter 24 examines the differences in gender-blind and gender-aware approaches when it comes to dealing with small-scale chicken production. The chapter includes three case studies focused on control of Newcastle Disease and Highly Pathogenic Avian Influenza (HPAI). As it shows, a gender-blind intervention during the HPAI outbreak adversely affected women smallholder farmers raising chickens whilst gender-aware approaches were more effective.

Part 1

Women smallholders and global agricultural production

Chapter 1

The role of women smallholder farmers in achieving global food security: an overview

Caroline Muchiri, Els Lecoutere and Nicoline de Haan, GENDER Impact Platform – CGIAR, Kenya

1 Introduction

The United Nations Sustainable Development Goal (SDG) 2 focusing on Zero Hunger aims to end hunger, improve nutrition and promote resilient and sustainable agricultural systems by 2030. This is important because having reliable physical, social and economic access to sufficient, safe and nutritious food at all times is a fundamental human right for every individual (CFS, 2023).

http://dx.doi.org/10.19103/AS.2024.0148.01

This food should meet the dietary needs and food preferences of people so they can lead active and healthy lives (IFPRI, 2021).

Global food security accounts for the efficient flow of food resources through agricultural production, trade, distribution, consumption, policy and governance both nationally and internationally, and significantly influences the food security status of individuals and households. The *Global Food Security Index 2022* report indicates that food security is also vulnerable to negative disruptions caused by global crises including climate change, natural disasters, economic shocks, violence and social conflict leading to displacement, and the lingering socioeconomic impacts resulting from pandemics (Economist Impact, 2022). These undesirable disruptions are increasingly impeding the achievement of global food security through the pervasive vulnerability to hunger, food insecurity or malnutrition inequalities across all regions and countries (FAO, IFAD, UNICEF, WFP and WHO, 2023; FSIN, 2023; FAO, 2022).

The *Global Report on Food Crises* (2023) showed that the number of people experiencing acute food insecurity and requiring urgent food, nutrition and livelihood assistance rose from 193 million people in 53 countries in 2021 to nearly 258 million in 58 countries in 2022. Evidence shows that the challenge of food insecurity is also gendered. According to the *State of Food Security and Nutrition in the World* by FAO (2022), food security among men improved in all global regions except Africa, while it worsened among women in all regions except Asia. It also suggests that unless immediate and effective strategies to address this gender gap in food insecurity are implemented, approximately 670 million individuals around the world will still be grappling with hunger by the year 2030.

This underscores the critical link between gender equality and global food security, as the gender gap in food insecurity reflects disparities in access to resources, opportunities and decision-making power between women and men. By ensuring the equal participation and decision-making power of women smallholder farmers in farming, food processing, marketing and ensuring household food security and nutrition, we can create more equitable and sustainable food systems that support their role in achieving global food security.

The agricultural sector has proven to be two to four times more impactful in increasing incomes and reducing poverty compared to other sectors, with women smallholder farmers playing a pivotal role in fostering its growth (World Bank, 2015). This is because more women than men participate in and rely on agriculture as an important means of livelihood, with women smallholder farmers and adolescent girls constituting well over half of the global agricultural labour force (FAO, 2023).

Due to structural gender inequalities, women face higher vulnerability to climatic shocks and stressors than men do, which limits the women farmers' capacity to respond, adapt to, or mitigate subsequent impacts (FAO, 2021b). While this heightened vulnerability is highly contextual, it is attributed to reliance on rain-fed agriculture and natural resources prone to fluctuations and environmental degradation and less-diversified livelihoods (Koo et al., 2022).

Women's vulnerability is also impacted by the out-migration of men in farming households to urban cities and industrial centres in search of remunerative off-farm employment (FAO, 2021c; UN Women, 2019; HLPE, 2020; Choithani, 2020). In this regard, valuing the diverse roles women smallholder farmers play, listening to their concerns and needs and ensuring they are involved in finding food security solutions is crucial for improving their well-being and that of their families and communities. It is also essential in the design and implementation of targeted policies and interventions aimed at enhancing women's access to resources and financial services, building their capacity through knowledge and securing more acceptable equitable opportunities and outcomes (FAO, 2021c; IFPRI, 2021).

In this chapter, we provide a synthesis of the critical role of women smallholder farmers in achieving global food security. It draws on recent research and scientific evidence to provide an overview to address these four research questions:

- What roles do women smallholder farmers play in agriculture and food systems and their contribution to global food security?
- What are the existing inequalities that limit their contribution?
- What potential strategies can be implemented to address these inequalities and constraints, leading to improved global food security outcomes?
- How do aspirations and motivations influence women smallholder farmers' roles in achieving food security?

Ways of approaching these questions are discussed in the next section.

2 Analysing the role of women smallholder farmers and global food security

We acknowledge that women smallholder farmers are not confined to rural areas whereby they engage in food production only for home consumption and local markets; they are also present in peri-urban or urban areas, cultivating small backyard plots or gardens to support their families or generate income. However, for this chapter, we will consider women smallholder farmers as rural participants in agriculture and focus on the pivotal roles they play that advance the achievement of global food security through their efforts at the household and community level.

While our criteria to guide the literature search for this overview were not rigid, they served as a strategic framework to ensure the relevance and quality of the literature sources. We used various search terms relating to women smallholders in agriculture and their contributions to food security, including 'smallholder women', 'women farmers', 'women's roles in agriculture', 'women's roles in agrifood systems', 'women's roles in food security', 'gender and food security', 'women's participation in agriculture', and 'gender and food security'. Then, we screened electronic publications retrieved from web search engines and scholarly databases, including Google Scholar, ScienceDirect, the CGSpace Repository and CAB Abstracts, prioritising those dated between 2013 and 2023. We focused on including studies that explored or directly addressed the involvement of women smallholder farmers in agriculture, the challenges and opportunities they face and the potential strategies to empower and support women in achieving food security at the national or local level.

It is important to note that this overview may not cover every perspective or study conducted around the roles and contributions of women smallholder farmers to global food security. The conceptual framework presented in Fig. 1 which presents the overarching synthesis of the literature review on the role of women smallholder farmers in achieving global food security.

The conceptual framework structuring our overview explains the relationships between the roles that women smallholder farmers play in the larger agriculture and food systems and their contributions to individual, household and global food security and nutrition outcomes (Fig. 1). The framework also illustrates how inequality, gender and social norms as well as aspirations may determine the extent to which the roles of women farmers in agriculture and food systems and the ensuing food security outcomes are impacted.

The conceptual framework lays out how addressing constraints that collectively hinder rural women's full and equal participation can transform existing agriculture and food systems into agriculture and food systems that enhance women's potential and contribution to global food security (as shown by the blue dotted arrows in Fig. 1). The green dotted paths show how gender-responsive agriculture and food systems influence women-empowerment outcomes and ultimately enhance the roles of women smallholder farmers.

In what follows, we address the research questions based on the results of our review of selected literature. We structured our discussion accordingly and first addressed ways in which women smallholders participate in agriculture and food systems as shown in the conceptual framework. Secondly, we lay out the prevailing constraints and drivers of women smallholder farmers' roles in achieving food security. Thirdly, we discuss potential strategies to address constraints and enable drivers for women's roles in agriculture and food systems.

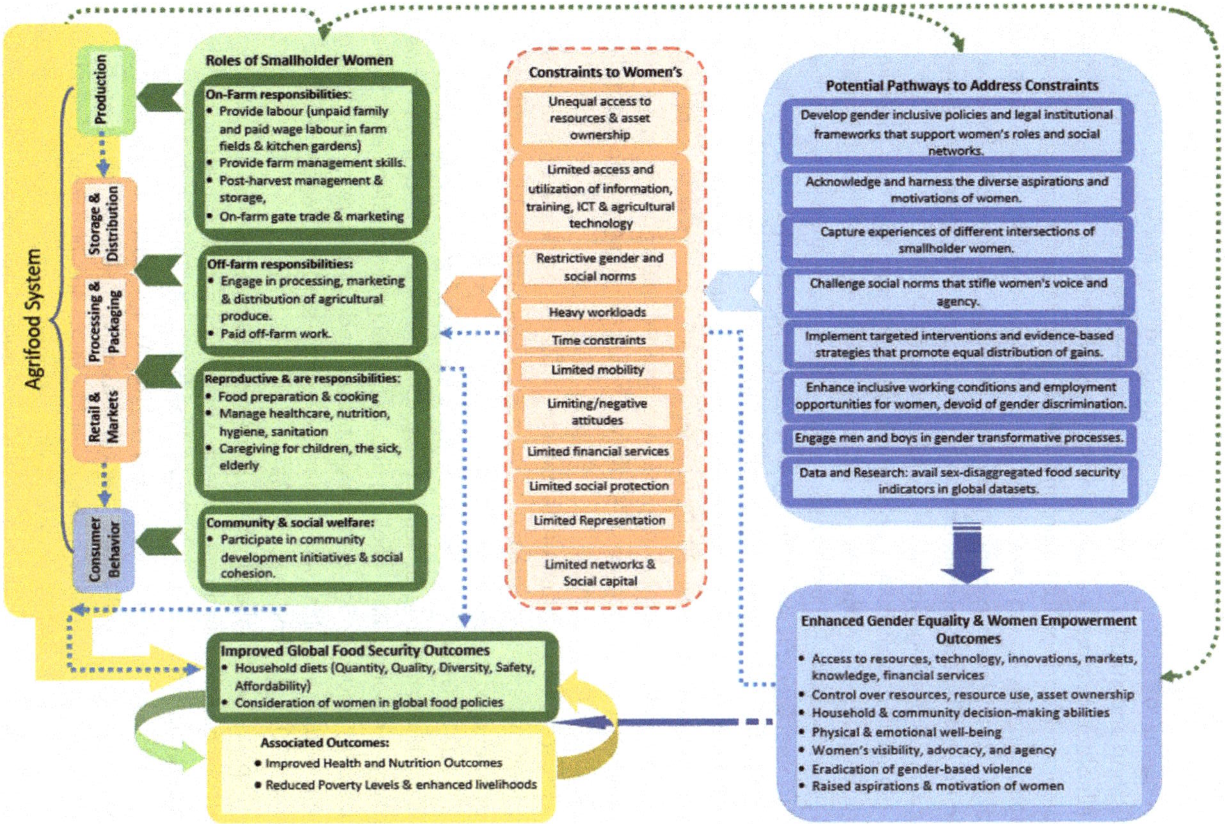

Figure 1 Conceptual framework: the role of women smallholder farmers in global food security.

3 Roles of women smallholder farmers in agriculture and food systems for global food security

Through their multiple roles as food producers and traders of agricultural produce, women smallholder farmers actively contribute to the foundational dimensions of global food security including the availability, accessibility, utilisation and stability of food in households and local communities. Women smallholder farmers are also decision-makers in their households' dietary choices and food consumption, and they play the role of managing the stability of food supplies during crises that decrease food availability (FAO, 2021a; Akter et al., 2020). While these multiple roles contribute to achieving food security, women smallholder farmers encounter challenges across the interrelated and mutually reinforcing dimensions of food security.

Regarding availability, opportunities for cultivating food are often limited. Regarding access, gendered norms can lead to reduced financial means to purchase food and limited transportation options for market access. Regarding use, women's nutritional requirements vary, especially during pregnancy and breastfeeding, and regarding stability, they bear a disproportionate burden during food shortages, frequently reducing their intake more significantly than others (Glazebrook et al., 2020). Women living in rural, remote, conflict-prone and climatically vulnerable regions face greater challenges in these aspects of food security, making up 60% of the chronically hungry global population (FAO, 2020a; HLPE, 2020).

The extent and nature of women smallholders' participation in agriculture and food systems vary worldwide, across different communities and regions, and depend on the structure of the economy and gender norms (Quisumbing et al., 2021). Usually, participation includes growing and managing crops, tending kitchen gardens, rearing livestock, obtaining food from farms or markets, preparing and cooking food, ensuring food safety and hygiene, processing and storing farm produce and trading and retail marketing of farm produce, as well as working on-farm or off-farm for wages (FAO and UN Women, 2021; Malapit et al., 2020).

In parts of sub-Saharan Africa, women smallholder farmers are responsible for growing more than three-quarters of the region's food, managing two-thirds of marketing farm produce and overseeing at least one-half of food storage and livestock-rearing activities (Faiez, 2018). As agriculture and food systems evolve, women smallholders are increasingly assuming decision-making responsibilities in the daily management of farms and households, cultivating crops and taking on tasks traditionally assigned to men. They are exploring diverse roles within the agriculture and food system by venturing into off-farm work, which offers better returns and opportunities than traditional agricultural labour.

Published by Burleigh Dodds Science Publishing Limited, 2025.

Women farmers who are not self-employed crop or livestock producers may provide unpaid labour on farms belonging to their spouses or other family members without receiving direct remuneration for their efforts (Kabeer, 2018). Additionally, women in indigenous communities concurrently hold pivotal roles as custodians and transmitters of traditional knowledge, which not only contributes to their livelihoods and resilience but also plays a fundamental role in ensuring the survival and integrity of their culture (ILO, 2019).

The roles and decision-making of women smallholder farmers regarding household food security are significantly influenced by gender norms, which vary across different contexts (Qanti et al., 2022; Galiè and Kantor, 2016; Twyman et al., 2015). Detailed case studies conducted by Gennovate (https://gennovate.org) and Align (www.alignplatform.org) show that within agricultural and fishing communities in sub-Saharan African and Southeast Asian countries (Bergman-Lodin et al., 2019; Locke et al., 2017), as well as across both rural and urban communities in low- and middle-income countries (Boudet et al., 2013), prevailing social norms continue to designate men as primary earners and household heads while assigning women the responsibility for domestic tasks.

In such communities where male household heads are the primary earners responsible for providing money for food and other necessities, the decisions of women to purchase and prepare food are closely tied to the amount and frequency of financial support received from their husbands. This affects household dietary choices and nutrition outcomes. Despite often facing insufficient financial support, some of these women empower themselves by leveraging their own enhanced purchasing capacity through selling agricultural products or earning income from off-farm activities (Farnworth et al., 2023).

Other women may prefer to take a back seat in decision-making around expenditure for food to avoid potential marital conflicts with the men who fund and provide food for the home. Still, the women are expected to make nutrition-related decisions based on the food made available to them by their husbands (Galiè, 2019). Such gendered customary norms that impose or normalise limitations on decision-making for women ultimately hinder their ability to fulfil the role of providing proper nutrition for their families.

3.1 Care responsibilities that nurture households and communities

Women smallholders in most regions of the world spend a disproportionate amount of time carrying out the socially ascribed responsibility of unpaid care and reproductive duties in households. These gendered responsibilities include primary caregiving roles for children, the elderly and sick members in the household and community, as well as tasks such as breastfeeding, complementary feeding of infants, managing health care and many other domestic chores essential to human well-being (CFS, 2023; Seedat and Rondon, 2021; UN Women, 2020; Kabeer, 2018).

Although the demanding nature of these tasks and their associated drudgery vary greatly, these activities can constrain women's time, movement and physical and emotional energy to fully participate in agricultural and related income-generating activities. This reduced participation diminishes their productivity and decision-making influence. Research indicates that globally, the unequal allocation of care and domestic work results in women working longer hours than men, as women dedicate nearly three times the daily hours to unpaid care and domestic work compared to men (Data2XILO, 2020; UN Women, 2020).

However, evidence shows that although income-related differences within countries exist, women in low- and middle-income countries devote more time to unpaid work than women in high-income countries. It also suggests that although the demanding nature of unpaid work among poorer women may cause substantial fatigue and stress, the relational aspect of unpaid work, for example playing with children, could be stress reducing and provide a sense of fulfilment (UN Women, 2020).

Surprisingly, although these unpaid responsibilities contribute significantly, amounting to about 35% of Gross Domestic Product in 90 countries, they are often not officially recognised as work in national and global datasets (UN Women, 2020). Addressing these disparities by equitably redistributing care work within families remains a prerequisite for realising substantial gender equality. It allows women smallholder farmers to allocate more time and resources to agricultural activities, thereby fostering increased food production, household income and overall well-being.

3.2 Life cycle and aspirations: link to roles of women smallholder farmers

Women smallholder farmers come from a variety of backgrounds and have fundamentally different experiences, desires, ambitions and aspirations as they move through the different stages of their lives. The homogenisation of women farmers and their roles can lead to inaccurate assumptions because it does not capture intragender differences such as age and marital status, which create power relationships that, in turn, influence women's participation in production and decision-making processes (Tavenner and Crane, 2019; ISF et al., 2020).

The influence of life cycle-related responsibilities is evident from studies by UN Women (2020b) indicating significant disparities in time and income poverty rates based on gender and age, with women between the ages of 25 and 34 being the most affected. This age range represents a critical phase in women's lives when they commonly establish families and undertake childcare responsibilities. This period is also associated with higher household expenses related to childcare (education, health care and other essential needs of children).

This explains why young adult and middle-aged women smallholders aspiring to have greater control over the income they earn need the flexibility to balance their productive, reproductive and childcare roles. This flexibility allows them to pursue seasonal on-farm or off-farm paid employment opportunities that can be easily combined with unpaid care and domestic work responsibilities. Often, these part-time employment opportunities available to rural women are mostly informal, lower-quality and lower-wage work (FAO, 2020b; UN Women, 2020).

As women's care responsibilities reduce over the course of their lives, they may regain some time and flexibility to engage in other economic opportunities such as growing and managing a business (Halim et al., 2021; ILO, 2019, 2020; Van den Broeck and Kilic, 2019). A study on cross-border bean trade in Kenya revealed that older women, whose children need less attention, were more actively involved and experienced in the trading business compared to men and younger women. These women mentioned that during their younger years, they prioritised caring for their young children. As they advanced in age and experienced reduced family obligations, they gained more flexibility and opportunities to participate in cross-border trading activities (Odera et al., 2016).

The attitudes and perceptions of women smallholder farmers about their roles and possibilities in achieving food security vary and are shaped by their upbringing, education, exposure to agriculture, gender norms and their stage of life. These factors play a crucial role in shaping women's aspirations and values concerning their future work in agriculture. Crossland et al. (2021) found that the aspirations of rural women who perceived limited prospects in unpaid or low-wage farm work were to seek employment in other segments of the agricultural supply chain.

Understanding the highly contextual, complex and multifaceted aspiration nuances of women smallholders is necessary because they are central to their sense of having a fulfilling life. Aspirations and motivations also directly influence women's decisions on agricultural resource use, investment in technology, engagement in other segments of the value chain and use of agency to engage in other future-oriented behaviours to achieve well-being (Hernandez et al., 2023; DeJaeghere et al., 2022; Nandi and Nedumaran, 2021; Crossland et al., 2021).

Aspirations also matter for women's agency in agriculture and food systems. In Kyrgyzstan, for example, women smallholders with higher aspirations for their agricultural work exhibited greater agency. These women were more actively involved in decision-making about their responsibilities, had increased control over their earnings, had a stronger voice in household matters and were less likely to experience domestic violence (Kosec et al., 2018).

However, few studies are focusing on the aspirations for success and growth for women smallholder farmers, how they may shift during the different stages of their life cycle and their significance for food security outcomes. This is because women experience dynamics in their household structure and family responsibilities. This influences their physical capacity to undertake certain roles, and consequently their level of engagement in agricultural activities (Kosec et al., 2018; Verkaart et al., 2018).

These shifts can be influenced by factors such as age, life experiences, education, family socioeconomic status, belief in their capabilities and skills, desire for recognition and improved social status through agricultural work. Other external factors that influence aspirations and ambitions can either fuel their drive to thrive or dampen their enthusiasm. These factors include cultural norms and social relations deeply rooted in gendered informal and formal institutions, access to resources, the presence of economic opportunities, capabilities to move freely, local agro-ecological conditions and the extent of out-migration within agriculture (Lecoutere et al., 2023; Nandi and Nedumaran, 2021; Elias et al., 2018).

According to Rietveld et al. (2020), young women in Uganda rarely expressed aspirations to make farming their primary livelihood strategy, even if they engaged in various farming activities. Their involvement in farming was primarily driven by the social obligation to provide food for the households. Similarly, a global comparative study conducted by Elias et al. (2018) found that, whereas some young men aspired to engage in knowledge-intensive or 'modern' agriculture, young women did not express such interest.

Studies conducted by Nandi et al. (2022) in South India and Verkaat et al. (2018) in rural Kenya showed that in both regions some parents held agricultural aspirations for themselves but discouraged their sons and daughters from considering careers in agriculture. Despite this lack of agricultural aspirations for their children's future, there was a significant gap in outcomes, particularly among young women, who ultimately found themselves working on their family farms.

Other findings showed that increased male out-migration and the persistence of social norms restricting women's movement out of rural life led women smallholders in Makueni County in Kenya to increasingly take on the roles of farm managers. This change in the agricultural opportunities and the new realities for women shifted their aspirations towards commercialising their farm activities (Crossland et al., 2021). These examples indicate that the aspirations of different women smallholders evolve throughout their life cycle, ultimately influencing how they engage in agriculture. Elias et al. (2018) suggest that while aspirations may not always align realistically with the available opportunities, thereby creating an aspiration-attainment gap, this gap can offer valuable policy insights, shedding light on the personal and structural constraints faced by young women and men in rural areas.

4 Constraints on women smallholders' roles in achieving global food security: structural gender inequalities

The goal of achieving global food security with sustainable and equitable food security outcomes is hindered not only by insufficient food production but also by inefficient connections between production and consumption, often compounded by gender biases and barriers. According to the 2023 *FAO Status of Women in Agricultural Food Systems Report*, women smallholder farmers are increasingly transitioning away from full-time agricultural primary production and participating in other segments of agricultural value chains, including value-adding processing, storage and marketing of agricultural products.

Like in primary production, women smallholders face barriers that block access to virtually every aspect of their off-farm roles because of prevailing structural gender inequalities. These inequalities include unequal access and control rights over productive resources such as land, natural resources and physical capital. Women smallholders also face constraints related to accessing complementary resources, technologies and services such as networks and social capital, information and communication technology, agricultural technology, training, extension and advisory services, financial services and social safety nets. Societal norms further restrict women's decision-making and leadership, confining them within traditional roles. The lack of time-saving infrastructure and market access also hinders their potential contributions to food security.

These constraints collectively hinder women from fully engaging and realising their potential in enhancing food and nutrition security outcomes (Kosec et al., 2023; Njuki et al., 2023; Quisumbing et al., 2021). Literature suggests that empowering women smallholders and increasing their participation in decision-making processes can improve food security and other associated welfare outcomes such as health, nutrition and education (Kosec et al., 2018).

Women smallholder farmers around the world encounter multifaceted dimensions of gender inequalities that significantly undermine their capacity to fully engage in agricultural activities and benefit from their contributions towards food security (Botreau and Cohen, 2020; FAO, 2020b; UN Women, 2020). These structural inequalities are deeply rooted in discriminatory social norms and gender-blind policies, laws and institutions that shape and prescribe societal expectations, gender roles and aspirations (FAO, 2023).

Unequal access to essential resources and services challenges women smallholder farmers' ability to engage effectively in productive activities and seize economic opportunities. Additionally, prevailing power imbalances lessen the voice and agency of women within decision-making processes by marginalising their participation in shaping policies and strategies and making life choices that directly affect them. Gender-based violence also undermines women's potential contributions to food security and sustainable livelihoods by compromising their safety and well-being, both within and outside their domestic and agricultural activities.

Structural gender inequalities encountered by women smallholders in their roles carry economic, environmental and social costs (HLPE, 2023). Addressing these inequalities and the associated costs not only contributes to enhanced food security and improved nutrition outcomes but also fosters healthier, more resilient and empowered communities (CFS, 2023; FAO, 2021a).

4.1 Discriminatory social and gender norms

The extent of roles, opportunities and capacities available to women smallholder farmers in different communities and countries can be determined by contextual social institutions such as cultural gender roles, practices and beliefs. These informal social institutions can also reinforce and mould the privileged positions held by men in formal institutions (FAO, 2023, 2020). Recent research suggests that the slow progress in achieving gender equality in agriculture and food systems is partly due to unsupportive policies, institutions and discriminatory social norms.

In this context, discriminatory norms profoundly affect food security outcomes, as they restrict the avenues through which women smallholder farmers can enhance agricultural production and increase their incomes, thereby perpetuating their vulnerability. This often occurs by assigning women to roles in unpaid care and domestic work, placing restrictions on their mobility and participation in public spaces and constraining their access to markets and control over assets and income (Lecoutere et al., 2023; Petesch and Badstue 2020). Such gender norms contribute to women accounting for a larger share of labour at the low-skilled and labour-intensive nodes of agricultural value chains, as the opportunities for value-added agricultural activities such as processing and marketing and secure off-farm economic opportunities are severely limited. Their exclusion from segments where more value is generated and earned persists due to discriminatory social norms, barriers to knowledge, assets, resources and social networks, and pervasive gender discrimination in wage employment (Van den Broeck and Kilic, 2019).

While development interventions focusing on household food security and nutrition have a proven potential to improve dietary diversity and decrease

Published by Burleigh Dodds Science Publishing Limited, 2025.

undernutrition among women and young children (Akter et al., 2020; Ragasa et al., 2021; Rakotomanana et al., 2020), disregarding gender norms that may undermine their effectiveness is detrimental to improving diets and nutrition. Solely improving women's access to nutrition education may not translate into changes in feeding practices if gender norms and relations, such as the decision-making roles of husbands, grandmothers and community members and the associated power imbalances, are disregarded (Ulaya, 2021; Ragasa et al., 2019; HKI, 2015; Bezner Kerr et al., 2013).

Food insecurity crises arising from climatic, economic, humanitarian or systemic shocks are globally becoming more intense and frequent. Rural women smallholder farmers, particularly those dependent on rain-fed systems and who function as net food buyers, are mostly impacted by food stresses linked to prices and market access rather than by production shocks (Botreau and Cohen, 2020). During and after these stresses, structural barriers compound the underlying gender inequalities by further limiting the capacity of women smallholders and girls to respond to shocks and cope with food crises (CFS, 2023; Koo et al., 2022; FAO, 2020b).

These women and girls are more likely to be burdened with additional care work, lose economic opportunities, be forced to deplete assets and forego meals to feed their children. They are also more exposed to increased insecurity, restricted mobility, sexual exploitation and abuse, as well as gender-based violence (Busingye and Ibrahim, 2021; FAO, 2020b). Placing substantial care obligations alongside other primarily unpaid agricultural and domestic responsibilities imposes heavier workloads and time constraints on women smallholders and girls, negatively impacting their physical and emotional wellness. It also leaves them with limited time for education and training and reduces their participation in decision-making and opportunities for economic empowerment (FAO, 2023; Lecoutere et al., 2023; AGRA 2021; ILO, 2019; HKI, 2015).

4.2 Unequal access to physical, financial and social resources

A recent report by the World Bank found that despite women producing over half of the world's food, they are still significantly disadvantaged in securing equal legal rights to own land and control its use because of discriminatory inheritance laws and social norms that prioritise male inheritance (World Bank, 2021; ISF et al., 2020). Studies have demonstrated that countries with inclusive and well-defined state policies and legal frameworks that advance equal land rights through women's involvement in land-policy reforms have increased access to land for rural women (Santpoort et al., 2021; FAO, 2020b).

Increased access to land has been found to have positive implications for food security at the household level. Quisumbing et al. (2021) found that wage-earning households, typically without any land of their own, tend to have the least wealth or assets compared to other households. Women smallholder farmers from such households often resort to providing cheap labour for landowners, which hinders their ability to purchase or produce food (Galiè, 2019).

Women's access to income and financial services, including the full benefits of social transfers tend to be constrained as well. Discriminatory norms and policies are the most common barriers to women's access to income and financial services (Kosec et al., 2023; Galiè, 2019). Improving women's access to financial services has often been seen as an approach to support their economic empowerment by facilitating substantial investment in agricultural technology, allocation of more resources to nutrition, health care and education, and breaking intergenerational poverty cycles (FAO, 2020b; Kabeer, 2018).

There is evidence that providing women smallholders with access to financial services has a positive effect on their contribution to household food security. Van den Bold et al. (2013), however, note that, in some cases, women who access rural microfinance are burdened with repaying money that is often controlled by other family members or is used for food and other household expenses. In such cases, women often shoulder the consequences of defaulting payments by themselves.

Women's access to networks and social capital outside the household through producer cooperatives, self-help groups and community organisations is limited by care and domestic workloads and mobility constraints. The ability to move freely varies across contexts and may be exacerbated by poor infrastructure, societal norms and disapproval from family members, particularly within the household. Barriers to free movement can discourage women from joining or actively participating in networks in ways that improve their access to productive and complementary resources. Consequently, women's ability to contribute towards achieving food security may be hindered (Kosec et al., 2023).

4.3 Limited voice and agency

Women smallholder farmers can exercise agency as individuals, collectively, within the family and through their participation in markets, politics and other formal and informal networks. These aspects of agency allow them to effect the changes they want, shape and achieve their own goals and achieve food security for themselves and their communities (Quisumbing et al., 2023; Elias et al., 2021).

Sociocultural biases that limit individual agency embedded within can cause women smallholders to lack a strong sense of self-worth and confidence about their capabilities to choose what they do (FAO, 2020b). They also dampen their aspirations and motivations for agricultural and non-agricultural work. According to Lecoutere et al. (2023), these dampened aspirations can both indicate and serve as structural barriers to women's agency at the individual level. Such internal barriers hinder the participation of women smallholders in agricultural and market activities, active engagement in intrahousehold decision-making and strategic life choices that enable them to achieve food security.

Similarly, the absence of instrumental agency – which unlocks external resources, opportunities, knowledge, skills and capabilities that give women smallholder farmers the power to exercise choice, take control over their lives and accomplish their goals – limits their capacity to enhance productivity and respond and adapt to shocks that impact food security outcomes. Moreover, weak women's collective agency prevents women smallholder farmers from leveraging group-based social networks to achieve common food security goals (Elias et al., 2021).

It is more pronounced in contexts where civil society is weak or there are restrictions on civic spaces that support women in claiming their rights (UN Women, 2020). Women's ability to join in, actively participate, publicly voice their opinions, and acquire meaningful social and leadership influence, particularly in mixed-gender groups and networks, is also hindered by the inadequate representation of women's leadership in resource management and governance and insufficient consultation of women in decision-making processes.

These and other institutional barriers (such as identification requirements and fees) increase their exclusion from strategic actions that facilitate greater access to markets and inputs and hampers their ability to advocate for policy changes and initiatives that could improve food security for themselves and their communities (FAO, 2023).

4.4 Gender-based violence

Gender-based violence presents a widespread threat impacting women's autonomy, choices, overall well-being and that of their children. It manifests in different forms (physical, sexual, economic and psychological) that are rooted in discriminatory, exploitative and coercive gender norms and power imbalances. In many food-producing communities in the global South, this extreme human rights violation acts as a catalyst for individual, household and community-level food insecurity. Studies indicate that the physical and psychological distress caused by gender-based violence weakens the ability of women smallholder farmers to focus on food production, as well as the capacity of mothers to offer adequate care for their children.

Additionally, gender-based violence might threaten women's ability to access higher-value markets as it makes it difficult for them to travel freely and safely to markets. Where this happens, women are more likely to grow crops that can be sold locally, thus remaining in lower-value markets. These disruptions undermine the access, consumption and utilisation of affordable, culturally appropriate and nutrient-dense food (Okpara and Anugwa, 2022; Botreau and Cohen, 2020).

However, Galiè (2019) found that increasing women's decision-making power and control over assets and income positively impacts their intrahousehold bargaining power and reduces their vulnerability to domestic violence. In addition, reducing time-consuming tasks such as fetching water and fuelwood for domestic purposes in areas with limited infrastructure reduces the exposure of women and girls in many low-income countries to the risk of violence and sexual abuse (CFS, 2023).

5 Constraints on women smallholders' roles in achieving global food security: life cycle-related constraints

Life cycle-related constraints affect the roles and responsibilities of women smallholders, negatively impacting their ability to contribute to easing the severity of the prevailing food insecurity. Women's roles evolve throughout their life cycle from childhood to adolescence, early adulthood, middle age and later adulthood.

In many low- and middle-income countries, patriarchal gender norms disproportionately restrict the agency, opportunities, aspirations and social networks of adolescent girls more as compared to boys. Alwab et al. (2022) demonstrate that while institutional and structural barriers may have gradually eased in Ethiopia, persistent gender- and age-related challenges, some stemming from beliefs, norms, traditions and relational dynamics, continue to hinder the advancement of adolescents' achievement of capabilities and aspirations. This hindrance is consistently more pronounced for girls than boys.

These patriarchal gender norms also impose increased household responsibilities and caregiving duties, along with early marriage and childbearing. It has been shown that the experiences of adolescent girls in both their productive and reproductive domains significantly influence their experiences in their adulthood and old age (ISF et al., 2020). To prevent young women from experiencing poverty and unachieved aspirations in later life, effecting policies that enhance economic rights, capabilities and opportunity spaces during their younger years is critical (Nandi et al., 2022; ISF et al., 2020; Kabeer, 2018; Elias et al., 2018).

Insecure land rights, coupled with increased family and care responsibilities, constrain the participation of women of reproductive age in productive and off-farm income-generating activities, more so than of older women with more secure land rights and fewer family responsibilities (UN, 2019; Kabeer, 2018). Evidence shows that women of reproductive age (15–49 years in most surveys) suffer from macronutrient and micronutrient deficiencies with long-term negative developmental impacts, as well as physical and sexual violence (Botreau and Cohen, 2020; Kabeer, 2018; FAO, 2023). According to ISF et al. (2020), the time-consuming and resource-intensive care burden seems to stall the ability of rural women to move up the resilience and agency ladder during this life stage.

As adult women transition to older age, they may assume more decision-making responsibilities on the farm and within their communities and aim to secure autonomous use of land for agricultural production. However, in situations where customary norms and laws discriminate against women inheriting property, those who experience separation, divorce or the death of a spouse in their old age face the risk of being marginalised by their in-laws. This can leave them voiceless and destitute, particularly if their livelihood solely relies on their farms (ISF et al., 2020).

Additionally, the economic capabilities of elderly people, including women smallholders in later adulthood, may be affected by age-related physical limitations, reducing their capacity for labour-intensive farming activities. Depending on the context, elderly women in good physical health may opt to remain economically active to provide financial support for their families or themselves. On the other hand, some women smallholders may find themselves compelled to keep working due to limited alternatives.

Overcoming these life cycle-related constraints that negatively impact the potential of women smallholders to contribute to easing the severity of the prevailing food insecurity (CFS, 2023; Lecoutere et al., 2023; Data2X and ILO, 2020) could also lead to increased global economic benefits that lift millions of people out of food insecurity (FAO, 2023). Policy interventions to support the processes of ending hunger and achieving global food security must also recognise and target constraints at specific stages of women smallholders' life cycles. This can create inclusive and supportive environments that empower and uplift women smallholder farmers, allowing them to fully participate in their roles, rather than being marginalised (Galiè, 2019; ISF et al., 2020).

5.1 Women's attitudes towards their roles

The roles of women smallholder farmers in agriculture may be influenced by internalised psychological constraints arising from their preferences, attitudes, values and beliefs (Kosec et al., 2023). With the rarity of role models of successful women agricultural producers and entrepreneurs, women engaged in agriculture might not necessarily identify themselves as producers or entrepreneurs (Farnworth et al., 2021). Limiting beliefs about their self-efficacy can vary widely among women, depending on factors such as social-cultural context, education, exposure and individual experiences. For example, Kosec et al. (2023) note that women believing that they have nothing useful to contribute to social groups or that it is inappropriate for them to pursue social networks prevents them from forming rich networks that support their roles in achieving food security.

Women smallholders are aware of the importance of their contributions to food security and their role in sustaining their households and communities. This is why they are more likely than men to invest in producing food for the household and to use their income to purchase food for their families (Reynolds et al., 2020; Galiè, 2019). However, those who consistently experience sociocultural constraints, limited opportunities or lack of recognition by others may develop negative attitudes towards their roles and capabilities, limiting their engagement in certain activities.

In Bangladesh, for instance, women and men in rural societies perceive that women are primarily confined to reproductive and domestic roles and that they cannot make substantial contributions beyond these domains. In many cases, men exclude women from crucial decisions concerning agricultural and livestock production, marketing and purchases. However, some women refrain from actively participating in household decision-making, expressing uncertainty regarding their capacity to independently make important decisions or engage in household-level choices (Galiè, 2019).

Additionally, women often find it challenging to engage in conversations with their husbands and other family members about the heavy domestic responsibilities that place time limitations on them (Yosef et al., 2015). This challenge is often tied to internalise social norms that affect women's bargaining power in households regarding these roles and ultimately hinder them from adopting optimal production and nutrition practices that contribute to overall household food security and well-being (HKI, 2015).

6 Potential strategies to address constraints: strengthening agency of women smallholders to achieve food security

Efforts to address the prevailing constraints faced by women smallholder farmers are of paramount importance for realising equitable and sustainable global food security. These constraints, rooted in gender-based norms, resource disparities and limited agency, undermine the full potential of women's contributions to food security and economic prosperity. Implementing targeted strategies to overcome these barriers not only promotes gender equality but also harnesses the untapped capacities of women in agriculture.

Published by Burleigh Dodds Science Publishing Limited, 2025.

Strengthening the individual and collective voice and agency of women smallholders plays a crucial role in driving change and addressing the constraints to their participation towards improved food security outcomes and socioeconomic well-being. However, there are still insufficient efforts that enhance women's visibility and voice, given their significant contribution to food and nutritional security (FAO, 2020a; Onyalo, 2019; Klugman et al., 2014).

6.1 Challenging discriminatory gender norms

Intervention efforts towards improving food security outcomes should offer both women and men as well as influential people such as power holders and traditional and religious leaders the opportunities to critically examine and challenge the customs, stereotypic values and beliefs, power relations, and exclusionary practices, laws and policies prevalent within their society that continue to restrict women's agency and opportunities (Lecoutere et al, 2023; FAO, 2020b; Galiè, 2019). For instance, achieving a more equal distribution of care and domestic tasks and childcare requires that men and boys are actively engaged in changing harmful gender norms that put a disproportionate burden of unpaid care and domestic work on women.

Addressing discriminatory gender norms enhances women's agency and bargaining power, fosters equitable time allocation, improves redistribution of unpaid care and domestic responsibilities, and inclusion in decision-making processes at the household, local community and national level (Lecoutere et al., 2023; Botreau and Cohen, 2020). This empowerment extends to women's ability to make agricultural production decisions, strategic livelihood choices, and engage in income-earning opportunities and market transactions (Bryan et al., 2022).

Relaxing or changing discriminatory norms over time can lead to positive gender norms that support the equal distribution of resources and decision-making power in agriculture and food systems. To support the role that women smallholders play in safeguarding household food and nutrition security in their households, development efforts focus on providing skills in dietary diversity and enhancing economic capacities to produce or purchase nutritious food. These efforts must strive to address prevailing harmful norms that disempower women smallholders by limiting their ability to produce and purchase sufficient and nutritious food. This in turn is essential for improving optimal health and nutrition outcomes for women and their families (Lecoutere et al., 2023; IFPRI, 2021; Mulema et al., 2019; Galiè and Kantor, 2016; Twyman et al., 2015).

6.2 Enhanced bargaining power through increased access to resources and services

Enhanced bargaining power in various aspects of agricultural production is an important empowerment outcome for women, as it positively impacts household food security (Botreau and Cohen, 2020). It facilitates women's improved access to resources, information and services, increased asset ownership, improved mobility, access to markets, and improved time use and agency. Improving women's education in nutrition alongside enhancing the voice of their bargaining power and agency in making food-related decisions within the household has been proven to increase control over budget allocations for household food expenses (Botreau and Cohen, 2020).

A review of women's land rights as a means to alleviate poverty by Meinzen-Dick et al. (2019) found that improved women's land rights were associated with increased bargaining power and decision-making capacity. It also discovered a strong correlation between women's land rights and favourable outcomes in agriculture and food systems, including food security and consumption. However, evidence on the connection between women's land rights and certain other outcomes, such as access to credit and technology, agricultural productivity and non-agricultural livelihoods, remains inconsistent and inconclusive (World Bank, 2021; Meinzen-Dick et al., 2019). Addressing the gender disparities in land rights requires raising awareness among both women and men about equal rights to land ownership, advocating for communities to extend land titles or tenure rights to women and ensuring that legal frameworks protect those rights (FAO, 2023).

Enhancing control over food budget allocations leads to improved food and nutrition security for women and their households. However, Tavenner et al. (2019) show that women's authority over agricultural and livestock production and marketing is reduced when intervention efforts primarily focus on increasing market-oriented smallholder production. This diminished control extends to activities that may have previously been within their control, yet concurrently strengthens men's influence over production benefits. Conversely, efforts that promote increased diversification of crops and livestock are likely to have a more favourable impact on women's control. As these are household-level decisions, gender-responsive approaches should recognise how gender norms and dynamics influence the implementation of household-level strategies by considering the roles of all household members, including men and boys (Galiè, 2019).

6.3 Participatory development of agricultural technologies and food policies

Research underlines the significance of allocating ample time for women smallholder farmers to engage and leverage their knowledge, skills and experiences in the early design, development and scaling of agricultural innovations (Thakur, 2023; Reynolds et al., 2020), as well as in the design of food and agricultural policies that promote gender-equitable outcomes (Botreau and Cohen 2020). However, without an enabling inclusive and supportive environment devoid of gender-based discrimination, women's ability to publicly advocate for their needs and influence decisions that affect their livelihoods and contributions to sustainable global food security is greatly hindered (Margolies et al., 2023).

The adoption of agricultural technologies has been shown to present a distinct gender dynamic influenced by the differing priorities of women and men. For instance, women smallholders with limited resources at their disposal and time constraints imposed by their domestic and caregiving roles often prefer food crops or crops that are easy to grow and demand less cooking time. They also prioritise nutritional value and palatability, reflecting their integral role in providing nutrition within their households (Reynolds et al., 2020).

For development programs that have the potential to effectively prevent and reduce time and income poverty among women to be inclusive and gender-responsive, they need to be aligned to address the distinct time, labour and mobility constraints that women smallholder farmers face (Van de Velde et al., 2020; UN Women, 2020). Similarly, efforts to enhance women smallholders' agency by growing their networks and social capital, creating leadership spaces or disseminating group training programs focused on agricultural skills, technologies and innovations should follow this approach while addressing the associated barriers in the civic space (UN Women, 2020). Inclusive approaches not only amplify women smallholders' voice and agency and increase productivity and incomes, but they also promote women's roles as agents of change in achieving better food security, overall well-being and more secure livelihoods (ISF et al., 2020).

Finally, while women smallholders and adolescent girls in low- and middle-income countries may face varied vulnerability to crises, we should avoid portraying women smallholders solely as victims, incapable of significantly contributing to solutions that address the complex shocks and stressors they encounter. Instead, we should acknowledge their agency in driving positive change through active participation and empowerment (IFPRI, 2023; Bryan et al., 2022; Quisumbing et al., 2008).

7 Potential strategies to address constraints: tailoring interventions to aspirations and life cycle of women smallholders

Addressing the constraints faced by women smallholder farmers requires a comprehensive approach that recognises the evolving roles and aspirations throughout their life cycle and considers the negative impact of harmful social and gender norms on those roles and aspirations. Policy and development initiatives aiming to empower women smallholders in agriculture should consider raising the aspirations that women set for themselves, their families and their communities.

Lecoutere et al. (2023) propose the inclusion and support of female role models who challenge gender stereotypes in influential roles such as mass communication campaigns, extension programs, leadership, school programs, academia and research bodies. This approach holds the potential to elevate the aspirations of women and girls while diminishing the aspirational gaps in various agricultural stages of their lives. It promotes their active involvement in realising global food security and empowers them to make informed decisions regarding how they allocate and invest their time and limited resources within various segments of the agriculture and food system.

Understanding the present socioeconomic context, aspirations in agricultural work and future desires of women smallholder farmers is crucial. This understanding facilitates the effective tailoring of food security policies and programs and supports targeted interventions that acknowledge and tackle the specific challenges experienced by women in different life stages (Kosec et al., 2021; Nandi et al., 2022). Such interventions can support their agency and reduce gender inequality in households (DeJaeghere et al., 2022; Kosec et al., 2021, 2018).

This could be done by providing women with relevant information about agricultural opportunities, knowledge and skills, and the supportive environments they need to achieve their aspirations (Nandi and Nedumaran, 2021) as well as developing markets for crops that women tend to produce (Botreau and Cohen, 2020). Moreover, the economic successes or failures related to their engagement in agriculture can influence their children and future generations to continue the agricultural family legacy or move away from it (Nandi et al., 2022). To achieve global food security, it is crucial to cater to the unique requirements of different women in various contexts, and stakeholders must be held accountable to women at every level of the food system (FAO, 2020b; Botreau and Cohen, 2020).

8 Potential strategies to address constraints: providing data to support women smallholders' roles towards global food security

In the past decade, progress has been made in availing sex- and age-disaggregated qualitative and quantitative data and research tools to measure gender equality and women's empowerment. However, significant challenges and gaps persist in the availability and utilisation of both quantitative and

Published by Burleigh Dodds Science Publishing Limited, 2025.

qualitative data to effectively measure and analyse gender equality and women's empowerment over time and across the various segments of agriculture and food systems.

Adopting an intersectional perspective to fill these gender equality and women's empowerment data gaps is useful for designing more rigorous evidence-based research and targeting effective gender-responsive food security programs and policies (Lecoutere et al., 2023). It involves the explicit recognition of multiple and often overlapping and intersecting aspects such as age, gender, ethnicity, health, disability and socioeconomic, marital and migration status, and how they combine to create different modes of discrimination, social exclusion and privilege. This data could provide useful insights that help us understand the impact of overlapping sources of discrimination on women smallholders' empowerment. With data, we can also challenge the implementation of policies and interventions that continue to perpetuate inequality around the roles of women smallholders in achieving global food security by monitoring whether women benefit from them or not (Tavenner and Crane, 2019).

Recognising and quantifying the economic value of women's unpaid work and ensuring its inclusion in national statistics can provide the evidence needed to address the persistent gender inequalities undermining the untapped potential of women smallholder farmers to enhance food security and nutrition outcomes (CFS, 2023). This data can support the scope for mobilising additional resources and implementing gender-responsive budgeting to alleviate the burden of unpaid care and domestic work on women. Investing in time-saving public services such as childcare, enhanced transport and other basic infrastructure, including household-level access to water and energy, can help reduce and redistribute the responsibilities of unpaid care and domestic work within communities (UN Women, 2020; ILO, 2019).

While the importance of measuring women smallholders' work and contribution towards food security is significant, empirically quantifying and verifying the share they produce can be complex due to the limitations in data collection. This is particularly apparent in households where food production is a venture among household members, both male and female, and involves a range of joint resources and inputs that cannot be readily assigned by gender (Doss et al., 2018). Nevertheless, the implementation of rigorous gender-research methods and data-collection tools can provide valuable insights for effective recognition of the actual contributions of women smallholders in food production, storage, marketing and consumption. It can also provide legitimate data for evaluating the impact of interventions on women smallholder farmers and for monitoring the progress on gender equality and women's empowerment around their roles in achieving food security (FAO, 2023).

A continuing data weakness is the absence of consistency in the capture, analysis and publication of accurate data on the gender-based dimensions of food security and nutrition (FAO, 2020b). Widely used global food security datasets lack detailed sex-disaggregated food indicators and have inadequate representations of women smallholder farmers. Statistics on women's yields and trade volumes along agricultural value chains are dispersed in many different microlevel studies but are rarely reported globally.

To address these data needs, methods of collecting and incorporating sex-disaggregated data in national and global datasets must make the challenges that women encounter in their roles visible. Otherwise, food security policies will overlook the need to include specific measures to address gender equality and women's empowerment issues around food security. This is illustrated by a recent agricultural policy assessment by FAO (2023) that revealed that while more than 75% of the policies recognised roles and/or challenges of women in agriculture, only 19% had explicit objectives to achieve gender equality and/or enhance women's rights in agriculture.

Another data-gap concern in global datasets is the absence of comprehensive sex-disaggregated data on secure land tenure. It creates an obstacle to fully appreciating the participation of women smallholder farmers in the agricultural workforce as it diminishes women's 'visibility', voice and support systems. Recognising women smallholders as independent farmers rather than mere labourers with limited decision-making power is crucial. Dismantling institutional barriers that render women smallholders 'invisible', including the lack of secure, documented land rights, will enhance their access to other essential productive resources (Van de Velde et al., 2020). This data gap results in inadequate monitoring of women's land rights and resource availability, presenting challenges to designing effective food security programs and policies that cater to the needs of these often overlooked 'invisible farmers'.

9 Conclusion

Women smallholders fulfil the roles of producers, traders of agricultural goods and decision-makers in shaping their households' dietary preferences and consumption patterns. These roles play an integral part in supporting the fundamental aspects of global food security – food availability, accessibility, utilisation and stability within households and local communities. To effectively support and harness their full potential in these significant roles, we must recognise and value their contribution by prioritising efforts that ensure equitable access to agricultural resources, services, technologies, output markets, off-farm income opportunities and increased decision-making.

Addressing the formal and informal structural constraints faced by women smallholders and implementing gender-responsive policies and institutions

Published by Burleigh Dodds Science Publishing Limited, 2025.

are priorities on the agenda for achieving global food security. This will foster an enabling environment where women can voice their needs and preferences and participate in changing discriminative social norms into positive norms. Engaging men, boys and community leaders in challenging discriminatory norms and practices also fosters women smallholders' ability to participate in advocating for inclusive policies, enhanced access to resources and stronger agency.

Moreover, it is crucial to recognise the life cycle of women smallholder farmers and understand how it influences their diverse roles and aspirations. The aspirations and motivations in agriculture among women farmers directly influence women's decisions on agricultural resource use, investment in technology and engagement in other segments of agriculture and food systems value chains, as well as how they use agency to engage in other tasks to achieve well-being. Roles and aspirations may evolve as women age and experience changes in their family dynamics, making it necessary to promote different empowerment strategies tailored to different women.

To improve our collective understanding of what women smallholders need, we have to shift away from generalising women and instead identify distinct typologies of women according to their aspirations and life stage. This approach focuses on understanding the constraints they face through listening to their concerns and building on their knowledge and practices, which provides insights into how they participate in agriculture. This recognition also nurtures effective policy planning and the delivery of tailored interventions that empower and support women smallholders in achieving their goals at the various stages of their lives as they contribute to global food security.

10 Where to look for further information

The CGIAR GENDER Impact Platform hosts a resource hub - https://gender.cgiar.org/. The hub features cutting-edge studies on women smallholder farmers, gender-responsive agricultural innovations, and inclusive food system governance. Researchers, policymakers, and practitioners can leverage these resources to deepen their understanding of gender dynamics in global food security and drive impactful change. It also provides toolkits and methodologies that are useful for analytical research.

11 References

AGRA. 2021. Women in agribusiness value chains in Africa. A white paper on constraints and opportunities for developing a gender-responsive agribusiness sector.

Akter R, Yagi N, Sugino H, et al. (2020) Household engagement in both aquaculture and horticulture is associated with higher diet quality than either alone. *Nutrients* 12(9): 2705.

Alwab BE, Els Lecoutere E and Jones N (2022) Adolescents' capabilities and aspirations across gender and generations in Amhara. *Ethiopia, Journal of Eastern African Studies* 16(3): 472–494.

Bergman-Lodin J, Tegbaru A, Bullock R, et al. (2019) Gendered mobilities and immobilities: women's and men's capacities for agricultural innovation in Kenya and Nigeria. *Gender, Place & Culture* 26(12): 1759–1783.

Bezner Kerr R, Shumba L, Dakishoni L, et al. (2013). *Participatory, Agroecological and Gender-Sensitive Approaches to Improved Nutrition: A Case Study in Malawi.* Submission to the FAO Expert Meeting 'Nutrition-Sensitive Food and Agriculture Systems' in Preparation for ICN+21.

Botreau H and Cohen MJ (2020) Gender inequality and food insecurity: a dozen years after the food price crisis, rural women still bear the brunt of poverty and hunger. *Advances in Food Security and Sustainability* 5: 53–117.

Boudet AMM, Petesch P and Turk C (2013) *On Norms and Agency: Conversations about Gender Equality with Women and Men in 20 Countries.* Washington, DC: World Bank.

Bryan E, Alvi M, Huyer S and Ringler C (2022) *Addressing Gender Inequalities and Strengthening Women's Agency for Climate-resilient and Sustainable Food Systems.* CGIAR GENDER Impact Platform Working Paper 013. Nairobi: CGIAR GENDER Impact Platform.

Busingye M and Ibrahim H (2021) CARESOM rapid gender analysis multiple shocks brief. Retrieved from file:///C:/Users/cmuchiri/Downloads/CARESOM-Rapid-Gender-Analysis-Report-August-2021.pdf

CFS (Committee on World Food Security) (2023) *Voluntary Guidelines on Gender Equality and Women's and Girls' Empowerment in the Context of Food Security and Nutrition.* Final version as agreed by OEWG on 14 June 2023. Rome: Food and Agriculture Organization (FAO).

Choithani C (2020) Gendered livelihoods: migrating men, left-behind women, and household food security in India. *Gender, Place and Culture* 27(10): 1373–1394.

Crossland M, Paez Valencia A, Pagella T, et al. (2021) Women's changing opportunities and aspirations amid male outmigration: insights from Makueni County, Kenya. *The European Journal of Development Research* 33: 910–932.

Data2X and ILO (International Labour Organization) (2020) *Making Women's Work Visible: The 19th International Conference of Labour Statisticians (ICLS) Standards, Purpose, and Progress.* Department of Statistics. Edited here and in the text. Retrieved from https://data2x.org/wp-content/uploads/2020/11/MakingWomensWorkVisibleReport_FINAL.pdf

DeJaeghere J, Wiger NP, Le H, et al. (2022) Why do aspirations matter for empowerment? Discrepancies between the A-WEAI domains and aspirations of ethnic minority women in Vietnam. *World Development* 159: 106057.

Doss C, Meinzen-Dick R, Quisumbing A and Theis S (2018) Women in agriculture: four myths. *Global Food Security* 16: 69–74.

Economist Impact (2022) The Global Food Security Index report 2022. https://impact.economist.com/sustainability/project/food-security-index/download-the-index

Elias M, Cole SM, Quisumbing A, et al. (2021) Assessing women's empowerment in agricultural research. In: Pyburn R and Eerdewijk A (eds) *Advancing Gender Equality through Agricultural and Environmental Research: Past, Present, and Future.* Washington, DC: International Food Policy Research Institute, pp. 329–364.

Elias M, Mudege N, Lopez D, et al. (2018) Gendered aspirations and occupations among rural youth, in agriculture and beyond: a cross-regional perspective. *Journal of Gender Agriculture and Food Security (Agri-Gender* 3(1): 82–107.

Faiez HE (2018) The role of women in household food security among small scale farmers in Gedarif and Rahad localities of Eastern Sudan. *Journal of Social and Humanities Sciences Research* 5(25): 2103–2117.

FAO (2020a) *Gender Equality and Women's Empowerment in the Context of Food Security and Nutrition. A Scoping Paper.* Rome: FAO.

FAO (2020b) *Gender Equality and Women's Empowerment in the Context of Food Security and Nutrition. A Scoping Paper*. CFS Forum 2017. Rome: FAO.

FAO (2021a) *NENA Regional Network on Nutrition-Sensitive Food System. Technical Brief: Empowering Women and Ensuring Gender Equality in Agriculture and Food Systems to Achieve Better Nutrition*. Rome: FAO.

FAO (2021b) *The State of Food and Agriculture 2021. Making Agriculture and Food Systems More Resilient to Shocks and Stresses, In Brief.* Rome: FAO.

FAO (2021c) *The State of Food Security and Nutrition in the World 2021. Transforming Food Systems for Food Security, Improved Nutrition and Affordable Healthy Diets for All.* Rome: FAO.

FAO (2022) *The State of Food Security and Nutrition in the World 2022. Repurposing Food and Agricultural Policies to Make Healthy Diets More Affordable*. Rome: FAO.

FAO (2023) *The Status of Women in Agriculture and Food Systems*. Rome: FAO.

FAO and UN Women (2021) *National Gender Profile of Agriculture and Rural Livelihoods, Bosnia and Herzegovina*. Rome: FAO.

FAO, IFAD, UNICEF, WFP and WHO. 2023. In Brief to The State of Food Security and Nutrition in the World 2023. Urbanization, agrifood systems transformation and healthy diets across the rural–urban continuum. Rome, FAO. https://doi.org/10.4060/cc6550en

Farnworth CR, Jafry T, Bharati LB, et al. (2021) From working in the fields to taking control. Towards a typology of women's decision-making in wheat in India. *The European Journal of Development Research* 33(3): 526–552.

Farnworth CR, Jumba H, Otieno PE, et al. (2023) Gender roles and masculinities in leveraging milk for household nutrition: evidence from two districts in Rwanda. *Food Policy* 118, 102486.https://doi.org/10.1016/j.foodpol.2023.102486.

FSIN and Global Network Against Food Crises. 2023. GRFC 2023. Rome. Retrieved from https://www.fsinplatform.org/sites/default/files/resources/files/GRFC2023-hi-res.pdf

Galiè A (2019) Leveraging gender for food and nutrition security through agriculture. In: Ferranti P, Berry EM, and Anderson JR (eds) *Encyclopedia of Food Security and Sustainability* 3: 426–431.

Galiè A and Kantor P (2016) From gender analysis to transforming gender norms: using empowerment pathways to enhance gender equity and food security in Tanzania. In: Njuki J, Parkins J, and Kaler A (eds) *Transforming Gender and Food Security in the Global South*. London: Routledge.

Glazebrook T, Noll S and Opoku E (2020) Gender climate change, gender bias, and women's farming in the global South and North. *Agriculture* 10(7): 267.

Halim D, Perova E and Reynolds S (2021) *Childcare and Mothers' Labor Market Outcomes in Lower- and Middle-Income Countries World Bank Group*. Research Working Paper 9828. East Asia and the Pacific Region Gender Innovation Lab & Gender

Global Theme Policy. Retrieved from https://documents1.worldbank.org/curated/en/450971635788989068/pdf/Childcare-and-Mothers-Labor-Market-Outcomes-in-Lower-and-Middle-Income-Countries.pdf

Hernandez MA, Alarcon C, Berrospi ML, et al. (2023) Cultural and economic barriers and opportunities for the participation of women in agricultural production systems: a case study in Guatemala. *Frontiers in Sustainable Food Systems* 7. https://doi.org/10.3389/fsufs.2023.1185756

HKI (Helen Keller International) (2015) *Nurturing Connections – Adapted for Homestead Food Production and Nutrition*. Bangladesh: HKI.

HLPE (High-Level Panel of Experts on Food Security and Nutrition Report #15) (2020) *Food Security and Nutrition: Building a Global Narrative Towards 2030*. Rome: Food and Agriculture Organization of the United Nations (FAO).

HLPE (High-Level Panel of Experts on Food Security and Nutrition Report #18) (2023) *Reducing Inequalities for Food Security and Nutrition*. Rome: Food and Agriculture Organization of the United Nations (FAO).

IFPRI (International Food Policy Research Institute) (2021) *Global Food Policy Report 2021: Transforming Food Systems after COVID-19*. Washington, DC: IFPRI.

IFPRI (International Food Policy Research Institute) (2023) *Global Food Policy Report 2023: Rethinking Food Crisis Responses*. Washington, DC: IFPRI.

ILO (International Labour Organization) (2019) *Empowering Women in the Rural Economy. Decent Work in The Rural Economy*. Policy Guidance Notes. Retrieved from https://www.ilo.org/sites/default/files/wcmsp5/groups/public/%40ed_dialogue/%40sector/documents/publication/wcms_601071.pdf - ISBN: 9789220308028 (web PDF)

ISF Advisors and the Mastercard Foundation Rural and Agricultural Finance Learning Lab (2020) *Pathways to Prosperity: Understanding Women's Rural Transitions and Service Needs*. Washington. Retrieved from https://isfadvisors.org/understanding-womens-rural-transitions-and-service-needs/

Katrina K, Cecilia HM, Emily S, et al. (2021). Perceptions of relative deprivation and women's empowerment, World Development, Volume 138.

Kabeer N (2018) *Gender, Livelihood Capabilities and Women's Economic Empowerment. Reviewing Evidence Oover the Life Course*. Gender and Adolescence: Global Evidence (GAGE) Programme Office, Overseas Development Institute. https://www.gage.odi.org/publication/gender-livelihood-capabilities/

Klugman J, Hanmer L, Twigg S, et al. (2014) *Voice and Agency: Empowering Women and Girls for Shared Prosperity*. Washington, DC: World Bank Group.

Koo J, Azzarri C, Mishra A, et al. (2022) *Effectively Targeting Climate Investments: A Methodology for Mapping Climate–Agriculture–Gender Inequality Hotspots*. CGIAR GENDER Platform Working Paper #005. Nairobi.CGIAR GENDER Platform. Retrieved from https://cgspace.cgiar.org/server/api/core/bitstreams/5c1e74cc-93e9-4f16-b459-864bad60257c/content

Kosec K, Akramov K, Mirkasimov B, et al. (2018) *Aspirations and Women's Empowerment Evidence from Kyrgyzstan*. IFPRI Discussion Paper 01786, December. Washington, DC: Development Strategy and Governance Division.

Kosec K, Hidrobo M, Gartaula H, et al. (2023) *Making Complementary Agricultural Resources, Technologies and Services More Gender Responsive*. CGIAR GENDER Impact Platform Working Paper #010. Nairobi: CGIAR GENDER Impact Platform.

Lecoutere E, Achandi E, Ampaire EL, et al. (2023) *Fostering an Enabling Environment for Equality and Empowerment in Agri-Food Systems*. CGIAR GENDER Impact Platform Working Paper #009. Nairobi: CGIAR GENDER Impact Platform.

Locke C, Muljono CM and Morgan M (2017) Innovation and gendered negotiations: insights from six small-scale fishing communities. *Fish and Fisheries* 18(5): 943–957.

Malapit HJ, Meinzen-Dick RS, Quisumbing AR, et al. (2020) Women: transforming food systems for empowerment and equity. IFPRI book chapters, in: *2020 Global Food Policy Report: Building inclusive food systems,* chapter 4, pp. 36–45. International Food Policy Research Institute (IFPRI). https://doi.org/10.2499/9780896293670_04

Margolies A, Colantuoni E, Morgan R, et al. (2023) The burdens of participation: a mixed-methods study of the effects of a nutrition-sensitive agriculture program on women's time use in Malawi. *World Development* 163, 106122.

Meinzen-Dick R, Quisumbing A, Doss C, et al. (2019) Women's land rights as a pathway to poverty reduction: framework and review of available evidence. *Agricultural Systems* 172: 72–82.

Mulema AA, Jogo W, Damtwe E, et al. (2019) Women farmers' participation in the agricultural research process: implications for agricultural sustainability in Ethiopia. *International Journal of Agricultural Sustainability* 17(2): 127–145.

Nandi R and Nedumaran S (2021) Understanding the aspirations of farming communities in developing countries: a systematic review of the literature. *The European Journal of Development Research* 33: 809–832.

Nandi R, Pratheepa CM, Swamikannu N, et al. (2022) Farm parent and youth aspirations on the generational succession of farming: evidence from South India. *Frontiers in Sustainable Food Systems* 5. https://doi.org/10.3389/fsufs.2021.804581

Njuki J, Eissler S, Malapit H, et al. (2023) A review of evidence on gender equality, women's empowerment, and food systems. In: Braun J, Afsana K, Fresco LO, et al. (eds) *Science and Innovations for Food Systems Transformation*. Springer, Cham. https://doi.org/10.1007/978-3-031-15703-5_9

Odera EV, Mshenga PM, et al. (2016) Analysis of common bean cross border trade in East Africa: the case of Kenya. *Journal of Agriculture* 3(2). Retrieved from https://researchjournali.com/pdf/2457.pdf

Okpara UT and Anugwa IQ (2022) Harms to community food security resulting from gender-based violence. *Land* 11: 2335.

Onyalo PO (2019) Women and agriculture in rural Kenya: role in agricultural production. *International Journal of Humanities and Social Science* 5: 1.

Petesch, P. and Badstue, L. (2020). Gender Norms and Poverty Dynamics in 32 Villages of South Asia. International Journal of Community Well-Being. 3. 10.1007/s42413-019-00047-5.

Qanti SR, Peralta A and Zeng D (2022) Social norms and perceptions drive women's participation in agricultural decisions in West Java, Indonesia. *Agriculture and Human Values* 39: 645–662.

Quisumbing A, Cole S, Elias M, et al. (2023) *Measuring Women's Empowerment in Agriculture: Innovations and Evidence*. Background Paper for *The Status of Women in Agriculture and Food Systems 2023*. Nairobi: CGIAR Gender Platform.

Quisumbing A, Heckert J, Faas S, et al. (2021) The pro-WEAI for Market Inclusion Study Team 2021 Women's empowerment and gender equality in agricultural value chains: evidence from four countries in Asia and Africa. *Food Security* 13: 1101–1124.

Quisumbing AR, Meinzen-Dick RS, Bassett L, et al. (2008). *Helping Women Respond to the Global Food Price Crisis*. IFRPI Policy Brief. Washington, DC: International Food Policy Research Institute.

Ragasa, C., Aberman, N. L., and Mingote, C. A. (2019). Does providing agricultural and nutrition information to both men and women improve household food security? Evidence from Malawi. *Global Food Security*, 20, 45–59.

Ragasa C, Malapit H, Rubin D, et al. (2021) *It Takes Two: Women's Empowerment in Agricultural Value Chains in Malawi*, IFPRI Discussion Paper 02006. Washington, DC: International Food Policy Research Institute.

Rakotomanana H, Hildebrand D, Gates GE, et al. (2020) Maternal knowledge, attitudes, and practices of complementary feeding and child undernutrition in the Vakinankaratra Region of Madagascar: a mixed-methods study. *Current Developments in Nutrition* 4(11).

Reynolds TW, Tobin D, Otieno G, et al. (2020) Differences in crop selection, resource constraints, and crop use values among female- and male-headed smallholder households in Kenya, Tanzania, and Uganda. *Journal of Agriculture, Food Systems, and Community Development* 9(4): 65–92.

Rietveld AM, van der Burg M and Groot JCJ (2020) Bridging youth and gender studies to analyse rural young women and men's livelihood pathways in Central Uganda. *Journal of Rural Studies*. 75: 152–163. https://doi.org/10.1016/j.jrurstud.2020.01.020

Santpoort R, Steel G, Mkandawire A, et al. (2021) The land is ours: Bottom-up strategies to secure rural women's access, control and rights to land in Kenya, Mozambique, Senegal and Malawi. *Frontiers in Sustainable Food Systems* 5: 697314.

Seedat S and Rondon M (2021) Women's well-being and the burden of unpaid work. *BMJ* 374: n1972. doi:10.1136/bmj.n1972.

Tavenner K and Crane TA (2019) Beyond "women and youth": applying intersectionality in agricultural research for development. *Outlook on Agriculture* 48(4): 316–325.

Tavenner K, Wijk M, Fraval S, et al. (2019) Intensifying inequality? Gendered trends in commercializing and diversifying smallholder farming systems in East Africa. *Frontiers in Sustainable Food Systems* 3. https://doi.org/10.3389/fsufs.2019.00010

Thakur N (2023) Women farmers and technologies in agriculture: a review of current practices. In: Munshi S and Singh M (eds) *Women Farmers: Unheard Being Heard. Sustainability Sciences in Asia and Africa*. Singapore: Springer. 169–182.

Twyman J, Muriel J and García MA (2015) Identifying women farmers: informal gender norms as institutional barriers to recognizing women's contributions to agriculture. *Journal of Gender, Agriculture and Food Security* 1(2): 1–17.

Ulaya F (2021) *Agriculture and Nutrition in Bangladesh: Mapping Evidence to Pathways Effect of Participation in Orange-fleshed Sweetpotato focused Nutrition Education Intervention on Nutrition Knowledge, Infant and Young Child Feeding Practices, and Women Dietary Diversity in Zomba, Mulanje and Chikwawa*. MSc Thesis, Lilongwe University of Agriculture and Natural Resources, Malawi.

UN (United Nations) (2019) *Guidelines for Producing Statistics on Asset Ownership from a Gender Perspective*. Studies in Methods, Series F. New York: Department of Economic and Social Affairs Statistical Division.

UN Women. (2019). Progress of the World's Women 2019–2020: Families in A Changing World. United Nations Entity for Gender Equality and the Empowerment of Women. Retrieved from https://www.unwomen.org/sites/default/files/Headquarters/

Attachments/Sections/Library/Publications/2019/Progress-of-the-worlds-women-2019-2020-en.pdf

UN Women (2020) World survey on the role of women in development: Report of the Secretary-General (2019): Why addressing women's income and time poverty matters for sustainable development. Rome: United Nations.

Van de Velde P, Stanley V and Stickler MM (2020) Invisible farmers: why recognizing and supporting women farmers is key to food and nutrition security. *World Bank Blogs*. World Bank. Blog retrieved from https://blogs.worldbank.org/en/developmenttalk/invisible-farmers-why-recognizing-and-supporting-women-farmers-key-food-and

Van den Bold M, Quisumbing AR and Gillespie S (2013) Women's empowerment and nutrition: an evidence review, IFPRI Discussion Paper 01294. Available at SSRN: https://ssrn.com/abstract=2343160 or http://dx.doi.org/10.2139/ssrn.2343160

rVan den Broeck G and Kilic T (2019) "Dynamics of Off-farm Employment in Sub-Saharan Africa: A Gender Perspective," *World Development, Elsevier* 119(C): 81–99. .

Verkaart S, Mausch K and Harris D (2018) Who are those people we call farmers? Rural Kenyan aspirations and realities. *Development in Practice* 28(4): 468–479.

World Bank (2015) *Ending Poverty and Hunger by 2030: An Agenda for the Global Food System*. Washington, DC: World Bank.

World Bank (2021) *Women, Business and the Law 2021*. Washington, DC: World Bank.

Yosef S, Jones AD, Chakraborty B and Gillespie S (2015). Agriculture and Nutrition in Bangladesh: Mapping Evidence to Pathways. *Food and Nutrition Bulletin*. 36(4):387–404. doi:10.1177/0379572115609195

Chapter 2

Women smallholder farmers and the achievement of the UN Sustainable Development Goals

Rhiannon Pyburn[1] and Julie Newton[2], KIT Royal Tropical Institute, The Netherlands

1 Introduction

Since 2016, the Sustainable Development Goals (SDGs) and, before them, the Millennium Development Goals (2000–2015) have been a reference point for governments, non-governmental organizations, research for development organizations, civil society, and private companies, internationally. The SDGs are ambitious political aspirations to guide policy and practice across 17 domains and focus not only on low- and middle-income countries (LMICs) but also on so-called developed and emerging economies. They are globally agreed by all 196 member states of the United Nations (UN) and are a time-bound agenda to guide political action and development efforts until 2030. Most of the 169 targets for the SDGs relate in one way or another to food and farming (Terlau et al., 2018). But what do these SDGs mean for smallholder women farmers? This question is particularly salient given that agrifood systems are

1 Gender Team Lead, KIT Royal Tropical Institute.
2 Senior Advisor, KIT Royal Tropical Institute.

http://dx.doi.org/10.19103/AS.2024.0148.02

an even more important livelihood for women than men in many countries: comprising 66% of women's employment in sub-Saharan Africa compared to 60% for men and 71% in southern Asia versus 47% for men (FAO, 2023). Women also account for a greater share of agricultural employment than men – well over 50% in many sub-Saharan countries and about half in some Southeast Asian nations (ibid.). Globally 36% of working women and 38% of working men are employed in agrifood systems (ibid). This chapter explores the impact of the SDGs on smallholder women's lives and how smallholder women are contributing to the achievement of these goals.

The chapter is based on a narrative review that sought to identify the evidence available related to smallholder women farmers and the UN SDGs. The guiding questions were two-directional: *What is the relevance of the SDGs for smallholder women?* and *what do smallholder women contribute to the achievement of the SDGs?* A narrative review is appropriate for these questions as our intent is to provide an overall summary of the (limited) literature, providing interpretation and critique (Sukhera, 2022). Initial literature searches were done in June 2023[3] using different combinations of the terms – SDGs, women farmers, women smallholders, farming, and contribution – in the search engine Google Scholar. The initial searches identified 35 sources – 25 research papers, 5 books and book chapters, and 5 reports. Of those sources, 13 were most directly related to the topic. From the references in these publications, further relevant sources were drawn out. In addition, we analyzed the latest 'Gender Snapshot' of the SDGs (UN Women, 2023d) to understand not only where gender equality, women's empowerment, or sex disaggregation is clear in targets and indicators but specifically to narrow down *where* smallholder agriculture and smallholder women farmers come out or are implicated across the 17 SDGs. We analyze where stronger links to smallholder agriculture and smallholder women farmers could be articulated.

This chapter begins by exploring the visibility of women smallholder farmers in the SDGs, highlighting what the SDGs bring to women working in this sector and key omissions (Section 2). We then turn to what smallholder women farmers contribute to the achievement of the SDGs (Section 3). On that basis we articulate limits and critiques of the SDGs in relation to smallholder women farmers and smallholder farming (Section 4). We conclude with propositions to better engage smallholder women farmers as agents and as beneficiaries as we move into the remaining 6 years of SDG implementation (Section 5) (Fig. 1).

3 The authors acknowledge and thank Ekaterina Gualoto (KIT Intern May/June 2023) for her contribution to the initial literature scan.

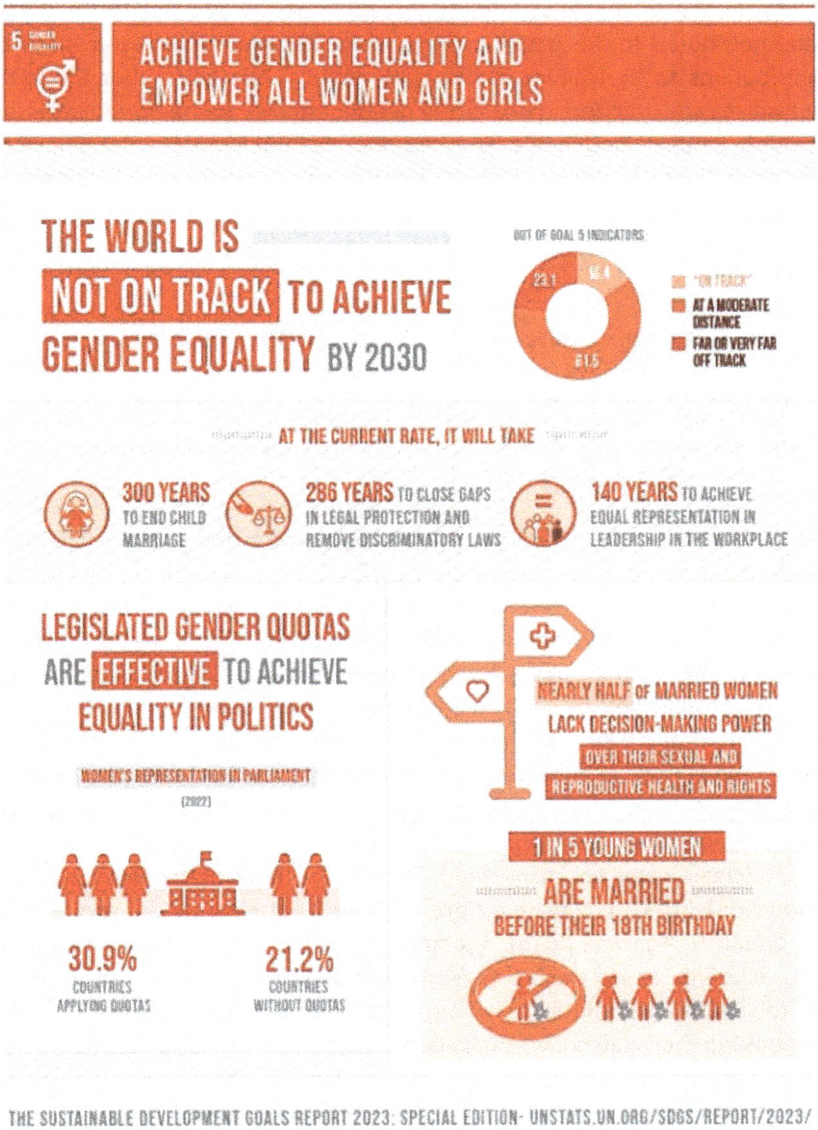

Figure 1 Overview of progress on Sustainable Development Goal 5.

2 Smallholder women farmers within the Sustainable Development Goals

The SDGs have reinforced the importance of women farmers on international agendas (Doss et al, 2018; Agarwal, 2018; FAO, 2023) and have helped to highlight the importance of smallholder farmer productivity and agrarian gender justice as central to ending global hunger (Cohen, 2019). They have also contributed to the growing demand for reliable indicators of women's contributions to agriculture (Doss et al., 2018; FAO, 2023). But the SDGs themselves are criticized in relation to gender equality for being too universal, ignoring context, and not acknowledging the feminization of agriculture[4] (Glazebrook and Opoku, 2020). Further, despite the gains gleaned from having an SDG focus specifically on gender equality, according to the UN Women's *SDG Gender Snapshot 2023*, progress is very limited and strikingly disappointing.

> *The world is not on track to achieve gender equality by 2030. At the global level, none of the 18 indicators 'met or almost met' the targets and only one is 'close to target'. At the current rate of progress, it is estimated that it will take up to 286 years to close gaps in legal protection and remove discriminatory laws, 140 years for women to be represented equally in positions of power and leadership in the workplace, and 47 years to achieve equal representation in national parliaments.*[5]

The SDGs represent broad domains. A review of farming and smallholder farming vis-à-vis the SDGs (see Terlau et al., 2018) finds that smallholder farmers have a double function in relation to the SDGs: as beneficiaries of the SDGs and as agents or promoters of sustainable development (Terlau et al., 2018). As noted by other authors, SDGs 1 and 2 on ending poverty and ending hunger, will only be achieved by supporting farmers: '*smallholder farmers are written into the SDGs and their respective targets .... the SDGs focus first on supporting and empowering smallholder farmers and second on how empowered smallholder farmers and family farmers can help achieve the SDGs*' (Terlau et al., 2018). In fact, smallholder farming is strongly linked to many SDGs, including SDG 6 (clean water and sanitation), SDG 12 (responsible consumption and production), SDG 13 (climate action), SDG 14 (life below water), and SDG 15 (life on land) (Agarwal, 2018). But smallholder farmers fall short of potential as a backbone for sustainable development due to socioeconomic limitations and related needs, including new organizational forms (cooperative and start-ups through the value chain), financial support, access to farming information, suitable technological solutions and innovations, and an enabling institutional environment (Terlau et al., 2018). The same authors conclude that smallholder farmers can only fulfill their potential as agents of sustainable development if their basic needs are satisfied.

4 Note: the 'feminization of agriculture' is a highly debated and broadly used term. See Kawarazuka et al., 2022, for a debunking and contextualizing of the myths surrounding the term. It is important to note that the most recent FAO report on the status of women in agrifood systems found that '*women do not constitute the majority of agricultural workers globally, nor is their share in agriculture increasing in most regions pointing to little evidence of "feminization of age" at the global or regional level*' (FAO, 2023: 30).

5 https://knowledge-hub.circle-economy.com/frameworks/629/1719

Extending the critique of the SDGs vis-à-vis smallholder farmers to the reality of smallholder *women* farmers, another set of issues arises (see Table 1). Remarkably, only SDG 2 *explicitly* highlights smallholder women farmers. SDG 2 includes targets aimed at doubling agricultural productivity and incomes of small-scale food producers, particularly women. This is monitored through the only indicator across all the SDGs to refer directly to smallholder women farmers: *'2.3.2. Average income of small-scale food producers, by sex and indigenous status'*. Sex-specific indicators are included across nine of the SDGs – namely SDG 1, SDG 2, SDG 3, SDG 4, SDG 5, SDG 8, SDG 10, SDG 11, and SDG 16. They are missing in the remaining eight, namely SDG 6, SDG 7, SDG 9, SDG 12, SDG 13, SDG 14, SDG 15, and SDG 17 (UN Women, 2023d). However, not all of the sex-specific indicators are relevant to women smallholder farmers. Table 1 offers an overview of all SDGs, notes those that have indicators referring explicitly to women as identified by *Gender Snapshot 2023*, and provides analysis as to the relevance of each for smallholder women farmers, specifically.

Table 1 clearly illustrates the limited reference to smallholder women farmers across the SDGs, as well as the lacuna that this implies. But whether or not they are referred to explicitly, smallholder women farmers are implicit as either agents or beneficiaries throughout many of the SDGs. Beyond SDG 5 (on gender equality), women smallholder farmers' livelihoods are embedded in the problems many SDGs seek to solve and smallholder women could be, and in many cases are, pivotal to their realization. We consider each SDG in relation to smallholder women farmers in the paragraphs that follow.

SDG 1 (No poverty) is highly relevant for smallholder women farmers. One in ten women live in extreme poverty (10,3%), most of whom are in sub-Saharan Africa (UN Women, 2023d). Gender gaps are highest amongst women aged 25–34 (ibid.). Farming is an important livelihood and a pathway out of poverty for women. Access to social protection across different intersectional markers like gender, age, and position in lifecycle (child, pregnant/lactating, school going, older) is a key strategy to build resilience against poverty and provide more flexibility in the choice of livelihoods and flexibility in adapting to climate change which affects smallholder agriculture food systems. To achieve zero poverty for all women and girls by 2030, there is a need for simultaneous action on access to land, health care, family planning, education and labor market, and gender-responsive social protection (ibid.). All SDG 1 sex or gender-specific indicators are directly relevant for smallholder women farmers as they aim to track those living below the poverty line, social protection coverage and access to secure tenure land rights.

SDG 2 (No hunger) is highly relevant for small-scale agriculture and particularly for women farmers who are widely recognized as key players in ending hunger and supporting more resilient and sustainable food production

Table 1 Most relevant Sustainable Development Goals vis-à-vis women smallholder farmers

SDG	Indicators directly or indirectly relevant for smallholder women farmers
	All gender-specific indicators are directly relevant for smallholder women farmers helping to track those living below the poverty line, social protection coverage, access to secure tenure land rights, and so on. 1.1.1 Proportion of the population living below the international poverty line by **sex**, **age**, employment status, and geographic location (urban/rural). 1.2.1 Proportion of population living below the national poverty line, by **sex** and **age.**1.2.2 Proportion of men, **women,** and children of all ages living in poverty in all its dimensions according to national definitions. 1.3.1 Proportion of population covered by social protection floors/systems, by **sex**, distinguishing children, unemployed persons, older persons, persons with disabilities, pregnant women, newborns, work-injury victims, and the poor and the vulnerable. 1.4.2 Proportion of total adult population with **secure tenure rights to land**, (a) with legally recognized documentation, and (b) who perceive their rights to land as secure, by **sex** and type of tenure.
	One indicator directly relevant to smallholder women farmers. 2.3.2 **Average income** of **small-scale food producers,** by **sex** and **indigenous status.**
	No mention of smallholder women farmers but could be used to support analysis of the **health** of smallholder women farmers across different ages and access to health care. Also, links could be made to the importance of healthy and sufficient food, which contributes to good health and well-being and which smallholder women produce.
	4.3.1 Participation rate of youth and adults in formal and non-formal education and training in the previous 12 months, by **sex.** 4.5.1 **Parity indices** (**female**/male, **rural/urban**, bottom/top wealth quintile and others such as disability status, indigenous peoples and conflict-affected, as data become available) for all education indicators on this list that can be disaggregated. 4.7.1 Extent to which (i) global citizenship education and (ii) education for sustainable development are mainstreamed in (a) national education policies; (b) curricula; (c) teacher education; and (d) student assessment. No direct mention of smallholder women farmers, but potential to explore their access to formal/non-formal education as pathway out of agriculture; and access to agricultural extension education.

5 Gender equality	No direct mention of smallholder women farmers, but various indicators directly relevant relating to secure ownership of agricultural land, legal frameworks for security of land tenure; time spent on unpaid care roles. Also relevant links to other legal frameworks supporting equality and tracking of public expenditure towards this goal. 5.1.1 Whether or not legal frameworks are in place to promote, enforce and monitor equality and non-discrimination on the basis of **sex.** 5.4.1 Proportion of **time spent** on unpaid domestic and care work, by **sex**, age and **location.** 5.a.1 (a) Proportion of total agricultural population with **ownership or secure rights over agricultural land**, by sex; and (b) share of women among owners or **rights-bearers of agricultural land**, by **type of tenure.** 5.a.2 Proportion of countries where the **legal framework (including customary law)** guarantees **women's equal rights to land ownership and/or control.**
6 Clean water and sanitation	No gender-specific indicators. However, there is a clear need to track smallholder women farmers' access to clean drinking water and for agriculture use.
7 Affordable and clean energy	No gender-specific indicators.
8 Decent work and economic growth	No direct mention of smallholder women farmers, yet relevant to look at wages for women farmers across intersectional markers (age, migrant status) working as producers in more commercial agricultural value chains at different nodes; and labor rights and safety. No indicators to analyze access to different forms of financial input. 8.3.1 Proportion of **informal employment** in total employment, by sector and sex. 8.5.1 Average **hourly earnings** of employees, by sex, age, occupation and persons with disabilities. 8.8.2 Level of **national compliance with labor rights** (freedom of association and collective bargaining) based on International Labor Organization (ILO) textual sources and national legislation, by **sex** and **migrant status.**
9 Industry, innovation and infrastructure	No gender-specific indicators and not relevant.
10 Reduced inequalities	No direct link for smallholder women farmers unless working in commercial agriculture, where indicators could be useful to look at income and perceptions of discrimination. 10.2.1 Proportion of people living below 50% of **median income**, by sex, age, and persons with disabilities. 10.3.1 Proportion of population reporting **having personally felt discriminated against** or harassed in the previous 12 months on the basis of a ground of discrimination prohibited under international human rights law.

(Continued)

Table 1 (Continued)

SDG	Indicators directly or indirectly relevant for smallholder women farmers
11	No direct links to women smallholder farmers. However, useful information about access to transport which could be important for further access to the market, and inputs necessary for production.
12	No direct links to smallholder women farmers but could be useful for understanding the enabling environment for urban smallholder women farmers.
13	No indicators for smallholder women farmers, yet considerable scope to add around their exposure to climate change risk, ability to access support to mitigate climate change effects and coping strategies.
14	No gender-specific nor direct indicators on smallholder women, yet potential scope to explore how smallholder women farmers who diversify in multiple livelihood options and may also be engaged in fisheries at different nodes of the value chain.
15	No gender-specific indicators on smallholder women, yet considerable relevance. These include access and management of forestry, wetland, and dryland natural resources; exposure to climate change; quality of soil for farming; exposure to climate change risk (i.e. desertification).
16	No specific indicators on smallholder women, but could be relevant for smallholder women farmers working in conflict-affected areas as victims of violence, human trafficking, or killed, also in experiences of discrimination.
17	No direct indicators on smallholder women. Yet, would be useful to see proportion of aid dedicated to smallholder agriculture that benefits smallholder women farmers directly or indirectly (related to changes in policy).

systems. Smallholder farming produces diverse food for sustenance (own consumption) and to supply local and regional markets, boosting food and nutrition security. Women (in all their diversity) are recognized as key actors in small-scale agriculture, but a striking proportion (49%) receive little or no pay because they work as family workers (UN Women, 2023d). Access to resources (land, knowledge, financial resources, market opportunities) is a critical gender gap and an important lever to increase both productivity and incomes: '*Limited access to assets and agriculture inputs generates a gender gap in land productivity, reaching 24% between female and male managed farmers of the same size …*' (UN Women, 2023d). Land ownership and secure tenure rights over agricultural land in 40 of the 46 countries with available data were found to be less likely for women than men (UN Women, 2023d). An important aspect, but overlooked in the SDG, is that norms influence food consumption practices within the household that systematically undermine the nutritional status of women and girls with consequences on their well-being (see Farnworth et al., 2023; Ahmed et al., 2023). Women's unequal status in agrifood systems (and in societies more broadly) amplifies vulnerability to hunger: '*Addressing these disparities would reduce food insecurity and boost global GDP by 1% ($1 trillion)*' (UN Women, 2023d). Just one indicator related to SDG 2 targets directly mentions smallholder women farmers: it pertains to average income by gender and indigenous status.

SDG 3 (Good health and well-being) is less directly relevant to smallholder women farmers. The targets and related indicators for this SDG focus on maternal mortality, neonatal mortality, communicable and non-communicable diseases, substance abuse, access to sexual and reproductive health services, and universal health coverage. While they are not specifically mentioned, smallholder women farmers can also be pregnant women, so are implicit in and loosely connected to this SDG. Weak and fragile health-care systems render pregnant smallholder women farmers at higher risk with indirect implications for productivity and the agricultural labor force. While the indicators do not mention smallholder women farmers, they could be used to support analysis of the health of this demographic and their access to health care across the life cycle.

SDG 4 (Quality education) is less directly relevant to smallholder women farmers. While the SDG makes no reference to smallholder women farmers, a loose argument can be made that women and girls end up farming as a livelihood by default. This is because formal and non-formal education opportunities are less accessible, in part due to gender norms around, for example, unpaid domestic work, the care economy, and early pregnancies.

SDG 5 (Gender equality) is clearly directly and highly relevant for and inclusive of smallholder women farmers. As the only goal focused on women, SDG 5 has numerous targets and gender-specific indicators that affect smallholder women farmers as well as women, more broadly (see Annex for details on SDG 5 targets and indicators). These include indicators related to time spent on unpaid and domestic work (see SDG 1) where farming work is very often unpaid and unrecognized. There are also a number of indicators that capture access to the key resource for agricultural production (land). The indicators also refer to institutional frameworks that aim to monitor women's rights and equality more broadly and to track public expenditure toward this goal. This is important for the enabling environment for smallholder women farmers. While there is no direct mention of smallholder women farmers, various indicators are directly relevant relating to secure ownership of agricultural land, legal frameworks for the security of land tenure, and time spent on unpaid care roles.

However, progress on SDG 5 has been very limited to date.

None of Goal 5 indicators and sub-indicators are at 'target met or almost met', 2 are 'close to target', 8 are at a 'moderate distance to target', 4 are 'far or very far from target' and 4 lack sufficient data to assess at global level. No country is within reach of eradicating intimate partner violence (indicator 5.2.1) and only 27 countries have comprehensive systems to track and make budgetary allocations for gender equality and women's empowerment (indicator 5.c.1). The world is at a moderate distance to the target for achieving legal frameworks that promote, enforce and monitor gender equality. Much more action is needed to deliver all 2030 Agenda promises to women and girls, in all their diversity. (UN Women, 2023d)

This lack of progress is attributed to broader institutional challenges that include deeply rooted biases against women, unequal political representation, unequal access to sexual and reproductive health, economic disparities, and lack of legal protection (UN Women, 2023d). Further, lack of data continues to be impedimented to tracking SDG 5 with the data needed to track this SDG lacking in many countries (UN Women, 2023d). Critically, gender equality is not only about women, and not about all women. An intersectional approach that differentiates women and includes other stakeholders – particularly men and boys – is needed.

SDG 6 (Clean water and sanitation) is highly relevant for smallholder women farmers who need access to clean drinking water as well as water for irrigation. However, this SDG has no gender-specific indicators at all. This is a major omission as water is a critical resource for agricultural production where access to irrigated water is a key agricultural input. Moreover, women are often responsible for obtaining water for their households. As water becomes less accessible due to climate change, contamination, and land grabbing – to list just a few of the many reasons for this – distances to reach safe and clean water are increasing (UNICEF & WHO, 2023). These can negatively impact women's time availability for agricultural tasks as they have to walk longer distances to attain water. The rights of water and the living entities that rely on it have often been protected and advocated by women, particularly indigenous women (UN Women, 2023b). Yet, women and indigenous groups still remain under-represented or excluded in water management bodies due to discriminatory gender norms: 'only 26% countries achieved high or very high levels of gender mainstreaming in water resource management' (UN Women, 2023d).

SDG 7 (Affordable and clean energy) is less directly relevant for smallholder women farmers and has no gender- or sex-specific indicators. Access to electricity is one loose link as this could help smallholder women farmers to access more agricultural information (i.e. extension advice) to improve agricultural practices and help with the division of labor by reducing time for cooking. Furthermore, smallholder women's health could be improved through access to clean cooking fuels and technologies that reduce indoor air pollution.

SDG 8 (Decent work and economic growth) is highly relevant for smallholder women farmers, though they are not specifically mentioned. This SDG refers to decent working conditions (safety), and wages allowing analysis of gender disparities across some intersectional markers (including age, disability, migrant status, and occupation). This is relevant to smallholder women farmers who may choose to work beyond production to processing, value addition activities, and the marketing of agricultural products. This kind of livelihood diversification and improvement through upgrading value chain activities to increase productivity and profitability can lead to entrepreneurship opportunities for women and women's economic empowerment within and beyond small-scale agriculture. Financial capital is a critical input for smallholder farmers; however, women farmers, in particular, face considerable barriers to accessing it: '*women's economic agency depends on their ability to protect their money and build savings. Yet, women continue to face barriers accessing and using diverse financial institutions. In 2021, in developing regions, 68.5% women had access to financial accounts compared to men (74.2%)*' (UN Women, 2023d). While the targets for this SDG acknowledge the importance of access to finance institutions, this is unfortunately not captured in the indicators.

SDG 9 (Industry innovation and infrastructure) is not relevant for smallholder women farmers. There are no gender- or sex-specific indicators. This omission is a shame given the recognized barriers that constitute obstacles to better representation of women in science, technology, engineering and math (STEM) jobs internationally (UN Women, 2023d). Were this SDG to have a broader frame that included agricultural innovations and technologies the relevance for smallholder women would be significant. In the agricultural sector, the cultural linking of technology and masculinity underpins gender norms that constrain both women and men (McDougall et al., 2021).

SDG 10 (Reduced inequalities) is relevant for smallholder women farmers. While the targets and indicators do not refer to smallholder agriculture, they could be relevant for commercial agriculture. Potential links include access to fair wages for smallholder women farmers who may also be working as laborers in commercial agriculture. It could also capture smallholder women farmers' experiences of discrimination based on sex and how these intersect with other social markers.

SDG 11 (Sustainable cities and communities) is less directly relevant. Indicators make no reference to women smallholder farmers or urban agriculture. This is a significant omission given that smallholder farmers are important actors in urban agriculture.

SDG 12 (Responsible consumption and production) is less directly relevant, has no gender-specific indicators and no clear links to smallholder women farmers. That said, given that the SDG includes production, this may be an oversight.

SDG 13 (Climate action) is highly relevant for smallholder women farmers but lacks gender- or sex-specific indicators. Although the SDG does not directly mention smallholder farming or smallholder women farmers, there are many critical links. It is well-documented that women and girls are disproportionately affected by climate change, amongst the most vulnerable and often hardest hit (UN Women, 2023d; Sultana, 2014). They are also pioneers of climate-smart agricultural initiatives and are at the forefront of climate adaption measures and coping strategies and adaptation responses to climate stresses (Rao, 2017; Glazebrook et al., 2020). However, as UN Women (2023d) points out: *'A recent review of national climate action plans identified that only 55 have specific climate adaption measures referring to gender equality, only 23 recognize women as agents of change accelerating progress on climate commitments.'*

SDG 14 (Life below water) is relevant for smallholder women farmers; however, there are no gender- or sex-specific indicators. Smallholder women farmers have diverse livelihoods, which can include engagement in fisheries. Fisheries are a key natural resource that smallholder farmers rely upon, and which also have gender dimensions where women play key roles within the fisheries value chain. Women farmers, and particularly indigenous women play a critical role in water and land protection and conservation efforts, often at the forefront of activist environmental movements (UN Women, 2023a). There is scope to explore how smallholder women farmers who diversify in multiple livelihood options may be engaged in fisheries and at different nodes of the related value chains.

SDG 15 (Life on land) is highly relevant for smallholder women farmers though there are no gender- or sex-specific indicators for this SDG, and no direct mention of smallholder agriculture or smallholder women farmers. Yet, conservation and restoration of terrestrial and inland freshwater ecosystems are key natural resources on which smallholder farmers depend. Smallholder farmers are pioneering climate-smart agriculture and aquaculture practices which have the potential to combat climate change and protect natural resources. Relevance includes access to and management of forest, wetland, and dryland natural resources; exposure to climate change; quality of soil for farming; and exposure to climate change risk (i.e. desertification).

SDG 16 (Peace, justice, and strong institutions) is less directly relevant for smallholder women farmers and smallholder farming is not mentioned in the targets or indicators. However, it may be relevant for smallholder women farmers who are engaged in agriculture in locations prone to conflict. Conflict can result in death, higher exposure to violence, food insecurity and increases in intimate partner violence (UN Women, 2023d) not to mention the effect on agricultural production and ability to engage in value chain activities.

SDG 17 (Partnerships for the goals) is relevant for smallholder women farmers, despite the absence of gender- or sex-specific indicators. This goal influences the resources dedicated to supporting institutional environments that enable or restrict smallholder women farmers' access to resources, services and ultimately the benefits from agriculture, as well as the possibilities to diversify livelihoods beyond agriculture. This SDG speaks to resource allocation and aid allocations. UN Women (2023d) point out: '*An urgent boost in finance is needed where gender equality lags furthest behind ... the percentage of aid dedicated to gender equality and women's empowerment from 2020-2021, was only 4% of total bilateral aid, a decrease from 5% ...*'. It would be useful to see policy change and a related proportion of aid dedicated to smallholder agriculture that benefits smallholder women farmers directly or indirectly. SDG 17 is important from an institutional perspective in relation to funding for gender equality initiatives, and policy change to enable smallholder women farmers.

3 Smallholder women farmers' contributions to the achievement of the Sustainable Development Goals

Whilst the role of women farmers in smallholder agriculture and agrifood systems more broadly is increasingly well-documented (see, e.g. FAO, 2023; Njuki et al., 2022; FAO, 2010–2011), there is surprisingly little research that comprehensively explores how smallholder women farmers *specifically* contribute to the achievement of *all* SDGs in a holistic way. To date, most of the discussion of smallholder women farmers' contribution to SDGs tends to fall under the umbrella of SDG 5 or in reference to sex-specific indicators across the other SDGs (Doss et al., 2018; Agarwal, 2018; see Table 1). The only specific mention of smallholder women farmers is in SDG 2 (zero hunger). Agarwal (2018) argues that women smallholder farmers' contribution to the

SDGs beyond SDG 5 is visible from their role in the household and national food security, which links to SDG 1 (No poverty) and SDG 2 (No hunger). She expands that they play a critical role as food producers, household food managers, and consumers and highlights their contributions to food systems in fisheries and agroforestry, which is overlooked in SDG 14 (Life below water) and 15 (Life on land). Analysis of agroforestry's contribution to SDGs (Mbow et al., 2014) supports this, highlighting women's contribution as managers of trees, gatherers of fuelwood, and in agroforestry enterprises. Other authors indirectly allude to women smallholder farmers' contributions with reference to specific sectors. For example, Wong (2017) discusses women's critical contribution to Small-scale Poultry (SSP) given that chickens are often under women's control. He maps the contribution of SSP across multiple SDGs,[6] most notably, its importance for vulnerable populations as a source of income.

Part of the difficulty in determining women smallholder famers' contribution to SDGs is insufficient quality data on women (and gender equality) in agriculture (Doss et al., 2018; UN Women, 2023d), though the evidence base is slowly building up (see FAO, 2023). Since the first FAO report on the *State of Food and Agriculture* that focuses on women in agriculture in 2010–2011 (FAO, 2011), significant improvements have been made in terms of the availability of sex-disaggregated data, gender statistics, and research on the nature and drivers of women's empowerment and gender equality, including country-level data on key dimensions of women's empowerment such as access to land and national household and labor-force surveys (FAO, 2023). Over the past decade, significant advances have been made in measuring women's agency and empowerment in agricultural development with the release of the Women's Empowerment in Agriculture Index (WEAI) and its iterations such as the Abbreviated WEAI, Project WEAI (Pro-WEAI), the Women's Empowerment in Fisheries and Aquaculture Index (WEFI[7]), and the Women's Empowerment in Livestock Index (WELI[8]) (Alkire et al., 2013; Twyman & Ambler, 2021; Quisumbing et al., 2023; Pyburn et al., 2023; Newton, 2020). Nonetheless, good evidence remains critical, especially given some persistent and deeply engrained myths (see Box 1).

6 Specifically SDG 2, 3, 1, 8, 5, 4, 12, 15.

7 Women's Empowerment in Fisheries and Aquaculture Index (WEFI) Guidance Note https://digitalarchive.worldfishcenter.org/bitstream/handle/20.500.12348/5107/e38075a3971911324c424ccc55a66c61.pdf?sequence2 (all accessed February 7, 2024).

8 Women's Empowerment in Livestock Index (WELI) https://www.ilri.org/womens-empowerment-livestock-index-weli (all accessed February 7, 2024).

Box 1

Myths skewing evidence related to women in agriculture (extracted from Doss et al., 2018 and Kawarazuka et al., 2022)

Myths about women in agriculture (Doss et al., 2018)	**Myths about the feminization of agriculture** (Kawarazuka et al., 2022)
1. 70% of world's poor are women	1. Feminization of agriculture is the predominant global trend
2. Women produce 60–80% of the food	2. Women 'left-behind' are passive victims and not farmers
3. Women own 1–2% of the land	3. Feminization is bad for agriculture
4. Women are intrinsically better stewards of the environment	4. All women farmers face similar challenges in relation to the feminization of agriculture

While aiming to bring much-needed attention to women's significant contributions to and challenges in agricultural development, these myths obfuscate the possibility to better understand and address gender inequalities in the agricultural sector. Myths make it more difficult to see the nuance in the data by focusing only on women and overlooking gender relations – including the contributions of men farmers – and the diversity of women's positions in households and communities. The 'kernels of truth' embedded in each myth confirm that gender inequalities in agrifood systems matter and underscore the important contributions that women are making. That these myths are often used to highlight women's critical role in the achievement of the SDGs is problematic for several reasons.

First, the data on which these claims are made are inaccurate, sometimes missing important contributions. For example, the myth that *women produce 60–80% of the world's food* is arguably one of the most cited statistics to demonstrate why smallholder women farmers' contributions to food security and dietary diversity are crucial.[9] This myth is problematic on several counts. Namely, it is difficult to accurately calculate what proportion of food is produced by women because we do not have good data. Moreover, this myth relies on labor input indicators, which do not include women's work in kitchen gardens,

9 Websites referring to the SDGs citing this statistic, including the OECD, FAO, United Nations, or Fairtrade International: https://www.oecd-ilibrary.org/sites/35ec6754-en/index.html?itemId=/content/component/35ec6754-en; https://www.fao.org/3/x0171e/x0171e02.htm; https://www.un.org/sustainabledevelopment/blog/2023/06/press-release-world-leaders-urged-to-prioritize-womens-land-rights-at-the-un/; https://www.fairtrade.net/issue/sdg5 (all accessed February 7, 2024).

care of small livestock nor food processing and cooking; activities that are often not counted as 'agricultural work'. Hence the data not only overlooks women's roles within the care economy but also fails to capture how gender norms around gender division of labour prevents women from benefitting from new opportunities (innovations, market opportunities) in food production. The lack of accurate data reinforces stereotypes that overlook women's contribution and undervalues the breadth of women's labor contributions.

Second, the data tend to homogenize women as a universal group which has implications for policymaking and targeting, overlooking intersectional differences between and amongst women and opportunities to work toward joint resource management efforts with men. The myth that women are intrinsically better stewards of the environment, for example, is used to justify a focus on smallholder women farmers for targeting and policy implementation because data shows that women farmers (in both the global North and South) embrace more sustainable goals and values compared to contemporary industrial agriculture (Glazebrook et al., 2020). This myth is based on a common misrepresentation of discourses of women in development rather than misleading data. In other words, it is not the data per se that is the problem, but rather how it is interpreted. The evidence base is mixed. On the one hand, there is research that shows that women farmers in both Global South and North practice more sustainable and organic practices, often driven by different circumstances, one by default due to poverty and others by choice (Glazebrook et al, 2020). Moreover, there is a 'kernel of truth' in the myth as Doss et al. (2018) acknowledge, in the fact that because of women's historical roles in agriculture they would be more likely to be engaged in environmental movements to protect resources when they are threatened. What is more problematic is how the myth perpetuates essentialist arguments that claim women have a special link to nature which serves to obscure the diversity amongst women. The latter is also a myth related to the feminization of agriculture: that all women farmers face similar challenges (Kawarazuka et al., 2022). By recognizing the diversity amongst women, it would allow for a stronger analysis of how other factors influence conservation activities such as access to tenure, information, finance, and so forth.

Third, these myths perpetuate worldviews of gender and development that overlook how different women are entrenched in power relations with men in households and in their communities (Doss et al., 2018). Whilst well-intentioned, these myths can be harmful because they undermine the credibility of gender research and have serious implications for analysis and evidence on interventions aimed at promoting gender equality (Doss et al., 2018).

In spite of the nuanced data on how and to what extent women smallholder farmers contribute to SDGs, there is agreement that there is a limit to what is possible due to broader structural constraints. In other words, women smallholder farmers' contributions to SDGs could be much greater. One of the biggest structural constraints for women smallholders' contribution to the SDGs is the lack of access to land (Asadullah & Kambhampati, 2021; Agarwal, 2018), which is a fundamental resource for agriculture. Currently, women smallholder farmers are marginalized when it comes to access to land, due to gender norms around customary tenure and inequitable family laws linked to land inheritance and martial property. Agarwal (2018) highlights that a major shortfall of SDG 5 is that it not only fails to capture how gender norms affect women's ability to access and claim land, but it also ignores all the other necessary inputs women require to increase land productivity. Women lack access to a range of other key agricultural resources and assets that could help to increase productivity, which is well-documented (Glazebrook et al, 2020; Kawarazuka et al., 2022; Coles & Mitchell, 2011; Ihalainen et al., 2021). They also often lack access to capital and credit, financial institutions, market and trade networks, agricultural technology and extension (Bragdon, 2019). Moreover, the SDGs do not mention all the other constraints that women farmers face related to gender norms around household work reproductive domestic roles (and related time burden).

The structural constraints facing women smallholder farmers are immense. If addressed, the impact on empowerment pathways would be significant, with positive effects on productivity and ripple effects to other SDGs (Asadullah & Kambhampati, 2021). For example, improving access to assets, inputs, and services (extension, financial services, technologies) would increase women's yield by an estimated 20–30%, which would then lead to an overall increase in global agricultural food production by 4%, which would in turn reduce people living in hunger by 17% (Cohen, 2019:184). This highlights the knock-on effect of SDG 1. Other scholars working on agriculture-nutrition-empowerment pathways also point to the knock-on effects of SDG 3. For example, interventions focused on empowerment (e.g. gender transformative interventions looking at decision-making and changing gender norms, targeting both women and men) would lead to greater gains in food and nutrition security (Asadullah & Kambhampati, 2021; Bragdon, 2019; Newton, 2020). Asadullah and Kambhampati (2021) argue that for women smallholder farmers to make a further contribution to food security, they need better access to agricultural trade networks, more equity in market access and bargaining power and asset ownership. Terlau et al. (2018) conclude that '*strengthening the role of smallholder farmers as agents of change will deliver many co-benefits, from biodiversity conservation and climate change to ensure economic growth and human wellbeing*'. Given the prominence of women farmers within this demographic, it is clear that they are important actors who play a vital role in the SDGs. Yet, the lack of data on their contribution, their diverse experiences depending on where they are located, and constraints vis-à-vis men, and in different contexts remains a critical gap.

4 Limits and critiques of Sustainable Development Goals relating to women smallholders

The limitations of the SDGs vis-à-vis gender equality and for smallholder women farmers come out strongly in critiques. Overall, SDG 5 put the spotlight on women and has helped to reinforce the importance of looking at gender equality in its own right. Most notably, SDG 5 includes indicators that pay attention to women's access to land and natural resources, which is key to the achievement of household food security and includes dedicated indicators on the ownership of agricultural land (Agarwal, 2018). A significant omission across all the SDGs is that production constraints that women face are overlooked. This is evident from the dearth of relevant indicators that could have been captured under SDG 13 (climate change) and SDG 15 (life on land), which look at issues related to the state of natural resources including soil quality, desertification, and sustainable use of terrestrial and inland freshwater ecosystems. Moreover, there are no gender- or sex-specific indicators related to access to water, which is needed for irrigation (SDG 6). The impact of climate change on smallholder women farmers is also critical. A recent gender analysis of climate change on poverty, productivity, and food security exploring different scenarios concludes that under our current scenario, the goal of reaching zero extreme poverty among women and girls by 2030 will not be achieved. If unchecked, climate change will push 15.3 million women and girls into poverty by mid-century (UN Women, 2023d). The report concludes that areas heavily dependent on agriculture will need more policy aid to support adaptation capabilities. Sustainable Development Goals 13 (climate change) misses gender dimensions and overlooks the critical roles played by smallholder women farmers in climate change adaptation.

Further lost opportunities include synergies with SDG 8 (decent work and economic growth) where an analysis of smallholder farmers' participation in other nodes of value chain, their working conditions, and access to fair pay could have been analyzed. Another important gap is the lack of sex-disaggregated data on smallholder women's access to different forms of finance. A glaring omission of smallholder women farmers concerns their reference to SDGs 14 and 15 on fisheries and forests, wetlands, mountains, and drylands (Agarwal, 2018). Yet there are no gender-specific indicators in either of these SDGs. There are also missed opportunities to highlight smallholder women farmers in other SDGs. For example, SDG 16 (peace, justice, and strong institutions) could also pertain to smallholder women farmers living and working in conflicted affected areas. Whilst there are gender-specific indicators on victims and deaths, there is no effort to explore who of these women are smallholder women nor the impact of conflict on production. Last but not least, there is no mention of smallholder women in SDG (17), which could be used to track partnerships and resource mobilization dedicated to improving the resilience of smallholder women.

Some authors consider having one dedicated SDG (5) on gender equality as limiting and silo-ing gender equality. They argue for much better integration of gender indicators overall and throughout all SDGs (Glazebrook and Opoku, 2020). A paper examining the potential and limitations of SDG 5 in achieving household food security underscores the connectedness of the issues the SDGs seek to address (Agarwal, 2018). The SDGs are interconnected and interlocking and gender dynamics are, likewise, interwoven – often implicitly or silently. Agarwal (2018) calls for a bold interpretation of SDG 5 and establishing linkages and synergies with other SDGs as a way forward: including both SDGs that acknowledge the importance of gender equality already (SDG 1, 2, 13) and those silent on the issue (SDG 14, 15) (Agarwal, 2018).

5 Conclusion

This chapter has highlighted how SDGs have helped to bring the critical role of women and gender equality into the spotlight. However, smallholder women farmers are still largely in the shadows. Based on the literature reviewed, we have endeavored to illustrate how women smallholder farmers also play critical roles – or could potentially play more critical roles – in the realization of other SDGs. However, there are major shortcomings.

First, we have surfaced and underscored the significant data gaps around women smallholder farmers' contributions. Where data exists, there has been a tendency to perpetuate gender myths rather than generate sound data. Moreover, these data gaps are missed opportunities to understand important interlinkages and gather further evidence to support policy coherence and demonstrate smallholder women farmers' contributions to the SDGs. There is still work to be done to better track progress and challenges related to gender equality not only in agriculture but more broadly for the SDGs. The critiques of UN Women (2023a, 2023d) are very clear on the shortcoming. The growing use of the WEAI is a promising response to the data gap.

Second, there are challenges around how the SDGs are organized and where women smallholder farmers' roles are emphasized. On the one hand, having a dedicated SDG (5) on gender equality has helped to legitimize gender equality as its own right. In doing so, it has helped to bring attention to gender beyond instrumentalist debates about why gender equality matters. However, putting gender equality in a silo as one SDG is having mixed results. Critical synergies with other SDGs are missed both for mapping impact and improving implementation. Critical omissions include women smallholder farmers' contribution to climate action (SDG 13), forestry (SDG 15), and fisheries

(SDG 14), amongst others. The value of both a stand-alone goal related to gender equality and a thorough and thoughtful integration of gender indicators across and through sustainability goals is a challenge, but one to embrace.

Third, there has been a broader interrogation of whether SDGs as a global instrument are sufficient when there are bigger issues that affect the realization of these goals in practice. Whilst SDGs are important for building and reinforcing political will, they need stronger policy coherence and support of constituencies to be truly effective. Further, the deeply engrained and biased gender norms (c.f. Pyburn & Hallin, 2023, for a discussion of gendered social norms in agrifood systems) that prevail not only in the agricultural sector, but broadly, in many societies, need attention in their own right.

As we enter into the final 6 years of SDGs, the focus will shift to thinking beyond the SDGs (e.g. next steps). We identify several priorities for consideration in the way forward.

5.1 Prioritize explicating links between climate change and smallholder women

A significant omission in the SDGs is in relation to SDG 13 (climate action). No indicators or targets specifically mention smallholder women farmers, yet there is considerable scope to elaborate on their increased exposure to climate change risk, ability to access support to mitigate climate change effects, and coping strategies.

5.2 Prioritize policies that tackle structural constraints to women's agricultural capacities

As highlighted by many, SDGs will not achieve their potential unless there is policy coherence (Cohen, 2019) around addressing the structural barriers to women smallholder farmers. This means investing in access to capital and technology and addressing gender asset gaps and requires institutional change in land rights and investments in system-wide changes to agricultural extension, financial institutions, and educational institutes. Policy change requires changes in prevailing gender norms that obstruct the potential of smallholder women farmers. In other words, it requires systemic change within the institutions that act as gatekeepers to access by different types of women smallholder farmers to key resources that would allow them to participate in and benefit equitability from agriculture-based livelihoods. Gender transformative approaches and interventions will be critical to advance this. In this case, we are referring to gender transformative change at the institutional level, within the mindsets and practices of policymakers.

5.3 Prioritize the collection of better data

Better data is needed in order to truly progress on the ambitious agenda that is the SDGs and to build a body of evidence on the impact of the SDGs on (smallholder) women (farmers) as well as their contributions to them. As Doss et al. (2018) argue, we need more reliable sex-disaggregated data to generate evidence that supports a more accurate understanding of gender relations that is both comprehensive and systematic. Without such understanding, we cannot assess the impacts of policies and programmes and measure change over time. Both quantitative and qualitative data are required to understand what men and women experience, do, and value. In particular, we need data on women farmers' capacity to support gender transformative change. This includes data on both women and men smallholder farmers' constraints and barriers, and we need more data on women smallholder farmers in urban and peri-urban agriculture. The recent advances in measuring women empowerment outcomes relying on mixed methods and different iterations of the WEAI provides promising metrics to achieve this and to measure change over time.

5.4 Prioritize using an intersectional lens

The SDGs run the risk of homogenizing women. A critical question is *which women and which smallholder women specifically do the SDGs address and miss?* For example, women in male-headed households and women in female-headed households face different challenges; women left farming as men migrate may face different challenges than those who choose to farm; economically challenged versus successful farmers; and all live different realities. Intersectional analysis of the '*multiple and often overlapping factors such as age, gender, ethnicity, health and disability and socioeconomic, martial and migration status combine to create different mode of discrimination, social exclusion and privilege*' (FAO, 2023) is critical, as is an analysis of structural dimensions of inequality and systemic disadvantage. Efforts to collect better data need to engage with intersectional issues to move beyond homogenous views of women smallholder farmers (FAO, 2023).

5.5 Engage civil society, including a diversity of women farmers to articulate post-Sustainable Development Goals' priorities

Finally, there is a serious issue of thinking about how to move beyond the growth paradigm of agriculture, which is clearly doing harm. Given what we know about smallholder women farmers' existing agricultural practices, they have much to contribute toward a more holistic understanding of sustainability that values ecosystem services, distributional justice, and care. This speaks to SDG 17 and partnerships to realize the SDGs (or what comes next). Smallholder women farmers need to be engaged as stakeholders and agents with voices and perspectives that can make a difference in the realization of the global sustainability agenda. Further, the challenges, systemic bias, and structural constraints facing smallholder women need to be articulated and addressed as part of this global agenda if results are to be realized.

6 Annex: Sustainable Development Goals 5: gender equality and women's empowerment – targets and indicators

Target 5.1: **End all forms of discrimination against all women and girls everywhere**

- *Indicator 5.1.1*: Whether or not legal frameworks are in place to promote, enforce, and monitor equality and non-discrimination on the basis of sex.

Target 5.2: **Eliminate all forms of violence against all women and girls in the public and private spheres, including trafficking and sexual and other types of exploitation**

- *Indicator 5.2.1:* Proportion of ever-partnered women and girls aged 15 years and older subjected to physical, sexual, or psychological violence by a current or former intimate partner in the previous 12 months, by form of violence and by age.
- *Indicator 5.2.2:* Proportion of women and girls aged 15 years and older subjected to sexual violence by persons other than an intimate partner in the previous 12 months, by age, and place of occurrence.

Target 5.3: **Eliminate all harmful practices, such as child, early, and forced marriage and female genital mutilation**

- *Indicator 5.3.1:* Proportion of women aged 20–24 years who were married or in a union before age 15 and before age 18.
- *Indicator 5.3.2:* Proportion of girls and women aged 15–49 years who have undergone female genital mutilation/cutting, by age.

Target 5.4: **Recognize and value unpaid care and domestic work through the provision of public services, infrastructure, and social protection policies and the promotion of shared responsibility within the household and the family as nationally appropriate**

- *Indicator 5.4.1:* Proportion of time spent on unpaid domestic and care work, by sex, age, and location.

Target 5.5: **Ensure women's full and effective participation and equal opportunities for leadership at all levels of decision-making in political, economic, and public life**

- *Indicator 5.5.1:* Proportion of seats held by women in (*a*) national parliaments and (*b*) local governments.
- *Indicator* 5.5.2: Proportion of women in managerial positions.

Target 5.6: **Ensure universal access to sexual and reproductive health and reproductive rights as agreed in accordance with the Programme of Action of the International Conference on Population and Development and the Beijing Platform for Action and the outcome documents of their review conferences**

- *Indicator 5.6.1:* Proportion of women aged 15–49 years who make their own informed decisions regarding sexual relations, contraceptive use, and reproductive health care.
- *Indicator 5.6.2:* Number of countries with laws and regulations that guarantee full and equal access to women and men aged 15 years and older to sexual and reproductive health care, information, and education.

Target 5.7: **Undertake reforms to give women equal rights to economic resources, as well as access to ownership and control over land and other forms of property, financial services, inheritance, and natural resources, in accordance with national laws**

- *Indicator 5.7.1:* (*a*) Proportion of total agricultural population with ownership or secure rights over agricultural land, by sex, and (*b*) share of women among owners or rights-bearers of agricultural land, by type of tenure.[10]
- *Indicator 5.7.2*: Proportion of countries where the legal framework (including customary law) guarantees women's equal rights to land ownership and/or control.

10 Men have greater ownership or secure tenure rights over agricultural land than do women in 40 of 46 countries reporting on SDG Indicator 5.a.1. (FAO, 2023: xxvi).

Target 5.8: **Enhance the use of enabling technology, in particular information and communication technology, to promote the empowerment of women**

- *Indicator 5.8.1*: Proportion of individuals who own a mobile, by sex.

Target 5.9: **Adopt and strengthen sound policies and enforceable legislation for the promotion of gender equality and the empowerment of all women and girls at all levels**

- *Indicator 5.9.1:* Proportion of countries with systems to track and make public allocations for gender equality and women's empowerment.

7 References

Agarwal, B. (2018). Gender equality, food security and the sustainable development goals. *Current Opinion in Environmental Sustainability,* 34, 26–32.

Ahmed, A., et al. (2023). *Agriculture, Nutrition and Gender Linkages (ANGeL) Evaluation Results*. Bangladesh: IFPRI.

Alkire, S., et al. (2013). The women's empowerment in agriculture index. *World Development,* 52, 71–91.

Asadullah, M. N. & Kambhampati, U. (2021). Feminization of farming, food security and female empowerment. *Global Food Security*, 29, 100532.

Bragdon, S. H. (2019). Living links connecting the United Nations sustainable development goals: small-scale farmers and agricultural biodiversity. *San Diego International Law Journal*, 21, 155.

Cohen, M. J. (2019). Let them eat promises: global policy incoherence, unmet pledges, and misplaced priorities undercut progress on SDG 2. *Food Ethics*, 4(2), 175–187.

Coles, C. & Mitchell, J., (2011). *Gender and agricultural value chains: a review of current knowledge and practice and their policy implications* (ESA Working Paper No. 11-05). Agricultural Development Economics (ESA), The Food and Agriculture Organization of the United Nations, Viale delle Terme di Caracalla, Rome, Italy.

Doss C., et al. (2018). Women in agriculture: four myths. *Global Food Security,* 16, 69–74.

FAO. (2010–2011). *State of food and agriculture: Women in agriculture – Closing the gender gap for development*. Rome. https://www.fao.org/3/i2050e/i2050e.pdf

FAO. (2023). *The status of women in agrifood systems*. Rome: FAO.

Farnworth, C.R., et al. (2023). Gender roles and masculinities in leveraging milk for household nutrition: Evidence from two districts in Rwanda. *Food Policy,* 118, 102486.

Glazebrook, T., et al. (2020). Gender matters: climate change, gender bias, and women's farming in the global South and North. *Agriculture*, 10(7), 267.

Ihalainen, M., et al. (2021). Promise and contradiction: value chain participation and women's empowerment. In R. Pyburn & A. van Eerdewijk (Eds.), *Advancing gender equality through agricultural and environmental research: Past, present and future*. International Food Policy Research Institute, p. 40.

Kawarazuka, N., et al. (2022). Myths about the feminization of agriculture: implications for global food security. *Global Food Security* 33, 100611.

Lopez, D., et al. (2023, in press). Advancing a feminist intersectional lens in gender transformative work with women smallholder farmers. In Sachs & Castellanos

(Eds.), *Women and smallholder farmers: addressing global inequities in agriculture*. Burleigh Dodds Science Publishing.

Mbow, C., et al. (2014). Knowledge gaps and research needs concerning agroforestry's contribution to sustainable development goals in Africa. *Current Opinion in Environmental Sustainability*, 6, 162–170.

McDougall, C., et al. (2021). Toward structural change: Gender transformative approaches. In R. Pyburn & A. van Eerdewijk (Eds.), *Advancing gender equality through agricultural and environmental research: past, present and future*. Washington, DC: 365-389, IFPRI.

Newton, J. (2020). Gender, nutrition, and food system approaches. In C. E. Sachs, L. Jensen, P. Castellanos, & K. Sexsmith (Eds.), *Routledge handbook of gender and agriculture*. New York: Routledge, pp. 85–100.

Njuki, J., et al. (2022). A review of evidence on gender equality, women's empowerment and food systems. *Global Food Security*, 33. https://www.sciencedirect.com/science/article/pii/S221191242200013X?via%3Dihub.

Pyburn, R., et al. (2023). Gender dynamics in agrifood chains: advance in research and practice over the last decade. *Global Food Security*, 39, 100721.

Pyburn, R. & Hallin, R. (2023). *Gendered social norms in agrifood systems: a conceptual framework*. Nairobi: CGIAR Gender Equality Initiative (HER+). IITA and IFPRI.

Quisumbing, A., et al. (2023). Measuring women's empowerment in agriculture: innovations and evidence. *Global Food Security*, 38, 100707.

Rao, N. (2017). Assets, agency and legitimacy: towards a relational understanding of gender equality policy and practice. *World Development*, 95, 43–54.

Sukhera, J. (2022). Narrative reviews: flexible, rigorous, and practical. *Journal of Graduate Medical Education*, 14 (4), 414–417.

Sultana, F. (2014). Gendering climate change: geographical insights. *The Professional Geographer*, 66(3), 372–381.

Terlau, W., et al. (2018). Smallholder farmers as a backbone for the implementation of the Sustainable Development Goals. *Sustainable Development*, 27(3), 523–529.

Twyman, J. & Ambler, K. (2021). *Tools and methods for gender research and integration in agricultural value chain, market and entrepreneurship projects*. CGIAR GENDER Platform Working Paper #004. Nairobi, Kenya.

UNICEF & WHO. (2023) *Progress on household drinking water, sanitation and hygiene 2000-2022: special focus on gender*. New York: United Nations Childrens Fund (UNICEF) AND World Health Organization (WHO).

UN Women. (2023a). *Data-Driven Insights: the effects of climate change on gender and development: UN Women's gender and climate data works: extracted compilations*. https://www.unwomen.org/sites/default/files/2023-11/data-driven_insight_the_effects_of_climate_change_on_gender_development.pdf

UN Women. (2023b). *From commodity to common good: a feminist agenda to tackle the world's water crisis*. New York, USA. 57pp.

UN Women. (2023c). *Gendered analysis of the impact of climate change on poverty, productivity and food insecurity*. A technical report. New York, USA. pp86.

UN Women. (2023d). *Progress on the sustainable development goals: the gender snapshot 2023*. UN Women and United Nations Department of Economic and Social Affairs, Statistics Division 2023. New York, USA. 35pp.

Wong, J. T., et al. (2017). Small-scale poultry and food security in resource-poor settings: a review. *Global Food Security*, 15, 43–52.

Chapter 3

The adoption of climate change mitigation strategies by women smallholder farmers

Margaret Alston, University of Newcastle and Monash University, Australia; and Zhaoen Pan, Monash University, Australia

1 Introduction

There are significant issues facing agriculture in general and women on smallholder farms in particular. These include a significant rise in catastrophic climate-related events, the need for adaptation and mitigation efforts to address the resultant changing circumstances, a lack of adequate accounting of small-farm numbers and niche farm industries making national accounting and support for smallholder farmers difficult, problematic access for women to land ownership through traditional routes such as inheritance or family support to purchase a farm, changing expectations about women's work on and off farms and gender politics in general. In the Australian context, this chapter outlines the disturbing and long-term neglect of women's contributions to agriculture in Australia, a lack of public accounting relating to the numbers of women operating industries on smallholder farms, low levels of farm ownership and all against a background of continuing and disturbing catastrophic climatic events.

For women on farms in the current Australian context, issues of gender politics are overlaid with the significant impacts of catastrophic climate

http://dx.doi.org/10.19103/AS.2024.0148.03

change-related events and the need for innovative mitigation actions to be part of the farm cycle. For much of the twentieth century, Australians have experienced a range of extreme climate-induced disasters each affecting agricultural production in some way. These include a devastating 10-year drought that covered the continent from the late 1990s to 2009 and beyond and which disrupted industries, reduced water available for irrigation and devastated crops and livestock production, causing significant damage to ecosystems including the river systems (Heberger 2012). Further events this century have included the Black Saturday fires of 2019/2020 which resulted in the deaths of 33 people, an estimated loss of 56 000 livestock and 3 billion native animals (ABC News 2020) and the devastation of farmlands and infrastructure across vast areas of Victoria, NSW and Queensland. These fires were equally and instantly overwhelming and have been followed most recently by history making floods across several areas of Queensland and New South Wales which resulted in, among other things, major losses of farm infrastructure, crops and pastures (ABC News 2023; Foote et al. 2023). As a result of these and other significant climate events this century, their regularity and their devastating consequences, it has become increasingly evident that farm production units must adapt to circumstances never previously experienced if they are to survive into the future. Because of the constancy of these events, those working in agriculture are particularly aware of the need to undertake mitigation efforts to address the impacts at farm levels. While these impacts on farm family units are widely understood, there is less understanding of the gendered impacts and even less of the mitigation efforts adopted by women on smallholder farms.

Assessing climate-related challenges and their gendered consequences is made more complex because of the low numbers of women who are owners and operators of farms (Sheridan et al. 2023) and by the lack of any useful national accounting of the numbers of small-scale farmers and niche industries. This chapter addresses these issues and challenges by drawing on studies undertaken since 2015 with women who co-own and operate farms, and women who run their own small-scale enterprises on their jointly managed or solely owned farms. The case study outlined below draws on one particular research project undertaken in 2022 with Australian rural women identified through national agricultural award schemes as leaders in their industries. These interviews give a significant insight into the attitudes of women farmers to climate change, the consequences of climate-induced events and the need to adopt mitigation efforts.

This chapter also draws on additional studies undertaken with women and men engaged in agricultural production in Australia over the last decade including a study in the Murray–Darling Basin in eastern Australia (Alston et al. 2017; 2018). This area, traditionally known as the food bowl of Australia (COALA 2021) lies at the heart of Australia's irrigated agriculture and has been subject to drought conditions and declining water availability. The disaster conditions over the last two decades have led to governments stepping in to address the over-allocation of water licences and this has had a significant impact on both the social and ecological systems that make up the Basin communities.

A further study was undertaken in 2021 with men and women engaged in agriculture and who were impacted by various climate disasters (Alston et al. under review). This study was conducted in three diverse agricultural areas, that had experienced a range of devastating climate events. It illustrates the impacts of these events on women and men on farms. Because of the focus of this book, this chapter focuses only on the issues raised by women farmers. While undertaken at different times and across a diverse area and range of industries, these studies shed light on the emerging influence of women in agricultural production in Australia, the significance of their contribution to agricultural production and, in the context of this chapter, the attention given by women farmers to climate change mitigation strategies and sustainability measures.

2 Agriculture in Australia

Agriculture together with mining dominates the Australian export economy and is therefore critical to our ongoing prosperity. Yet Australian farmers are not subsidized, resulting in volatility in the agricultural sector particularly when catastrophic conditions occur. There are currently approximately 90 000 farm units in Australia, a large majority of which are family farm production units (National Farmers' Federation, 2017). The industry is dominated by a neoliberal mantra which, despite current conditions, sees an expansionist approach as the optimal way forward in what Newsome (2020) describes as a 'get big or get out' mentality. The widespread expectation on Australian family farms across industry and area is that attention to climate changes and resulting income stress is best served through these changes. In the dairy industry for example there is evidence of expansion of farms to increase economies of scale. This strategy can have unintended consequences including a reduction in the local population, fewer children in the schools, business decline, fewer off-farm employment opportunities and fewer people available to fill on-farm labouring positions.

The competitiveness and indeed ruthlessness of the approach have led to a decline in the numbers of small farms as these are swallowed by expanding larger farms producing food and fibre for the global market. It has also made agriculture less attractive to younger generations with many younger farm family members turning their backs to agriculture as a long-term viable career. Consequently, the number of farms overall has steadily decreased, especially in the small to medium size range, at the same time as the size of farms has increased. Indicating the significant expansion of farms at the larger end of the scale, approximately 10% of Australian farms produce 90% of agricultural export products (Hamblin 2009). The continued expansion of a corporate approach to agricultural production would appear to run counter to sustainable practices and, therefore, to mitigation efforts. Not surprisingly it is farms at the lower end of the scale that appear to be more attuned to sustainable practices and it is here that women are more visible in land management practices and sustainable enterprises. As one participant in our women farm leaders research noted:

> *The statistics show that it's the small dairy farmers, it's the 200 cows and under, which are really falling off a cliff. The big guys are fine, they've got economies of scale – I did a study with small dairy farmers when I was formulating xxxx, and more than 40% of small dairy farmers told me that they thought the long-term prospect of their farm was poor to very poor. So, I think, there's that feeling out there, of feeling a little bit desperate, but not really knowing how to turn that around and how to ask for help either (participant 5).*

3 Women in agriculture in Australia

Australian women working in agriculture were a silent presence until the mid-1900s when, tired of their invisibility, farm women rallied to change their status (Liepins 1995). Previously their contribution had been dismissed and subsumed under domestic, or reproductive labour, and census data records did not enable women to register or be recorded as farmers (Alston 2000; Butcher n.d.). Consequently, the historical record of agricultural production and sustainability in Australia is manifestly incomplete and has failed to record women's long history of engagement. In research conducted in the 1990s (Alston 1995), it was evident that women's small-scale enterprises had been operating at least since colonization and included the production of eggs, butter, chickens and indeed wine. Writing about this in 2003, I noted that women were the 'new entrepreneurs' (Alston 2003, p. 163) because the research revealed that women had been deeply engaged in small sustainable enterprises. Respondents reported that they had run significant enterprises and these had included running up to 1000 hens and/or turkeys for egg production and meat, producing cream and butter, with produce being sold in the local community, to hotels and to marketing boards (Alston 2003). Women also reported having large and thriving vegetable gardens and being skilled seamstresses, making a large proportion of the family's clothes. What overturned these thriving enterprises at the time were aggressive government regulations including restrictions placed on the number of hens per farm to 25 and introducing draconian health regulations, which were viewed by women as acting in the interests of large production companies and against women's small scale enterprises. Consequently, for some decades, women's small-scale enterprises were largely abandoned.

In more recent times, women's enterprises have again expanded to include cheese production, winemaking, bee keeping and other small on-farm industries that comply with existing regulations (Alston et al. unpublished). Indeed, women have continued their entrepreneurial activities, but now in an environment where their industries are taken seriously and their production output significant. At the same time, women dominate university agricultural programs with 55% of undergraduates in Australian universities being female (Sheridan et al. 2023). Despite their dominance in university programmes, their entry to farming remains tenuous. For example, inheritance practices have traditionally worked against them. Yet Sheridan et al. (2023) note that daughters are now challenging inheritance practices, a position supported in our research in several farming communities as daughters argue their case (Alston et al., 2017, 2018). Nonetheless, the best current estimates suggest that less than 10% of farms are inherited by women (Newsome 2020). Yet Sheridan et al. (2023) note that inheritance practices are slowly being overturned as young people reject the harsh and uncertain farming life and as farms across Australia are being split amongst siblings and often sold in the interest of fairness and equity (Sheridan et al. 2023).

Despite a lack of ownership, somewhat surprisingly, given their invisibility in dominant agricultural discourses, official statistics also reveal that women make up approximately 34% of the Australian agricultural workforce (ABS 2016) and produce 48% of real farm income (Sheridan and Haslam-McKenzie 2009) through their activities on and off farm. Nonetheless, because farm families dominate agricultural production units and because men traditionally are still more likely to inherit farms, women's most evident point of entry is through marriage with the result that their position in the farm unit is never entirely clear and their input is overshadowed. What we do know is that the dominant units of farm production remain farm families and that women's entry to farming through marriage results in them being subsumed into a secondary role carefully scripted around care work and community involvement.

Previous research suggests that farm roles are reshaping around a farmer/manager role occupied by the male farmer and a 'directed worker' role occupied by the female partner and that women's directed-worker roles have expanded considerably into on-farm labour and off-farm income generation (Alston et al., 2017, 2018). Further, the on-farm roles are more arduous and the need for farm labour is more evident. What then does this mean for agriculture and sustainable practices? Our work over several years with dairy families (Alston et al. 2018) and in 2021–2022, with farm family members in a number of agricultural industries (Alston et al. under review) allowed us to examine the understanding respondents had about climate changes and the mitigation efforts undertaken on farms. This work indicates that farm families are adopting a number of mitigation strategies to address increasing climate uncertainty.

4 Mitigation efforts on family farms

In our study of farm families in three diverse areas of eastern Australia (Alston et al. under review), respondents noted they had experienced extreme conditions including drought, floods and bushfires. The mitigation efforts adopted had included preparing land and infrastructure for future fires, selling land, reducing livestock numbers, reducing staff and household spending and seeking grants. Additional mitigation efforts included reducing hazard risks, focusing more on animal health, using minimum tillage for crop planting, ensuring more shade areas were available on farms for livestock, storing silage and changing milking times to later to avoid heat stress. Critically adaptive measures also included a range of technological solutions including solar water troughs, automated irrigation, soil sensors, GPS on tractors, phone apps to source information and solar panels on the farmhouse and other buildings on the farm (Alston et al. under review). These measures are significant not only because of their innovation but also because of their capacity to reduce the challenges of heavy work on farms, making farming tasks more accessible to women. These innovative measures also demonstrate the significant investment in adaptive responses to climate changes being undertaken at the farm level in Australia.

Of most interest in the context of this chapter is that the most reliable forms of sustainability on these farms identified by the participants and researchers were both the co-option of a female partner to increase their hours worked on the farm and the increasing expectation that women would draw an off-farm income to support the farm. In fact, an off-farm income is now viewed as one of the most essential mitigation responses in periods of uncertainty (Alston et al. 2018). So too is reducing hired labour and replacing this with female partners working expanded hours on farm tasks (Alston et al. 2018). This is particularly evident in dairy farms where the twice-daily milking requires significant hands-on labour. This was demonstrably evident in the dairy communities of the Murray–Darling Basin which were the context for previous research (Alston et al. 2018). The expansion of farms in response to the 'get big or get out' mantra (Newsome 2020) had led to males adopting a more managerial role on the farm and women a more hands-on labour role attending to milking and to other physical tasks such as calf rearing.

One female respondent from a dairy farm in the Murray–Darling Basin study noted:

> *'my workload – it's a lot. The bookwork is insane. It's crazy ... like the wages side is just nuts .. and you have to understand the safety perspective ... but it's the time to do it all ... just milking cows alone is 35 hours, so 35 hours of my week I'm just milking cows before I do anything else, before I do the bookwork, before I do anything on the farm, and I'm still doing stuff on the farm – so its just a lot ... and the kids ... so just everything'.*

Women in the dairy industry noted the stresses associated with milking at the same time as they are looking after young children. Others noted the impacts on their health of working so hard in physically demanding work. The following quote is from a woman in the dairy industry who is working both on the farm and off.

> *'It's taken a toll on our health, its put huge financial pressures and emotional pressures ... it hasn't been easy. We've never had enough cash to pay all the bills each month, it's a huge pressure, a huge stress ... when we first started I milked seven days a week and then did the farm work in the middle of the day and then did all the bookwork at night, and now I only fill in the gaps ... I don't milk anymore because I have [an injury]* (female dairy farmer quoted in Alston et al., 2018)'.

A survey of 128 farming people undertaken in 2017 as part of the Murray–Darling study revealed that in the previous 12 months, 45% of women had increased their farm work roles in the previous 12 months, 15% of women had increased their off-farm work hours; and 45% noted they would be seeking additional off-farm work in the next 5 years. This study and these quotes reveal several trends including adaptation strategies being introduced to farms, extensive changes in work patterns on family farms and little diminution of structured gender relations.

5 Women and sustainable practices

In summary, the factors identified as barriers to women's greater involvement in agriculture – climate disasters, the reshaping of mitigation strategies to deal with disasters, the 'get big or get out' mantra, patrilineal inheritance practices, the lack of statistics on women's involvement in farm ownership and sustainability measures, the historical lack of recognition of women's small-scale enterprises and the failure to view women as farmers in their own right, make it difficult to predict trends in women's involvement in sustainable development practice in Australia. While Alston et al. (2016) note that women in a US context are moving into sustainable industries on their farms and producing niche products, we are less certain of the extent of this in an Australian context. Nonetheless, anecdotal evidence is emerging (see for example Newsome 2020) that when women do take control of their own future on farms, they are more likely to introduce niche industries on farms and to move away from a commitment to expansionist agriculture. In an Australian context, Newsome (2020, p. 57) notes 'female producers are seeking alternatives to hegemonic agriculture due to its gendered financial and social constraints'. Yet their limited access to land, labour and capital makes farming in their own right difficult to establish.

5.1 Case study

While we do not have national-level trend statistics, we can draw on small-scale studies to highlight the issues raised above. Our recent study with female national agricultural award recipients gives us an opportunity to assess women's growing involvement in sustainable agriculture (Alston and Pan).

In 2022, we conducted research with 16 women who had been recognized through national award schemes as exemplary leaders in their industries and/or in agri-politics. Of the sample of 16, 9 come from a farming background and 6 of these now farm with their husbands and three are involved in agricultural industries or politics. Altogether, 12 of the sample are farming. 6 of the 16 respondents are on high level boards, two in very senior positions.

These women are all recognized leaders either because of their significant contribution to their industry and/or because of the leadership roles they have undertaken in their various representative bodies. They recognize the significance of women's leadership in agriculture in the context of disasters and note that women are 'hyper aware of climate change' R12) and are 'challenging norms' (R4).

The sample of women typifies the precariousness of women's entry into farming in Australia. Nine of the participants were raised on farms but did not inherit the family farm. One still works with her father on the family farm, and one works with her mother on her farm but has her own successful niche product. One participant bought her family's farm when it was sold at auction and one, together with her husband bought the farm from her parents. Two bought small acreage farms with their partners and have established small niche industries. None of this group of recognized leaders in agriculture inherited their farms.

Commenting on inheritance practices, women noted that the tradition of patrilineal practice had been strong in their experience and had had a significant impact on their agricultural careers. R3 noted that she had been denied any possibility of having the farm passed on to her.

> *Dad said, 'well the boys get the farm ... Of course I absolutely flipped my lid on it. 'Bloody hell, I've worked as hard as these buggers,' and there was nothing ... it forced us to make our own way.*

The sense of injustice these farm inheritance practices in their own families had, often spurred women on to prove themselves. Nonetheless alongside this view is a growing acceptance that there is a reluctance amongst young people in general to take over the farm.

> *A lot of the older generation aren't able to look after their farm anymore, but a lot of the kids don't to take over the farm because its such a high risk business and you really do have to be passionate about it.* R13

R6 also noted that the growth of *the big corporate players'* has seen the *small family farms disappearing.*

Women were asked about a number of issues shaping their success including their attitudes to farm management and environmental consciousness. R5 noted that it was a trip to an island nation that changed her perception about Australian farmers' poor environmental record.

> *I saw food systems and ag systems done much differently. When I came back I was like – 'what the hell are we doing here?'*

Respondents suggested that women are *'the brains behind the operation'* (R10), that they are better at managing staff, at looking after livestock, at marketing and at keeping the books in order.

> *Women are better at rearing calves, ... at managing the cows, they're more sensitive, they're more in tune.'* (R 4)

This affinity with livestock had also been noted by women in our dairy study (Alston et al. 2017; 2018). Their more sensitive awareness was viewed as a reason why women were more likely to be more attuned to the environment.

Significantly in the context of this chapter, respondents noted that women are good at assessing risks, taking action, keeping things together, finding innovative solutions and focusing on sustainability. In commenting on women's contributions to addressing climate change, respondents noted that '*women are absolutely hyper aware of climate change*' (R6) and that women are promoting '*sustainable and regenerative agriculture because of their nurturing perspective*' ... and are '*challenging the norm regarding what has previously been done*' (R4). '*Women think about the whole business and the whole community, and they think and they see climate change in a more systematic way whereas men see it more lineally*' (R10).

Participant 1 commented on the significant interlinking of women with the environment – '*I do know it tends to be women who have a longer future legacy focus because, if they have kids... women are a bit more sustainable.*'

Participant 4 also linked women with environmental consciousness noting:

> *Regenerative agriculture comes from a female nurturing perspective. I think there's a lot more women getting behind it and pushing it ... they're challenging the norm, they're challenging what previously has been done ... a lot of that comes from a female influence ... the more women coming into agriculture the more there is around regenerative agriculture and looking after the land.*

As with several of the respondents, R15 noted how she and her partner were introducing mitigation strategies on their dairy farm to build sustainability. '*we are starting to get batteries that are capable of holding [energy] so we will be working towards being more neutral in our energy consumption. We recycle water … and just doing the right thing is creating our social licence … and what the community expects from us and how important that is.*

Respondents noted that, with climate changes, uncertain market conditions and fluctuating fortunes, innovative measures had to be introduced to farming. The old ways of *'treating land as a resource' (R5)* have to change in the context of sustainable farming in favour of land being viewed as something to be understood. These innovative leaders noted that agriculture is now a 'sophisticated, tech focused, and innovative' industry (R15), attracting 'strong women and men' (R14).

R5 noted that *'there will always be a need for bigger farms, but I think the middle-sized and smaller farms, the dynamic nature of these businesses and their ability to shift and move and be flexible, and react quickly to what they need to react to, is what we really need to focus on in agriculture.'*

Respondents noted a number of gendered issues including not being recognised or treated as the principal in their enterprise. R16 noted that she attended a trade show to assess a piece of machinery she was looking to buy for her small farm-based enterprise.

> *I was discussing something that I needed … and the person stood there and talked to [my husband] the entire time even though it was me asking the questions.*

This process of being publicly ignored was also noted by others including, for those who were members of public boards, having their advice ignored until it was mentioned by a male member; and in a similar vein being ignored when raising a point in a workplace situation. R3 noted that her point was only discussed when a male colleague made the same comment and was taken seriously. R3 noted:

> *I've been in leadership roles and the thing that's still happening is the 'invisibleness'. You can sit there and have an idea and you get gaslit [and I wonder] did I actually speak or did I imagine that I spoke?*

6 Discussion

The critical issues that emerge from this study include a need not only to consider sustainable practices in the context of climate change, and the need for mitigation efforts to be introduced, but also to be critically aware of gender politics which is thwarting the capacity of women farmers to be innovative. I have argued that women are working very hard on farms, most often in partnerships where their work on and off the farm is critical. These extensive roles have led to a critical overload for many women. Why then would we be arguing for greater involvement of women? The answer lies in evidence that women are more attuned to sustainable agriculture. But there are barriers to women adopting sustainable practices in large, commercial farms where they may not hold any ownership rights. Critical to incorporation of women's perspectives are for women's leadership to be recognized and for land ownership pathways for women to be facilitated, for capital to be accessible, for there to be a greater recognition of women's contribution to agriculture including adequate accounting of their numbers, their input to agriculture, the barriers that prevent them entering agriculture in their own right and the niche industries in which women are engaged. More generally, a broader understanding of sustainable agriculture in the context of climate changes and ongoing climate disasters is critical. Women's entry to agriculture remains largely via marriage thereby reducing their ability to enter agricultural production in large numbers despite their dominance of agricultural university degrees. This expertise is largely being lost to agriculture because of outdated views on patrilineal inheritance and patriarchy and the blinkered view that women's roles on the farm are bound by care work, 'helping' on farm and generating an off-farm income. Outdated attitudes specifying the need for strength as a limiter for women are easily eroded by the steady march of technology and the production of technologically enhanced solutions. As our studies illustrate, items such as automated dairies, solar water troughs, sheep lifters, GPS tracking and other instruments are all operating to reduce the arduous back-breaking work of farming. As Sheridan et al. (2023) note, these are not only reducing the need for physical strength, they are also shifting the power imbalance. Whether this can result in an increase in the number of women engaged in sustainable practices remains unclear.

Currently, we have a number of outstanding women being recognized through national awards schemes, schemes that have been developed largely through the activism of women. Their example brings a wealth of insight into the work women are doing on farms and small acre holding. But they are not a comprehensive assessment of women on farms, their commitment to sustainable practices or their mitigation efforts. More detailed research on the increasing trend of women engaged in farming and their knowledge of sustainable agriculture is needed.

7 Conclusion

Research suggests that women may well be more attuned to environmental stewardship, particularly on small acre enterprises but also in their increasing involvement on larger family farms. At the same time, their input to farm production both on and off the farm is now critical to the survival of the family farm mode of production in Australia. For some decades, research has indicated the increasing visibility of women in agriculture and their fight to address uneven gender politics. There is much to be done if we are to foster women's industry and input to agriculture. Supporting women's ambitions to own and operate their own farms through a breakdown in traditional inheritance practices and more ready access to capital is critical. Other factors include supporting and nurturing female university graduates to remain in agriculture, making the work and contribution of women on family farms more visible and addressing the portrayal of farming as a male occupation.

To adequately assess mitigation efforts undertaken by women smallholder farmers will require a great deal more research to open up our understanding of the numbers of women small farmers, their enterprises, their access to land and capital, the mitigation efforts they are undertaking and the barriers preventing them from entering agriculture and building sustainable practices. Without that, at least in an Australian context, we are not able to adequately assess women's input nor indeed to understand the mitigation practices underway in small enterprises. Ongoing climate changes and the resultant catastrophic disasters would now appear inevitable, and the need for mitigation practices is critical. Breaking down barriers to women's entry into agriculture and fostering the link between women and sustainable agriculture would seem a no-brainer.

8 References

ABC News (2020). Three billion animals killed or displaced in black summer bushfires study estimates. Data accessed on September 26 from https://www.abc.net.au/news/2020-07-28/3-billion-animals-killed-displaced-in-fires-wwf-study/12497976

ABC News (2023). Farmers still grappling with the impact of the 2022 flood crisis. Data accessed on September 26 from https://www.abc.net.au/news/2023-04-29/nsw-farmers-grappling-with-flood-crisis-hopes-for-winter-crop/102240454

Alston, Margaret (1995). *Women in Agriculture: the hidden heart of rural Australia*. UNSW Press, Sydney.

Alston, Margaret (2000). *Breaking through the Grass Ceiling: women, power and leadership in Australian agriculture*. Harwood Publishers, England.

Alston, Margaret (2003). Women in agriculture: the 'new entrepreneurs'. *Australian Feminist Studies*, 18(41), 163–171.

Alston, Margaret, et al. (2017). Gender relations, livelihood strategies, water policies and structural adjustment in the Australian dairy industry. *Sociologia Ruralis*, 57(S1), 752–768. https://doi.org/10.1111/soru.12164

Alston, Margaret, et al. (2018). Contemporary feminist analysis of australian farm women in the context of climate changes. *Social Sciences*, 7(2), 16. https://doi.org/10.3390/socsci7020016

Alston, Margaret, et al. (under review). Gender and transformative resilience in the context of Australian agriculture. *Journal of Rural Studies*, 2023.

Australian Bureau of Statistics (ABS) (2016). Census data.

Butcher, Sophie (n.d.). A brief history of Aussie women in agriculture. Data accessed on September 26, 2023 from https://farmbuy.com/post/a-brief-history-of-aussie-women-in-agriculture

Copernicus for Sustainable Agriculture in Australia (COALA) (2021). Australia's Food Bowl and Murray-Darling basin. Data accessed on October 7, 2023 from https://www.coalaproject.eu/blog/australias-food-bowl/

Foote, Wendy, et al. (2023). Women's Leadership and a community saving itself. Learning from disasters, University of Newcastle, Australia.

Hamblin, Ann (2009). Policy directions for agricultural land use in Australia and other post-industrial economies. *Land Use Policy*, 26(4), 1195–1204.

Heberger, Matthew (2012). *Australia's millenium drought*. Impacts and Responses, Springer. https://doi.org/10.5822/978-1-59726-228-6_5

Liepins, Ruth (1995). Women in agriculture: advocates for a gendered sustainable agriculture. *Australian Geographer*, 26(2), 118–126. https://doi.org/10.1080/00049189508703140

National Farmers' Federation (NFF) (2017). Food, fibre and forestry facts. Data accessed on September 26, 2023 from https://nff.org.au/wp-content/uploads/2020/01/171116-FINAL-Food-Fibre-Food-Facts.pdf

Newsome, Lucie (2020). Beyond 'get big or get out': female farmers' responses to the cost-price squeeze of Australian agriculture. *Journal of Rural Studies*, 79, 57–64.

Sheridan, Alison and Haslam-McKenzie, Fiona (2009). Revisiting Missed Opportunities: growing women's contribution to agriculture. Rural Industries Research and Development Corporation. RIRDC publication number 09/083. Data accessed on January 27, 2019 from https://www.agrifutures.com.au/wp-content/uploads/publications/09-083.pdf

Sheridan, Alison, et al. (2023). Changing scripts: gender, family farm succession and increasing farm values in Australia. *Journal of Rural Studies*, 100, 103024.

Part 2

Sociocultural factors and their effects on women smallholder farmers

Chapter 4

Socio-cultural factors and women smallholder farmers' empowerment in Uganda

Brenda Boonabaana, The University of Texas at Austin, USA; Florence Kyoheirwe Muhanguzi, Losira Nasirumbi Sanya, Susan Namirembe Kavuma, Grace Kyomuhendo Bantebya and Pius Okello, Makerere University, Uganda; Laura Meinzen-Dick, Villanova University, USA; and Nargiza Ludgate, University of Florida, USA

1 Introduction

Agriculture remains an important sector for most African economies, accounting for 14% of total GDP in sub-Saharan Africa (SSA). It is the primary source of food for human beings and fodder for animals, as well as a source of raw materials for agro-based industries (Food and Agricultural Organization [FAO] et al., 2020). The agricultural sector employs a sizable proportion of the rural population, the majority of whom are women, ranging from 19.9% in Botswana, 76.4% in Malawi (World Bank, 2023) and 81.2% in Uganda (FAO, 2023). In low-income landlocked countries such as Chad, farming makes up over 50% of GDP with over 75% of the population in subsistence farming. Crop production accounts for more than 75% of the total agricultural production on the African continent, with the main crops grown being wheat (mainly in North Africa), tubers and plantains (Central and West Africa), and maize (mainly in Southern

http://dx.doi.org/10.19103/AS.2024.0148.05

Africa). In 2021/2022, agriculture accounted for about 24.1% of GDP and 33% of export earnings (International Trade Administration [ITA], 2022). Around 80% of agricultural households engage in crop and livestock production both for their own consumption and income generation. The major food crops grown in Uganda are maize (55%), beans (54%), bananas (47%), cassava (29%), cereals (sesame, soybean, sunflower, millet and sorghum), sweet potatoes, vegetables, groundnuts, plantains and Irish potatoes (Uganda Investment Authority [UIA], 2023). The main cash crops grown in Uganda include coffee, tea, sugar, cotton and tobacco (Uganda Bureau of Statistics [UBOS], 2018; UIA, 2023).

As farming continues to loom large in the development possibilities of Africa, women constitute one vital human resource for the African agricultural sector (Amenyah and Puplampu, 2013). As primary drivers of African economies and local communities, women constitute more than 40% of the global agricultural labour force (Matthew et al., 2022; JICA, 2023). In Sub-Saharan Africa, 66% of women's employment is in agri-food systems compared to 60% of men (FAO, 2023). In countries such as Ethiopia, Malawi, Niger, Nigeria, Tanzania, Zambia and Uganda, women's share of labour in crop production ranges between 40% and 78% (World Bank, 2023). In Uganda, over 80% of the women derive their livelihoods from agriculture (Civil Society Budget Advocacy Group [CSBAG], 2022; UBOS, 2018; UN Women, 2018), depicting the critical role of agriculture to African women's wellbeing.

However, women in agriculture experience a myriad of challenges that constrain their contributions and benefits. These include their limited participation in decision-making over crops grown, activities to be undertaken and control of income from high-value cash and food crops (Carter et al., 2017; Peralta, 2021). Cash crops such as cotton, cocoa, coffee and cereals including maize, barley and wheat are often controlled by men, while food crops which are largely for household consumption such as plantains (bananas), sweet potatoes, sorghum, millet and vegetables, remain typically 'women's crops' (Alinyo and Leahy, 2012; Ali et al., 2015). Fruit tree species with a high market potential are also controlled by men, while women mainly control fruit tree species with a lesser market potential but offer products for domestic use like firewood and medicine (Twinorugyendo, 2019). Further, men are much more likely to engage in market-oriented work (agribusiness) and to obtain off-farm employment, while women face significant constraints to these opportunities, including being unpaid or less-paid workers (UBOS, 2014). As consumers, women are more likely to be food insecure than men in every region of the world; and as producers, rural women particularly face much greater constraints than their men in accessing essential productive resources and services, technology, market information and financial assets (FAO, 2018).

To date, an understanding of women's role in livestock production in developing countries has been limited to cultural biases that underestimate women's contribution with a tendency to view livestock as male enterprise. Yet evidence shows that women have more contact with animals and play a key role in livestock production particularly in the aspects of milk and meat production, hide processing, caring, feeding, birthing, processing of products, cleaning udders, milking, preparing milk for sale, selling milk products, cleaning cattle sheds, watering cattle, harvesting fodder and other daily dairy-related activities (Barnes, Mumma and Cumming, 2018; Tangka, Jabbar and Shapiro, 2020; Bonis-Profumo et al., 2022; Galiè et al., 2022, 2023). Despite, the key roles women play, they occupy the less lucrative nodes of the livestock value chain (Allan et al., 2023) and often have limited control over the process of decision-making, particularly over the disposal of animals and animal products for income (IFAD, 2013; Bonis-Profumo et al., 2022). The limited technical skills in animal care, technology, markets and marketing skills, and financial services limit women's opportunities to access, control and expand their small ruminant stock and production (Bikaako et al., 2022). In many African societies, large animals such as cattle are owned by men while smaller stock/animals such as goats, sheep, pigs and backyard poultry kept in the homestead are more a woman's domain (Allan et al., 2023; Zole, 2011). However, when small animals become an important source of family income, ownership, management and control are often turned over to the man (IFAD, 2013).

Apart from women-led value chains, participation of women in upper levels of the value chains, such as lead farmers and aggregators is restricted, with men dominating the high-value positions in processing, marketing and trading activities (Rubin, Boonabaana and Manfre, 2019; Rubin and Manfre, 2014). This is due to gender constraints around financial access, mobility, decision making and input (Alliance for a Green Revolution in Africa [AGRA], 2018). For instance, African women perform most of their agricultural work using rudimentary practices and technologies (Leon-Himmelstine et al., 2021) while men make use of labour-saving technologies (machinery) (FAO, 2018; Rubin, Boonabaana and Manfre, 2019). In Uganda, women are unable to scale up to the level required for high-value crops due to constraints associated with small plot sizes, plot quality and/or ownership (Mukasa and Salami, 2015). In addition, men have more access to and use of improved seeds than women (UBOS, 2012). Women perform most of the post-harvest activities while men dominate the decision-making roles of the activities (Mwinuka and Hyera 2022). Access to resources is often determined by social rules, norms and institutions that govern the distribution and exchange of these resources (Golla et al. 2011).

Further, women's heavy workload at the household level constrains their engagement in multiple agricultural enterprises associated with crop and livestock production (AGRA, 2018; Bikaako et al., 2022). In most African societies, women are discouraged from taking on traditional roles, especially if they threaten male power and superiority (Guloba et al., 2018). Identifying and supporting women's roles as livestock owners, processors and users of livestock products while strengthening their decision-making power and capabilities, would be critical for women's economic and social empowerment. Such discriminatory gender positioning and norms are rooted in religious and patriarchal kinship systems (Yisehak, 2008; Zole, 2011; IFAD, 2013) that define most African societies, including Uganda.

Sexsmith et al. (2017) note that gender discrimination in credit markets makes it more difficult for women farmers to acquire labour-saving and innovative production inputs. Evidence in Africa shows a substantial gender gap in mean yield of around 20–30% in favour of male farmers (Ali et al., 2015). A report by FAO (2011) demonstrates that if African smallholder women farmers had equal access to land, labour, information, technology, fertilizer, water and equal opportunity to use these resources effectively, agricultural productivity would increase by 20% in the continent. Similarly, UN Women, UNDP and World Bank (2015, p. 5) have noted that 'closing the gender gap in agricultural productivity could potentially lift as many as 238,000 people out of poverty in Malawi, 80,000 people in Tanzania, and 119,000 people in Uganda.'

This chapter draws on Kabeer's (1999, 2001) conceptualization of empowerment, defined as the ability to make strategic life choices, and to put those choices into desired actions and outcomes (Kabeer 1999). It enables women to participate on equal terms with men in society, especially, equal access to resources and decision-making power. As a multidimensional process, empowerment describes the different facets of agency as intrinsic *(power within)*, instrumental *(power to)* and collective agency *(power with)* (Malapit et al., 2019; Kabeer 1999; Mulema, et al., 2021). Drawing on a mixed methods approach and document review, focused on Uganda, this chapter interrogates social-cultural influences of women smallholder farmers' empowerment and agricultural opportunities. In the first part, we present the methodology, followed by findings on the status of rural women's empowerment, local understanding and perceptions about women's empowerment and the social-cultural dynamics, and implications for women's agricultural opportunities. It ends with conclusions and recommendations.

2 Methodology

This chapter utilized literature from studies conducted on the African continent and primary data from a mixed methods baseline survey undertaken in two regions of Uganda. This is part of an ongoing research titled 'strengthening women smallholders' resilience to agricultural shocks for enhanced income diversification and empowerment in Uganda.' Data were collected from a sample of 1280 women smallholder farmers and their main decision makers in their household. For quantitative participants, we used systematic random

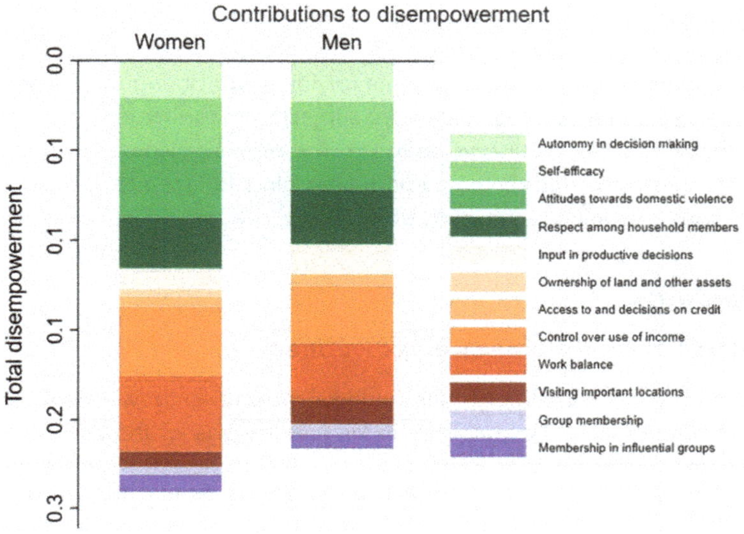

Figure 1 Indicator contributions to men and women's disempowerment.

sampling to select married women smallholder farmers that belonged to selected savings groups, aged between 18 and 64 years. For qualitative data, a total of 32 individual interviews (Life Histories), 19 Key Informant Interviews (KIIs), 32 male and female focus group discussions (FGDs), 8 mixed-sex community profiles and 8 seasonality pattern group discussions, in the two study districts – Isingiro in South Western Uganda and Alebtong in Northern Uganda. The participants were selected using snowball and purposive sampling procedures.

Quantitative data analysis utilized STATA software to compute the Project Level Women Empowerment in Agriculture Index (pro-WEAI) based on three domains of empowerment (3DE) colour-coded in *Fig.* 1: intrinsic agency (green), instrumental agency (orange) and collective agency (purple). The 3DE considers 12 indicators whereby, for each indicator, an individual who is considered adequate (empowered) receives a score of 1 while one considered inadequate (disempowered) receives a score of 0. The 12 scores are then equally weighted for an average 3DE score and a Gender Parity Index (GPI)[1] with a weight of 10%, to give a composite empowerment score that ranges between 0 and 1. Individuals with a score above 75% (adequate in 9/12 indicators) are considered to be empowered while those with a score below 75% (inadequate in 4 or more indicators) are categorized as disempowered (Malapit et al., 2019).

1 The GPI is computed by comparing the empowerment scores of an individual woman with her spouse/male respondent in each household (dual-adult households). It reflects: (1) the percentage of women who enjoy gender parity; a woman enjoys gender parity if: she is empowered or if her empowerment score is equal to or greater than the empowerment score of the primary man/spouse in her household; (2) the empowerment gap – the average percentage shortfall that a woman without parity experiences relative to the man in her household. For details about computation of GPI, see Malapit et al. (2019).

Qualitative data were analyzed following a rigorous and reflexive process drawing on the thematic approach. The first step involved field-based daily briefs to identify the emerging insights for follow-up during the subsequent interviews. After fieldwork, all data were subjected to further in-depth analysis using the ATLAS.ti22 qualitative software. The process involved transcription of audio interview recordings, coding, searching for themes, patterns and connections in relation to the study objectives.

3 Findings

3.1 Status of rural women's empowerment

Based on the Pro-WEAI scores, the findings show an empowerment level of 0.78, computed from the weighted individual average of the 3DE score for women (0.76) and the gender parity index (GPI) score at 0.92. According to the pro-WEAI, individuals with a 3DE score above 0.75 (adequate in 9/12 indicators) are considered to be empowered. In our sample, only 42% of women met this threshold compared to 48% of the men, with this difference statistically significant at 10% level. The GPI showed that 63% of households achieved gender parity with an average empowerment gap of 22% for women who did not achieve gender parity as men in their households. Findings further show that the mean adequacy (3DE) score for the disempowered women is at 0.58, meaning that these women achieved adequacy in an average of 58% of the indicators compared to the men whose mean adequacy (3DE) score was at 0.59. This indicates that they achieved adequacy in an average of 59% of the indicators, denoting a small and statistically insignificant gender difference in terms of indicator adequacy. However, these aggregate measures mask interesting patterns within the indicators of empowerment. *Figure* 1 shows the absolute contribution of each indicator to disempowerment (1-3DE score) for both women and men. Interestingly, the largest six contributors to disempowerment (work balance, attitudes about domestic violence, control over use of income, self-efficacy, respect among household members and autonomy in income) were similar for both women and men. However, attitudes about domestic violence play a larger role in women's disempowerment than men.

The findings indicate that indicators associated with intrinsic agency such as autonomy in decision-making, self-efficacy, attitudes towards domestic violence, and respect among household members are substantial contributors to women's disempowerment (*Fig.* 1). Other substantial contributors to women's disempowerment are aligned to instrumental agency indicators, particularly, control over use of income, work balance and input in productive decisions (*Fig.* 1). In contrast, a small proportion of women (18.7%) were considered to be

Table 1 Overall adequacy for men and women

Variable	*N*	(1) Men (mean/SE)	*N*	(2) Women (mean/SE)	*t*-test Difference (1)-(2)
Ownership of land and other assets	640	1.000 [0.000]	640	0.938[0.010]	0.063***
Control over the use of income	638	0.433[0.020]	637	0.386[0.019]	0.046*
Access to and decisions on credit	640	0.869[0.013]	617	0.911[0.011]	−0.042**
Input in livelihood decisions	640	0.758[0.017]	640	0.816[0.015]	−0.058**
Membership in influential groups	640	0.855[0.014]	640	0.784[0.016]	0.070***
Group membership	640	0.900[0.012]	640	0.895[0.012]	0.005
Autonomy in income	640	0.613[0.019]	640	0.641[0.019]	−0.028
Visiting important locations	639	0.793[0.016]	608	0.887[0.013]	−0.093***
Attitudes about domestic violence	640	0.689[0.018]	640	0.383[0.019]	0.306***
Respect among household members	640	0.519[0.020]	640	0.566[0.020]	−0.047*
Self-efficacy	640	0.470[0.020]	640	0.503[0.020]	−0.033
Work balance	616	0.458[0.020]	631	0.187[0.016]	0.271***

The values displayed for *t*-tests are the differences in the means across the groups.
***, ** and * indicate significance at the 1, 5 and 10 percent levels, respectively.

Source: Field data, 2022.

adequately empowered in the domain of work balance compared to 45.8% of men (Table 1). Concerns around work balance are connected with the labour-intensive nature of subsistence farming, the main livelihood source, especially for women. The production season in these communities begins in February with planting, followed by weeding and harvesting with women contributing substantial labour as a female key informant (Agricultural Officer from Alebtong District) mentioned: 'the women are the ones who put in a lot of effort from planting, wedding, harvesting. It is the women who are really involved there'.

Such farm labour performed by women comes on top of their socially established domestic responsibilities, including chores (washing dishes, cooking) and management of children's wellbeing. Although there are narratives that men focus on income-generating activities, and this gendered division of activities is perceived to be 'natural' to men and women, there can be a double burden of work balance for women as further demonstrated by the following voices:

> In our culture as Lango there are things that are automatic and everybody's role at home is clearly defined like washing clothes is a woman's responsibility unless in some cases where a man sees that madam is tired or has got a lot to do that is when he will help but this role has existed even before we were born a woman knows her role and a man knows his role like taking livestock to graze is a man's role but in other households even women participate in grazing animals but that depends on

> arrangement in that particular household. (FGD, Male, Abia Subcounty, Alebtong District, 2022)

Considering control over the use of income, only 38.6% of women reported adequate control over the use of income as compared with 43.3% of men (Table 1), a difference that was statistically significant. Women in the group discussions and interviews acknowledged that women have limited access to finance, but particularly stressed the gendered control of income. The women stressed that for the sake of peace in the home, they feared to challenge men's absolute control over income:

> Let truth be told. Men do not want to work and yet they want to spend like men who work. The only way they can get money is to sell what the woman has produced from her sweat. It can be livestock; goats, chicken at home. At home also they do not want to work and they want to show that they still have authority. They know if the woman gets money, they will be less of men without their own money, so, they take away the woman's money to keep her in a begging state. Men have killed women when it comes to the point of selling, so, if you want your life you leave whatever you have harvested to go, and you retain peace. So, men want to be in charge of money so that women keep begging from them, they limit women. Women will never look smart, women will never eat good sauce at home and to him, that is what makes him a man. He even tells fellow men that at his home, it is only him to authorize the use of a piece of soap, if he does not buy salt no one can in his home and he feels respected that way. (Seasonality Calendar, Masha Subcounty, Isingiro District, 2022)

Men held a view that control over money and assets defines one's social position in society, with the corollary that women should rely on their husbands for financial support:

> They know [meaning men] if the woman gets money, they will be less of a man without their own money, so they take away the woman's money to keep her in a begging state. So, men want to be in charge of money so that women keep begging from them, they limit women, women will never look smart, women will never eat good sauce at home and to him that is what makes him a man. He even tells fellow men that at his home, it is only him to authorize the use of a piece of soap, if he does not buy salt no one can in his home and he feels respected that way. (Seasonality Calendar, Masha Subcounty, Isingiro District, 2022)

In contrast to this, almost all women in the study (91.1%) indicated they were adequately empowered in terms of access to and control over credit, a higher proportion than that for men (86.9%) (Table 1). However, this is likely to be an artifact of our sample recruitment process, which selected households through women's membership in savings groups, many of which operate rotating credit funds. These groups, which can be understood as a collective empowerment strategy, were often mentioned by women, as demonstrated in some of their narratives below:

> ...there is a group that has started an association, and people can borrow from them. (Mixed sex Community Profile, Abia Subcounty, Alebtong District, 2022)
> We have groups, where we save our money. Now when times of shocks come, you go to those groups, borrow money from them which have small interest and do not need collateral like in the bank. We use that money to pay school fees, to build, to take patients to the hospital, to buy food for family. (Mixed sex Community Profile, Kikagate Subcounty, Isingiro District,2022)

Collective agency indicators, consisting of group membership and membership in influential groups, contribute relatively little to both women's and men's disempowerment in our study. Much as with credit, through savings organizations, women are increasingly engaged in other groups such as religious organizations as expressed below:

> Even in churches you find the majority are women praying and attending those discussion groups at church that encourage harmony in homes and are hardworking. All this is because women want to earn some money and become empowered, women are educating their children because men have stopped minding. At home, women are the ones who wake up very early and leave their drunk husbands in bed to go and work for their families. Men are getting unempowered because they are engaged in useless things like alcohol. (Women FGD, Masha Sub county, Isingiro District, 2022)

Table 1 presents the pro-WEAI measures of empowerment for both men and women, with the final column indicating differences between men and women and whether the differences are statistically significant.

3.2 Local understanding of empowerment and social-cultural dynamics

Qualitative data from focus group discussions and life histories indicate that men and women had varied definitions of empowerment most aligned to the three domains of empowerment (3DE) especially, instrumental agency.

3.2.1 Positive perceptions and characteristics of empowered women

Both men and women expressed their understanding of empowerment in terms of having access to finances and ability to make decisions over productive resources especially land, buildings, livestock, transportation and business. Financial capabilities were valued as empowering due to the desire to meet household needs. Having access to (or ownership of) productive resources such as land, livestock (goats, cows), income generating activities and a big farm was considered a key characteristic for an empowered woman, across all study sites. Women with such resources command fear and respect in their communities as expressed below:

> When a woman cultivates a garden and after harvesting the man allows her to sell the produce and then she buys a goat for the home which later becomes a cow. There you can do big things that everyone can see.... (FGD, Female, Mbaare Subcounty, Isingiro District, 2022)
>
> Most times people fear such a woman. People will say she is an empowered woman and the one who will develop such an area. She will go, make some money that can be used to develop the home so people are also respected. This happens because she is not just any woman who sits down unemployed. (FGD, Male, Abako Subcounty, Alebtong District, 2022)
>
> Okay, the way I see an empowered woman is that woman who participates in small businesses to raise income to cater for the family. (Empowered Male, Omoro Subcounty, Alebtong District, 2022)

Therefore, being empowered means being hardworking and having the capability to earn an income that meets essential household basic needs including children's school fees and household food. Having a formal job or business was considered respectful, a description related to a culturally ideal woman as expressed by a woman from Abako sub-county, Alebtong district: 'works so hard that when you see her home you just know that home is for an empowered woman.' The above perceptions were attributed to women because of their increased exposure to education that accords them an opportunity to make private money: '...because they are educated, they make money on their own...' (Female KII Alebtong). Other descriptions related to owning better homes (developed, good looking, with basic needs to better care for their families, especially children). Better home care, especially children's good discipline, is considered a key characteristic of an empowered and well-behaved woman as explained below:

> Such an empowered woman is the one who brings an opinion in her home and that opinion is considered. Her children look good, are educated, well disciplined, and are not smugglers of goods to the neighbouring country. Also, such a woman takes part in the decisions of her family with her husband. She is never left behind, their home looks good, she dresses decently and she is confident.... Any woman who decides on something and she does it, she is empowered.... (Empowered Female Life History, Kikagate Subcounty, Isingiro District, 2022)

Findings further show that being a leader or simply an advisor in the community was regarded as empowering. An empowered woman from Abako subcounty in Alebtong district said: 'Yes, I am empowered because most of the women come to me for help and advice and I help them'. Empowered women were described to have decision-making power and authority with regard to the ability to participate in economic, production decisions and other decisions within the household either alone or jointly with a partner/husband. Such women are considered to have the ability to plan for their lives and to put their plans into action. For others, being empowered is being in a position to make decisions without consulting a husband, a description related to a culturally ideal woman, as depicted below:

> That is a woman that does what she feels like doing. She is a woman that does not consult the husband at home when she is going to do something. That is a woman that never agrees with the husband to do anything. Each one in the family does her, or his own things.... (Empowered Female, Life History, Ngarama Sub-county, Isingiro District, 2022)

Empowered women were also defined by their ability to make decisions. Such women tend to be valued as community role models as the following voices depict:

> In village, for a woman that makes decision together with her husband, some people are actually going to give him a lot respect because you will find that such a man has good plans for his future and the family and he also has a well-established household. (Male, FGD Abako Subcounty, Alebtong District, 2022)
>
> We admire a woman that cooperates with the husband and does one thing as a family. They sit as family and agree as a family.... (Empowered Female Life History, Ngarama Subcounty, Isingiro District, 2022)
>
> A woman who decides on something and puts it into action is a hero. And men think that such a woman is the head of the family.... (FGD, women, Mbaare subcounty, Isingiro District, 2022)
>
> When other women see such women that make decisions with their husbands, they admire her, she is their role model.... (Female KII, Ngarama subcounty, Isingiro District, 2022)

Independent decision-making especially on economic matters (loosely termed 'economic independence') including the ability of women to provide for her family independently, and having access to everything was critical in defining an empowered woman by men. Such a woman is seen as an 'income source' and is seen to be a husband's financial saviour as expressed below:

> We respect this woman because she is not over reliant and dependent on the man for provisions but she has her own things going and such women make a home a better place to be because she relieves her man from some pressure to provide for the family. (FGD, Male, Aloi sub-county, Alebtong district, 2022)

Interestingly, some men expressed admiration and support for them to progress towards gaining a higher social-economic status. This was evidenced by some men who wished to have close connections with such women as their mothers, wives or mothers-in-law. Therefore, we found positive perceptions by both men and women about an empowered woman. However, there were negative views as well that presented ambivalences and tensions as below.

3.2.2 Complex ambivalences and tensions about an empowered woman

At the same time, an empowered woman was associated with negative perceptions that depicted complex contradictions grounded in the desire to preserve traditional gender relations, and male superiority/patriarchy. Empowered women were perceived as witches, proud, and incapable of managing and sustaining a good marriage life, including being unsuitable 'marriage material.' Empowered women were labeled as 'prostitutes', disrespectful, uncultured, or un-submissive women that are not worthy of marrying. There were narratives around the tendency of some men abandoning such women, divorcing, or separating from them while others just fear getting close to them. Further, they were equated to being 'too high to be under a man' and 'unable to take care of their marriages' with examples of such narratives are depicted below:

> The first perception will be that such women are the ones ruling their families, whereas others would say that the women could have bewitched their husbands.... (Male KII, Agricultural Officer, Kikagate Subcounty, Isingiro District, 2022)
> ...Others think she gives orders in the home and the man follows them. Others think a man for such a woman is bewitched.... (FGD, women, Mbaare subcounty, Isingiro district, 2022)
> ...they ask her what she gave her husband. Here, there is still a belief that a woman bewitched the man.... (Female KII, Ngarama subcounty, Isingiro district, 2022)

As such, while empowered women are expected to work hard and provide for their families, which is a positive perception, at the same time, they are thought about negatively. They are therefore, meant to 'watch' their moves and actions especially in regard to how they express themselves before their husbands and the public, in order to keep within their socially confined roles and behaviours. Women expressed the negotiations around the interconnectedness of marriage values and income control dynamics as below:

> ...In most cases when a woman earns more money than a husband, it is a bit challenging. The man feels inferior, he feels less empowered. So, the woman has to mind the words she uses in a home or when you are making decisions, because the moment you try to be at the forefront, the tone that you use, the man might think that you are doing that because you are earning more than him and you have your money....(Female KII, Ngarama Subcounty, Isingiro District, 2022)
> Once the public knows that the woman earns more than the man, they keep thinking that the woman is ruling the man in the home, she is the head of the family, she is the one that decides in a home. But it is the man and the woman who know how they manage... As a woman you need to know how to handle it if you value your marriage, and if the man does not torture you for earning more than him.... (Female KII, Ngarama Subcounty, Isingiro District, 2022)

To navigate such complexities, women's narratives depict how they often have to let men take charge and control their lives and income, which eventually limits their agency and potential. Most women mentioned having to negotiate such contradictory practices in terms of what they prefer to do vis-à-vis their 'acted' actions, for the sake of keeping peace in their marital relations. These acts were regarded as 'wise moves' by the women if they were to navigate their survival in their patriarchal communities. Further, some men expressed displeasure with empowered women by associating them to being competitive with their husbands and other men to 'overthrow' their social positions. As such, such women were labelled as 'fellow men' something that men were countering by working hard to 'beat' such 'competition.' Below are some examples of such negative stereotypes:

> They think that the woman has taken over now, so a man is under her control because he is doing domestic work. (Female FGD, Abako Subcouty, Alebtong District, 2022)
> ...The community also becomes resentful of such households where women decide together with their husbands. Men in such a household will be perceived as being controlled by their wives.... (Male KII, Agricultural Officer, Kikagate Subcounty, Isingiro District, 2022)
> Some people will say such a man is overpowered by the woman because the woman runs all the affairs of his home yet it's actually their decision to do so. (Male, FGD Abako Subcounty, Alebtong district, 2022)

The findings on the local perceptions of empowerment present narratives that demonstrate mixed feelings about what characterizes an empowered woman. While they are partly admired, they are also demonized when their behaviours fall out of the traditional expectations, associated with a socially ideal and submissive woman. The findings thus revealed contradictory expectations and contestations about the empowered woman vis-à-vis an ideal woman. In this regard, women empowerment was viewed as a threat to male position, power and control, and thus as a danger to marriage/family stability and peace.

3.3 The varied social image of an empowered woman: what did we learn?

While women who work and earn their own income were considered to be empowered, findings indicate community tensions that relegate such women to negative stereotypes due to their non-conformity to the social ideals of a submissive woman. The local narratives about an empowered woman remain contradictory, characterized by positive descriptions of empowered women within the context of a rigid gender system and a strong observance of the gender hierarchy. While the benefits of women's empowerment are appreciated, preference to work within the traditional gender system remains apparent with women's/men's empowerment viewed in the context of a proper/ideal woman and man, and any deviations creating tension in the family and the broader society/community.

Economically independent women were perceived to be uncultured and less submissive, competitors with their husbands and dangerous to marriage/family stability and peace. These perceptions are largely aligned to the traditional expectations of a married woman and keeping within the social gender hierarchy and associated norms. Yet empowered women were also appreciated as key role models and heroes.

Overall, the qualitative findings reveal contradictory expectations and contestations about the empowered woman vis-à-vis an ideal woman. In this regard, women's empowerment was viewed as a threat to male position, power and control, thus a danger to marriage/family stability and peace. Therefore, empowerment was closely linked to the social cultural gender system and patriarchal power. Such resistance to change slows the process of empowerment and continues to create hurdles for women to exploit their potential, within the agricultural sector. As Kabeer (1999, 2001) argues, women empowerment should address gender constraints around (i) resources that not only include access but also material, human and social resources; (ii) agency including processes of decision-making, negotiation and manipulation); (iii) achievements (wellbeing outcomes).

4 Social-cultural gender dynamics and women's empowerment: implications for gender-equitable agriculture in Uganda and beyond

4.1 Main contributors of women's disempowerment and gender inequality

As the findings depict, the limited women's control and autonomy over agricultural income remains an overarching constraint in Africa. This resonates with Sexsmith's et al. (2017) work that documents the five dimensions to gender inequality in African agriculture as: (i) land rights; (ii) productive resources; (iii) unpaid work and employment; and (iv) decision-making power. Aligned to this, the UN Women (2018) indicates that although women constitute a major agricultural labour force, only 13% own agricultural land. Many scholars across the continent (including Nigeria and Uganda) have also documented the difficulty for women in to access land and engage in more productive and lucrative agricultural opportunities (Chekene, and Kashim, 2018). In Uganda, majority of women express joint ownership of land with their husbands but only a few have formal documentation of the land (Chekene and Kashim, 2018).

Quisumbing, Meinzen-Dick and Njuki (2019) highlight the importance of women's control over income for their own empowerment, and reinvestment in crops and value chains that are important for their wellbeing, food and nutrition security, and poverty reduction outcomes. Evidence shows that if women had equal access to skills, resources and opportunities as men, they would become powerful agents in the fight against hunger, malnutrition and rural poverty (Quisumbing, Meinzen-Dick and Njuki, 2019; UN Women, UNDP and World Bank, 2015).

FAO (2018) denotes that in many developing countries, including Africa, the agricultural sector is underperforming, and one reason is attributed to women's unequal access to resources and the necessary opportunities to be more productive. In relation to this, IFAD (2012) draws our attention to the urgent need to promote gender-equitable agricultural opportunities in order to level the ground. It has been postulated that successful efforts to promote women's economic empowerment not only impacts incomes but also builds self-confidence, enhances women's agency, and contributes to improved education, health and security outcomes for families (Maestre & Thorpe, 2016). If a rural woman is empowered, she is able to make choices about assets and services – such as land, rural credit, technology, irrigation, labour, knowledge and extension – that contribute towards nutrition (UN Women, 2018).

Further, the findings depict women's disempowerment in terms of work–life balance, due to the heavy burden of work and amidst strict gender rules. In many African societies, women and girls have been socialized to undertake unpaid domestic and care roles such as fetching water, cooking, family farming/gardening, collecting firewood, washing clothes, caring for the sick, elderly, children and other time-consuming domestic tasks (Maestre & Thorpe, 2016; UN Women, 2018). Women and girls are responsible for water collection in 80% of households that have no access to water in their premises (UN Women, 2018). In sub-Saharan Africa, an average of 85.7% of households rely on unclean fuels for cooking and heating, with devastating effects on women and children's health who often spend more time working around the home (UN Women, 2018; Maestre & Thorpe, 2016). Globally, women perform unpaid care and domestic work 2.6 times more than men (UN Women, 2018). Yet such contributions are often unrecognized, unvalued and unrewarded (Grassi, Landberg and Huyer, 2015; FAO, 2018; Maestre and Thorpe, 2016). Power relations and social norms shape the provision of care and allocate more responsibilities to women than men (Maestre and Thorpe, 2016; Ferrant, Pesando and Nowacka, 2014). The socialization process in many African societies enhances the gender division of roles, including care roles, and reinforces its continuity. Ferrant et al. (2014) links unequal distribution of care responsibilities to discriminatory social institutions and gender role stereotypes.

While care work is critical for individuals, families, societies and economies to survive and thrive from generation to generation (UN Women, 2018), it is also challenging for women's productivity and wellbeing (Ali et al., 2015; Grassi et al., 2015). Broadly, while women in low-income countries provide 43% of the agricultural labour force, they mainly work as unpaid family labourers (UN Women, 2018). Many studies clearly show women's heavy workloads compared to men in Uganda (Kavuma et al, 2018; Acosta et al., 2015). This further limits women's mobility to access lucrative markets to sell their produce and to tap into agricultural income. Maestre and Thorpe (2016) have noted that often, men will easily have access to the market where they sell produce or buy inputs, compared to women, due to constrained mobility. Women also lack access to information regarding the choice and use of inputs that negatively affect their farming activities and associated outcomes (Maestre and Thorpe, 2016).

A qualitative study conducted in three districts of Uganda (Kaabong, Kabale and Kampala) in 2018 shows that both urban and rural women dominate the unpaid care work at an average of 32h for women and 6.8h for men, a week. This was found to be closer to the national estimate of 30 h spent on unpaid domestic care work by women, and 12 h by men per week (UBOS, 2017). Culture in these districts and much of Ugandan and African societies defines a good woman as a good cook who grows, harvests, peels, and sorts the food, fetches water and firewood, as well as serves the food with care and respect, and cleans the utensils (Guloba et al., 2018). Women are stereotyped as compassionate nurturers and therefore 'most suited' to perform care roles. On the other hand, men were stereotyped as mobile, impatient, and with limited time to perform care duties (Guloba et al., 2018). Cultural expectations associated with being a good wife and mother comes with women's time constraints that limit their participation in profitable agricultural opportunities

Recognising care is the first step towards change. This should begin with understanding gendered roles and responsibilities, household dynamics and community or other social group dynamics that affect women's time, mobility and agency (Maestre and Thorpe, 2016). They further note that 'where women are unable to leave their house due to their care responsibilities, such as preparing meals at specific times of the day, and participate in marketing crops, they may lose control over the money that is earned from farming' (Maestre and Thorpe, 2016, pp. 4–5). Given that unpaid care restricts women's involvement in the labour market, it also affects overall productivity, economic growth and poverty reduction. In Uganda, child-care activities constrain women's ability to either supervise or perform labour activities on their plots (Ali et al., 2015), while in Mozambique, women respond to the multiple roles and time constraints by multi-tasking (Arora, 2014). Arora (2014) has argued that the difficult trade-offs between agricultural work, income-generating activities, household maintenance and care activities limit women's leisure time in Mozambique. Studies have shown that reducing women's share of unpaid care work could increase agricultural labour productivity by 15%, and capital productivity by as much as 44% in certain countries (OECD, 2008).

When adult women do less care work because of other responsibilities, the work is usually transferred to paid workers, often, other women (daughters or grandmothers), or the quality of the care is compromised. Adolescent girls are particularly affected and may have to drop out of school to help with care work, perpetuating a cycle of inequality (Maestre and Thorpe, 2016). Social norms determine culturally acceptable roles for men as well as women, and men (and women) can face a backlash for contravening norms. Men who try to undertake more care tasks or women who try to undertake fewer may be considered 'unnatural'. Widespread evidence shows that criticism, shaming and violence are mechanisms used to enforce these 'natural' gender roles (Maestre and Thorpe, 2016).

One of the key steps to this constraint is for agricultural development programmes to carry out sex-disaggregated gendered market analysis and collect sex-disaggregated data that enables the identification of gendered household and community dynamics, such as roles and responsibilities. This will enable us to build evidence towards understanding women and men's time use as well as their access to and control over resources (Maestre and Thorpe, 2016). It is also important to recognize women's unpaid labour contributions and ensure equal pay, working conditions and benefits in agriculture. This requires overcoming the marginalization of poor rural women that stems from gender-based socio-cultural norms and practices rooted in capitalist and patriarchal systems at various scales (FAO, 2018).

The key finding around the disempowerment of women manifested through their higher acceptance of domestic violence (even higher than men's acceptance) and this needs urgent attention in Uganda and Africa as a whole. Domestic violence manifests in different forms: physical, emotional and sexual. For instance, while 26% of men reported that it is justified for a husband to beat his wife across five scenarios, including burning food or leaving children unattended, 61% of women were supportive of the vice.

4.2 Prioritization of economic/income resources: access and control rights

Finally, in tandem with the quantitative findings, qualitative data depicts that both women and men small-scale farmers prioritize access to finances and assets such as land, livestock (goats and cows), as well as the ability to make decisions over these resources, as the desirable ingredients for enhancing their empowerment. As findings depict, in patriarchal Ugandan societies, there is a tendency to privilege men to access and control over productive resources through inheritance regimes. By contrast, women's access to and joint ownership of land is often at the mercy of the men/husbands, for as long as the marriage is stable (Gibson et al., 2014; Doss, Meinzen-Dick and Bomuhangi, 2014). UBOS (2018) indicates that only 7.7% of women in Uganda own agricultural land individually while 39.9% own agricultural land jointly with someone else (UBOS, 2018). The limited access to property and assets, such as land, not only limits the productive capacities of women but also prevents them from accessing financial resources such as credit (using land as collateral) to enable them to engage in productive investments, such as re-investing in their farms to increase production, add value to their products or invest in off-farm income diversification. A survey by the Financial Sector Deepening (2018) indicates that only 4% of women had access to credit from formal financial institutions, 1% from formal and informal lenders, 23% relied on informal sources (savings groups, burial societies), 15% relied on family and friends and over a half of the women (57%) did not have access to credit. Therefore, women have less access to assets and services that would enable them to increase their productivity and reduce their workload (IFAD, 2012).

Across all study sites, participation in decision-making (independently or jointly), and having autonomy over income was considered critical. However, women's seeking and owning of productive resources was only perceived to be socially acceptable if they combined this with their traditional gender roles, including care roles, else they wouldn't be respected and their marriages would be at stake. This shows that income ownership for women is not enough unless they keep performing their socially expected roles and behaviours. Increased women's economic independence is perceived to weaken men's authority. Challenging men's overall authority over household income comes with reprimands such as gossip, domestic violence and loss of respect for the woman. Such reprimands further work to reinforce such unequal and unfair gendered practices.

5 Conclusion and recommendations

We conclude that in Uganda, and many other African societies, women's disempowerment and gender inequality are closely linked to social and gender norms and identities, rooted in the capitalist-patriarchal system that privileges men and boys, and disadvantages women and girls. Our mixed-methods empirical evidence and complementary literature attests to the ongoing tensions and ambivalences around women empowerment in Africa that need intentional and targeted attention if we are to enable gender-equitable and sustainable agricultural systems in Africa. The local narratives about an empowered woman remained contradictory, characterized by positive recognition and perception of empowered women but situated within contexts of rigid gender systems that strongly observe unequal gendered power relations. The high preference for empowered women to operate within their traditional gender system remains problematic for them to realize actual empowerment. The complexities of empowerment intricately connected to an ideal woman offer critical deviations from the global understanding of empowerment. These are seemingly creating hurdles for women to exploit their potential, including within the African agricultural systems.

We also conclude that the main contributors to women's disempowerment are aligned to the two domains of empowerment, namely, instrumental (work–life balance, control over use of income and autonomy in income) and intrinsic agency (self-efficacy, attitudes about domestic violence and respect among household members). This demonstrates the need to pay attention to these indicators if we are to promote gender equality and strengthen women's empowerment in through the agricultural sector. Some key recommendations are to enhance women's access to and control over resources, recognize, value, reduce and redistribute unpaid care work to free women's time to exercise their potential, and create a gender-responsive policy environment for women to function in informed by locally grounded gender evidence. An intentional gender transformative agenda will go a long way to address the structural gaps that limit women from engaging meaningfully in the agricultural sector in Uganda and Africa as a whole.

6 Acknowledgement and disclaimer statement

This research was funded by the Feed the Future Innovation Lab for Markets, Risks, and Resilience and made possible by the generous support of the American people through the United States Agency for International Development (USAID). The contents of this publication are the responsibility of the authors and do not necessarily reflect the views of USAID or the United States Government.

7 Where to look for further information

For further information on the subject, please read the following publications:

- Food and Agriculture Organization of the United Nations. 2011. The State of Food and Agriculture; Women in Agriculture: Closing the Gender Gap for Development. (FAO, Rome, 2011); Link: https://www.fao.org/3/i2050e/i2050e00.htm

- FAO. 2018. Empowering Rural Women, Powering Agriculture. Food and Agriculture Organization. Link: https://www.fao.org/3/CA2678EN/ca2678en.PDF
- FAO. 2023. *The Status of Women in Agrfood Systems*. Rome: Food and Agriculture Organization of the United Nations. doi:10.4060/cc5343en
- Muhanguzi, F.K., Boonabaana, B., Sanya, L.N., Kavuma, S.N., Kyomuhendo, G.B., Ludgate, N. and Meinzen-Dick, L., 2023. The meanings of resilience in climate justice: women smallholder farmers' responses to agricultural shocks in Uganda under the spotlight. *Agenda*, *37*(3), pp.106-123; Link: https://www.tandfonline.com/doi/full/10.1080/10130950.2023.2245844
- Quisumbing, A., Meinzen-Dick, R.S. and Njuki, J., 2019. *2019 Annual trends and outlook report: Gender equality in rural Africa: From commitments to outcomes*. International Food Policy; Link: https://ebrary.ifpri.org/utils/getfile/collection/p15738coll2/id/133470/filename/133685.pdf

Key research in this area can be found at the following organizations:

- Food and Agricultural Organization (FAO).
- International Food Policy Research Institute (IFPRI).

8 References

Acosta, M., Ampaire, E. L., Okolo, W. and Twyman, J. 2015. *Gender and climate change in Uganda: effects of policy and institutional frameworks*. CCAFS Info Note. CGIAR Research Program on Climate Change, Agriculture and Food Security (CCAFS), Copenhagen, Denmark.

Alinyo, F. and Leahy, T. 2012. Designing food security projects: Kapchorwa and Bukwo, Uganda. *Development in Practice* 22(3), 334–346.

Allan, F., Wong, J., McLeod, A., Galie, A. and Miller, B. 2023. Building a community: the gender and livestock data community of practice. *Agriculturists*. Available at: https://agrilinks.org/post/building-community-gender-and-livestock-data-community-practice.

Alliance for a Green Revolution in Africa (AGRA) 2018. *The Business Case for Financial Inclusion of Female Smallholder Farmers*. Final Report, Mastercard Foundation. Available at: https://agra.org/wp-content/uploads/2020/10/The-Business-Case-for-Financial-Inclusion-of-Female-Smallholder-Farmers.pdf .

Ali, D., Bowen, D., Deininger, K. and Duponchel, M. 2016. Investigating the gender gap in agricultural productivity: evidence from Uganda. *World Development* 87, 152–170.

Amenyah, I. D. and Puplampu, K. P. 2013. *Women in Agriculture: as Assessment of the Current State of Affairs in Africa*. The African Capacity Building Foundation, Harare.

Barnes, A. N., Mumma, J. and Cumming, O. 2018. Role, ownership and presence of domestic animals in peri-urban households of Kisumu, Kenya. *Zoonoses and Public Health* 65(1), 202–214.

Bikaako, W., Kabahango, P., Mugabi, K., Yawe, A., Stallon, K., Kyewalabye, E., Tukahirwa, L., Kusiimakwe, D., Stanley, M., Miller, B., Mugisha, A., Rosenbaum, M. H. and Amuguni, H. 2022. Breaking institutional barriers to enhance women's participation in and benefit from the Peste des Petits Ruminants and Newcastle disease vaccine value chains for Sembabule District of Uganda. *PLoS One* 17(10), e0270518.

Bonis-Profumo, G., do Rosario Pereira, D., Brimblecombe, J. and Stacey, N. 2022. Gender relations in livestock production and animal-source food acquisition and consumption among smallholders in rural Timor-Leste: A mixed-methods exploration. *Journal of Rural Studies* 89, 222–234.

Carter, N. A., Humphries, S., Grace, D., Ouma, E. A. and Dewey, C. E. 2017. Men and women farmers' perceptions of adopting improved diets for pigs in Uganda: decision-making, income allocation, and intra-household strategies that mitigate relative disadvantage. *Agriculture and Food Security* 6(1), 1–24.

Chekene, M. B. and Kashim, I. U. 2018. Gender equality: women in agriculture or gender in agriculture. *Agricultural Research and Technology* 18(5), 1–4.

Civil Society Budget Advocacy Group (CSBAG) 2022. *Assessing National Funding for Women's Economic Empowerment in Uganda*. Available at: https://africa.unwomen.org/sites/default/files/2022-09/Assessing-National-Funding-For-WEE-Uganda.pdf.

Doss, C., Meinzen-Dick, R. and Bomuhangi, A. 2014. Who owns the land? Perspectives from rural Ugandans and implications for large-scale land acquisitions. *Feminist Economics* 20(1), 76–100.

FAO 2018. *Empowering Rural Women, Powering Agriculture*. Food and Agriculture Organization, Rome. Available at: https://www.fao.org/3/CA2678EN/ca2678en.PDF.

FAO 2023. *The Status of Women in Agrfood Systems*. Food and Agriculture Organization of the United Nations, Rome. doi:10.4060/cc5343en.

Food and Agriculture Organization of the United Nations. 2011. *The state of food and agriculture; women in agriculture: closing the gender gap for development*. Food and Agriculture Organization, Rome. Available at: https://www.fao.org/3/i2050e/i2050e00.htm.

Ferrant, G., Pesando, L. M. and Nowacka, K. 2014. *Unpaid Care Work: the Missing Link in the Analysis of Gender Gaps in Labour Outcomes*. OECD Development Center, Boulogne Billancourt.

Financial Sector Deepening [FSDU] and the Government of Uganda. 2018. *FinScope Uganda: Topline Findings Report*. FSDU and the Government of Uganda. Available at: https://fsduganda.or.ug/wp-content/uploads/2018/10/FinScope-Uganda-Survey-Report-2018.pdf, Kampala.

Galiè, A., Njiru, N., Heckert, J., Myers, E. and Alonso, S. 2022. Gendered barriers and opportunities in Kenya's informal dairy sector: enhancing gender-equity in urban markets. *Gender, Technology and Development* 26(2), 214–237.

Galiè, A., Njiru, N., Wanyoike, F., Tawanda, M., Boonabaana, B., Bisikwa, J., Njuguna-Mungai, E., Jones, C. S. and Baltenweck, I. 2023. Exploration of gender dynamics in the production and marketing of forage technologies in Kenya. *Frontiers in Animal Science* 4, 1113243.

Gibson, S., Kabuchu, H. and Watkins, F. 2014. *Gender equality in Uganda: a Situational Analysis and Scoping Report for Gender Development Partners Group*. Commissioned by the UK Department for International Development (DfID) and Irish Aid. The IDL Group, Coffey International Development Ltd, UK.

Golla, A. M., Malhotra, A., Nanda, P. and Mehra, R. 2011. *Understanding and Measuring Women's Economic Empowerment: Definition, Framework and Indicators*. International Center for Research on Women (ICRW), Washington, DC.

Grassi, F., Landberg, J. and Huyer, S. 2015. *Running out of Time: the Reduction of Women's Work Burden in Agricultural Production*. Food and Agricultural Organization of the United Nations, Rome.

Guloba, M., Katunze, M., Ssewanyana, S., Ahikire, J., Musiimenta, P., Boonabaana, B. and Ssennono, V. 2018. *Gender Roles and the Care Economy in Ugandan Households: the Case of Kaabong*. Oxfam, Kabale and Kampala Districts.

IFAD 2012. *IFAD Policy on Gender Equality and Women's Empowerment*. International Fund for Agricultural Development. Available at: https://www.ifad.org/documents/38711624/39417906/genderpolicy_e.pdf/dc871a59-05c4-47ac-9868-7c6cfc67f05c?t=1507215182000.

IFAD 2013. *Gender and Livestock: Tools for Design*. Available at: https://www.ifad.org/documents/38714170/39148759/Gender+and+livestock.pdf/67c6dca9-4a11-4f53-931e-2ccb46105a3c.

International Trade Administration (ITA) 2022. *Uganda: Country Commercial Guide*. Available at: https://www.trade.gov/country-commercial-guides/uganda-agricultural-sector.

JICA 2023. *Reference Material for Gender Mainstreaming in Agricultural and Rural Development*. Available at: https://www.jica.go.jp/english/our_work/thematic_issues/gender/c8h0vm0000f3jmj6-att/gender_mainstreaming_07.pdf.

Kabeer, N. 1999. Resources, agency, achievements: reflections on the measurement of women's empowerment. *Development and Change* 30(3), 435–464.

Kabeer, N. 2001. Reflections on the measurement of women's empowerment. In: Sisask, A. (Ed.). *Discussing Women's Empowerment: Theory and Practice*, pp. 17–57. Swedish International Development Cooperation Agency, Stockholm, Sweden.

Kavuma, S. N., Muhanguzi, F. K., Bogere, G. and Cunningham, K. 2018. *Supporting Business Opportunities for Rural Women in East and Southern Africa: A case study of Uganda*. Action for Development (ACODE). Policy Research Series No. 87.

Leon-Himmelstine, C., Phiona, S. and Löwe, A. 2021. *Young Women in the Agricultural Sector in Uganda: Lessons from the Youth Forward Initiative*. Available at: https://odi.org/en/publications/young-women-in-the-agricultural-sector-in-uganda-findings-fromyouth-forward/.

Maestre, M. and Thorpe, J. 2016. *Understanding Unpaid Care Work to Empower Women in Market Systems Approaches*. BEAM Exchange, London. Available at: https://www.beamexchange.org.

Malapit, H., Quisumbing, A., Meinzen-Dick, R., Seymour, G., Martinez, E. M., Heckert, J., Rubin, D., Vaz, A., Yount, K. M., Gender Agriculture Assets Project Phase 2 (GAAP2) Study Team and Team, S. 2019. Development of the project-level women's Empowerment in Agriculture Index (pro-WEAI). *World Development* 122, 675–692.

Matthew, O., Osabohien, R., Lakhani, K. H., Aderounmu, B., Osadolor, N. E., Adediran, O., Mabinuori, O. and Igharo, A. E. 2022. Women engagement in agriculture and human capital development in developing countries: an African sub-regional analysis. *Plos One* 17(12), e0277519.

Mukasa, A. N. and Salami, A. O. 2015. *Gender Productivity Differentials among Smallholder Farmers in Africa: A Cross-Country Comparison*. African Development Bank, Abidjan.

Mulema, A. A., Boonabaana, B., Debevec, L., Nigussie, L., Alemu, M. and Kaaria, S. 2021. Spiraling up and down: mapping women's empowerment through agricultural interventions using the community capitals framework in rural Ethiopia. *Community Development* 52(1), 113–130.

Mwinuka, L. and Hyera, E. O. 2022. Effect of socio-cultural factors on gendered decision-making in the adoption of improved maize storage technologies. *Cogent Food and Agriculture* 8(1), 2132849.

OECD 2008. *Gender and Sustainable Development: Maximising the Economic, Social and Environmental Role of Women*. OECD Publishing, Paris. doi:10.1787/9789264049901-en.

Peralta, A. 2021. The role of men and women in agriculture and agricultural decisions in Vanuatu. *Asia and the Pacific Policy Studies* 9(1), 59–80.

Quisumbing, A., Meinzen-Dick, R. S. and Njuki, J. 2019. *Annual Trends and Outlook Report: Gender Equality in Rural Africa: From Commitments to Outcomes* (vol. 2019). International Food Policy Research Institute. doi:10.2499/9780896293649.

Rubin, D., Boonabaana, B. and Manfre, C. 2019. Building an inclusive agriculture: strengthening gender equality in agricultural value chains. In: *Annual Trends and Outlook Report: Gender Equality in Rural Africa: From Commitments to Outcomes*, pp. 83–96. doi:10.2499/9780896293649_06.

Rubin, D. and Manfre, C. 2014. Promoting gender-equitable agricultural value chains: issues, opportunities, and next steps. In: *Gender in Agriculture: Closing the Knowledge Gap*, pp. 287–313. doi:10.1007/978-94-017-8616-4_12.

Sexsmith, K., Smaller, C. and Speller, W. 2017. *How to Improve Gender Equality in Agriculture*. International Institute for Sustainable Development (IISD). Available at: https://genderinsite.net/sites/default/files/iisd%20brief5.pdf.

Tangka, F. K. 1999. *Gender Roles and Child Nutrition in Livestock Production Systems in Developing Countries: A Critical Review*. Socio-economics and Policy Research Working Paper 27. ILRI International Livestock Research Institute, Nairobi, Kenya.

Twinorugyendo, P. 2019. Women farmers in Uganda: constraints and ways forward. In: *Final AgriFoSe2030 Report 21. Theme 1 Social and Economic Dimensions of Smallholder-based Agriculture and Food Security*. Swedish International Development Agency. ISBN: 978-91-576-9694-6. Available at: https://www.slu.se/globalassets/ew/org/andra-enh/uadm/global/agrifose/phase-1/outputs/reports/agrifose2030-report-21.pdf, vol. 2019).

Uganda Bureau of Statistics 2017. *Women and Men in Uganda: Facts and Figures 2016*. Government of Uganda, Kampala.

Uganda Bureau of Statistics (UBOS) 2018. *The Annual Agriculture Survey 2018 Statistical Release*. Governement of Uganda, Kampala.

Uganda Bureau of Statistics (UBOS) 2018. *Uganda National Household Survey (UNHS)* (vol. 2016/2017). Governement of Uganda, Kampala.

Uganda Investment Authority (UIA) 2023. *Agriculture Value Addition; Opportunities in Agriculture Value Addition*. Available at: https://www.ugandainvest.go.ug/why-uganda/opportunities/agriculture-value-addition/.

UN Women, UNDP, UNEP, and the World Bank Group 2015. *The Cost of the Gender Gap in Agricultural Productivity in Malawi, Tanzania, and Uganda*. UN Women, the United Nations Development Programme–United Nations Environment Programme Poverty-Environment Initiative (UNDP-UNEP PEI) Africa, and the World

Bank. Available at: https://africa.unwomen.org/sites/default/files/Field%20Office%20Africa/Attachments/Publications/2015/10/Costing%20Gender%20Gap.pdf.

UN Women 2018. *Turning promises into action: gender equality in the 2030 agenda for sustainable development: UN. Women, New York*. Available at: https://www.unwomen.org/sites/default/files/Headquarters/Attachments/Sections/Library/Publications/2018/SDG-report-Gender-equality-in-the-2030-Agenda-for-Sustainable-Development-2018-en.pdf.

World Bank 2023. *Women, Agriculture and Work in Africa*. Available at: https://www.worldbank.org/en/programs/africa-myths-and-facts/publication/women-agriculture-and-work-in-africa.

Yisehak, K. 2008. Gender responsibility in smallholder mixed crop–livestock production systems of Jimma zone, South West Ethiopia. *Livestock Research for Rural Development* 20(11), 12.

Zole, E. W. 2011. *Gender relations and their effect on livestock management in Uganda: A case study of Ciforo Sub-county, Adjumani District* (Doctoral dissertation, Makerere University).

Chapter 5

Sociocultural factors affecting women smallholders: the case of ethnic minority women in the highlands of Southeast Asia

Nozomi Kawarazuka, International Potato Center, Vietnam; Nguyen Thi Van Anh and Vu Xuan Thai, Formerly Institute for Social Development Studies, Vietnam

1 Introduction

Women smallholder farmers in Asia are highly diverse in culture, religions, languages, and agricultural landscapes, especially among ethnic minority people who are concentrated in the highlands of Southeast Asia, northeastern India, and southwest China (e.g. Michaud, 2009, 2017; Scott, 2009). This upland region is distinguished from low-land areas in terms of agricultural, cultural, and political landscapes, forming unique research agendas in area studies. Farming conditions and gendered experience of these minority men and women farmers, including the young generation, are very different from those of the rice-based lowland region.

However, agricultural technologies (e.g. mechanization, hybrid seeds, and new cash crops for the global market) are developed from the majority

http://dx.doi.org/10.19103/AS.2024.0148.06

ethnic groups' perspectives in response to the needs of low-land agriculture, especially with men's interest. Women with minority backgrounds face double discrimination by the majority-centered and male-oriented agricultural technology design, policy, and interventions. Hence, it is deeply problematic to generalize farmers' interests and challenges in agriculture in Asia without considering gendered, cultural, and geographical differences.

The major narratives of agricultural development in the upland region focus on boom export crops such as cassava, coffee, rubber, and spices. The literature on political economy highlights the negative consequences of those mono-crop investment for the minority and poor farmers through environmental degradation, land grabbing, exploitative labor, and, most importantly, uneven and unequal distribution of benefits over space and time (Akram-Lodhi, 2005; Castella et al., 2023; Évrard and Baird, 2017; Hall, 2013; McElwee et al., 2023). In terms of gender, studies suggest that women are often losers of the shift from diverse farming systems to mono-crop cultivation, because their involvement in household decision-making over crop management is reduced, as those new cash crops require interactions with (male) outsiders and decisions on high investment, which are considered as men's domains (Bonnin and Turner, 2014; Douangphachanh et al., 2021; Kusakabe and Chanthoumphone, 2021).

Although women still play significant roles in agricultural production both for cash crops and home consumption, less attention is paid to the contribution of women's unpaid labor to overall household economy. This oversight of women's unpaid labor contribution is also discussed in feminist critiques of capitalist ideologies that women's unpaid labor for domestic work and care is neglected in their account of capital accumulation (Curty, 2020; Mies, 2007). Since farming is unpaid work for ethnic minority women in the Southeast Asian uplands, their contribution is under-recognized in economics-oriented agricultural research and its frameworks where women's roles in agriculture are classified as reproduction activities along with domestic work and care. Similarly, the narratives of women's empowerment in general in the global south have focused on women's involvement in paid labor work in the agricultural sector and elsewhere, neglecting women's unpaid work in reproduction and its relations to economic activities (see critiques: Calkin, 2015; Cornwall, 2018; Hickel, 2014; Wilson, 2015). Current research trends focusing on the economic aspect of agriculture are thus gender biased, contributing to reproducing gender inequality in knowledge production. This chapter addresses this challenge.

The concepts of feminist political economy and political ecology allow researchers to uncover women's hidden but important roles in unpaid work by exploring how unpaid labor functions the household's operation of economic activities (Collard and Dempsey, 2020; Morinville and Van Lier, 2021; Philipps, 2008; Power, 2004). In this book chapter, we apply this concept in the analysis of women's unpaid agriculture through the case study drawing on the experience of ethnic Thai young married women and men.

We demonstrate how women's unpaid labor in agriculture influences and is influenced by household economic activities that are mainly performed by men. The two contrasting components of unpaid and paid labor, women's and men's work, and on-farm and off-farm activities are inseparable from both conceptual and analytical viewpoints of explaining the lives of young married couples under disadvantaged conditions. The key message we bring in this chapter is the importance of integrating the often-forgotten aspects of women's unpaid labor in farming into research frameworks to position the ethnic minority women as key actors in agriculture. By so doing, we seek to contribute to producing pluralistic knowledge in agricultural and development scholarship. We believe that this is a first step to discuss how development policy and interventions can support those marginalized women in the uplands in transformative ways.

2 The narratives of agricultural development in the upland regions of Southeast Asia

As discussed in the introduction, less attention is paid to the roles of women in agricultural development in this region due to the narrow positioning of agriculture as an economic activity. There are limited studies on the roles of women in other aspects of agriculture such as small-scale farming and livestock production for both home consumption and local trading. A single model of agrarian change with cash crops masks not only the complexities and diversities in agriculture (Vicol et al., 2018) but also women's significant roles in sustaining local food and environmental systems and kin-based safety nets (Turner et al., 2022).

Ethnic minorities' livelihood strategies are much more complex than a mono-crop model. For example, income sources are diverse with not only cash crops but also labor migration and other non-farming incomes from men and/ or women (Cramb et al., 2009; Tugault-Lafleur and Turner, 2009; Turner et al., 2015; Vicol et al., 2018). Meanwhile, subsistence farming remains an integral part of agriculture for household food security with diverse crop varieties associated with their cultural value. As a result, high genetic diversities are observed in the upland region, such as rice varieties (Vanlalsanga et al., 2019), minor horticultural crops (Deka et al., 2012), maize, banana, root crops (Son et al., 2021), and indigenous pigs (Ho et al., 2022).

Historically, swidden agriculture (shifting cultivation) played a major role in farming among ethnic minorities in the region, and women are the ones who manage and control swidden agriculture (Erni, 2015; Sengupta, 2015). As swidden agriculture has been restricted in many countries in this region, farming systems among ethnic minorities become more diverse with so-called 'composite systems' which include swidden agriculture, permanent wet-rice farming, home gardens, perennial crop farming, and enduring three-crop cultivation in which women play central roles (Cairns and Brookfield, 2011). Composite systems allow ethnic minorities to ensure household food security meanwhile they invest in cash crops for local and regional markets.

In patriarchal societies in the upland region, women face a vicious circle of disadvantages throughout their life in male-privileged gender norms, such as girls' lower attainment than boys in both primary and higher education, early marriage, and having many children (Inui, 2022; Jones et al., 2018; Thi et al., 2023). As a consequence, women and girls often have limited physical mobilities and are more likely to stay at home and work for family farms and livestock production as (unpaid) family laborers (Jones et al., 2018; Kawarazuka et al., 2019; Le et al., 2023). This is very different from low-land areas where both married and unmarried women's labor force is absorbed in emerging industrial zones as well as in the informal sector in cities (Huijsmans, 2014; Kusakabe, 2020; Phouxay and Tollefsen, 2011; Resurreccion, 2005). Although ethnic minorities in this region cannot be generalized as they are extremely diverse linguistically and ethnologically, they must have certain commonalities in the challenges they face due to cultural, geographical, and political marginalization. In order to recognize women's roles and opportunities and constraints in agriculture among minority ethnic groups, it is essential to explore social organizations of households and women's positions as a wife, daughter-in-law, and mother.

3 Revisiting women's roles in (unpaid) agriculture labor work

The lack of attention to women's unpaid roles in agriculture research appears to be related to the dominant capitalist concepts employed in various disciplines. Women's unpaid labor in farming activities is classified as the domain of reproductive activities rather than that of productive activities. Productive and reproductive activities are then separated in theoretical, conceptual, and analytical frameworks with reproduction being the secondary importance in agricultural research and elsewhere. This is partly because the original theories were developed by men from men's perspectives: unlike women's, men's productive activities are usually independent from reproductive activities (Ahl, 2006; Ahl and Marlow, 2012; Little, 2016). Recent feminist approaches propose integrating the two domains into one framework and explore relationships between the two domains and women's roles therein (Collard and Dempsey, 2020; Gustavsson, 2021; Kawarazuka et al., 2023; Morinville and Van Lier, 2021). This is a starting point to appropriately recognize the complexities of livelihoods and women's roles therein.

Another concept relevant to this theme is the concept of the conjugal contract (Jackson, 1995; Whitehead, 1981), which provides useful insights to explore the role of women in unpaid work including subsistence farming. The conjugal contract explains that in contexts where women do not have direct access to resources, they are able to ensure security via spouses by fulfilling gendered obligations (unpaid and intangible) for their husbands, children, and other family members (Jackson, 2007). In exchange for their (unpaid) contribution, women could interpret their husbands' resources as theirs and access resources such as land, crops, and food (Hanrahan, 2015; Jackson, 2007). A conjugal contract is flexible and can be continuously reinterpreted and manipulated in women's favor (Moore, 1986). Obligations that women fulfill through conjugal contracts can also improve their bargaining position over the life course (Kandiyoti, 1998). In old age, women's roles turn from providing labor to receiving support, and in Asia especially, supporting old parents is an expression of filial piety for adult children (Eklund, 2018; Göransson, 2013; Knodel, 2014). In this way, a conjugal contract is interwoven by an intergenerational contract (Croll, 2006; Göransson, 2013) as women's interdependent relationships often shift from their husbands to their sons and daughters. The notion of a conjugal contract also emphasises the highly gendered nature of obligations and rewards in the life cycle in interdependent and intergenerational family relationships. Conjugal contracts thus significantly influence women's strategic choices for productive and reproductive activities.

Some studies on youth confirm that providing unpaid labor for family farming and domestic work – especially by women – is a means of receiving future support from their parents. In the matrilineal context of Laos, some young unmarried women choose not to migrate and instead to support family farms and domestic work. In exchange, they receive significant support from their mothers for childcare after marriage (Huijsmans, 2014). Huijsmans (2014) points out that young women's decisions to stay home to follow their parents' wishes rather than to migrate to a city to pursue their own aspirations is a strategic choice because women's inter-dependent relationships with their parents more strongly influence (positively and negatively) their life cycles compared with those of young men. Such young women's strategic choices are similar to those made by married women to contribute unpaid labor for access to their husbands' resources (Jackson, 2007; Moore, 1986). Hence, both married and unmarried women's unpaid labor for family farms should not simply be viewed as the issue of unemployment or women's immobility.

Building on concepts in forementioned literature, the following section describes a case study to explore ethnic minority women's lived experience in managing family farms.

4 Case study: 2018 Thai ethnic minority commune in Dien Bien, Vietnam

4.1 Methods

This case study is based on a qualitative study conducted in October 2018 in a Thai ethnic minority commune in Dien Bien province in Vietnam. The study began with a series of participatory activities to establish a rapport with the villagers and to obtain background information on the research context through participatory photos and videos and then held an event to receive feedback on these. Thereafter, data collection was undertaken through key informant interviews (KIIs), focus group discussions (FGDs), and in-depth interviews (Table 1). In-depth interviews were conducted with 12 young women and men under 30 years of age. The stories described on this paper mainly draws on these conversations in in-depth interviews (Table 2).

4.2 Study context

The study commune is located approximately 400 km northwest of Hanoi and has a population of around 4800 people, 80% of whom are members of the Thai ethnic minority group. This Thai ethnic group is a patrilineal society, and it is common for three generations to live together. According to information from KIIs, most children in this commune complete primary school education, while girls have slightly higher enrollment rates for secondary and high school education than boys (around 50% complete high school). Girls who drop out of school tend to do so for marriage, while boys drop out to migrate for labor. Thai women and men in our study site usually marry in their mid-to-late teens and early twenties, respectively. Marriage is considered a very important and an expensive life event: although simplified over the past 10 years, the average

Table 1 The lists of participants

	Men	Women	Total
Participatory video (film making)	6	6	12
Key informant interviews*	4	2	3
FGD (above 30 years old)	6	6	12
FGD (under 30 years old)	6	6	12
In-depth interviews	6	6	12
The community events (in two villages)	18	28	46

*A village leader, a farmer's Union leader, and a women's Union leaders.
FDG, focus group discussion.

Published by Burleigh Dodds Science Publishing Limited, 2025.

Table 2 The lists of respondents

	*Name	G	Age	**Economic Status	Living with parents or in-laws	Migration destination	Paid work
1	Dung	F	25	Poor	Yes	Hanoi***	Construction
2	Phuong	F	22	Poor	Yes	No experience	
3	Hanh	F	24	Poor	Yes	Hanoi***	Construction
4	Thuy	F	25	Better off	No	Hanoi	Factory
5	Mai	F	22	Poor	Yes	Hanoi***	Construction
6	Lan	F	28	Poor	Yes	No experience	
7	Nam	M	29	Poor	Yes	Within the province	Rock mining
8	Hieu	M	27	Poor	No	Five provinces	Construction
9	Sang	M	29	Poor	Yes	South-central coast	Construction
10	Minh	M	20	Better off	Yes	Hanoi	Construction
11	Vien	M	22	Poor	Yes	Hanoi	Construction
12	Duc	M	23	Poor	Yes	Hanoi, Dien Bien	Construction

*Proxy names.
**Self-reported, based on personal perceptions.
***These women go to Hanoi with their husbands and work as construction assistants for a short time during the winter season.

cost of a wedding runs between VND[1] 60 million and 70 million (USD 2500–3000) for the groom's household and VND 10 million (USD 420) for the bride's household. The money is provided by parents and relatives and the groom's own savings (if he has any).

The commune's poverty rate is 42% for the Thai ethnic group and less than 20% for a small number of Kinh people (the ethnic majority in Vietnam). Agriculture and livestock are central to lives and livelihoods. However, engaging in farming, including animal raising, is often not considered as an economically viable occupation for men. The commune used to grow mountain rice, but then shifted first to cassava and then maize in response to rapid soil erosion and price declines for cassava. Indigenous species of buffalos, cattle, pigs, and chicken were originally highly valued for cultural events, such as weddings and Tet (New Year cerebrations). However, many families now raise livestock as an income source and a hybrid cattle variety has been introduced to poor households through the government's poverty-reduction schemes. Cattles and buffalo are taken to the mountain areas for grazing.

Women of all ages and marital status have responsibility for feeding animals, growing vegetables and root crops in their home gardens, and contributing labor in farm fields. In contrast, most young men – married and

1 Vietnamese dong (VND): one US dollar equals approximately 23 000 VND at the time of fieldwork.

unmarried – work outside the village throughout the year. Around 15% of the commune population (700 people) are regularly absent from the village, working outside the province and sending remittances home. During winter, married women and middle-aged men also go to work in Hanoi or neighboring districts and provinces, while children are left in the village with their aunts or grandmothers. Building a house, purchasing new motorbikes, and/or agricultural machines are symbols of success in male labor migration. No cases of international migration were recorded in our study site.

5 Findings: gender pathways, dependencies, and opportunities

5.1 Gendered transition pathways from schooling to marriage

In our study site, we found a clear gender difference in the transition process from schooling to marriage, contributing to women's concentration in (unpaid) farming work. Young men typically spend at least a few years earning money outside the village before marriage, while young women tend to marry immediately after leaving or completing school, between 15 and 18 years old. We provide short case studies, turning first to the experience of young men.

Sang (29-year-old man) stopped studying in the 11th grade because his parents could not afford the school fee of VND 800 000 (USD 34). His father tried unsuccessfully to borrow money from a moneylender to enable Sang to graduate to high school. Observing the difficult situation, Sang decided to leave school to support his younger siblings still in school. He found casual labor work in various locations through social connections, going as far away as 700 km near the north-central coast. Through the experience, he acquired welding skills which helped him increase his salary over time from VND 180 000 (USD 7.7) per day to VND 400 000 (USD10.79) per day. He said, 'I learned a lot during my first four years of work … I learned about the way Kinh people live, work and save money as well as their dialects and culture. I learned more as I moved far away to the South.' In addition to his young siblings, his remittances to the family are also used to support his sister studying in college – and Sang takes great pride in this. He married at 19 and, by virtue of his work, was able to purchase some items considered essential for young men such as a motorbike and a mobile phone.

Similarly, Nam (29-year-old man) left school after completing the sixth grade at 15. In his first 3 years, he worked in road construction in a neighboring district. Through social connections, he shifted to work in mining and found consistent job offers for more work. He married at 19 and continues to migrate to work in the mines while his wife lives at home with his parents. Hieu (27-year-old man) also left school at 15, and worked as a laborer on national road and infrastructure projects in the northern mountainous region. He used to give some money to his parents when he returned home, but he kept most of his income for personal consumption including clothes, drinking, a smart phone, and a motorbike. He married at 23 and is now working in road construction within the district.

The narratives illustrate that work experience allows young men to have certain autonomy and freedom in spending money during their transition between schooling and marriage. At the same time, however, they also face some discrimination. Being ethnic minorities and of a young age with limited education, they run the risk of being cheated by their employers in Hanoi or industrial zones. Minh another 20-year-old man went to Hanoi when he was 15. He said he was cheated many times, either not paid wages owed or paid less than promised. From this experience, he said he learned he could not trust people in Hanoi. Sang reported that he had run away from some construction work at a foreign-invested company because he was assigned extremely dangerous tasks without training and safety equipment. For these reasons, men in our study site reported preferring to work with employers who were connected to their relatives or village neighbours – even though the wage was likely to be lower.

In contrast, since young women are less likely to travel seeking work, they have fewer opportunities for learning new skills, interacting with people living outside the village, and earning income. Rather, marriage is central to their youth experience in their late teens as the following examples demonstrate.

Lan (28-year-old woman) stopped her studies for marriage at 15 after completing the ninth grade. Her marriage was arranged by her parents and her husband's parents: 'I did not know what love is. My husband and I had no feeling of love for each other, but we had to sleep together. I just felt very sad and under considerable pressure. I was too young to marry. I used to weep alone at night, but gradually I got used to it. Now my husband is very sympathetic. We live together with happiness.' While arranged marriages have become less common in our study site, leaving school for marriage is still common. Hanh (24-year-old woman) also stopped her studies when she completed ninth grade as she did not find value in studying further even though her family could afford it. She helped her mother with cooking, farming, and animal husbandry for 2 years until her marriage at 17. When she lived at home not schooling or married, she recalls feelings of weakness: 'I did not have any power in the family. I just followed what my parents asked me to do. I did not have money and did not go anywhere outside the village. The only thing I decided was who I was going to marry.' Her reflections clearly indicate her unpaid labor contributions to the family were not considered as 'work' and went unrecognized while her position in the household remains low. However, when women marry, this situation changes as they begin working for their in-laws, which we now discuss in the next sub-section.

5.2 Intergenerational dependency and the roles of agriculture

Young married men's incomes from casual labor depend on the demand for work, which is unstable. Sang told us, 'I need to provide VND 50 000 (USD 2.5) per week for school meals for my child, but many times I have failed, and it places me under a lot of pressure.' As such, it is essential for married men to have support from his parents and relatives. Both married men and women ask support by working as laborers on their parents' or in-law's farms. Support varies from offering accommodation, food, farmland, care services to grandchildren, and connections for finding paid work.

Vien (22-year-old man) works in Hanoi while his wife and their child live with his parents. He sends 80% of his income to his parents but he thinks this only meets 30% of his parents' expectation. His wife works on the farm and looks after three heads of cattle and one buffalo. He returns to the village during planting and harvesting seasons to help his father even though he loses potential income because the farm is his top priority. In exchange with unpaid labor to maintain family farms, Vien and his wife secure accommodation and food.

Minh is the second son in his family. His youngest brother will inherit his father's house, but Minh and his wife live with his parents' support at the moment because his income from casual labor is too low to build a house. In turn, he gives 75% of his salary (USD 130) to his wife who then purchases daily necessities for his whole family, including for his parents and younger siblings. In addition, his wife works on his father's farm as a laborer and looks after three heads of cattle and one buffalo. Thus, although his remittances are limited, his wife's labor contribution in agriculture allows the couple to receive support from Minh's parents for accommodation and food.

Young men also receive material and non-material support from the family of their wives. Hung (24-year-old man) lives with his in-laws and works in his father-in-law's farm full-time and is unpaid. His parents-in-law and brother-in-law provide food, accommodation, and basic necessities for him, his wife, and their child. In our study site, a man living with in-laws is not common, but it does occur. In Hung's case, his own family is too poor and the family land is too small to divide among four sons. On the other hand, his wife's father has a large farmland, but his only son works outside the province. Therefore, Hung's labor, rather than remittances, was needed in this circumstance. Hung grows rice and maize, while his father-in-law oversees farm management and sales of rice and maize. This arrangement allows Hung to live under the safety net of his wife's family while establishing his marriage.

Young men can also receive support from their wives' parents in exchange with their wives' labor contribution to their natal families. Vien has constant offers for wage labor work due to his father-in-law's social connections. In return, Vien sends his wife to work as a laborer during the peak farming season to help his father-in-law. Similarly, Minh (20-year-old man) was given a small plot of farmland from his father-in-law. In return, his wife returns to her home village to help her parents and relatives with farming and animal husbandry. Another case, Thuy's husband borrowed money from his father-in-law (Thuy's father) to renovate his parents' house so that the newly married couple could live together with his parents. They knew that her husband would not be able to return the money in his circumstance with unstable employment, so to keep showing her appreciation to her own parents for this arrangement, Thuy takes care of her grandmother who lives in a distant village twice per month.

Thus, young married men have substantial support from their parents and in-laws. In turn, they offer not only their own labor but also the labor of their wives. Although this opens up an opportunity for young married women to visit their natal village, their visit must be coordinated with their mother-in-laws who help with domestic work including child care during young women's absence. This indicates that the ways women's labor contribution turns to cash/goods/services are complex and diverse. We now move on to highlight women's cases.

Phuong (22-year-old woman) is in charge of all farming work including raising pigs and chicken. Her mother-in-law then sells the pigs and chicken in a local market and controls the money. Her husband sends remittances to Phuong and she hands them over to her mother-in-law who controls it and gives some to Phuong. She said, 'If I need to buy something such as clothes and shoes, then she can give me money I would need. Everything I have bought for my daughter has been purchased by my parents-in-law.' Her unpaid labor contribution is, however, under-valued by herself. She continued, 'I cannot make money as I am staying at home. I am always scared of being criticised by my parents-in-law. I have to work hard even when I feel very tired. I am also under pressure from my husband. He often asks me by phone where I go and why I go out for so long and so on. I work hard at home every day. That is only the way for me to avoid being criticised.'

Dung (25-year-old woman) manages the farm of her parents-in-law with whom she lives, feeding chickens, grazing two heads of cattle and one buffalo, cooking, cleaning, and caring three children, including one of her brother-in-law's children (whose wife is working in Hanoi with him). Dung's mother-in-law is too old to do farming work and she can only do some light domestic work when Dung is in the farm. Dung's husband returns to the village only twice per year during planting and harvesting seasons, and he sends his remittances to his parents. When the family has a financial shortage, she asks her own parents for food and/or some money (around USD 10–20 per request). In turn, she goes to her parents' house during the peak farming season to help with their farm work and some domestic chores to relieve her mother's burden. Thus, the reciprocal support from her parents is as important as those with in-laws as her labor to her parents' farm turns to cash in times of need.

Mai (22-year-old woman) desires to be independent from her in-laws both physically and financially. She thinks it will take a long time as her husband is still paying off debts from their wedding 5 years ago. To help with repayment, she works in Hanoi with her husband during the off-farm season in winter. Her additional income from working as a laborer in construction is used primarily for repaying the debt, but she also puts aside some money for when her mother-in-law needs to attend weddings and funerals. Mai does this because her mother-in-law looks after her children and livestock during her absence.

These stories of men and women confirm that farming and domestic work are central to young couples' safety nets. Their labor contribution justifies support from their parents and parents-in-law. Young wives have a joint obligation with their husbands to sustain reciprocal relations with their in-laws as well as their parents.

5.3 Young married men's and women's opportunities in agriculture

In this sub-section, we present findings on young Thai men and women's perceptions about agriculture as income sources. Although both young men and women work on family farms, many of them do not consider it as 'work'. Hung who lives with his in-laws says he feels less important than his other male family members because 'I am not a breadwinner. I am dependent on my father-in-law who earns incomes from farming and a brother-in-law who sends remittances. I am under a lot of pressure from them to work hard and behave properly. If I don't do so, they will criticize me. I don't have any decision-making power in this house where I live. I follow what they say.'

However, young men are able to negotiate with their father or father-in-law. Hung, for example, has built trust relations with his father-in-law and gained some degrees of autonomy and independence in farming. He said, 'Two months before we met, he had planted 400 orange trees.' The local government had distributed planting materials to two households with large farms to test orange tree growth – one of them was Hung's father-in-law. Orange groves would be a new addition in this area as most farmers grow only rice, maize, and cassava. Hung's father-in-law assigned him to look after the trees, and Hung was excited to see if the trees produced fruit.

Young men who live with their parents often interpret their fathers' farmland and livestock as shared assets in which they are co-owners although they have not yet inherited the farmland. This interpretation allows them to access their fathers' farmland. For example, Vien has not yet inherited his father's farmland, but he is interested in grafting plums after seeing a video on YouTube. Nobody in his village has tried plums. In the case of Nam, his father was given two cattle as part of a local government's poverty reduction scheme for supporting poor households. The cattle produced four calves, eventually bringing in VND 15 million (USD 652) as additional income to his family. Nam still owes the debt of 50 million (USD 2170) for his father's buffaloes which died before they began generating income. Furthermore, Nam's father is sick, and he must pay the hospital bills while his wife looks after Nam's father. With this interdependency, Nam has a notion of co-ownership for and control over the cattle as his asset, even though it is his wife who invests the labor to take care of them.

Women also actively seek opportunities for themselves. Phuong looks after pigs and chickens that were originally purchased by her parents-in-law when she married. While the income from the pigs is taken and controlled by her mother-in-law, Phuong has a certain notion of ownership and autonomy over the pigs she breeds:

> As a daughter-in-law, I have very limited decision-making power. Many decisions in everyday lives are made by parents-in-law or my husband, but raising pigs is an area where I can make decisions. In my household, I know the best how to raise pigs and which pigs are ready for selling.

Although Phuong is not physically taking any money from raising pigs, her work certainly gives her power and space for negotiating other things such as asking her mother-in-law to help with domestic work and childcare. Similarly, Hanh has control over raising chickens. She learned from a friend in her natal village how to raise chickens in a cage and using artificial feed instead of rice. Adopting these lessons, she learned that chickens grow faster. Her role model is her sister-in-law who is economically independent and earns income from breeding buffaloes, pigs, chickens, and ducks. Earlier, Hanh had thought that earning cash incomes in Hanoi would be better, but she and her husband were cheated by their employers and exhausted by dusty and overcrowded urban life. She decided not to go to Hanoi again with her husband. Instead, she is improving her skills in livestock production and learning to manage the business like her sister-in-law.

In the last case, Lan's situation is different from others who live with their in-laws. She has full control of farm and livestock activities (including the income) as she lives independently from in-laws while her husband is absent from the house earning an income from mining work. Her main income source is breeding cattle, which produce one or two calves per year and she has sold six calves over the past 4 years. She also raises other livestock such as pigs, chickens and fish but they are mostly for home consumption. She perceives cattle breeding to have great potential for generating income for women, as it is less labor- and time-consuming compared with raising pigs. Her case demonstrates that young women are capable of independently managing agriculture including livestock production although their contribution is less recognized by themselves and their families.

6 Conclusion

This book chapter has discussed the challenges of women farmers who have double disadvantages from current mainstream male-oriented and the low-land centered agricultural policy and research priorities and frameworks. The example case was drawn from ethnic Thai women in the upland region of Northern Vietnam. We particularly looked at women's unpaid labor for domestic work, care, and farming and its relationships with economic activities to uncover women's significant roles in agriculture in this marginalized upland region.

The findings revealed that young married women are responsible for meeting greater labor contributions than men in both family farming and domestic work, managing overall household economy within the patrilineal organization. In return for unpaid labor, women receive both material and non-material support from their kin in times of need. This mechanism of safety nets is particularly important in the context where men's off-farm incomes are precarious. Hence, women's unpaid labor contribution to family farming and care work is an integral part of maintaining household economic activities.

In the literature, women's reproductive roles are often viewed as support for other family members' participation in productive activities (Shah and Lerche, 2020) and/or constraining their opportunities and capacities to enter labor markets (Ferguson and McNally, 2015; Kofman and Raghuram, 2015). Our findings suggest that reproduction is an integral part of economies (Ahl, 2006; Collard and Dempsey, 2020; Little, 2016; Philipps, 2008) and is also a tool for bargaining access to cash, food, goods, and childcare services for themselves in times of need (Jackson, 2007). In this respect, women's hidden roles in agriculture should not be excluded from the analysis of livelihoods. Research frameworks and models for exploring cash crops (as a productive domain) are not sufficient to understand women's roles in agriculture as farming systems and gender and social systems are interwoven with each other.

Our study also confirms that married young women living with in-laws eventually gain bargaining power through their unpaid labor contribution to the family: a sort of conjugal contracts in which women ensure access to resources via their husbands through their contribution to reproduction (Jackson, 1995; Jackson, 2007; Whitehead, 1981). Hence, the effect of unpaid work on bargaining should not be overlooked in research on women's empowerment. Young married women build their skills and experiences in agriculture and seek income generating activities within a small domain where they have autonomy such as livestock rearing. In thecase of Hung who lives with his wife's parents, he had a similar experience as women who live with their husbands' parents. This confirms that it is the relational position of young men and women in their family as daughter-in-law or son-in-law that determines their bargaining power and roles in agriculture.

Furthermore, ethnic minority young married women's challenges are closely associated with the struggles of their husbands as minority young men, who face various discrimination and challenges in non-farm employment in the low-land region. This confirms that ethnic minority women and men are socially embedded in existing family, gender and cultural structures within which they seek to find advantageous entry points for incremental change. Understanding women in isolation from their social and family backgrounds results in a misdirected focus on individual characteristics such as lack of knowledge, skills, and employment opportunities. Given that, any discussions on agriculture for women's empowerment among ethnic minority communities in upland Asia need to start by looking at their positions as a mother and a daughter-in-law. Simply promoting individual women to participate in income-generating activities and assessing the level of women's empowerment based on their involvement in economic activities is very narrow and superficial.

Our findings also highlight methodological implications for research on gender and agriculture. Young married men and women often tell researchers they are 'unemployed' or 'just staying at home' and express their limited decision-making power. These modest assessments of their position in conversation might simply reflect their understanding of how outsiders might value their agricultural contributions, especially their apprehension of outsiders' value for and interest in agriculture as a source of income. Oftentimes, respondents' words hide their aspirations to, dependency on and interests in family farming. Careful attention is therefore needed to understand smallholder farmers' especially women's paradoxical contributions to and benefits from agriculture, a situation in which small innovation spaces can open up opportunities for both women and men.

7 Future trends in research

Recently, there has been a trend to understanding the complexities of agriculture consisting of farming and non-farming incomes and from farmers' perspectives to re-evaluate swidden cultivation in the upland region of Asia (Mertz and Bruun, 2017; Pritchard et al., 2019; Teegalapalli and Datta, 2016; Vicol et al., 2018). Farmers' agency is central to these studies, contributing to enriching our understanding of the dynamics and flexibility of agriculture. Incorporating women's perspectives is, however, still a challenge in such crop-focused research. Although the study is from Latin America, Turner et al. (2022) employs the concept of food sovereignty and gender to understand women's unpaid- and under-valued contribution to agriculture as well as their roles in a sustainable environment. Miller and Gibson-Graham (2019) also propose new concepts to explore relationships of household economy, environment, and non-material aspects of life (e.g. gifts and reciprocal support). The concept of degrowth (Dengler and Lang, 2022) may be another important theme in this area to challenge the paradigm of (unsustainable) economic growth maintained by the exploitation of women's unpaid labor. The shift in the focus of agriculture from the quantity (productivity) to quality (e.g. well-being, cultural value, and environmental sustainability) is an important aspect of new knowledge production in both agriculture and gender scholarship.

As mentioned earlier, another important new trend relevant to this topic is feminist approaches to political economy and political ecology (Cohen, 2018; Curty, 2020; Gustavsson, 2021; Mies, 2007). The approaches have been applied to various themes from the analyses of natural resource management (Collard and Dempsey, 2020; Morinville and Van Lier, 2021), entrepreneurship (Ahl, 2006), and more broadly women's experience during the coronavirus disease 2019 pandemic (Stevano et al., 2021; Sultana, 2021). Agriculture, especially for understanding the role of unpaid labor, could be the next area to apply the forementioned feminist approaches. This allows researchers to be free from the ideology of capitalism and its research frameworks, enabling us to think of the problems of women farmers as structural than individuals' lack of skills and capacities.

To better incorporate the minorities' social, gender, and cultural contexts into research on agriculture, anthropological and ethnographic studies are very important. However, seminal work on ethnic minorities in the upland Southeast Asia has been led by Western scientists (e.g. Bonnin and Turner, 2014; Kunstadter, 1967; Michaud, 2009; Scott, 2009; Turner et al., 2015). They are very rich and extremely valuable to knowledge production, and some of them critically address subjectivity and reflexivity of themselves as outsiders (Bonnin, 2010; Turner, 2010). The future trend we hope to see is more studies conducted by local researchers, especially minority backgrounds and women scientists. Research with insider viewpoints would strengthen not only agricultural scholarship but also anthropological, geographical, and regional studies and help develop appropriate agricultural policy and intervention strategies with stronger notion of ownership by local communities, especially women.

Studies on ethnic minority people's structural problems are also conducted in various other sectors such as public health, education, and nutrition. Inter-sectoral and interdisciplinary research in those sectors to identify and address common research questions could help improve our understanding of agricultural challenges in the upland region.

Lastly, to link science into practice to transform the communities on their own, applied research learning from (failure) cases in agricultural and/or multi-sector development projects would be important. This is the first step to offer tailored approaches for ethnic minority men and women with different steps to agricultural technologies and development interventions.

8 Where to look for further information

- IFAD indigenous people.
- https://www.ifad.org/zh/indigenous-peoples.
- Journal of Mountain Research and Development.
- https://www.jstor.org/journal/mounresedeve.
- Journal Asian Ethnicity.
- https://www.tandfonline.com/journals/caet20.

9 Acknowledgments

We thank all women and men young farmers for sharing their experiences and opinions with us. We are grateful to local authorities for their institutional support. The project was funded by CGIAR GENDER Impact Platform and CGIAR Research Program on Policies, Institutions, and Markets (PIM) led by IFPRI.

10 References

Ahl, H. 2006. Why research on women entrepreneurs needs new directions. *Entrepreneurship Theory and Practice*, 30, 595–621.

Ahl, H. & Marlow, S. 2012. Exploring the dynamics of gender, feminism and entrepreneurship: advancing debate to escape a dead end? *Organization*, 19, 543–562.

Akram-Lodhi, A. H. 2005. Vietnam's agriculture: processes of rich peasant accumulation and mechanisms of social differentiation. *Journal of Agrarian Change*, 5, 73–116.

Bonnin, C. 2010. Navigating fieldwork politics, practicalities and ethics in the upland borderlands of northern Vietnam. *Asia Pacific Viewpoint*, 51, 179–192.

Bonnin, C. & Turner, S. 2014. 'A good wife stays home': gendered negotiations over state agricultural programmes, upland Vietnam. *Gender, Place & Culture*, 21, 1302–1320.

Cairns, M. & Brookfield, H. 2011. Composite farming systems in an era of change: Nagaland, Northeast India. *Asia Pacific Viewpoint*, 52, 56–84.

Calkin, S. 2015. "Tapping" women for post-crisis capitalism. *International Feminist Journal of Politics*, 17, 611–629.

Castella, J.-C., et al. 2023. Beyond the boom-bust cycle: an interdisciplinary framework for analysing crop booms. *Global Environmental Change*, 80, 102651.

Cohen, J. 2018. What's "radical" about feminist radical political economy? *Review of Radical Political Economics*, 50, 716–726.

Collard, R. & Dempsey, J. 2020. Two icebergs: difference in feminist political economy. *Environment and Planning A: Economy and Space*, 52, 237–247.

Cornwall, A. 2018. Beyond 'empowerment lite': women's empowerment, neoliberal development and global justice. *Cadernos Pagu*, 52. https://doi.org/10.1590/180 94449201800520002

Cramb, R. A., et al. 2009. Swidden transformations and rural livelihoods in Southeast Asia. *Human Ecology*, 37, 323–346.

Croll, E. J. 2006. The intergenerational contract in the changing Asian family. *Oxford Development Studies*, 34, 473–491.

Curty, G. 2020. Rethinking capitalism, crisis, and critique: an interview with Nancy Fraser. *Critical Sociology*, 46, 1327–1337.

Deka, B. C., et al. 2012. Horticultural diversity in North-East India and its improvement for value addition. *Indian Journal of Genetics and Plant Breeding*, 72, 157–167.

Dengler, C. & Lang, M. 2022. Commoning care: feminist degrowth visions for a socio-ecological transformation. *Feminist Economics*, 28, 1–28.

Douangphachanh, M., et al. 2021. Agriculture transition and women's decision-making power in coffee-farming households in Lao PDR. *Journal of Southeast Asian Studies*, 26, 49–71.

Eklund, L. 2018. Filial daughter? Filial son? How China's young urban elite intergenerational obligations. *NORA - Nordic Journal of Feminist and Gender Research*, 26, 295–312.

Erni, C. 2015. *Shifting Cultivation, Livelihood and Food Security: New and Old Challenges for Indigenous Peoples in Asia*. Bangkok, Thailand: FAO, IWGIA and AIPP.

Évrard, O. & Baird, I. G. 2017. The political ecology of upland/lowland relationships in Laos since 1975. In: Bouté, V. & Pholsena, V. (eds.) *Changing Lives in Laos: Society, Politics, and Culture in a Post-Socialist State*. Singapore: NUS Press.

Ferguson, S. & Mcnally, D. 2015. Precarious migrants: gender, race and the social reproduction of a global working class. *Socialist Register*, 51, 1–23.

Göransson, K. 2013. Reassessing the intergenerational contract. *Journal of Intergenerational Relationships*, 11, 62–75.

Gustavsson, M. 2021. The invisible (woman) entrepreneur? Shifting the discourse from fisheries diversification to entrepreneurship. *Sociologia Ruralis*, 61, 743–758.

Hall, D. 2013. Land grabs, land control, and Southeast Asian crop booms. In: Peluso, N. & Lund, C. (eds.) *New Frontiers of Land Control*. London: Routledge.

Hanrahan, K. B. 2015. 'Mɔn' (to marry/to cook): negotiating becoming a wife and woman in the kitchens of a northern Ghanaian Konkomba community. *Gender, Place & Culture*, 22, 1323–1339.

Hickel, J. 2014. The 'girl effect': liberalism, empowerment and the contradictions of development. *Third World Quarterly*, 35, 1355–1373.

Ho, N.-N., et al. 2022. Indigenous pig production and welfare of ultra-poor ethnic minority households in the Northern mountains of Vietnam. *Environment, Development and Sustainability*, 24, 156–179.

Huijsmans, R. 2014. Becoming a young migrant or stayer seen through the lens of 'householding': households 'in flux' and the intersection of relations of gender and seniority. *Geoforum*, 51, 294–304.

Inui, M. 2022. Education access and continuity in northern Laos: a comparative study of the Hmong and Lanten minorities. *Hmong Studies Journal*, 24, 1–47.

Jackson, C. 1995. From conjugal contracts to environmental relations: Some thoughts on labor and technology. *IDS Bulletin*, 26, 33–39.

Jackson, C. 2007. Resolving risk? Marriage and creative conjugality. *Development and Change*, 38, 107–129.

Jones, N., et al. 2018. Intersecting inequalities: the impact of gender norms on Hmong adolescent girls' education, marriage and work in Viet Nam. In: Harper, C., ēt al. (eds.) *Empowering Adolescent Girls in Developing Countries*. London: Routledge.

Kandiyoti, D. 1998. Gender, power, and contestation: rethinking bargaining with patriarchy. In: Jackson, C. & Pearson, R. (eds.) *Feminist Vision of Development: Gender, Analysis and Policy*. London: Routledge.

Kawarazuka, N., et al. 2023. Social reproduction: the sidelined aspect in gender and agricultural research. *Frontiers in Sustainable Food Systems*, 7. https://doi.org/10.3389/fsufs.2023.1220486

Kawarazuka, N., et al. 2019. 'A bird locked in a cage': Hmong young women's lives after marriage in northern Vietnam. In: Sachs, C. E. (ed.) *Gender, Agriculture and Agrarian Transformations*. London: Routledge.

Knodel, J. 2014. Is intergenerational solidarity really on the decline? Cautionary evidence from Thailand. *Asian Population Studies*, 10, 176–194.

Kofman, E. & Raghuram, P. 2015. *Gendered Migrations and Global Social Reproduction*. London, UK: Palgrave Macmillan.

Kunstadter, P. (ed.) 1967. *Southeast Asian Tribes, Minorities, and Nations*. Princeton: Princeton University Press.

Kusakabe, K. 2020. Migrant workers and childcare at the place of destination: an overview of issues in the Mekong region, South East Asia. In: Huang, S. & Ruwanpura, K. N. (eds.) *Handbook on Gender in Asia*. Northampton: Edward Elgar Publishing.

Kusakabe, K. & Chanthoumphone, C. 2021. Transition from subsistence agriculture to rubber plantations in northern Laos: analysis of household livelihood strategies by ethnicity and gender. *SAGE Open*, 11. https://doi.org/10.1177/21582440211011463

Le, T. T., et al. 2023. Ethnic minority women's empowerment in agriculture in the central region of Viet Nam. *Plos One*, 18, e0287115.

Little, J. 2016. Gender and entrepreneurship. In: Shucksmith, M. & Brown, D. L. (eds.) *Routledge International Handbook of Rural Studies*. London: Routledge.

Mcelwee, P., et al. 2023. Climate precarity in rural livelihoods: agrarian transformations and smallholder vulnerability in Vietnam. *Journal of Agrarian Change*, 23, 661–686.

Mertz, O. & Bruun, T. B. 2017. Shifting cultivation policies in Southeast Asia: a need to work with, rather than against, smallholder farmers. In: Cairns, M. (ed.) *Shifting Cultivation Policies: Balancing Environmental and Social Sustainability*. Wallingford, UK: CABI.

Michaud, J. 2009. Handling mountain minorities in China, Vietnam and Laos: from history to current concerns. *Asian Ethnicity*, 10, 25–49.

Michaud, J. 2017. What's (written) history for?: on James C. Scott's Zomia, especially Chapter 6½. *Anthropology Today*, 33, 6–10.

Mies, M. 2007. Patriarchy and accumulation on a world scale revisited. *International Journal of Green Economics*, 1, 268–275.

Miller, E. & Gibson-Graham, J. K. 2019. Thinking with interdependence: from economy/environment to ecological livelihoods. In: Zournazi, M. & Bennet, J. (eds.) *Thinking in the World*. London and New York: Bloomsbury Publishing.

Moore, H. 1986. *Space, Text and Gender: An Anthropological Study of the Marakwet of Kenya*. Cambridge, Cambridge University Press.

Morinville, C. & Van Lier, N. 2021. On nature, degradation, and life-making in late capitalism. *Capitalism Nature Socialism*, 32, 43–61.

Philipps, L. 2008. Silent partners: the role of unpaid market labor in families. *Feminist Economics*, 14, 37–57.

Phouxay, K. & Tollefsen, A. 2011. Rural–urban migration, economic transition, and status of female industrial workers in Lao PDR. *Population, Space and Place*, 17, 421–434.

Power, M. 2004. Social provisioning as a starting point for feminist economics. *Feminist Economics*, 10, 3–19.

Pritchard, B., et al. 2019. The importance of non-farm livelihoods for household food security and dietary diversity in rural Myanmar. *Journal of Rural Studies*, 67, 89–100.

Resurreccion, B. P. 2005. Women in-between: gender, transnational and rural-urban mobility in the Mekong region. *Gender, Technology and Development*, 9, 31–56.

Scott, J. C. 2009. *The Art of Not Being Governed: An Anarchist History of Upland Southeast Asia*. New Have: Yale University Press.

Sengupta, M. 2015. Migrant narratives in state-led development: the Reang experience in Tripura (North-East India). *Journal of South Asian Development*, 10, 73–95.

Shah, A. & Lerche, J. 2020. Migration and the invisible economies of care: production, social reproduction and seasonal migrant labour in India. *Transactions of the Institute of British Geographers*, 45, 719–734.

Son, H. N., et al. 2021. Indigenous knowledge and the enhancement of community resilience to climate change in the Northern Mountainous Region of Vietnam. *Agroecology and Sustainable Food Systems*, 45, 499–522.

Stevano, S., et al. 2021. Hidden abodes in plain sight: the social reproduction of households and labor in the COVID-19 pandemic. *Feminist Economics*, 27, 271–287.

Sultana, F. 2021. Climate change, COVID-19, and the co-production of injustices: a feminist reading of overlapping crises. *Social & Cultural Geography*, 22, 447–460.

Teegalapalli, K. & Datta, A. 2016. Shifting to settled cultivation: changing practices among the Adis in Central Arunachal Pradesh, north-east India. *Ambio*, 45, 602–612.

Thi, H. D., et al. 2023. Socio-cultural norms and gender equality of ethnic minorities in Vietnam. *Journal of Racial and Ethnic Health Disparities*, 10, 2136–2144.

Tugault-Lafleur, C. & Turner, S. 2009. The price of spice: ethnic minority livelihoods and cardamom commodity chains in upland northern Vietnam. *Singapore Journal of Tropical Geography*, 30, 388–403.

Turner, K. L., et al. 2022. Food sovereignty, gender and everyday practice: the role of Afro-Colombian women in sustaining localised food systems. *The Journal of Peasant Studies*, 49, 402–428.

Turner, S. 2010. Challenges and dilemmas: fieldwork with upland minorities in socialist Vietnam, Laos and southwest China. *Asia Pacific Viewpoint*, 51, 121–134.

Turner, S., et al. 2015. *Frontier livelihoods: Hmong in the Sino-Vietnamese borderlands*. Seattle: University of Washington Press.

Vanlalsanga, et al. 2019. Rice of Northeast India harbor rich genetic diversity as measured by SSR markers and Zn/Fe content. *BMC Genetics*, 20, 79.

Vicol, M., et al. 2018. Rethinking the role of agriculture as a driver of social and economic transformation in Southeast Asia's upland regions: the view from Chin State, Myanmar. *Land Use Policy*, 72, 451–460.

Whitehead, A. 1981. I'm hungry mum: the politics of domestic bargaining. In: Young, K., et al. (eds.) *Of Marriage and the Market*. London: CSE Books.

Wilson, K. 2015. Towards a radical re-appropriation: gender, development and neoliberal feminism. *Development and Change*, 46, 803–832.

Chapter 6

Women smallholder farmers in South America, Central America and the Caribbean

Janet Momsen, University of California-Davis, USA

1 Introduction

These regions are those in which smallholder farmers have been mostly ignored in official statistics and the gender of their ownership rarely considered. Consequently, I have had to depend on field research for much of the information presented in this chapter. The nature of these smallholdings reflects the ethnographic mix and migration patterns of the region (Momsen, 2017). It has been strongly influenced by European colonialism:

- Spanish in most of South and Central America and the Greater Antilles;
- Portuguese in Brazil; and
- English, French, Dutch and Danish in the Lesser Antilles.

Following European settlement, export agriculture depended on forced labour originally supplied by indigenous peoples, and when these sources became inadequate, they were replaced by African slaves and Asian indentured labour. Local food production remained in the hands of women smallholder farmers.

http://dx.doi.org/10.19103/AS.2024.0148.07

Boserup (1970) was first to recognize male and female farming systems and their regional patterns in developing countries. These systems were defined by their gender roles, as to which gender contributed most to agricultural labour. Boserup suggested that where plough agriculture is found, as in Latin America, there is low female participation in agriculture. She considered the Caribbean region an aberration because women are numerically important as small farmers there (Momsen, 1988). Boserup explained this in terms of the predominance of Africans brought as slaves to the region during the seventeenth, eighteenth and part of the nineteenth centuries who carried with them their African attitudes to agriculture. In this way, she negated her own classification based on farming systems while introducing a focus on migration.

We now know that South America, Central America and the Caribbean contain many different types of agriculture, from shifting agriculture in the Amazon (Henshall, 1968; Henshall, 1982) to intensive irrigated viticulture in Argentina (Imburgia, 2019) and land-farmed in cooperatives and land settlements (Momsen, 1987a; Besson, 2007), all of which involve women farmers. There have also been demographic and social changes encouraging women to turn to farming as a livelihood. In Latin America and the Caribbean, some 11 per cent of households are now headed by a single parent, almost always a mother, a level much higher than elsewhere in the world (Economist, 2023). Recent data from the World Bank (2020) suggest that 78 per cent of single mothers in the region are in the workforce. Some of these single mothers are raising crops on small plots to feed their families and to sell. These home gardens have been found to have developed adaptive functions over time: producing plentiful supplies of food with relatively low labour input, and 'providing a "genetic backstop" during periods of crop failure' according to Lope-Alzina (2007). Yet home gardens are often ignored because of their physical proximity to the home, their strong association with women's decision-making and labour and their non-commercial cultural values (Howard, 2003).

The first section is based on my own field research in Mexico and Costa Rica and on fieldwork carried out in collaboration with graduate students. Differences related to ethnicity and migration status are shown by Momsen (2017) to be relevant in many areas. The following section considers women small farmers in South America, especially in Brazil, Argentina and Colombia. The final section includes the broad Caribbean region including the British, French, Spanish and Dutch Antilles. Here production is largely focused on food for the domestic market and some crops for inter-island trade. This section is based on fieldwork in Trinidad and Tobago, the Windward Islands, Nevis and St Kitts in the Leeward Islands, St Martin and Jamaica and shorter periods in the Dutch Antilles, Cuba and Puerto Rico, the Dominican Republic and Haiti.

Special emphasis is given to Barbados. Barbados has a higher proportion of women small farmers than anywhere else in the region (see Table 1). This

Table 1 Women small scale farmers in the Eastern Caribbean

Country	Year of survey	Sample size	% of women farmers
Barbados	1963	207	53.1
Barbados	1987	126	25.8
Grenada	1969	256	20.7
Grenada	1982	186	20.4
Martinique	1964	203	35.5
Martinique	1981	17919	20.0
Montserrat	1972	527	44.2
Montserrat	1985	136	27.9
Nevis	1979	91	30.8
Nevis	1985	407	39.3
St Lucia	1964	187	42.8
St Lucia	1980/1981	7520	23.0
St Vincent	1972	6862	46.2
Trinidad	1979	80	28.8

Source: Field surveys and census data adapted from Momsen (1993: 241).

proportion has varied over time and the determinants of this variation will be considered over the post-Emancipation period since 1834 (Momsen, 2007). Women smallholders have been studied during 60 years of fieldwork since 1963 using sample surveys and in-depth interviews. These surveys reveal the impact of economic modernization, post-colonialism and self-government, on smallholder farming.

2 Women smallholder farmers in Mexico

In this section, I look at women small farmers in commercialized farming areas of the Bajío north of Mexico City and in the milpa agriculture of the Yucatán. In both areas, maize is the main crop and commercialization of agriculture undermines women's roles on farms while male migration leaves women-headed households with responsibility for farming (Chambers and Momsen, 2007; Lopez-Alzina, 2007; Torres, 2011). Women usually have a small household plot where they grow plants needed for home consumption and often keep small animals such as chickens and pigs. In this way, these household plots are a source of genetic diversity.

In the Bajío where maize is both a commercial and a subsistence crop, there are home gardens and small plots, usually cultivated by women, side by side with the larger fields cultivated by men. The selection of maize cultivars differs by gender as women often grow varieties chosen for dietary reasons for domestic use rather than for commercial ease of sale as is the main determinant for men. As migration by men to the USA increases, women's roles on the farm expand and their knowledge of crops is important in protecting biodiversity (Chambers and Momsen, 2007).

Migration is also important in the Yucatán, but there it is often women who migrate for work in tourist poles (Torres, 2011; Lope-Alzina, 2007). This migration is usually short term, and contact is maintained with their home villages. Food for sale to tourists is also grown by women who move between farmwork and domestic work in tourism areas (Torres, 2011). Agriculture is mainly the traditional shifting cultivation of milpas growing corn and squash and carried out by men. Women maintain home gardens near their homes and sometimes on communal plots in the village as noted by Lope-Alzina (2007). On these plots, women grow food for the family and maintain a great diversity of crops. I interviewed one young woman who was growing a single cotton plant by her back door. She explained that she needed the cotton balls to clean her new baby. External agencies also work with women farmers and encourage entrepreneurship and the production of specialist crops for the tourist market (Momsen, 2002, 2003; Blanco-Gregory et al., 2020).

3 Women smallholder farmers in Central America: Nicaragua and Costa Rica

This is a region of banana and coffee production by both large- and small-scale farmers. It has been seriously affected by civil disturbances in recent years, with the exception of Costa Rica. This section looks at Nicaragua and Costa Rica.

3.1 Nicaragua

Under the Sandinista Revolution of 1979 in Nicaragua, it was government policy to put women's rights at the forefront of agricultural change. In the post-Sandinista era, the position of women farmers deteriorated but, through the intervention of many religious organizations, cooperatives were set up and the Fair Trade system was introduced. This system encourages women to join producer groups and take on leadership roles (Dilley and Momsen, 2013). Women especially appreciate the social premium provided by Fair Trade which enables them to educate their children and improve their living conditions as shown in Momsen (2009). A study of the SOPPEXCCA cooperative in Nicaragua revealed that within 10 years of its foundation in 1997, its members had an income 25 per cent above the local average. Its goal was to work with women coffee producers right from the start. After 10 years women made up about one-third of producers in each of the 15 sub-cooperatives, and the Director was a woman. The cooperative helped women gain access to farmland, but many had difficulty coping with producing high-quality coffee on top of their household duties which often included cooking for workers on their husbands' farms. About half the women interviewed sought help from their husbands for labour on the land but worked with their children in cleaning and sorting the coffee beans.

3.2 Costa Rica

According to the 2014 Agricultural Census of Costa Rica 12 634 women producers, 15.6 per cent of the total farmers, owned 106 500 hectares of land making up only 8 per cent of the agricultural land owned by individuals in the country. Women's land is mainly in the poorest parts of Costa Rica and in areas of forest conservation. Many women farmers are engaged in reforestation, ecotourism, and harvesting of non-traditional forest products (World Bank, 2020). In 2019, Costa Rica became one of the few countries in the world to prepare a Gender Action Plan guaranteeing financial resources, technical assistance and monitoring to enable women to be integrated into land use planning and forest protection. Women coffee farmers generally earn 40 per cent less than men because they lack access to direct markets but recently women have started two all-women coffee cooperatives and have set up new businesses such as beekeeping and organic agriculture (UNDP, 2020). In the south of the country, in Talamanca, the indigenous BriBri, a matriarchal society, practice traditional, largely subsistence, agroforestry and grow a variety of organic cocoa resistant to most diseases (Echavarría, 2020). Eighty per cent of these cocoa producers are women and owners of their lands (Pelliccia, 2021).

4 Women smallholder farmers in South America: Brazil, Argentina and Chile

Small farming by women has been much affected by the varied environment of the region, its complex history of immigration and recent political crises. Women small farmers exist throughout and have grown in importance in areas such as Colombia where recent violence has left widows forced to turn to farming (Parra-Fox, 2015) although in the past Colombian women, especially in frontier settlements, were housewives with no role in farming (Townsend, 1995).

4.1 Brazil

According to the 2017 Agricultural Census of Brazil, the proportion of farms run by women rose to 18.6 per cent, with almost a million women involved, compared with 12.7 per cent eleven years earlier (UN Women, 2019). Encouraged by feminist agricultural groups, new female farmers who are part of Brazil's Landless

Workers Movement, are embracing beekeeping and agroforestry on formerly unproductive land in more than 500 agrarian reform settlements (Gulon, 2022).

Earlier Japanese pioneers settled in the Amazon region and demonstrated how men and women could work together on small farms producing vegetables, eggs and black pepper for local markets (Henshall, 1968). Agricultural settlements and cooperatives have been established in many parts of Brazil (Henshall and Momsen, 1974; Henshall, 1982). Although men and women are technically equal members of such cooperatives, women's views are often ignored. Sometimes men will abandon the farm and move to a nearby town which appears to offer greater opportunities. The success of such deserted women as independent farmers largely depends on the presence of children old enough to provide labour and assist in marketing produce. If these women are unable to cultivate the land, they lose their membership in the cooperative. Today, increasing attention is being paid to women small farmers, encouraged by funding from overseas. Women are seen as the guardians of biodiversity and are being taught how to measure the economic value of their farm production by keeping detailed records (UN Women, 2019). In this way, women are becoming aware of the financial and dietary importance of their farm output to the family and are thus gaining status and confidence.

4.2 Argentina and Chile

In both these countries, agriculture is dominated by agribusiness and patriarchal attitudes. There is much rural poverty and migration to cities. In Chile, women make up only one-quarter of small farmers and own only 9 per cent of the land. More gender awareness, stimulated by international agencies, is slowly changing this situation. In Argentina, the government has launched a programme to help small farmers transition to more resilient agro-ecology based production with women as 40 per cent of the beneficiaries (OECD, 2022). In areas of intensive agriculture such as viticulture in the Mendoza area women farmers do participate. However, this type of agriculture is highly dependent on irrigation and women farmers on their own, despite having equal rights to land, find it is difficult to remain as farmers because they have higher labour costs, cultivate less land and earn less income from farming and so are financially least able to invest in the modernization of irrigation systems (Imburgia, 2019). Thus, there has been a decrease in female participation in small-scale farming in the area as the need to earn off-farm income has become more pressing. In Chile women, smallholders also are under increasing pressure to take off-farm jobs and are moving into tourist-related employment and handicraft production.

5 Women smallholder farmers in the Caribbean: Trinidad, Cuba, Barbados and the Eastern Caribbean

Women small farmers are more common here than elsewhere in the region but least so in the Hispanic Caribbean. Caribbean women generally exhibit a higher level of economic autonomy than is found in Central and South America. This may be attributed to three factors. Firstly, under slavery, women were more impacted by labour exploitation than slave men (Morrissey, 1989). Secondly, the restrictions on marriage enforced by slave owners and, after emancipation, male migration out of the region, left many women alone to bear the burden of caring for their children. Sex-specific migration has led to 35 per cent of rural households in the Eastern Caribbean being headed by women and two-thirds of children being born out of wedlock. A second enabling factor for women in the Anglophone Caribbean has been their legal equality with male siblings in access to land (see Table 1) while in the Francophone and Hispanic Caribbean patriarchal inheritance laws limit women's land ownership (Momsen, 1993). Thirdly, this region is characterized by a highly educated female population. Where men have sought economic opportunity through migration women have achieved it through education for themselves and their children. Yet, in most parts of the region, farms run by women tend to be smaller, less accessible and have fewer resources than male-operated farms (Henshall, 1981; Brierley, 1993).

5.1 Trinidad

This is the most southerly island of the Lesser Antilles. Crop cultivation was introduced from Venezuela about 300 BC. Shifting agriculture was practiced with men doing the land clearing and women the planting, weeding and harvesting. Women were instrumental in food production and this role increased the status of women in society (Newson, 1976). In the sixteenth century waves of European settlers bringing slaves and plantations led to a decline in the indigenous population and the beginning of commercial agriculture on the island with the introduction of cacao and tobacco (Momsen and Richardson, 2009). Women continued to grow all the food for domestic consumption on small plots but there was a dietary switch from root crops to maize (Newson, 1976) which may have been related to the new time demands of the export crops. After slave emancipation indentured labourers were brought from India. Following indentureship many Indians were given plots of Crown Land and became small-scale sugar growers, rice producers or market gardeners. Indra Harry (1993) in a survey of small farmers, found that many women farmers were proud to grow organic crops, and most had inherited land from deceased spouses. Both men and women irrigated their crops. Women farmers worked longer hours than men as they had to undertake household duties and childcare in addition to agriculture (Momsen, 1988).

5.2 Cuba

Cuba is the largest island of the Caribbean with over 11 million people, of whom unusually for the region only a small proportion, mainly Jamaican immigrants, are black and 27 per cent mulatto. It was settled by the Spanish in 1492 but became independent in 1902. Slavery was abolished in 1886 much later than elsewhere in the Caribbean. In 1959, Fidel Castro took over and Cuba became Communist with land reforms placing 70 per cent of farmland in large-scale holdings in the hands of the state. In the late 1960s, there were 250 000 individual farmers on 30 per cent of the farmland but by the mid-1980s, there were only 180 000 farmers on 15 per cent of the farmland (Stubbs, 1993). In the 1990s, a need for domestic food production led to the development of cooperatives strongly supported by women who liked working together. For the first time in Cuban agrarian history, women were encouraged to become farmers in their own right (Stubbs, 1993). Under the 1959 land entitlement programme, land security had been granted to male heads of households 'reflecting an agrarian pattern more characteristic of Hispanic America than that of the Caribbean' (Stubbs, 1993). However, women increasingly worked as members of tobacco-growing cooperatives and gradually took on leadership roles in many general cooperatives and by 1979 over a third of cooperative farmers were women. Women also began producing food crops on small peri-urban farms for urban markets and were especially involved in organic farming methods encouraged by the shortage of imported pesticides and fertilizers (Torres et al., 2007). Private producers earn more than their collective counterparts and so more women have moved into small farming and become strong advocates of such opportunities.

5.3 Barbados and the Eastern Caribbean

Women have long played a major role in agriculture on the smaller islands (Table 1) but the involvement of women has changed over time. From the 1640s until the slave trade ended in 1803, men and women were brought from West Africa as enslaved workers on the plantations. Under slavery, plantation owners often allowed slaves to use small plots of land on the edge of the plantations to grow their own subsistence crops and so support their families. Women became important in marketing this food throughout the island (Momsen, 1988).

Following slave emancipation in 1834, in Barbados, unlike the situation in most of the nearby islands (Brierley, 1993), there was no empty land available for the former slaves. Many planters gave land to their slave mistresses or to the children of such liaisons as a source of support. In this way, for a brief period, there were more women than men small farmers in Barbados (Momsen, 1995). Land ownership was vital for political suffrage as at this period a mere two acres could give voting rights. Information on the early period draws from a study of wills and changes in landholding structure in the period immediately following the end of slavery. Many women moved out of agriculture into urban occupations after 1875 and left their land to their sons. The predominance of women small-scale farmers re-occurred during male migration to Panama between 1880 and 1921 for work on the Panama Canal. Men would often send money home to be invested in land, often under the title and operation of their wives/partners. This further entrenched the culture of female-headed households and women small-scale farmers in Barbados (Momsen, 1987a). After 1921, the male deficit caused by migration patterns decreased but increased again in the 1950s prior to changes in the UK and US immigration laws. In 1946, there were some 11 400 women farmers in Barbados, and women made up 46 per cent of the agricultural labour force. Examination of the structural characteristics of farms operated by women reveals that such farms are generally smaller, have poorer quality land, are less accessible to markets and have less rented land than those operated by men. Women farmers are more likely to get financial support from family members living overseas than are male farmers and to see farming as an extension of social reproduction (Barrow, 1993).

As the importance of sugar production declined and women migrated overseas to work as domestics and nurses, the number of Barbadian women employed in agriculture fell from 9200 in 1960 to 5100 in 1970 when they constituted 37 per cent of the agricultural workforce. However, these figures do not include self-employed women farming on their own account. The 1989 Agricultural Census reported that 6714 holdings were operated by women. In 1992, the National Statistical Service counted 1900 women, 4 per cent of the economically active female workforce, as employed in agriculture. Women often do not identify themselves as farmers in official surveys but as housewives and so it has been estimated that there were probably 17 865 women involved in agriculture as own-account farmers or agricultural labourers in 1992.

There has been no Agricultural Census of Barbados since 1989 but various sources provide some more recent information. The Association of Women in Agriculture indicates that women food producers now occupy 39 per cent of the total number of agricultural holdings but these holdings comprise only 6 per cent of the total area under agricultural production (AWIA, 1919). Women are responsible for producing and marketing a significant proportion of the fresh vegetables and fruit and especially herbs and spices in Barbados. They also make prepared jams and chutneys. Some 72 per cent of women farmers are over 45 years of age and more than one-quarter do not own the land they cultivate (AWIA, 1919). Some 47 per cent of women small farmers cultivate crops or raise livestock on less than one acre of land while more women than men, 41 per cent versus 26 per cent, cultivate between two and five acres.

Farms of between five and ten acres are cultivated by 12 per cent of women and 13 per cent of men small farmers (Caribbean Policy Development, 2021).

More jobs are now available for women in tourism and manufacturing industries and women are generally better educated than men, so few younger women are going into farming. Some may combine small-scale farming with off-farm jobs especially when young with children to support (Momsen, 1993). The Chief Agricultural Officer of Barbados is now a woman, and she is providing funding for various initiatives to help women small farmers in Barbados (Bennett, 2022). COVID-19, with its associated decline in tourism, plus problems caused by climate change caused a recent increase in poverty throughout the Caribbean and South and Central America but at the same time did encourage a greater dependence on locally produced food.

6 Conclusion

The feminization of agriculture is occurring worldwide (Momsen, 1987b), but time constraints are especially hard for women farmers who must combine childcare and household responsibilities with farming. Migration is often sex-specific with male out-migration leaving women forced to cultivate land in order to feed their families. As in Mexico, this may result in women's farms having greater agrobiodiversity than men's farms. On the other hand, tourism often benefits women as it offers off-farm jobs as well as a market for locally produced food. The impacts of climate change and COVID-19 have been especially hard for women small farmers throughout the region. However, greater global awareness of sustainability and gender issues has led to international agencies providing funding and new approaches to supporting women small farmers in Latin America and the Caribbean and so overcoming traditional patriarchal attitudes.

7 References

Association of Women in Agriculture. (1919). 'Profile of rural women producers in Barbados'. CANROP, St. Michael, Barbados.

Barrow, C. (1993). 'Small farm food production and gender in Barbados'. In J.H. Momsen (Ed.), *Women and Change in the Caribbean*. Kingston, Jamaica: Ian Randle; Bloomington and Indianopolis, IN: Indiana Press; London: James Currey, pp. 181–193.

Bennett, R. (2022). 'Help coming for small farmers'. *Barbados Today*. https://barbadostoday.bb/2022/02/25/help-coming-for-small-farmers.

Besson, J. (2007). "Squatting" as a strategy for land settlement and sustainable development'. In J. Besson and J. Momsen (Eds.), *Caribbean Land and Development Revisited*. New York: Palgrave Macmillan, pp. 135–147.

Blanco-Gregory, R., et al. (2020). 'Agroecological entrepreneurship, public support, and sustainable development: The case of rural Yucatán (Mexico)'. *Land,* 9(401), 1–24.

Boserup, E. (1970). *Women's Role in Economic Development*. New York: St. Martin's Press.

Brierley, J. S. (1993). 'A profile of Grenadian women small farmers'. In J. H. Momsen (Ed.), *Women and Change in the Caribbean*. Kingston, Jamaica: Ian Randle; Bloomington and Indianopolis, IN: Indiana Press; London: James Currey, pp. 181–193.

Caribbean Policy Development Centre. (2021). *Gender and Climate and Disaster Risk, Finance and Insurance. A focus on Small-scale Farmers in Antigua and Barbuds, Barbados and Grenada*. Bridgetown, Barbados: Caribbean Policy Development Centre.

Chambers, K.J. and Momsen, J.H. (2007). 'From the kitchen and the field: gender and maize diversity in the Bajìo region of Mexico'. *Singapore Journal of Tropical Geography,* 28(1), 39–56.

Dilley, C. and Momsen, J. (2013) 'Women in Fair Trade Coffee production in Nicaragua'. In A. Buang and J. Momsen (Eds.), *Women and Empowerment in the Global South*. Bangi, Malaysia: Universiti Kebangsaan Malaysia, pp. 95–111.

Echavarría, T.R. (2020). 'Circulación de saberes y apropiación del conocimiento local alrededor del cultivo de cacao en Talamanca, Costa Rica'. *Anuario del Centro de Investigación y Estudios Políticas',* 11, 124–161.

Economist. (2023). 'Latin America's mothers: all the single ladies and their kids'. *Economist,* June 10, p. 40.

Gulon, I. (2022). 'Newcomer farmers in Brazil embrace bees, agroforestry and find success'. Mongebay Series: Agroecology, Latin America.

Harry, I.S. (1993). 'Women in agriculture in Trinidad: an overview.' In J. H. Momsen (Ed.), *Women and Change in the Caribbean.* Kingston, Jamaica: Ian Randle; Bloomington and Indianopolis, IN: Indiana Press; London: James Currey, pp. 205–218.

Henshall, J.D. (1968). 'Japanese pioneers in Brazil.' *Geographical Magazine,* XL(5), 136–137.

Henshall, J.D. (1981). 'Women and small-scale farming in the Caribbean'. In O. Horst (Ed.), *Papers in Latin American Geography in Honor of Lucia C. Harrison, Muncie, Indiana, Conference of Latin Americanist Geographers*, pp. 44–56.

Henshall, J.D. (1982). 'Agricultural colonization in Rondônia, Brazil'. *Luso-Brazilian Review,* 19(2), 169–186.

Henshall, J.D. and Momsen, R.P., Jr. (1974). *A Geography of Brazilian Development*. London: Bell and Hyman.

Howard, P. (2003). 'Women in the plant world: an exploration'. In P. Howard (Ed.), *Women and Plants: Gender Relations in Biodiversity Management and Conservation*. London: Zed; New York: Palgrave-Macmillan, pp. 1–48.

Imburgia, L. (2019). 'Irrigation and equality: A gender-analytic approach for water government with examples from Ethiopia and Argentina'. *Water Alternatives,* 12(2), 571–587.

Lope-Alzina, D.G. (2007). 'Gendered production spaces and crop varietal selection: Case study in Yucatán, Mexico'. *Singapore Journal of Tropical Geography,* 28(1), 21–38.

Momsen, J.H. (1987a). 'Land settlement as an imposed solution'. In J. Besson and J. Momsen (Eds.), *Land and Development in the Caribbean*. London and Basingstoke: Macmillan Publishers Ltd, pp. 46–69.

Momsen, J.H. (1987b). 'The feminization of agriculture in the Caribbean'. In J.H. Momsen and J. Townsend (Eds.), *Geography of Gender in the Third World*. London: Hutchinson Ltd., pp. 344–347.

Momsen, J.H. (1988). 'Gender roles in Caribbean agricultural labour'. In M. Cross and G. Heuman (Eds.), *Labour in the Caribbean*. Basingstoke, UK: Macmillan Publishers Ltd, pp. 141–159.

Momsen, J.H. (1993). 'Development and gender divisions of labour in the rural Eastern Caribbean', In J. Momsen (Ed.), *Women & Change in the Caribbean: A Pan-Caribbean Perspective*. Kingston: Ian Randle; Bloomington and Indianopolis, IN: Indiana University Press; London: James Currey, pp. 232–246.

Momsen, J.H. (1995). 'The marginalization of women farmers: gender and small-scale agriculture in Barbados, 1838-1935'. *Caribbean Geography*, 6(1), 52–60.

Momsen, J.H. (2002). 'NGOs, gender and indigenous grassroots development'. *Journal of Development Studies*, 14, 859–867.

Momsen, J.H. (2003). 'Participatory development and indigenous communities in the Mexican Caribbean'. In J. Pugh and R. B. Potter (Eds.), *Participatory Planning in the Caribbean: Lessons from Practice*. Aldershot, UK and Burlington, VT: Ashgate, pp. 155–172.

Momsen, J.H. (2007). 'The waxing and waning of land for the peasantry'. In J. Besson and J. Momsen (Eds.), *Land and Development in the Caribbean*. London and Basingstoke: Macmillan Publishers Ltd, pp. 175–188.

Momsen, J.H. (2009). 'Fair trade versus unfair trade: the Windward Island Banana Industry'. *Caribbean Geography*, 15(2), 130–141.

Momsen, J.H. (2017). 'Gender and rural migration in Mexico and the Caribbean'. In B. B. Bock and S. Shortall (Eds.), *Gender and Rural Globalization*. Wallingford, UK: CAB International, pp. 177–188.

Momsen, J.H. and Richardson, P. (2009). 'Caribbean cocoa, planting and production'. In L. E. Grivetti and H.-Y. Shapiro (Eds.), *Chocolate: History, Culture and Heritage*. Hoboken, NJ: John Wiley and Sons, pp. 481–491.

Morrissey, M. (1989). *Slave Women in the New World: Gender Stratification in the Caribbean*. Lawrence, KS: University Press of Kansas.

Newson, L.A. (1976). *Aboriginal and Spanish Colonial Trinidad*. New York: Academic Press.

OECD. (2022). *Agricultural Policy Monitoring and Evaluation 2022: Reforming Agricultural Policies for Climate4 Change Mitigation*. Paris: OECD Publishing.

Parra-Fox, A. (2015). 'La Ruta, the Pacific Way. Women for a negotiated solution to the armed conflict'. In A. Coles, L. Gray and J. Momsen (Eds.), *The Routledge Handbook of Gender and Development*. London and New York: Routledge, pp. 398–406.

Pelliccia, M. (2021). 'For Costa Rica's Indigenous Bribri women, agroforestry is an act of resistance and resilience'. Mongabay Series: Global Agroforestry, September 29. https://nws.mongabay.com/series/global-agroforestry/.

Stubbs, J. (1993). 'Women and Cuban smallholder agriculture in transition'. In J. Momsen (Ed.), *Women & Change in the Caribbean: A Pan-Caribbean Perspective*. Kingston: Ian Randle; Bloomington and Indianopolis, IN: Indiana University Press; London: James Currey, pp. 219–231.

Torres, R.M. (2011). 'Life between the two *milpas*: tourism, agriculture and migration in the Yucatan'. In R.M. Torres and J.H. Momsen (Eds.), *Tourism and Agriculture: New Geographies of Consumption, Production and Rural Restructuring*. Abingdon UK; New York: Routledge, pp. 47–71.

Torres, R.M., et al. (2007). 'Cuba's farmers' markets in the "special period", !990-1995'. In J. Besson and J. Momsen (Eds.), *Caribbean Land and Development Revisited*. New York and Basingstoke, UK: Palgrave Macmillan, pp. 53–66.

Townsend, J.G. (1995). *Women's Voices from the Rainforest*. London and New York: Routledge.

UNDP in Latin America. (2020). 'In Costa Rica, rural women grow their own businesses', United Nations Development Programme in Latin America, October 15, https://www.undp.org.-latin America.

UN Women. (2019). 'The simple scheme that's driving a quiet revolution for Brazil's family farmers'. *News*, July 15.

World Bank News. (2020). Feature Story 'For Costa Rica, the Future of Forests Hangs in the Gender Balance', Word Bank, Washington DC, October 7.

Chapter 7

The impact of sociocultural factors on women smallholder farmers: Middle East and North Africa

Dina Najjar, International Center for Agricultural Research in the Dry Areas, Morocco; and Bipasha Baruah Western University, Canada

1 Introduction

Gender inequality persists in the Middle East and North Africa (MENA) region because of an entrenched patriarchy from within the private domain of households to communities, to a more systemic patriarchy sustained by

http://dx.doi.org/10.19103/AS.2024.0148.08

the state and other key political and economic institutions. In agriculture, the largest employer is women in the MENA region, this is manifested in many ways, including:

- women's marginalization from land rights, and subsequently access to credit, information;
- seeing themselves and being labeled as helpers instead of farmers, including as housewives on their national IDs;
- gender wage gaps and discrimination for paid work; and
- having to contend with sexual harassment.

Although women exercise their agency to transgress these harmful norms, often by accentuating other norms, enforcing existing legislation and targeted gender transformative approaches to challenge these norms is a necessary first step to enable women to benefit from their engagement in agriculture.

2 Visible and invisible contributions of women to agriculture in the Middle East and North Africa region

Women's contributions to agriculture in the MENA region have grown dramatically in recent decades because migration – as a coping strategy in response to climate change, globalization, political instability, and other influences upon agriculture – is much more widely available to men than to women. Entrenched familial and societal gender norms ensure that women remain almost solely responsible for homemaking and caregiving even as they are increasingly participating out of choice and necessity in the paid economy. These multifaceted economic and domestic responsibilities render women less mobile than the men in their families and communities.

In dry areas, which characterizes the majority of the Arab World, a combination of climatic change and population growth has led to increased male and youth outmigration to cities and towns among rural populations who depend on rainfed agriculture (Abdelali-Martini and Hamza 2014; Afifi et al. 2016; Kristensen and Birch-Thomsen 2013; FAO 2011; Läderach et al. 2022). A report published by the International Centre for Agricultural Research in Dry Areas (ICARDA) of the proceedings from the International Conference on Food Security in Dry Lands held in Qatar in 2012 emphasized that populations that inhabit deserts and dry rural areas of the world are most likely to experience male outmigration to urban areas and to subsequently witness the 'feminization of agriculture,' a rise in the participation of women in agriculture, often borne of necessity (Pedrick et al. 2012; Baada and Najjar 2020). Other recent research affirms these trends (Abdelali-Martini and Hamza 2014; Gaurtala et al. 2010;

Pattnaik et al. 2018; de Brauw et al. 2008). Male outmigration strongly influences women's roles in agriculture, with associated implications for agricultural productivity and gender equity.

Although agriculture is the largest employer of women in the MENA region, women's contribution to the sector remains largely undervalued, if not invisible. Some estimates suggest that about 50% of women engaged in agrarian labour are either not counted at all in national surveys or classified as economically inactive (Kabeer et al. 2019; Hoodfar 1997; Jensen 1994; Larson 1991). This is especially true for women who provide unpaid labour on family farms and are assumed to be (and may even see themselves as) economically inactive or, at most, as helpers to male farmers instead of farmers. In some settings (see, for example, Najjar et al. (2017a) for findings from rural Egypt), even female wage workers in the agricultural sector are listed as 'housewives' on their national IDs while their male counterparts are listed as agricultural workers. This prevents women from being recognized, and often even from recognizing themselves, as workers who make significant contributions to their households, communities and the national economy.

Established patriarchal gender hierarchies play a powerful role in how women see themselves and are seen by others. There is an extensive literature on gender and agriculture in the MENA region that emphasizes the tendency for women to be classified as 'helpers' rather than as farmers regardless of their contributions, and this is identified as a major barrier to women's optimal participation in agriculture (Barnes 2013; Galie et al. 2013; Gouda 2013; Najjar et al. 2017a). There is both legal and social failure to recognize women as workers. Being categorized as 'helpers' to male workers instead of as workers has negative implications for women's rights, entitlements, well-being and agency.

3 Lack of land rights for women in the Middle East and North Africa region

The gap between women's and men's education and income levels is narrowing gradually all over the world, but the disparity in land and property ownership is stark and seemingly intransigent. Even in countries where women consistently outperform men in educational attainment, they remain extremely marginalized in land and property ownership (Baruah 2010). Consequently, there is growing acknowledgement that women are particularly discriminated against in their right to land and property (UNDP 1995; UNIFEM 2000; World Bank 2001). Women are disadvantaged in land ownership almost everywhere in the world.

At 5%, the MENA region has one of the lowest documented rates of women's land ownership in the world (FAO 1999). In 2007, the United Nations Economic Commission for Africa (UNECA) described women's access to land in Egypt as 'alarmingly low' and 'lagging behind'. More recent surveys conducted in Egypt in 2014 reveal an even lower rate of 2% (Ministry of Health and Population [Egypt], El-Zanaty and Associates [Egypt], and ICF International 2015). As in several other MENA countries, by law, women in Egypt are entitled to inherit only half of the natal property as their brothers, but ultimately most women do not receive even this limited share. Deep-seated cultural norms ensure that brothers control the shares of their sisters; thus, in practice, women's ownership is just notional (Najjar et al. 2020). The inability to access land rights at par with men renders unmarried, divorced and widowed women particularly vulnerable to poverty and climate insecurity.

If threats to a family's well-being and security are external to the household, it may be less important whether a woman has independent ownership of and control over an asset – be it land, houses, livestock, or jewellery. The entire household may be protected if someone within it owns and controls the asset. However, a very different scenario may present itself upon internal threats to family security such as death, divorce, separation or abandonment. In the latter scenario, a woman's well-being is likely to depend much more significantly on whether she can exercise ownership and control over assets. Thus, inability to access land and other asset rights at par with men renders widowed and divorced women particularly vulnerable.

Since women rarely own land, they face more challenges than men do in accessing loans and credit due to their inability to offer individual land titles as collateral. Women also had weaker access than men to extension services and training in skills deemed masculine, such as irrigation and other drought-mitigation strategies (Najjar et al. 2019a; Najjar and Baruah 2023). Women also have weaker access than men to markets for the goods they produce because of their lack of income and collateral and the higher burden of domestic responsibilities and higher time poverty.

4 Lack of decent and safe paid work for women in agriculture

Although women increasingly perform most of the agricultural work in the Global South, including in the MENA region, they are almost always paid less than men, experience more precarious working conditions and have limited bargaining power (Abdelali-Martini 2011; Abdelali-Martini and de Pryck 2015; de Pryck and Termine 2014; Mills 2003; Razavi 2009). Alongside women's growing participation in agriculture in MENA countries, there is widespread evidence of gender-based wage inequity in agricultural labour. The literature reveals that women systematically perform lower paid and manual tasks that are precarious and seasonal, while men perform more permanent, technologically sophisticated, and higher paying jobs although

Published by Burleigh Dodds Science Publishing Limited, 2025.

these stereotypes tend to break in conditions when the supply of male work, for example due to migration, is limited (Deere 2005; de Pryck and Termine 2014; Najjar et al. 2018).

Wages and working conditions are also reported to vary depending on migration dynamics, type of crops, farm and market (traditional versus more cash-generating crops). For example, export-oriented crops such as fruits and vegetables and large farms tend to have lower gender wage gaps and less precarious working conditions than smaller family-owned farms (Barrientos et al. 2003; de Pryck and Termine 2014; Maertens and Swinnen 2012). Although commercial farms tend to pay women better wages than family farms, the expansion of agricultural commercialization in many contexts in the Global South, including MENA countries, has simultaneously been found to be very dependent on the availability of cheap and easily replaceable female labour (Razavi 2009). Very little is known about longer-term economic and social effects and outcomes upon women's lives from their growing participation in commercial farming in MENA. For example, there is virtually no research available to help us understand how women in MENA countries who participate actively in commercial agriculture in their youth fare as they get older and become less productive as agricultural workers. There is no existing research on their needs, priorities, or their ability to transition into less physically demanding sectors of employment or into retirement.

Almost all countries in the MENA region have framed national legislation against wage discrimination based on gender. Most countries in the region are also signatories to international treaties and guidelines (such as the International Labour Orgnanization's decent work agenda) that stipulate equal pay for equal work, but major gaps remain in the implementation, enforcement and monitoring of such laws and policies, explaining the persistent gender wage gaps in many sectors of employment (ILO 2014). Enforcing existing legislation to ensure equal pay for women is an essential first step toward enabling women to benefit equitably with men from their agricultural labour contributions.

Most female agricultural wage workers in MENA countries come from families that are landless or from families that own plots of land that are too small to sustain their needs (Najjar et al. 2017a, 2018). Thus, while land ownership and inheritance practices remain extremely male-biased in this region, large numbers of men from poor families are also landless or otherwise disadvantaged in land ownership. Since their labour is the most important asset, the landless poor, in general, and women, in particular, have to offer, it is critical that due emphasis is given to improving wages and working conditions (Razavi 2007; de Pryck and Termine 2014).

The concept of decent work, which stipulates fair pay, equal opportunities and treatment, security, social protection, social dialogue, as well as rights at work, offers a valuable opportunity to improve the working conditions of wage workers (ILO 2014) even as decent work principles remain largely unenforced, especially for informal workers in the rural agricultural sector in the MENA context. The enforcement of existing gender equity wage legislation is crucial as is raising awareness among employers to ensure equal pay and better working conditions for women.

There have been significant advancements in strengthening social protection floors in many countries in Asia and Latin America recently, including both conditional and unconditional cash transfer programs that enable poor women to make priority decisions for themselves and their dependents (for example, Bolsa Familia in Brazil and Oportunidades/Prospera in Mexico). Structural inequality constrains individual ability to exercise rights and demand entitlements. The benefits women derive from wage employment in agriculture can only find optimal traction within the context of a wider and more comprehensive social security infrastructure.

Countries in the MENA region are well-advised to try to introduce and expand the types of social protection programs (pensions, maternity, basic income) that are already in place in other countries. It is difficult to improve wages and working conditions in agriculture when labour supply is abundant and when social security is weak. These considerations are particularly important in the context of land fragmentation and climate change rendering rainfed agriculture less reliable, and paid wage work in irrigated agriculture on larger commercial farms becoming a more important source of income, especially for women. In rural Morocco, male and female workers in rural areas, which are drought-prone but rainfed, migrate for agricultural wage work opportunities on irrigated commercial farms in large numbers (Najjar et al. 2017a).

4.1 Widespread evidence of sexual harassment in paid agricultural work

Sexual harassment has been systematically reported as a problem faced by women in the agricultural wage sector in MENA countries, often made worse by the seasonal and informal nature of much agricultural labour in the region and the lack of unionization (Maertens and Swinnen 2012; de Pryck and Termine 2014). Women were overwhelmingly the victims of sexual harassment in the agricultural sector (Najjar et al. 2017a; Najjar et al. 2018). Young unmarried women were particularly vulnerable to sexual harassment from labour supervisors or fellow male labourers. Because women's access to work often depended on their tolerance of sexual harassment, they were rarely able to resist unwanted advances or to demand respect at work (Najjar et al. 2017a, 2018). Women may also self-select into lesser paid prospects to avoid

Published by Burleigh Dodds Science Publishing Limited, 2025.

gender-based harassment (Najjar et al. 2018). Sexual harassment diminishes women's ability to participate optimally in the agricultural sector, which has implications for women's ability to build resilience to climate change via agricultural productivity and food and income security.

Women agricultural workers in Morocco often self-select to work in women-only groups to avoid harassment, despite that meaning that they would get lower pay (Najjar et al. 2018). Patriarchal gender norms are a key barrier to women taking advantage of opportunities arising from rural transformation (Kosec et al. 2020). Even when women provide most of the labour in milk processing and livestock rearing, they may not own livestock and may have limited decision-making power such as in South Jordan (Najjar et al. 2019b). Breaking down norms that allow such imbalances to persist is critical for empowering women.

5 Opportunities for women to organize and access to policy and decision-making structures

Many of the challenges and inequities women face in working in agriculture in MENA countries may have been addressed if women had opportunities to organize and mobilize for fairer wages and better working conditions. Farmers' unions are rare in the MENA context, and where they exist, their primary focus appears to be on improving access to subsidized fertilizer and water rights not on workers' rights, fair wages or decent working conditions (Najjar et al. 2017a). Male agricultural workers in MENA also fare quite poorly at bargaining collectively to improve their working conditions, but women were mostly unable to access any opportunities to bargain formally for better wages and working conditions.

Women in agricultural labour in MENA countries contend with erratic incomes, insecure work, lean seasons, work-related injuries, lack of health insurance and unaffordable or unavailable childcare. Increased mechanization of harvesting in both areas is leading to job losses for women (Najjar et al. 2017a; Abdelali-Martini and de Pryck 2015). 'Many women asked the interviewer to discourage efforts to mechanize their work, particularly the well-paid lentil harvesting' (Abdelali-Martini and de Pryck 2015, p. 13). In the absence of adequate social security nets, in most MENA settings, women must work well into old age to support themselves and their families.

Although there is some evidence of women organizing and mobilizing informally for better wages and working conditions in agriculture (Najjar et al. 2017a, 2018), there are presently very few opportunities for women to organize formally as workers via collectives such as unions, producer groups or cooperatives.

5.1 Access by women to policy and decision-making structures

The lack of opportunities for women in the MENA region to organize and mobilize as workers also translates into women having much weaker access than men to governance, policy and decision-making spaces. It is important to create more visibility and social acceptance for women in these spaces. The limited existing research on the topic points to some of what might be needed to enable women to participate effectively in decision-making spaces (for example, institutional support from non-governmental organizations (NGOs) and international or intergovernmental agricultural organizations, gender-based quotas or targets, and closer attention and responsiveness to women's domestic responsibilities). More research aimed at understanding how to enable women to participate optimally in public institutions in agriculture in the context of specific countries in the MENA region should be a priority.

6 Ways forward: legal recognition and land ownership

6.1 How best to enable women to gain legal and social recognition as farmers

Despite women's growing contribution to agriculture, sex-disaggregated data on wages and working conditions is extremely scarce in MENA countries. The lack of empirical evidence about the opportunities and constraints faced by women in agricultural labour make it difficult to design policy responses to improve their situation, including enabling them to be legally recognized as workers. The fact that women's contributions to agriculture in the MENA region are undervalued or invisibilized due to patriarchal laws and regressive social norms is well-established (Galie et al. 2013; Geha and Karam 2021; Najjar 2015; Najjar et al., 2017a, 2018; Ragetlie et al. 2021) as are the negative implications women's marginalization in agriculture have for their ability to adapt and build resilience toward worsening climate effects in the region (Najjar 2015; Najjar and Baruah 2023).

What is less well known is how best to go about enabling women to gain both legal and social recognition as farmers. In addition to advancing legal recognition via gender-disaggregated data collection and inclusion and analysis of formal and informal work, governments and donors may be able to play an important role in enabling women to gain social recognition as workers by supporting efforts to raise public awareness about the contribution female agricultural workers make to national economies, food security, the fight against climate change and to society at large.

6.2 How to advocate for reform of male-biased inheritance practices and for gender equality in land ownership

The MENA region has the lowest level of women's land ownership in the world, yet very little research has explored the specific barriers women face in land ownership in different countries in the region and how to go about addressing them in each specific context. Doss et al. (2015) found that gender land statistics were available in only 9 of 53 African countries, and they emphasize that there is virtually no data on gender and land ownership available for North African countries such as Morocco, Algeria and Tunisia.

The question of how best to advocate for gender equality in land ownership in MENA is complicated by findings from other regions, notably from South Asia and sub-Saharan Africa and emerging research in Egypt (see Najjar et al. 2020) which suggest that policy interventions such as advocating for titling land in women's names can only play a limited role in altering gender norms and social hierarchies in the absence of wider political consciousness and awareness about gender equality (Baruah 2010; Gammage et al. 2016). Even when women are potentially able to acquire land and property through either inheritance, purchase in the market or distribution by the state, they are unwilling or hesitant to assert the equal inheritance rights of sons and daughters. The simple targeting of resources to women does not always ensure equitable outcomes, since resources may then be allocated in biased ways to children under conditions of strong son preference. Therefore, legal literacy and initiatives that raise awareness among women as well as men about the benefits of greater equity and that address fears about undoing customary male privileges may be as crucial as policy reforms and state actions that protect women's interests and facilitate their agency. Gender equity goals can only be partially accomplished through legal measures and economic interventions, regardless of how well-intentioned and progressive they may be. Since inheritance practices remain so male-biased (and since women are themselves often complicit in making them so), it is also important to ask from a more practical perspective if assets other than land (houses, livestock, jewellery, savings and investments, for example) can also enable women in agriculture to build economic security and resilience to the effects of climate change.

7 Ways forward: asset ownership and access to training

7.1 Research on women's ownership of non-land assets

Though there have been a few studies addressing women's ownership of land assets, there has been very little research on non-land assets owned by women in the MENA region, including assets such as livestock, jewellery,

Published by Burleigh Dodds Science Publishing Limited, 2025.

private enterprises, cellphones, computers, guaranteed investment certificates, agricultural and irrigation equipment.

The existing research on gender and asset ownership points to land and homes as the most economically and socially valuable assets for women. Livestock and gold are also certainly useful assets for women, especially because they can be sold easily for cash (Moors 2013; Najjar et al. 2020). Because these latter assets are highly liquefiable, it is important for women to have control over their sales and use. Therefore, the existing research highlights the importance of implementing specific policies to enable and optimize land and property ownership by women even if they own other non-land assets such as gold, poultry and livestock.

Future research on gender and asset ownership should also consider various forms of sole and joint ownership (with, for example, spouses, family members and friends) through which women and men may own and control assets. However, researchers should be cognizant that joint ownership does not necessarily imply equal ownership (Doss et al. 2014; Jacobs and Kes 2015). In addition to enabling women to access and own land and non-land assets, it is important to visibilize, strengthen and validate women's roles in agriculture. This is critical since women often saw themselves and were seen by others in the community as 'helpers' rather than farmers.

7.2 How to enable women to access training in irrigation and other skills deemed 'masculine'

Parts of the MENA region that are dependent on rainfed agriculture are more vulnerable to climate effects of reduced precipitation (Schilling et al. 2012; Waha et al. 2017). Water scarcity has motivated various irrigation initiatives in this region, ranging from drawing water from aquifers, lakes and rivers to desalination of seawater (Waterbury 2013; Sowers et al. 2011). In irrigated areas, the expansion of commercial labour has led to a growing demand for hired agricultural labour. Women find themselves employed in such contexts in growing numbers, often because they are cheaper to hire than men labourers and because they are more willing than men to accept temporary, seasonal or otherwise precarious work. Large numbers of women migrate out of necessity from rural areas with rainfed family farms to other areas in the country with irrigated commercial farming due to the prospect of paid wage work (see Najjar et al. 2021 for findings from Morocco).

The existing scholarly and practitioner literature on women's participation in irrigation in MENA countries suggest that although women participate quite actively in irrigation and water management, their contributions are poorly understood and undervalued by landowners (typically men in their own families and communities) as well as by irrigation engineers and extension agents

Published by Burleigh Dodds Science Publishing Limited, 2025.

(Barnes 2014; Ibrahim 2006; Najjar et al. 2019a). Research aimed at identifying and describing women's contributions to the irrigation sector and making recommendations for improving women's access to irrigation technologies and participation in water governance is critical. Detailed and comparative research in specific countries and local contexts in MENA is important for ensuring equitable participation of women in irrigation especially since control of water resources is increasingly being transferred from centralized entities such as water boards and utilities to community institutions such as local Water Users Associations.

Gender norms also play an important role in determining the adoption of irrigation technologies. For example, while pedal pumps were deemed beneficial for men in sub-Saharan Africa, they were deemed inappropriate or even shameful for women to use in the same contexts (Theis et al. 2018). The literature on this topic suggests that the role played by gender norms in enabling or disabling women's ability to adopt irrigation technologies is under-researched and so are the benefits and constraints following the adoption of specific irrigation technologies.

Very few studies explore gendered patterns of irrigation technology adoption and women's ability to benefit from them. The existing research on gender and technology adoption is based in South Asia and sub-Saharan Africa (see, for example, Njuki et al. 2014; Theis et al. 2018). There is very little published research on this topic in the context of the MENA region.

Although gender norms play an important role in technology adoption, it is important to simultaneously emphasize that gender norms are never static. They are constantly being reproduced, adapted, contested and negotiated as part of everyday social interactions (Jackson 1998). Migration and resettlement, for example, may play an important role in changing gender norms. These are also important considerations for future research on gender and irrigation.

The existing evidence suggests that women are far more actively engaged in irrigation roles in the MENA region than is generally assumed and documented in the literature on the topic (Najjar et al. 2019a). Women transgress norms related to their participation in irrigation by calling peer men brothers and younger men sons, while also circumscribing to other norms related to propriety by taking their young sons with them to the field. Although women's invisibility in irrigation may be a consequence of their lower levels of land ownership and social status and stronger association with the domestic sphere, it may also be a consequence of the continuing association of irrigation with masculinity (Bossenbroek and Zwarteveen 2014; Carreta 2015; Cole et al. 2015). The limited existing research on gender and irrigation in MENA identify land ownership; educational attainment; institutional support from government, donors and NGOs; and access to training in irrigational technologies as factors that enable women to optimally undertake irrigation.

8 Ways forward: climate-smart practices and mechanization in crop and livestock production

8.1 Identifying opportunities for climate-resistant 'green' agriculture and rangeland cultivation

In addition to improving women's access to irrigation and other skills and training in agriculture that are still largely only offered to male farmers, it is important to explore and identify opportunities for climate-responsive or 'green' agriculture via drought-resistant crops such as barley and cactus and livestock such as hardier breeds of poultry, goats, sheep, camels and cattle (Najjar 2015; Najjar and Baruah 2023). Most of the existing training in MENA about drought management focuses on supplementary and alternative irrigation techniques and practices (Sowers et al. 2011). Other complementary drought mitigation and management strategies such as the introduction of cacti, including as livestock feed, and other drought-tolerant crops and animal breeds were not presently being explored by the agricultural extension and training programs in MENA countries (Haddad and Shideed 2013; Haddad et al. 2011). These are worth exploring in the future.

Because women are known to be more vulnerable to the effects of climate change, they also benefit more from access to risk mitigation strategies and tools (Bageant and Barrett 2017; Chanamuto and Hall 2015). Findings from rural Tunisia suggest that women are often unable to access innovations which may mitigate the effects of climate change at par with men (Najjar and Baruah 2021). Skills and training related to drought and irrigation, for example, were targeted almost exclusively to men. It is crucial that women gain access to drought management and adaptation training at par with men. Alongside increasing women's access to such training, it is important to create more visibility and social acceptance for women in roles such as irrigation, grazing and marketing that are deemed masculine. This will enable more women to participate in agriculture (including rangeland cultivation) and livestock rearing on a more equal footing with men and to voice their concerns and priorities in policy dialogues.

8.2 How to mechanize the agricultural sector responsibly

Broadly speaking, agriculture in MENA countries is still heavily reliant on human labour because the cost of large-scale mechanization presently far exceeds the cost of human labour, especially given the high levels of rural unemployment and the high availability of cheap migrant labour in the region (ILO 2014; ILO 2017). The existing research on the mechanization of agriculture in other regional contexts in the Global South suggests that the introduction of mechanization in agriculture will be gradual and incremental rather than sudden and absolute (Lewis et al. 2022). Mechanization will be heavily influenced by economic and social realities of the local context and will initially affect medium- and low-skilled workers, among whom women are the majority, the most (Kawarazuka et al. 2018). In some contexts, mechanization may also be driven by environmental, health and safety concerns.

The limited existing evidence on mechanization of agriculture from MENA countries confirms these broader trends. The mechanization of commercial agricultural enterprises in Egypt and Morocco has raised concerns because women typically have much weaker access to training and machinery and are more likely than men to lose work because of mechanization (Najjar et al. 2018, Najjar and Baruah 2020). Research and public policy should anticipate mechanization displacing wage work and train those affected in other skills.

Mechanization of commercial fruit, vegetable and flower farming in the MENA has also enabled more widespread use of chemical fertilizers and pesticides in agriculture. Research is needed to explore the health implications for workers of heavy chemical use in the agri-export industry (Bain 2010). The fact that agricultural labourers in many MENA contexts do not have health insurance makes this an especially urgent issue.

8.3 Women, rangeland management and livestock farming

Pastoralism, which is practiced by close to 200–500 million people and in more than 75% of countries, is projected to become more vulnerable to climate change (McGahey et al. 2014). Coupled with non-climate-related stressors, climate change is projected to affect pastoralism by lowering pasture and animal productivity, damaging reproductive function, reducing litter size and survival rates, and leading to biodiversity loss (IPCC 2019).

Livestock rearing is an important livelihood strategy for rural communities all over the world. Particularly in developing countries, livestock often serve as assets, capital and 'insurance policies' of rural people with poor access to formal financial institutions and high vulnerability to crop failures (Njuki and Sanginga 2013). Livestock rearing can be an important component of building household and community resilience (Dumas et al. 2018). Livestock rearing tends to be a significant (if not the primary) livelihood activity for rural households and communities located in areas of the MENA region that are dry or that have more extreme climates where crop cultivation is not as reliable (Archambault, 2016; Turner and Williams, 2002). Livestock may also be integral to crop production in some contexts and therefore to livelihood security (Debela 2016; Fisher et al. 2000).

Existing ecological issues such as overgrazing and desertification that are associated with livestock production have been exacerbated by climate change. An increasing number of dryland areas will be subjected to drought conditions in the future (Fraser et al. 2011). Dryland areas are particularly vulnerable to the effects of climate change because of their low adaptive capacity and higher sensitivity to changes in temperature and precipitation (ibid.). By rendering grasslands and rangelands less productive, climate change will adversely affect dairy, meat and wool production. Thus, the livelihoods of the 2.5 billion people who inhabit dryland areas are particularly vulnerable to climate change (ibid.). Since the effects and outcomes of climate change have been demonstrated to be borne more heavily by women, often as a result of the household and societal division of labour, improving their ability to adapt to changing environmental conditions is critical (Alexander et al. 2011; Carr and Thompson 2014). Climate change also interacts with structural barriers such as tenure insecurity, credit inaccessibility, limited access to agricultural inputs and improved technologies to the disadvantage of women (Carr and Thompson 2014).

Women's roles in and contributions to rangeland cultivation and the effects of climate change on their livelihoods are both under-researched topics (Nelson et al. 2002; Huyer 2016; Chanana-Nag and Aggarwal 2018; The CGIAR Impact Platform 2022). There is very little research on gender and pastoral livelihoods in the MENA. Almost all the existing research on the topic was conducted in sub-Saharan Africa. In order to enable women in pastoral communities and rangeland cultivation in MENA countries to adapt and build resilience towards climate change, we need more empirical research aimed at understanding the roles they play in their communities, and the challenges and opportunities they face.

9 Ways forward: empowering rental farmers and income diversification

9.1 How to empower renter farmers

Due to land fragmentation in all MENA countries, growing number of people must rent land in order to be able to farm. Renter farmers are a disadvantaged group regardless of gender, but women were more vulnerable because of the lower social status attributed to them because of their gender. Landowners, who are almost always men, prefer to rent their lands to men: 'Even if the woman [the wife] will be the one farming, I prefer to rent the land to a man', explained a landowner in Kafr Sheikh, Egypt (Najjar et al. 2020). Women had very weak bargaining positions with farm owners and often faced conflicting priorities between growing food crops (which renters preferred) and cash crops (which farm owners preferred) with very important implications for the ability to optimize returns from agriculture while preventing further soil erosion and degradation, and water overuse, thereby building resilience towards climate effects. Reframing and regulating renter–owner relationships to ensure more equity for renters should be a priority (see, for example, Bush 2019 for Egypt).

Published by Burleigh Dodds Science Publishing Limited, 2025.

Because female renters have even lower bargaining power than male renters, they will benefit tremendously from the reframing of renter–owner relations in MENA countries.

9.2 How to create and optimize livelihoods complementary with agriculture

It is increasingly rare for smallholder farming families in the MENA region to sustain themselves solely through agriculture. Members of farming families are compelled to diversify their livelihoods to respond not just to climate-induced stresses but also to other influences such as political conflict, globalization, urbanization, migration and return migration. It is not uncommon for different members of the same family, or even the same individual, to carry out a range of farming and non-farm-related economic activities. Since men are more mobile than women in the MENA region, they typically have more opportunities to participate in off-farm activities. More research is needed to understand how women might optimize incomes from agriculture (via activities such as agro-processing, meat, milk, manure, livestock sales) as well as non-agricultural economic activities such as producing handicrafts and other marketable goods and services such as carpets, leather goods, essential oils, and soaps. NGOs and other civil society organizations may be well placed to assist women in business training and marketing their goods physically and virtually.

10 Ways forward: organization and access to extension services

10.1 How best to organize and mobilize for rights, resources, entitlements (cooperatives, membership-based organizations, unions, etc.)

Members of cooperatives hold jobs in a co-operative producing goods and services. They participate in making major decisions concerning the cooperative. Cooperatives are not common in the MENA context, and the ones that exist are almost entirely staffed by men. The existing cooperatives are often not well-organized or well-run. Members had limited access to training, social protection and benefits via cooperatives, but they were able to access markets and export licenses through their participation in cooperatives. Barring a few examples such as argan cooperatives in Morocco, women's access to cooperatives and producer organizations was minimal in MENA (Perry et al. 2019; Biermayr-Jenzano et al. 2014). Significant research has demonstrated that women are able to increase their social and economic power through membership in co-ops and producer groups and that such organizations can help women mobilize resources and networks effectively (de Pryck and

Termine 2014). However, women were rarely able to join cooperatives in MENA countries. Women's only cooperatives in other settings have been known to empower women on social, economic and political fronts (the Self-Employed Women's Association in India is a good example, see Baruah, 2004, 2010), but they are rare in MENA. Research aimed at understanding the experiences of existing co-ops and producer groups in the MENA region as well as in culturally comparable regions such as South Asia and of replicating or adapting them for the MENA is much needed. Increased engagement of unions as organizations that lobby for improved wages and working conditions for workers rather than just on increasing subsidized access to water and fertilizer, which is presently the focus of most existing farmers' unions in MENA, is also important.

10.2 How to deliver extension services and information in-person and digitally

It is well-established that agricultural extension services do not serve farmers well in the MENA and that women are often completely overlooked by extension agents. Some researchers have recommended training more women as extension agents as a means of reaching more female farmers. Other researchers have noted that the gender of the extension agent is not as important as it is for them to think of women as farmers rather than as 'helpers' to male farmers, and for extension agents to reach out to women more proactively with agricultural extension support. Progress has been made on both these fronts (training more women as extension agents and encouraging extension agents of all genders to reach out to female farmers) in some contexts in MENA countries, but one of the biggest limitations remains that some governments in the region do not have the financial resources to offer extension support, especially in remote rural areas, with meaningful levels of training frequency and outreach. In Egypt, for example, since 2015, the annual budget for extension services has been slashed to 230 000 EGP, which as one official from the Ministry of Agriculture and Land Reclamation informed 'is not even enough to run 10 motorcycles, let alone 350 extension centres around the country' (Najjar et al. 2017b). If governments are unable to provide adequate funding and support for agricultural extension services, it may be worthwhile to explore the possibilities, which are already available in countries in other regions, of domestic and international civil society organizations, aid agencies and private companies offering such services.

With the expansion of access to the internet and to mobile technologies even in remote rural areas in MENA countries, the idea of providing extension services digitally is being explored both by governments and by intergovernmental agricultural organizations. Of course, the fact that women in MENA typically have much weaker access to Information and Communication

Technologies (ICTs) and digital resources than men remains a serious challenge (Najjar et al. 2021; Ragetlie et al. 2022) but research aimed at finding solutions to address this digital gap is important and must be pursued in the future as a means to improve equity in the provision of agricultural extension services. The limited research that is emerging on the topic has suggested that gender-responsive digital extension, accomplished via means such as phone distribution, radio and SMS messages, and sharing of information prompts can be useful for enabling women to learn about and adopt new agricultural inputs, techniques and technologies (see Ragetlie et al. 2022 for findings from northern Tunisia). Much more empirical evidence needs to be generated in specific countries and contexts in the MENA region to understand how best to deploy and maintain such services.

11 Ways forward: risk management, supporting refugees and youth

11.1 How to reduce risk for farmers

As the effects upon agriculture of climate change worsen, the conversation about how to reduce the risks and losses associated with farming via instruments such as crop, livestock and asset insurance and green bonds for agriculture is gaining more traction in research and policy circles. Other tools like payment for environmental or ecosystem services are also being proposed as payments to farmers or landowners who have agreed to take certain actions to manage their land or watersheds to provide an ecological service. However, systematic research on the feasibility and outcomes of such initiatives is lacking almost everywhere, including the MENA region. While such tools may be able to protect farmers from the vagaries of climate effects and may also enhance the adoption of improved agricultural and rangeland management practices (Louhaichi et al. 2016), other related thorny issues such as the rights of landowners, land users and renters have neither been researched nor received any policy attention. The gender implications, effects and outcomes (especially vis-à-vis community and private property rights regimes and women's weak access to both) are potentially profound and powerful but remain mostly unknown. Addressing them proactively via focused research and responsive public policy is critical to ensuring gender equity in climate mitigation, adaptation and resilience.

11.2 Research on refugees in agriculture in Middle East and North Africa

Four of the top ten refugee host countries in the world are in the MENA region: Lebanon, Jordan, Turkey and Iran (Al-Dajani and Marlow 2019). It is well known that large number of refugees living in rural areas in MENA are working in agriculture, both as formal and informal wage labour, but there is virtually no research on their experiences in farming, what challenges and opportunities they face and how they might be supported to stabilize their livelihoods and build resilience towards climate effects. Given the continued political strife and turmoil in the region and the scale of migration within and between MENA countries, research on this topic should be considered an urgent priority (see Al Zoubi 2022 on Syrian women migrants' agricultural coping strategies in rural Lebanon). Women and girls represent almost half of the world's migrants and refugees. While refugees of all genders are likely to face challenges working in agriculture in MENA countries, women and girls may face additional problems given entrenched gender hierarchies and inequities. Specific attention must be paid to understanding and responding to them.

11.3 Better understanding of youth perceptions and interests in agriculture

There is strong evidence to suggest that youth in the MENA region are increasingly disinterested in agriculture as a livelihood strategy. The reasons for youth disengagement and disenchantment with agriculture are complex and may be attributable to the influences of globalization, wider exposure to urban lifestyles and professions, generational differences in material and ideological aspirations and the diminishing prospects of building economic security via sole reliance on agriculture (Bossenbroek et al. 2015; Najjar et al. 2018). As discussed previously, growing female participation in agriculture is often a direct outcome of male and youth migration out of, or diminishing contribution to, agriculture. Youth face significantly higher levels of unemployment in MENA and migrate from rural to urban areas more frequently than other age groups (Bossenbroek et al. 2015). The existing research on agricultural labour in MENA confirms that most of those between the ages of 15 and 24 only participate in agriculture when they have no other viable livelihood options. Among young women, the existing research confirms that agricultural wage labour is most often carried out by women from the poorest landless families as an option of last resort (Najjar et al. 2018). Diminishing returns from farming due to climate change may exacerbate these trends of youth departure from agriculture. More research is required in different contexts to better understand youth perceptions of agriculture. Research and responsive policies aimed at revalorizing agricultural labour to render it more compatible with the contemporary aspirations of youth are urgently needed. A deeper structural revalorization of the importance and necessity of the agricultural sector within MENA is also required so that agricultural labour does not continue to be perceived as an occupation of last resort. Agriculture remains the backbone of many national economies in the MENA region. Given the added stress that climate change places upon agricultural systems and productivity, the sector can

Published by Burleigh Dodds Science Publishing Limited, 2025.

ill afford to lose out on the labour, energy, enthusiasm and creativity that youth can bring to it. Governments, donors, intergovernmental organizations and NGOs can also play a role in revalorizing agricultural livelihoods by supporting efforts to raise public awareness about the contribution agricultural workers make to national economies and to society at large.

12 Ways forward: policy lessons from the global coronavirus disease experience

Coronavirus disease 2019 (COVID-19) led to an unprecedented global reckoning of what constitutes essential work and essential services in all countries around the world, leading to calls for comprehensive reforms of sectors such as basic income, pensions, paid sick leave, childcare, eldercare and other stronger social protections for those who work in such sectors (Doss et al. 2020; Najjar and Baruah 2020; Baruah 2021; Bossenbroek and Ftouhi 2021). Agriculture, food security, food accessibility and availability (grocery stores, vegetable stands, for example) were recognized worldwide as sectors and services that no country could do without. COVID-19 has also led to a broader conversation about the possibilities and benefits of providing social protection within a human rights framework and delinking social security from employment status. In recent years, there have been significant advancements globally in expanding and strengthening social protection policies as more countries transition towards developing welfare systems.

Some strategies that are being tried in European, African, Asian and Latin American countries include basic income schemes as well as conditional and unconditional cash transfer programs designed specifically to enable poor women to make priority decisions for themselves and their dependents. Programs like Brazil's Bolsa Familia, Mexico's Prospera, Mali's Social Cash Transfer initiative, Togo's Novissi scheme (a fully digital cash transfer scheme designed and deployed during COVID-19 for workers in the informal sector) and India's basic income pilot are important developments given that structural inequality constrains individual ability to exercise rights and demand entitlements. Since so many countries around the world rolled out new social protection schemes (basic income and paid sick leave, for example) or strengthened and expanded existing schemes during the COVID-19 pandemic, there is now a clear opportunity to design research to understand how such schemes might be introduced or adapted to benefit the agricultural labour force in MENA and other contexts in the Global South.

13 Conclusion

Gender inequality persists in the MENA region because of an entrenched patriarchy from within the private domain of households, to communities, to

Published by Burleigh Dodds Science Publishing Limited, 2025.

a more systemic patriarchy sustained by the state and other key political and economic institutions (Geha and Karam 2021; Kandiyoti 1988; Moghadam 2020; Ragetlie et al. 2021). Geha and Karam (2021) and Ragetlie et al. (2021) emphasize that MENA governments focus almost exclusively on skills development as a means to strengthening women's economic participation, but they rarely implement policies that address the root causes of gender inequality, such as patriarchal inheritance laws, male-biased land rights and structural gender inequality in democratic and inclusive governance.

Recognizing women as workers, framing and enforcing equal pay legislation for women and men in agriculture, strengthening social protection for all workers, increasing funding and support for agricultural extension services (including hiring more female extension agents), and providing gender-inclusive digital extension services are essential first steps that will enable women in MENA countries to benefit more optimally from working in agriculture and building resilience towards climate change. Investing in more public awareness about women as agricultural workers rather than helpers may also eventually shift more deeply entrenched social norms.

14 Where to look for further information

The following publications provide a good overview of the subject:

- Baruah, B. & Najjar, D. (2023). Priorities for Research on Gender Equality, Climate Change, and Agriculture in the MENA Region: A Policy Brief (https://hdl.handle.net/20.500.11766/68298)
- Baruah, B. & Najjar, D. (2023). Priorités de recherche sur l'égalité de genre, le changement climatique et l'agriculture dans la région MENA : note de politique (https://hdl.handle.net/10568/130021)
- Baruah, B. & Najjar, D. (2023). أ ولويات البحث حول المساواة بين الجنسين وتغير المناخ- والزراعة فيمنطقة الشرق األوسط وشمال إفريقيا: موجز السياسة(https://hdl.handle.net/10568/130020)
- Baruah, B. & Najjar, D. (2022). Gender Equality, Climate Change and Agriculture in the MENA region: Priorities and Possibilities. Beirut, Lebanon: International Center for Agricultural Research in the Dry Areas (ICARDA). (https://cgspace.cgiar.org/handle/10568/126987)
- Najjar, D., Nyantakyi-Frimpong, H., Devkota, R., and Bentaibi, A. (2023). A feminist political ecology of agricultural innovations in smallholder farming systems: Experiences from wheat production in Morocco and Uzbekistan. *Geoforum, 146*, 103865.
- Najjar, D. and Baruah, B. (2023). "Even the Goats Feel the Heat:" Gender, Livestock Rearing, Rangeland Cultivation, and Climate Change Adaptation in Tunisia. *Climate and Development*, 1–14.

- Ragetlie, R. Najjar, D., and Oueslati, D. (2022). "Dear Brother Farmer": Gender-Responsive Digital Extension in Tunisia during the COVID-19 Pandemic. *Sustainability*, *14*, 4162. 10.3390/su14074162
- Ragetlie, R., Najjar, D., and Baruah, B. (2021). Paying 'lip service' to gender equality: The hollow implementation of gender mainstreaming in Jordan. In special issue "Challenging Power and Inequality: Gender and Social Justice in the Middle East". Civil Society Review, November (5): 17-45. https://dx.doi.org/10.28943/CSR.005; https://hdl.handle.net/20.500.11766/66536
- Baada, J. and Najjar, D. (2020). A Review of The Effect of Migration on the Feminization of Agrarian Dryland Economies. *Journal of Gender, Agriculture and Food Security*, 5(2), 1-12.
- Najjar, D., Baruah, B., and Al Garhi, A. (2020). Gendered Patterns of Asset Ownership and Control in Egypt. *Feminist Economics*, 26(3), 119-143.
- Najjar, D., Baruah, B., and Al Garhi, A. (2019). Women, Irrigation and Social Norms in Egypt: "The more things change, the more they stay the same?". *Water Policy,* 21 (2), 291-309.
- Najjar, D., Baruah, B., Aw-Hassan, A., Bentaibi, A. and Kassie, G. (2018). Women, Work and Wage Equity in Agricultural Labour in Saiss, Morocco. *Development in Practice*, 28(4)
- Najjar, D., Baruah, B. and Garhi, A. (2018). Women and Land Ownership in Egypt: Continuities, Contradictions, and Disruptions. Chapter 3, p. 59-86. In Emerich, Y. and Saint-Pierre, L. (Eds). Access to Land and Social Issues: Precarity, Territoriality, Identity. McGill-Queen's University Press (MQUP).
- Najjar, D. (2015). Women's Contributions to Climate Change Adaptation in Egypt's Mubarak Resettlement Scheme through Cactus Cultivation and Adjusted Irrigation. Chapter 8, p. 141–161. In Buechler, S. and Hanson, AM. (Eds). A Political Ecology of Women, Water and Global Environmental Change. London and New York: Routledge.

15 References

Abdelali-Martini, M. (2011). *Empowering Women in the Rural Labor Force with a Focus on Agricultural Employment in the Middle East and North Africa (MENA).* Algeria: ICARDA. https://www.un.org/womenwatch/daw/csw/csw56/egm/Martini-EP-9-EGM-RW-Sep-2011.pdf

Abdelali-Martini, M. & de Pryck, J.D. (2015). Does the feminisation of agricultural labour empower women? Insights from female labour contractors and workers in Northwest Syria. *Journal of International Development* 27(7): 898–916.

Abdelali-Martini, M. & Hamza, R. (2014). How do migration remittances affect rural livelihoods in drylands? *Journal of International Development* 26: 454–470.

Afifi, T., et al. (2016). Human mobility in response to rainfall variability: opportunities for migration as a successful adaptation strategy in eight case studies. *Migration and Development* 5(2): 254–274.

Al-Dajani, H. & Marlow, S. (2019). *Poverty Alleviation and Arab Women Refugees in Lebanon: Empowerment through Grassroot Micro Entrepreneurship?* Project Report. Amman: King Hussein Foundation. https://haqqi.info/en/haqqi/research/poverty-alleviation-and-women-refugees-middle-east-empowerment-through-grassroots

Alexander, P., et al. (2011). The link between climate change, gender and development in Africa. *The African Statistical Journal* 12: 119–140.

Al Zoubi, S. (2022). When coping strategies become a way of life: a gendered analysis of Syrian refugees in Lebanon. *Oxford Development Studies.* https://doi.org/10.1080/13600818.2022.2096210

Archambault, C.S. (2016). Re-creating the commons and re-configuring Maasai women's roles on the rangelands in the face of fragmentation. *International Journal of the Commons* 10(2): 728–746.

Baada, J.N. & Najjar, D. (2020). A review of the effects of migration on the feminization of agrarian dryland economies. *Journal of Agriculture Gender & Food Security* 5: 1–12.

Bain, C. (2010). Structuring the flexible and feminized labor market: global gap standards for agricultural labor in Chile. *Signs: Journal of Women in Culture and Society* 35(2): 343–370.

Bageant, E.R. & Barrett, C.B. (2017). Are there gender differences in demand for index-based livestock insurance? *The Journal of Development Studies* 53(6): 932–952.

Barnes, J. (2013). Who is a water user? The Politics of gender in Egypt's water user associations. In: Harris, L.M., et al. (eds) *Contemporary Water Governance in the Global South: Scarcity, Marketization, and Participation*. Abingdon: Routledge. pp. 185–198.

Barnes, J. (2014). Mixing waters: the reuse of agricultural drainage water in Egypt. *Geoforum* 57: 181–191.

Barrientos, S., et al. (2003). A gendered value chain approach to codes of conduct in African Horticulture. *World Development* 31(9): 1511–1526.

Baruah, B. (2004). Earning their keep and keeping what they earn: a critique of organizing strategies of South Asian women in the informal sector. *Gender, Work and Organization* 11: 605–626.

Baruah, B. (2010). *Women and Property in Urban India*. Vancouver: UBC Press.

Baruah, B. (2021). *What does degrowth say about gender equality and social justice?* SSHRC Evidence Brief. https://www.sshrc-crsh.gc.ca/society-societe/community-communite/ifca-iac/evidence_briefs-donnees_probantes/earth_carrying_capacity-capacite_limite_terre/pdf/SSHRC%20KSG%20Evidence%20Brief_Baruah%20Bipasha_FinalE.pdf

Biermayr-Jenzano, P., et al. (2014). *Understanding Gender and Poverty Dimensions of High Value Agricultural Commodity Chains in the Souss-Masaa-Draa Region of Southwestern Morocco.* Amman, Jordan: ICARDA (International Center for Research in Dry Areas).

Bossenbroek, L. & Zwarteveen, M. (2014). Irrigation management in the pamirs in Tajikistan: a man's domain. *Mountain Research and Development* 34(3): 266–275.

Bossenbroek, L., et al. (2015). Broken dreams? Youth experiences of Agrarian change in Morocco's Saïss region. *Cahiers Agricultures* 24(6): 342–348.

Bossenbroek, L. & Ftouhi, H. (2021). The plight of female agricultural wageworkers in Morocco during the COVID-19 pandemic. *Cahiers Agricultures* 30(40): 1–6.

Bush, R. (2019). *Economic Crisis and the Politics of Reform in Egypt*. London and New York: Routledge.

Carr, E.R. & Thompson, M.C. (2014). Gender and climate change adaptation in agrarian settings: current thinking, new directions, and research frontiers. *Geography Compass* 8(3): 182–197.

Carreta, M. (2015). Hydropatriarchies and landesque capital: a local gender contract analysis of two smallholder irrigation systems in East Africa. *The Geographical Journal* 181(4): 388–400.

Chanana-Nag, N. & Aggarwal, P.K. (2018). Woman in agriculture, and climate risks: hotspots for development. *Climatic Change* 158: 13–27.

Chanamuto, N.J.C. & Hall, J.G. (2015). Gender equality, resilience to climate change, and the design of livestock projects for rural livelihoods. *Gender and Development* 23(3): 515–530.

Cole, S., et al. (2015). Exploring the intricate relationship between poverty, gender inequality, and rural masculinity: a case study from an aquatic agricultural system in Zambia. *Culture, Society, and Masculinities* 7(2): 154–170.

Debela, B.L. (2016). Factors affecting differences in livestock asset ownership between male and female-headed households in Northern Ethiopia. *European Journal of Development Research* 29(2): 328–347.

de Brauw, A., et al. (2008). Feminization of agriculture in China? Myths surrounding women's participation in farming. *China Quarterly* 194: 327–348.

de Pryck, J.D. & Termine, P. (2014). Gender inequalities in rural labor markets. In: Quisumbing, A. et al. (eds) *Gender in Agriculture: Closing the Knowledge Gap*. Dordrecht: Springer. pp. 343–370.

Deere, C.D. (2005). *The Feminization of Agriculture? Economic Restructuring in Rural Latin America*. UNRISD Occasional Paper 1. Geneva: UNRISD.

Doss, C., et al. (2014). Who owns the land? Perspectives from rural ugandans and implications for large-scale land acquisitions. *Feminist Economics* 20(1): 76–100.

Doss, C., et al. (2015). Gender inequalities in ownership and control of land in Africa: myth and reality. *Agricultural Economics* 46(3): 403–434.

Doss, C., et al. (2020). The potential intersections of Covid-19, gender and food security in Africa. *Journal of Gender, Agriculture and Food Security* 5(1): 41–48.

Dumas, S.E., et al. (2018). "Men are in front at eating time, but not when it comes to rearing the chicken": unpacking the gendered benefits and costs of livestock ownership in Kenya. *Food and Nutrition Bulletin* 39(1): 3–27.

FAO. (1999). Gender and Land Rights Database. http://www.fao.org/gender-landrights-database/data-map/statistics/en/.

FAO. (2011). *The State of the World's Land and Water Resources for Food and Agriculture (SOLAW) – Managing Systems at Risk.* Rome: FAO.

Fisher, M.G., et al. (2000). Gender and agricultural change: crop-livestock integration in Senegal. *Society & Natural Resources: An International Journal* 13(3): 203–222.

Fraser, E., et al. (2011). Assessing vulnerability to climate change in dryland livelihood systems: conceptual challenges and interdisciplinary solutions. *Ecology and Society* 16(3): 3.

Galiè, A., et al. (2013). Women's identity as farmers: a case study from ten households in Syria. *NJAS-Wageningen Journal of Life Sciences* 64: 25–33.

Gammage, S., et al. (2016). Voice and agency: where are we now? *Feminist Economics* 22(1): 1–29.

Gaurtala, H.N., et al. (2010). Feminisation of agriculture as an effect of male out-migration: unexpected outcomes from Jhapa District, Eastern Nepal. *International Journal of Interdisciplinary Social Sciences* 5(2): 407–412.

Geha, C. & Karam, C. (2021). Whose feminism? Gender-inclusive policymaking in the Arab Middle East and North Africa. *SAIS Review of International Affairs* 41(1): 23–31.

Gouda, D.M. (2013). *The impact of Social Capital on the operation of Water Users' Organizations in Egypt's Old Lands*. PhD Dissertation, University of Sheffield, UK.

Haddad, N., et al. (2011). The potential of small-scale rainfed agriculture to strengthen food security in Arab countries. *Food Security* 3(1): 163–173.

Haddad, N. & Shideed, K. (2013). Mainstreaming adaptation to climate change into the development agenda. In: Sivakumar, M.V.K. et al. (eds.) *Climate change and food security in West Asia and North Africa*. Dordrecht: Springer. pp. 301–315.

Hoodfar, H. (1997). *Between Marriage and the Market: Intimate Politics and Survival in Cairo*. Los Angeles: University of California Press.

Huyer, S. (2016). Closing the gender gap in agriculture. *Gender, Technology and Development* 20(2): 105–116.

Ibrahim, S. (2006). *Brief Overview on the Current Situation on Gender and Water Management in Egypt*. Paper presented at the 3rd Arab Water Regional Conference: Research Advancement in Managing Limited Water Resources, Cairo, Egypt, December 2006.

ILO. (2014). *Women's Entrepreneurship Development Brochure*. Geneva: ILO.

ILO. (2017). Current guidelines on the status in employment. http://www.ilo.org/global/statistics-and-databases/statistics-overview-and-topics/status-in-employment/current-guidelines/lang--en/index.htm.

IPCC. (2014). *Climate Change 2014: Impacts, Adaptation, and Vulnerability. Part A: Global and Sectoral Aspects*. Contribution of Working Group II to the Fifth Assessment Report of the Intergovernmental Panel on Climate Change. Cambridge: Cambridge University Press.

IPCC. (2019). Climate change and land: an IPCC special report on climate change, desertification, land degradation, sustainable land management, food security, and greenhouse gas fluxes in terrestrial ecosystems. https://www.ipcc.ch/srccl/.

Jackson, C. (1998). Gender, irrigation, and environment: arguing for agency. *Agriculture and Human Values* 15: 313–324.

Jacobs, K. & Kes, A. (2015). The ambiguity of joint asset ownership: cautionary tales from Uganda and South Africa. *Feminist Economics* 21(3): 23–55.

Jensen, K. (1994). Who carries the load? Who carries the cash? Work and status among Egyptian farm women. *Frontiers: A Journal of Women Studies* 15(2): 133–152.

Kabeer, N., et al. (2019). *Women's Access to Market Opportunities in South Asia and the Middle East & North Africa: Barriers, Opportunities and Policy Challenges*. London, UK: Department of International Development, London School of Economics and Political Science.

Kandiyoti, D. (1988) Bargaining with patriarchy. *Gender and Society* 2(3): 274–290.

Kawarazuka, N., et al. (2018). *Gender in Agricultural Mechanization: Key Guiding Questions*. Lima, Peru: CGIAR Gender Impact Platform.

Kosec, K., et al. (2020). *Gender and Rural Transformation*. Washington, DC: International Food Policy Research Institute.

Kristensen, S. & Birch-Thomsen, T. (2013). Should I stay or should I go? Rural youth employment in Uganda and Zambia. *International Development Planning Review* 35(2): 175–201.

Läderach, P., et al. (2022). *Strengthening Climate Security in the Middle East and North Africa Region*. Position Paper No. 2022/3. CGIAR FOCUS Climate Security.

Larson, B. (1991). Women's work and status in Rural Egypt. *NWSA Journal* 3(1): 38–52.

Lewis, D., et al. (2022). Rural mechanization for equitable development: disarray, disjuncture, and disruption. *Development Policy Review*. https://doi.org/10.1111/dpr.12612

Louhaichi, M., et al. (2016). Financial incentives: possible options for sustainable rangeland management? *Journal of Environmental Management* 180(1): 493–503.

Maertens, M. & Swinnen, J. (2012). Gender and modern supply chains in developing countries. *Journal of Development Studies* 48(10): 1412–1430.

McGahey, D., et al. (2014). *Pastoralism and the Green Economy – A Natural Nexus?* Nairobi: IUCN and UNEP.

Mills, M.B. (2003). Gender and inequality in the global labor force. *Annual Review of Anthropology* 32: 41–62.

Ministry of Health and Population [Egypt], El-Zanaty and Associates [Egypt], and ICF International. (2015). *Egypt Demographic and Health Survey 2014*. Rockville, MD: Ministry of Health and Population and ICF International.

Moghadam, V. (2020). Gender regimes in the Middle East and North Africa: the power of feminist movements. *International Studies in Gender, State & Society* 27(3): 467–485.

Moors, A. (2013). Wearing gold, owning gold: the multiple meanings of gold jewelry. *Etnofoor* 25(1): 79–89.

Najjar, D. & Baruah, B. (2020). *The vital contribution of women to livelihoods resilience during COVID-19*. https://www.icarda.org/media/drywire/vital-contribution-women-livelihoods-resilience-during-covid-19.

Najjar, D. (2015). Women's contributions to climate change adaptation in Egypt's Mubarak resettlement scheme through cactus cultivation and adjusted irrigation. In Buechler, S. & A.M. Hanson (eds) *A Political Ecology of Women, Water, and Global Environmental Change*. London: Routledge. pp. 141–161.

Najjar, D., et al. (2017a). *Climate Change, Gender, Decision-making Power, and Migration into the Saiss Region of Morocco*. Economic Research Forum Working Papers (No. 1102). Cairo: ERF.

Najjar, D., et al. (2017b). *Women, Decent Work and Empowerment in Rural Egypt*. Beirut, Lebanon: ICARDA.

Najjar, D., et al. (2018). Women, work, and wage equity in agricultural labour in Saiss, Morocco. *Development in Practice* 28(4): 525–540.

Najjar, D., et al. (2019a) Women, irrigation and social norms in Egypt: "The more things change, the more they stay the same?" *Water Policy*, 21(2): 291–309.

Najjar, D., et al. (2019b). *Decision-Making Power of Women in Livestock and Dairy Production in Jordan*. Beirut, Lebanon: International Center for Agricultural Research in the Dry Areas (ICARDA).

Najjar, D., et al. (2020). Gender and asset ownership in the old and new lands of Egypt. *Feminist Economics*. https://doi.org/10.1080/13545701.2020.1743877

Najjar, D., et al. (2021). Climate-induced migration, women, and decision-making power in the agricultural wage sector in Saiss, Morocco. In Eastin, J. & Dupuy , K. (eds) *Gender, Climate Change and Livelihoods: Vulnerabilities and Adaptations*. Oxfordshire, UK: CAB International. pp. 185–195.

Najjar, D. & Baruah, B. (2023). "Even the goats feel the heat:" gender, livestock rearing, rangeland cultivation, and climate change adaptation in Tunisia. *Climate & Development*, 1–14. https://doi.org/10.1080/17565529.2023.2253773

Nelson, V., et al. (2002). Uncertain predictions, invisible impacts, and the need to mainstream gender in climate change adaptations. *Gender and Development* 10(2): 51–59.

Njuki, J., et al. (2014). A qualitative assessment of gender and irrigation technology in Kenya and Tanzania. *Gender, Technology and Development* 18(3): 303–340.

Njuki, J. & Sanginga, P.C. (2013). Gender and livestock: key issues and opportunities. In Njuki, J. & Sanginga, P.C. (eds) *Women, Livestock Ownership and Markets: Bridging the Gender Gap in Eastern and Southern Africa*. London: Routledge. pp. 1–8.

Pattnaik, I., et al. (2018). The feminization of agriculture or the feminization of Agrarian distress? Tracking the trajectory of women in agriculture in India. *Journal of the Asia Pacific Economy* 23(1): 138–155.

Pedrick, C., et al. (2012, November). Strategies for combating climate change in drylands agriculture. In International Conference on Food Security in Drylands. Beirut, Lebanon, International Center for Agricultural Research in the Dry Areas.

Perry, W., et al. (2019). Argan oil and the question of empowerment in rural Morocco. *The Journal of North African Studies* 24(5): 830–859.

Ragetlie, R., et al. (2021). Paying "lip service" to gender equality: the hollow implementation of gender mainstreaming in Jordan. *Civil Society Review* 5: 18–45.

Ragetlie, R., et al. (2022). "Dear brother farmer": gender-responsive digital extension in Tunisia during the COVID-19 pandemic. *Sustainability* 14(7): 1–22.

Razavi, S. (2007). Liberalisation and the debates on women's access to land. *Third World Quarterly* 28(8): 1479–1500.

Razavi, S. (2009). Engendering the political economy of Agrarian change. *The Journal of Peasant Studies* 36(1): 197–226.

Schilling, J., et al. (2012). Climate change, vulnerability and adaptation in North Africa with focus on Morocco. *Agriculture, Ecosystems & Environment* 156: 12–26.

Sowers, J., et al. (2011). Climate change, water resources, and the politics of adaptation in the Middle East and North Africa. *Climatic Change* 104(3–4): 599–627.

The CGIAR Gender Impact Platform. (2022). *Effectively Targeting Climate Investments: A Methodology for Mapping Climate–Agriculture–Gender Inequality Hotspots*. CGIAR GENDER Platform. Working Paper #005.

Theis, S., et al. (2018). What happens after technology adoption? Gendered aspects of small-scale irrigation technologies in Ethiopia, Ghana, and Tanzania. *Agriculture and Human Values* 35(3): 671–684.

Turner, M. & Williams, T. (2002). Livestock market dynamics and local vulnerabilities in the Sahel. *World Development* 30(4): 683–705.

UNDP. (1995). *Human Development Report 1995*. Oxford: Oxford University Press.

UNIFEM. (2000). *Progress of the World's Women*. New York: UNIFEM.

Waha, K., et al. (2017). Climate change impacts in the Middle East and Northern Africa (MENA) region and their implications for vulnerable population groups. *Regional Environmental Change* 17(6): 1623–1638.

Waterbury, J. (2013). *The political economy of climate change in the Arab Region*. Arab Human Development Report Research Paper Series. UNDP.

World Bank. (2001). *World Development Report*. Washington, DC: World Bank.

Part 3

Gender negotiations for women smallholders

Chapter 8

Rebranding and reconstructing masculinity in agriculture: an overview

Angie Carter, Michigan Technological University, USA

1 Introduction

Children's songs familiar to those growing up in settler states in Global North countries, such as 'The Farmer in the Dell' or 'Old MacDonald Had a Farm,' clearly define the farmer as a man who 'takes' a wife, nurse, cow, dog, cat, mouse, cheese' (in the former) and 'has' a farm, cows, chicks, and pigs (in the latter). In a catchy sing-song, these songs initiate children into colonized norms of agriculture as a relationship of dominance between the farmer – a man – and *his* family, animals, and land. These songs also establish farming as a solitary or individualized act. Thus, the songs are successful not only in championing hierarchies of gender, sexuality, and species but also in erasing the Indigenous cultural knowledge systems found world over in which the provisioning and cultivation of food is not an action of dominance but rather one of community, done in relationship with (rather than power over) human and more-than-human communities. Even as the modes of agriculture championed by agribusiness and prioritized by governments and NGOs in much agricultural development work rely on the adoption of ever-evolving technologies, these traditional colonial tropes of 'the farmer' continue to remain central to the project of capitalist agricultural maintenance and expansion.

This framing of masculinities then helps to identify how it is that without attention to gendered power relations, transitioning to new modes of agricultural production will not create greater gender equity; rather, new modes of production or adoption of technologies risks replicating existing gendered

http://dx.doi.org/10.19103/AS.2024.0148.10

hierarchies (Peter et al., 2000). The social construction of masculinities in agriculture is central to the expansion of capitalist agriculture as a neocolonial project; it follows, then, that gender equity in agriculture requires attention to masculinity.

From chattel slavery to the Green Revolution to contemporary land grabbing, the expansion of capitalist agriculture remains consistent across time and space in its aim to centralize control of land, water, and people. The success of this project depends upon the exclusion of women, children, non-dominant men, and non-binary people from power (Mies and Shiva, 1993; Sachs, 1996; Leslie et al., 2019). How agribusiness employs masculinity to further its agenda is of importance well beyond the Global North where the majority of attention to it has been conducted; transnational corporations' success in the Global North shapes agricultural research, trade, and aid internationally.

Even as racialization takes different forms at different times and in different cultural contexts, capitalist agribusiness is white just as it is masculine – the expansion of new markets, new technology adopters, and new land depends upon a very specific model of agriculture that has proved largely successful as a settler colonial project in the Global North. In the US, for example, genocidal federal policies such as the Homestead Act of 1862 complimented forceful removal policies and false treaties with Indigenous nations and enabled the expansion of white colonists westward. Through their 'improvements' of the land, these white colonists became landowners, establishing white men as the rightful stewards of agricultural land in the eyes of both US law and settler colonial narratives about how land is to be used, and by whom. Here, we see how the social constructions of race, motivated by ownership of property, intersect with social constructions of gender and sexuality (Delgado and Stefncic, 2012; Hoffelmeyer, 2021; Leslie et al., 2019). The continuation of this patriarchal and colonial hegemony persists in the US today and is exported from the Global North through trade agreements, developmental aid and programs, and international agribusiness. These gendered and racialized exclusions privilege the power of dominant men such that their leadership in agriculture becomes hegemonic, sang by children, celebrated in popular culture, and embraced even by scholars, scientists, and policymakers who resort to how to 'empower' women to compete and be successful in this world of dominant men as a means of agricultural 'equity.'

Yet, as critical scholars teach us, these norms do not come from nowhere and the children's songs are anything but innocent. In this paper, I will provide an overview of the importance of masculinity (re)construction to the work of agricultural equity. First, I will discuss how contemporary capitalist agriculture strategically engages a hegemonic agricultural masculinity – one that commodifies land and bodies and emphasizes dominance and control of 'property' for personal profit-making and power-holding – in order to normalize a certain type of land use and land decision-maker to maintain global agricultural 'progress' and expansion. Then, I will discuss how agribusiness strategically rebrands this masculinity to maintain and expand the reach of capitalist agriculture. I will conclude with an overview from the literature summarizing proposed interventions to inform the work of gender and agriculture scholarship and policy if we are to reach transformative change.

2 The importance of masculinities to the expansion of capitalist agriculture as a settler colonial project

Masculinity refers to the cultural acceptance of specific behaviors, beliefs, expressions, and styles of social interactions in a specific place and time (Bird, 2008:5). Critical study of masculinities (rather than the study of men merely as a social group or demographic category) makes visible how masculine power works to maintain the status quo of agriculture (Campbell and Bell, 2000). Masculinities are situated within hegemony; the hegemonic patterns of masculine power are embedded in specific environments that are shaped by conceptualizations of manhood at local, regional, and global scales (Connell, 2016). The characteristics of dominant masculinity emphasize 'the currently most honored way of being a man' and are relational, thus requiring others define themselves in comparison to the hegemonic ideal and creating hierarchies as they do so (Connell and Messerschmidt, 2005:832). The dominant agricultural masculinity championed by mainstream agricultural production aligns with what Connell (2016:312) defines as a 'transnational corporate masculinity' sustained by the continued acceptance of the global economy by the most powerful states, the continued legitimacy and organizational control of a masculine self-selecting elite, and the continued extraction from the periphery states.

In capitalist agriculture, the mode dominant in the Global North and exported world over under the guise of development and aid, agricultural masculinity is characterized by 'physical strength, control of nature, tenacity, hardship, toughness, independence and individualism' (Pini, 2008:34), as evidenced in the children's songs or in the representation of agricultural masculinity in advertisements (Brandth, 1995; Bell et al., 2015; Carter and Lopez, 2019). This conceptualization of dominant agricultural masculinity in the Global North has been further developed by feminist and rural scholars to include the control of land and animals (Liepins, 2000), a fascination with machines (Peter et al., 2000), heterosexuality and heteronormativity (Carter and Lopez, 2019; Hoffelmeyer, 2021; Leslie et al., 2019), as well as the ability to endure long hours of extreme labor in harsh weather (Little, 2002). These hierarchies exist among men whose class, race, or sexuality do not align with those of dominant men, as well as between dominant men and women, non-binary people, land, and animals. As a result, those whose identities belong to excluded groups – women, LGBTQIA, racialized minorities – find their identities othered in contemporary agricultural education, programs, and in larger culture. This 'brand' of masculinity contributes to the invisibility of women (Sachs, 1983; Campbell and Bell, 2000), as well as to the invisibility of non-dominant men and non-binary people (Leslie et al., 2019).

The theory of social constructionism can aid in understanding how capitalist agricultural expansion both depends upon and strategically engages the reconstruction of agricultural masculinity to successfully embed itself as the dominant form of agricultural production, especially in the face of resistance. Social construction identifies how it is that certain conceptualizations of masculinity – such as who is respected as an agricultural decision-maker or whose land access is accepted as legitimate – become dominant cultural narratives (Berger and Luckmann, 1966). These narratives carry with them 'formulas' detailing acceptable practices, e.g. how one might 'become' a farmer, as well as the symbolic codes detailing the expectations, responsibilities, rights, and norms that make one's enactment of this story successful (Loseke, 2007). Dominant cultural narratives are defined by those in power through the use of cultural symbols that are then reified through interactional processes (Greider and Garkovich, 1994:6). Bell et al. (2015), drawing upon Pini (2008), note that in Western contexts, control of nature, individualism, and physical strength characterize the dominant narrative of agricultural masculinity. The dominance of these cultural narratives and the power required to maintain their control results in a loss of diversity across the agricultural landscape, from the farmer to the crops grown; the monocultures of the mind replicate themselves on the ground (Shiva, 1993).

Burton's (2004) conceptualization of a 'good farmer' in an industrialized agricultural system emphasizes the importance of the appearance of the farmer's crop and/or livestock, their productivity, and the farmer's compliance with community farm standards and their longevity/history on the farm. The embodied capital of farm longevity in colonized agricultural systems, such as the US, depends upon one's proximity to whiteness and economic capital. In a study of white women who engage in conventional farming in the US Corn Belt, respondents shared how they could overcome the gendered barriers of entry in the mainstream US agricultural system if they were able to engage the cultural capital of longevity on the farm (Nichols and Carter, 2023). Of course, this cultural capital is only available to those permitted access by the power holders, white men.

The agricultural formula stories driving today's agricultural expansion prioritize technology adoption, increased productivity using inputs, and expert knowledge. New practices and technologies are not neutral, but carry with them embedded symbolic valuations and meanings that prioritize productivist agricultural approaches (Burton, 2004). Those in power control agricultural resources – e.g. land, knowledge, access to capital – by (re)defining the biophysical environment and cultural symbols to best serve the status quo (Loseke, 2007; Greider and Garkovich, 1994). Social interactions reify these definitions, revising them as needed to adapt to changing economic, social, and environmental contexts. (Re)definitions of how land is to be used and by whom take place over time, becoming normative and even hegemonic (Saugeres, 2002). Multiple studies of agricultural technology highlight how participants in these formula stories revise them to accommodate technological changes even as these changes may seem, on the surface, to run counter-to-cultural norms (Bell et al., 2015; Brandth, 1995; Carter and Lopez, 2019; Saugeres, 2002). For example, Bell et al. (2015) interviewed seed salespeople at the second largest indoor farming show in the US. These salespeople relayed how in visiting farmers, often riding with them in their tractors during planting season, they provided consultation on products and, importantly, what was or was not working well for farmers' neighbors. The dependence of farmers upon these salespeople for agronomic advice and farmers' reluctance to consult with other farming neighbors may seem to maintain the individualism of the farmer; however, farmers' dependence upon industry and lack of solidarity with other farming neighbors stems from a cultural manipulation by which farmers accept dependence upon the agribusiness consultants as a 'partnership' even as doing so furthers farmers' own deskilling and loss of autonomy.

As rural sociologists and rural studies scholars have argued, hegemonic masculinity serves agriculture by defining what 'good' or acceptable farming and farmers look like and establishes norms to which farmers are expected to conform in order to be accepted as farming men (Bell et al., 2015; Brandth, 1995; Liepins, 1998). Heteronormative norms permeate agriculture no matter the paradigm of practice. To use a US-based example, the 'family farm' is such a powerful social construct that sustainable agricultural movements advocate for its preservation even as it re-enforces heteropatriarchal, white settler, and white supremacist norms (Keller, 2015; Leslie et al., 2019; Hoffelmeyer, 2021). Even within sustainable agriculture movements in New England, queer farmers face expectations to conform to norms of monogamy and marriage (Leslie, 2017). The result is that the status quo of capitalist agriculture has long gone unquestioned (Leslie et al., 2019; Horst and Marion, 2019). Indeed, the participation in the expectations of and conformity to hegemonic masculinity provides men privileged, insider access and resources unavailable to others (Carter and Lopez, 2019).

Beyond the Global North, studies of agricultural restructuring programs and land grabs emphasize the dangers of failing to critically engage with constructions of masculinity. As the expansion of capitalist agriculture redefines agricultural masculinity to align with the needs of transnational agricultural markets, it devalues women's work and renders it invisible and dangerous

(Sachs and Alston, 2010). Tavenner et al. (2021) document how a low emissions dairy market intensification program in Kenya resulted in a 'win–lose' as women engaging in informal milk markets faced gender-based sanctions. Programs to increase food security exacerbated household food insecurity and increased women's household labor in India's Kumaon Hills (Nichols, 2015). As masculinities are redefined and repackaged to be of use to industry, so too are femininities rendering the labor of women to a 'largely invisible reserve army of corporate agriculture' (Sachs and Alston, 2010:280).

Drawing on Collins' (2002) and Crenshaw's (1991) definitions of intersectionality as an approach to making evident axes of power, Norgaard et al. (2017:3) argue that the environment must be included in sociohistorical analyses of power and identity, as how one relates to the environment is a key component of racialization in settler colonialism and racism. Intersectional theory identifies how race, class, gender, sexuality, and other categorizations of domination, e.g. able-bodiedness or citizenship, maintain dominant hierarchical power structures (Collins, 2002; Crenshaw, 1991). Applying intersectionality to the understanding gender in agriculture 'is a de-hierarchizing approach and examines power relations that produce gendered subjectivities in particular contexts' (Leder and Sachs, 2019). To this work, queer, critical masculinity, and Indigenous theory contribute needed attention to how capitalist agricultural production engages heteropatriarchy and cissexism to protect, defend, and expand settler colonialism through control of land, knowledge, technology, and bodies.

Rather than a gender binary, many Indigenous peoples in what is now North America lived and continue to live according to place-based gendered responsibilities, relationships, and knowledge that sustained and sustain gender diversity and egalitarianism (Vinyeta et al., 2015). Anishinaabe scholar Kyle Whyte's (2018) discussion of gender in relation to environmental injustice and settler colonialism explains that Anishinaabe traditions emphasize gender diversity and fluidity rather than a gender binary. Whyte (2018:134) discusses the 'ecology of settler colonial domination' as ecological violence and environmental injustice. Settler colonialism, as defined by Whyte (2018:134–135), 'refers to the complex social processes in which at least one society seeks to move permanently onto the terrestrial, aquatic, and aerial places lived in by one or more other societies who already derive economic vitality, cultural flourishing, and political self-determination from the relationships they have established with the plants, animals, physical entities, and ecosystems of those places,' thereby transforming 'Indigenous homelands into settler homelands.' Capitalist agricultural expansion, whether through the push of petrochemical dependence, land theft, or co-optation of knowledge, continues to intentionally rupture the relationships among Indigenous communities in the Global North and Global South today.

By continually revising and adapting masculinity to adapt to the needs of agribusiness, agricultural corporations normalize agricultural practices, decision-making, and leadership. It follows then that targeting masculinity is an opportunistic site for redefining and remaking power relations in agriculture. Alienation, deskilling, and domination as a result of colonial ecological violence (as in the case of the Karuk tribal fishermen in Norgaard et al.'s 2017 study) or the successful rebranding of white agricultural masculinity (as documented by Brandth, 1995; Bell et al., 2015; Carter and Lopez, 2019 and others) identify how capitalist agriculture, through the control of symbolic and material resources, targets masculinity as a tool for settler colonialism.

3 Agribusiness and the rebranding of agricultural masculinity

There is little more unnatural than the tens of thousands of acres of genetically similar corn and soybeans that now spread across the once most diverse biome of North America, the tall grass prairie, yet the US farmer standing along his field of corn is held up as an idealized form of agricultural manhood in the Global North. Connell (1995:74) defines hegemonic masculinity as the version of masculinity occupying 'the hegemonic position in a given pattern of gender relations' rather than something fixed; the malleability of masculinity is important in understanding how rebranding masculinity works across local and global scales to further agricultural expansion. As a project of settler colonialism, the rebranding of agricultural masculinity supports capitalist reproduction – not only the products of capitalist agriculture, such as the crops or the technologies sold/consumed, but also its social relations. The expansion of capitalist production describes the waged relation in which one class owns the means of the production and the other class labors (Friedmann, 1978:548). As Friedmann (1978:555) argues, 'reproduction is both social and technical' and can enable a transformation in production in which an old form of production is undermined and a new form of production becomes dominant.

Drawing on Butler's (1999:33) definition of gender as 'always a doing' or a performative set of acts, Chung (2020:724) argues that gender, in a postcolonial world, is deeply intertwined within colonial power relations, including race, class, sexuality, and age. Her ethnographic study of the EcoEnergy Sugar Project's evolution in Tanzania from a 'liminal' land deal to a 'militarised, masculinised and necropolitical space' (737) identifies how the biopolitics of the project's training programs were shaped by the existing domestic politics of social reproduction (Chung, 2020:730). She cautions that 'the reproduction of subjects does not occur outside or separate from existing matrices of power' (727).

Capitalist agriculture's rebranding of masculinities to align with the goals of global agribusiness results in the deskilling and alienation of the farmer (Bell et al., 2015), but it begins with the control and maintenance of a mythology – that of the farmer as a 'man of the Earth' – while also a consumer of the most up-to-date technologies needed to maximize yield per acre and, ultimately, profit. In its maintenance of this mythology, corporate agriculture constructs cultural narratives in the Global North that both celebrate settler nostalgia while also co-opting the farmer for continued expansion of ever-evolving markets, thus negating the dissonance of the US Corn Belt farmer driving his corn crop to the ethanol plant even as his truck sports a 'Farmers Feed the World' bumper sticker (Carter and Lopez, 2019). This deskilling of the farmer – who is now less reliant upon place-based or generational knowledge because of dependence upon experts and inputs – facilitates continued control of who is/is not a farmer by agribusiness (Bell et al., 2015). And so, new types of manhood emerge as new technology is consumed (Brandth, 1995; Carter and Lopez, 2019). Even as this redefinition of masculinities centers cisgender white man bodies, it is more importantly using specific masculinities to advance particular organizational, occupational, political, industrial, and/or institutional dispositions, reifying the dominant gender order even as it seems to present something new (Carter and Lopez, 2019:43).

Bell et al. (2015) remind us that Brandth (1995), in her study of Norwegian tractor advertisements, first theorized how the expansion of agricultural technology and increased reliance upon external inputs might result in a more 'professional' masculinity to replace the rugged individualism of the dominant agricultural masculinity found throughout capitalist agriculture in the Global North. Since that early study, others have confirmed that a more 'professional' or 'businessman' identity indeed is replacing the rugged yeoman farmer (Bell et al., 2015; Brandth and Haugen, 2005; Carter and Lopez, 2019; Saugeres, 2002). Brandth's (1995) study of the relationship between agricultural technology transfer and the construction of masculinities identifies a hegemonic masculinity characterized by men's dominance over technology. Building on Brandth's (1995) study of tractor advertisements, Bell et al. (2015) document how US agribusiness facilitates agricultural deskilling through the strategic construction of masculinities. As farmers become more reliant upon outside inputs and knowledge sources, they become more dependent upon the products of agribusiness. To sell these outside inputs and knowledge sources, Bell et al. (2015) document a shift in emphasis from the rugged individualism of the farmer to the farmer as a businessman. Advertisements portray this 'businessman farmer' as wearing clean, crisp button-down shirts and khakis rather than overalls and consulting with industry experts for knowledge regarding agronomic decision-making (Bell et al., 2015). This echoes the trends identified by Little's (2002) review as to how this hegemonic masculinity shifts as new types of managerial and entrepreneurial skills are being used in relation to land management.

Agribusiness shifts the construction of the 'farmer' to incorporate settler nostalgia while also incorporating advanced technology adoption, expert knowledge acquisition, and the increased consumption of both (Carter and Lopez, 2019). The production of the farmer as an 'entrepreneurial subject within modern capitalist society' facilitates the farmer's deskilling, dependence upon the agribusiness representatives selling inputs, and, ultimately, disconnection from the land, enabling further corporate control over agricultural knowledge production (Bell et al., 2015:309). As a rebranding project, this reconstruction of the farmer allows the seeming contradictions – farmers are connected to and care the land and animals, yet farmers use the best technology to maximize yield – to persist (Carter and Lopez, 2019). The project is further successful in that this 'reconstituted brand of conventional masculinity' prevents farmers from building the solidarity together that is needed to resist this process because they view each other as competition rather than resources (Bell et al., 2015:310).

Rebranding is of course not new. Kroma and Flora (2003) traced the social construction of pesticide advertisements from the 1940s to 1990s, documenting shifts in eras of science, control, and environment. As they note, these changes are a strategic attempt by industry to co-opt cultural narratives for use in sustaining markets and corporate profits. The success of this project depends upon the adoption of products and cultural acceptance of revised narratives; it is to masculinity that the marketing appeals and it is through masculinity that the project succeeds.

4 Reimagining agricultural masculinity

The currency of capitalist agricultural production is a patriarchal masculinity, rooted in domination (Pini, 2008). Heteropatriarchal models of family and land control continue to be used as a means for colonial expansion through technology adoption today. Gender-based interventions will continue to be limited by these models. Scholars studying masculinity reconstruction among Global North farmers have identified the ongoing patterns of rebranding masculinity to promote and serve the purposes of agribusiness; there is a need to learn more from and with the farmers and rural people in the Global South so as to create new solidarities in the face of the continuing settler colonial project that is the expansion of capitalist agriculture. Peter et al. (2000) point to the importance of masculinities in preserving existing gendered hierarchies in agriculture and preventing sustainable practices from being more equitable – they argue that agricultural men need masculinities with more flexible boundaries that can make space for other types of farming and farmers. In addition to the calls for increased and more critical research studying gendered power and masculinities, more scholarship is also needed to study successful projects rebranding masculinity with justice in mind, such as agrarian social movements and Indigenous continuance.

Instead of a 'world gender order,' Connell (2016:313) argues there are 'conflicting hegemonic projects' in which 'different configurations of masculinity are at work, and come into conflict' and, despite the gender equity that has been gained, 'these changes have yielded only a little ground for democratic projects of change in masculinity.' Connell (2016:314) cautions there seems little threat to this 'corporate masculinity'; Connell's published research identifies 'few connections between masculinity reform efforts and union activism, landless people's movements, environmental activism, or other movements that offer a significant challenge to corporate or state power.' Connell (2016:304) stresses the need decolonize the study of masculinities, 'to learn not only from Western Europe and North America but also from the majority world.' Connell (2016:305-306) calls upon scholars to situate gender historically and contends that the conceptualization of hegemonic masculinity must be treated 'as a collective project for realizing gender hierarchy' and be 'understood on a world scale.' There is resistance, though, to this brand of colonial gender relations, and it can be found within both the Global North and the Global South.

Despite centuries of colonial expansion and extraction of resources, relational models rooted in Indigenous and Native worldviews persist and offer important alternatives. Norgaard et al. (2017) identify how the continued ecological violence of settler colonialism has resulted in the collapse of fisheries in northern California, presenting challenges to masculinity for Karuk tribal fisherman as well as the Karuk nation at large, and yet, the Karuk fisherman engage in Whyte's (2013) definition of 'collective continuance,' responsibility, and resistance. Some reconstruct their masculinity through environmental activism and management of the fisheries, creating new ways of embodying traditional masculine values of respect and responsibility to the fish and larger communities (Norgaard et al., 2017:6). The protests led by Indian farmers beginning in summer of 2020 against structural adjustments were not only about agrarian reforms but also in response to the larger financialization of India's economy and the consequences for agriculture and rural life (Baviskar and Levein, 2021). While similar to past periods of agrarian protest in India in that the most recent protests were mostly led by larger landowners of dominant castes, the protests organized in 2020 formed a much broader coalition across the caste and class divisions of the agricultural sector, including broader participation from women (Baviskar and Levein, 2021).

Relationship with community and creation of solidarity with others are points of differentiation in the rebranding of masculinity as corporate co-optation to continue the capitalist agriculture expansion machine and the rebranding of masculinity as resistance to or continuance in the face of the same. The masculinity dominant among white settler farmers in the face of increased consolidation, global competition, and environmental degradation resorts to increased reliance upon external inputs and expert knowledge in order to remain competitive and continue dominance over the land (Bell et al., 2015; Carter and Lopez, 2019). In doing so, the settler farmer exchanges possibilities of solidarity with other farmers, as well as economic, cultural, and environmental systems transformation, for acceptance as a global businessman even as he is beholden agribusiness. The Karuk fishermen recommit themselves in new ways to their communities by reconstructing the fisherman's identity and gendered practices through activism and fishery management, finding alternative means through which to enact the traditional values of reciprocity and responsibility (Norgaard et al., 2017). The Indian farmers protests created a new coalition across class, caste, gender, and geographic region to advocate for reforms benefitting rural livelihood (Baviskar and Levein, 2021). Contrary to the settler farmers, the Karuk fishermen or Indian farmers engage new modes of masculinity in protection of community livelihood and in resistance to cultural exploitation and threat; their rebranding of masculinity serves continuance rather than co-optation.

Sachs and Alston's (2010:278) introduction to a *Signs* special issue about women and agriculture asked how gender relations shift as the globalization of agricultural markets, declining subsidies, increased consolidation, and rising consumer demand for healthier food change the agricultural landscape. They directed questions towards new imaginaries and a reconceptualization 'of the intersections of gender, race, ethnicity, class, citizenship, and species.' The need for greater critical attention to these intersections continues. In her recent study of gender research in *Agriculture and Human Values*, the journal of the Agriculture, Food, and Human Values Society, Sachs (2023) noted that masculinities and sexualities still receive little scholarly attention in agricultural research. As Pyburn and Kruijssen (2020) caution, gender relations rather than gender roles should be the focus of gender-conscious interventions and policy changes – roles are fluid and can change over time, emphasizing them as static can reinforce stereotypes and work to further entrench gendered inequalities. Gender relations – how people engage with one another across space and time – provide a lens through which to see opportunities for change.

Critically engaging with constructions of masculinities and their strategic use by settler colonialism to normalize specific types of agricultural power contributes to larger theoretical and empirical projects. For example, increased study of agricultural masculinity construction, particularly within Indigenous and Global South contexts, has much to contribute to Bonds and Inwood's (2016:715) call for the need to develop 'geographically sensitive theorizations of white supremacy as the animating logic of racism and privilege' and their argument that settler colonialism is 'an ongoing mode of empire.'

Approaches extending beyond the transactional inclusion of the needs of women, non-binary people, non-dominant men, land, and animals must focus on the contributions and synergies each contribute to diversifying agriculture. Such approaches will be inclusive of different masculinities, offering interventions that are relational, as emphasized by Whyte (2018), rather than hierarchical. Increased study of how the masculinity emphasized through the expansion of capital agricultural influences gender relations in different contexts, and especially study of how this masculinity is resisted, contested, and revised in support of ecological, economic, and social equity, is needed to inform transformative, rather than additive, agrarian change.

5 References

Baviskar, A. and Levein, M. (2021), Farmers Protests in India: An Introduction to the JPS Forum. *Journal of Peasant Studies* 48(7): 1341–1355. DOI: 10.1080/03066150.2021.1998002.

Bell, S. E., et al. (2015), Manipulated Masculinities: Agribusiness, Deskilling, and the Rise of the Businessman-Farmer in the United States. *Rural Sociology* 80(3): 285–313. DOI: 10.1111/ruso.12066.

Berger, Peter L. and Luckmann, T. (1966), *The Social Construction of Reality: A Treatise in the Sociology of Knowledge*. Garden City, NY: Anchor Books, Doubleday & Company.

Bird, S. R. (2008), Masculinity Studies. In *International Encyclopedia of the Social Sciences*, 2nd ed., edited by William A. Darity, Jr. pp. 5–11. New York: McMillan Library Reference. See https://searchworks.stanford.edu/view/7692595

Bonds, A. and Inwood, J. (2016), Beyond White Privilege: Geographies of White Supremacy and Settler Colonialism. *Progress in Human Geography* 40(6): 715–733. DOI: 10.1177/0309132515613166.

Brandth, B. (1995), Rural Masculinity in Transition: Gender Images in Tractor Advertisements. *Journal of Rural Studies* 11(2): 123–133. DOI: 10.1016/0743-0167(95)00007-A.

Brandth, B. and Haugen, M. (2005), Text, Body, and Tools: Changing Mediations of Rural Masculinity. *Men and Masculinities* 8(2): 148–163. DOI: 10.1177/1097184X05277716.

Burton, R. J. F. (2004), Seeing Through the 'Good Farmer's' Eyes: Towards Developing an Understanding of the Social Symbolic Value of 'Productivist' Behaviour. *Sociologia Ruralis* 44(2): 195–215. DOI: 10.1111/j.1467-9523.2004.00270.x.

Butler, J. (1999 [1990]), *Gender Trouble: Feminism and the Subversion of Identity*. New York: Routledge.

Campbell, H. and Bell, M.M. (2000), The Question of Rural Masculinities. *Rural Sociology* 65(4): 552–546. DOI: 10.1111/j.1549-0831.2000.tb00042.x.

Carter, A. and Lazaro Lopez, A. (2019), Rebranding the Farmer: Formula Story Revision and Masculine Symbolic Boundaries in US Agriculture. *Feminist Formations* 31(3): 25–50. DOI: 10.1353/ff.2019.0029.

Chung, Y. B. (2020), Governing a Liminal Land Deal: The Biopolitics and Necropolitics of Gender. *Antipode* 52(3): 722–741. DOI: 10.1111/anti.12612.

Collins P. H. (2002), *Black Feminist Thought: Knowledge, Consciousness, and the Politics of Empowerment*. New York: Routledge.

Connell, R. W. (1995), *Masculinities*. Cambridge: Polity Press.

Connell, R. W. (2016), Masculinities in Global Perspective: Hegemony, Contestation, and Changing Structures of Power. *Theory and Society* 45(4): 303–318. https://www.jstor.org/stable/44981834.

Connell, R. W. and Messerschmidt, J. W. (2005), Hegemonic Masculinity: Rethinking the Concept. *Gender & Society* 19(6): 829–859. DOI: 10.1177/08912432052786.

Crenshaw, K. (1991), Mapping the Margins: Intersectionality, Identity Politics, and Violence against Women of Color. *Stanford Law Review* 43: 1241–1299.

Delgado, R. and Stafncic, J. (2012), *Critical Race Theory: An Introduction*. New York: NYU Press.

Friedmann, H. (1978), World Market, State, and Family Farm: Social Bases of Household Production in the Era of Wage Labor. *Comparative Studies in Society and History* 20(4): 545–586.

Greider, T. and Garkovich, L. (1994), Landscapes: The Social Construction of Nature and the Environment. *Rural Sociology* 59(1): 1–24.

Hoffelmeyer, M. (2021), "Out" on the Farm: Queer Farmers Maneuvering Heterosexism and Visibility. *Rural Sociology* 86(4): 752–776. DOI: 10.1111/ruso.12378.

Horst, M. and Marion, A. (2019), Racial, Ethnic and Gender Inequities in Farmland Ownership and Farming in the U.S. *Agriculture and Human Values* 36(2): 1–16. DOI: 10.1007/s10460-018-9883-3.

Keller, J. (2015), Rural Queer Theory. In *Feminisms and Ruralities,* edited by B. Pini, B. Brandth, and J. Little. Lanham, MD: Lexington Books, pp. 141–151.

Kroma, M. M. and Flora, C. B. (2003), Greening Pesticides: A Historical Analysis of the Social Construction of Farm Chemical Advertisements. *Agriculture and Human Values* 20: 21–35. DOI: 10.1023/A:1022408506244.

Leder, S. and Sachs, C. E. (2019), Intersectionality at the Gender-Agriculture Nexus: Relational Life Histories and Additive Sex-Disaggregated Indices. In *Gender, Agriculture and Agrarian Transformations: Changing Relations in Africa, Latin America, and Asia*, edited by C. E. Sachs. London: Routledge, pp 75–92.

Leslie, I. S. (2017), Queer Farmers: Sexuality and the Transition to Sustainable Agriculture. *Rural Sociology* 82(4): 747–771. DOI: 10.1111/ruso.12153.

Leslie, I. S., et al. (2019), Relational Agriculture: Gender, Sexuality, and Sustainability in US Farming. *Society & Natural Resources* 32(8): 853–874. DOI: 10.1080/08941920.2019.1610626.

Liepins, R. (1998), The Gendering of Farming and Agricultural Politics: A Matter of Discourse and Power. *Australian Geographer* 29(3): 371–388. DOI: 10.1080/00049189808703230.

Liepins, R. (2000). Making men: the construction and representation of agriculture-based masculinities in Australia and New Zealand. *Rural Sociology* 65(4): 605–620. DOI:10.1111/j.1549-0831.2000.tb00046.x

Little, J. (2002), Rural Geography: Rural Gender Identity and the Performance of Masculinity and Femininity in the Countryside. *Progress in Human Geography* 26(5): 665–670. DOI:10.1191/0309132502ph394pr.

Loseke, D. (2007), The Study of Identity as Cultural, Institutional, Organizational, and Personal Narratives: Theoretical and Empirical Integrations. *The Sociological Quarterly* 48(4): 661–688. DOI: 10.1111/j.1533-8525.2007.00096.x.

Mies, M. and Shiva, V. (1993), *Ecofeminism*. London: Zed Books.

Nichols, C. (2015), Time Ni Hota Hai: Time Poverty and Food Security in the Kumaon Hills, India. *Gender, Place & Culture* 10: 1404–1409. DOI: 10.1080/0966369X.2016.1160871.

Nichols, C. and Carter, A. (2023), It Doesn't Bother Me: An Intersectional Analysis of Discrimination Among White Women Farmers in the US Corn Belt. *Journal of Rural Studies* 101. DOI: 10.1016/j.jrurstud.2023.103054.

Norgaard, K. M., et al. (2017), How Environmental Decline Restructures Indigenous Gender Practices: What Happens to Karuk Masculinity When There Are No Fish? *Sociology of Race and Ethnicity* 4(1): 1–16. DOI: 10.1177/2332649217706518.

Peter, G., et al. (2000), Coming Back Across the Fence: Masculinity and the Transition to Sustainable Agriculture. *Rural Sociology* 65(2): 215–233.

Pini, B. (2008), *Masculinities and Management in Agricultural Organizations Worldwide*. London: Routledge.

Pyburn, R. and Kruijssen, F. (2020), Gender Dynamics in Agricultural Value Chain Development: Foundations and Gaps. In Routledge *Handbook of Gender and Agriculture*, edited by C. E. Sachs, L. Jensen, P. Castellanos, and K. Sexsmith. New York: Routledge, pp. 32–45.

Sachs, C. (1983), *The Invisible Farmers: Women in Agricultural Production*. Totowa, NJ: Rowman & Allanheld.

Sachs, C. (1996), *Gendered Fields: Rural Women, Agriculture, and the Environment*. Boulder, CO: Westview Press.

Sachs, C. (2023) Gender, women and agriculture in *Agriculture and Human Values*. *Agriculture and Human Values* 40: 19–24. DOI:10.1007/s10460-022-10391-6

Sachs, C. and Alston, M. (2010), Global Shifts, Sedimentations, and Imaginaries: An Introduction to the Special Issue on Women and Agriculture. *Signs* 35(2): 277–536. DOI: 10.1086/605618.

Saugeres, L. (2002), The Cultural Representation of the Farming Landscape: Masculinity, Power and Nature. *Journal of Rural Studies* 18(4): 373–384.

Shiva, V. (1993), *Monocultures of the Mind: Perspectives on Biodiversity and Biotechnology*. London: Zed Books.

Tavenner, K. et al. (2021), "Breaking Even" under Intensification? Trade-Offs for Women Milk Marketers in Kenya. *Rural Sociology* 86(1): 110–138. DOI: 10.1111/ruso.12345.

Vinyeta, K. et al. (2015), *Climate Change Through an Intersectional Lens: Gendered Vulnerability and Resilience in Indigenous Communities in the United States*. Gen. Tech. Rep. PNW-GTR-923. Portland, OR: U.S. Department of Agriculture, Forest Service, Pacific Northwest Research Station, p. 72.

Whyte, K. P. (2018), Indigenous Resurgence, Decolonization, and Movements for Environmental Justice. *Environment and Society* 9: 125–144.

Whyte, K. P. (2013), Justice Forward: Tribes, Climate Adaptation and Responsibility. *Climatic Change* 120(3): 517–530.

Chapter 9

Sexuality and small-scale farmers

Prisca Pfammatter, University of Bern, Switzerland; and Michaela Hoffelmeyer, University of Wisconsin-Madison, USA

1 Introduction

While gender has been a focal point of discussion in prominent international organizations such as the United Nations, sexuality has been insufficiently researched, leading to policies that do not or only superficially address sexualities, often only within the context of sexual violence (e.g. UNDROP, 2018; *Agriculture Development, Food Security and Nutrition: Resolution,* 2022). Furthermore, research has played a role in upholding the discriminatory nature of the dominant agricultural model. By failing to recognize how sexuality shapes agrifood systems, research has failed to address heteronormativity, thus perpetuating heterosexual, cisgender biases (Leslie, 2017). Research, in turn, translates into policies and programs that fail to recognize queer individuals as potential farmers and actors in building sustainable and just agrifood systems (Leslie, 2017, 2019). We raise the question of how policies and programs can be truly gender-sensitive, -inclusive, or even -transformative if they rarely, if at all, acknowledge and understand the full range of genders, sexualities, and sexual relations. We further consider how sexuality-mainstreaming might contribute to queer-inclusion in agriculture and gender-mainstreaming policies, research, programs, and activism.

In this chapter, we begin to draw out the relationship between gender and agriculture research, and sexuality and agriculture research thus far. While there is no clear dichotomy between these two bodies of scholarship, the first

http://dx.doi.org/10.19103/AS.2024.0148.11

has gained unprecedented attention over the last several decades as gender mainstreaming policies, initiatives, and programs have been and continue to be taken up widely across the world. This achievement must be celebrated for the advances in acknowledging the role that women smallholders[1] play in food production and environmental conservation. The mainstreaming of gender-based efforts in agriculture calls attention to the structures that inhibit or bolster women smallholder farmers' livelihoods and have provided meaningful shifts in resources and policies. We see the various areas of gender and agriculture research as necessary to the broader effort to bring about equity and justice in agriculture globally. However, we see much potential in what we might brand sexuality in agriculture research to advance the engagement with structures surrounding heterosexuality and cisheterosexism in the agrifood system and the implications for smallholder farmers, food security, and environmental sustainability globally.

Box 1: Heteronormativity

Heteronormativity refers to the belief that heterosexuality is the normal form of sexuality. It is rooted in and simultaneously co-constitutive of the binary and static understanding of gender, sex, and sexuality. As such, individuals are perceived and classified as either men or women, male or female, heterosexual or homosexual. Based on these categorizations, the expectations for gender norms and performance are linked to their assigned sex. For example, if a person is assigned male at birth on the basis of their biological and genetic characteristics, 'he' is socially expected and conditioned to identify as a man, perform masculinity, and be attracted to women. This framework excludes any notions beyond or in between these categories and dismisses the potential for changes over time (Barker, 2007). As such, queer theorizing and concepts such as heteronormativity challenge this interlocking of sex with gender identification, performance, and sexuality.

---

2 Reflecting on the modern/colonial gender system

Despite acknowledging the need to uproot patriarchal norms, we argue that gender in agriculture scholarship has, by and large, failed to see the 'colonial/modern gender system' (Lugones, 2007) as part of the enforcement of patriarchy. The modern/colonial gender system, as Lugones (2007) describes

1 We recognize that the category of smallholder and small-scale farmers varies greatly depending on geographic location and cultural norms. Here, we use these terms as short hands to draw potential parallels across the world while also recognizing that a small-scale farmer in the United States context would vary greatly from a smallholder in, for example, sub-Saharan Africa.

it, offers a historical account of the application of the European gender system to societies that were colonized. Lugones argues 'that gender itself is a colonial introduction, a violent introduction consistently and contemporarily used to destroy peoples, cosmologies, and communities as the building ground of the "civilized West"' (p. 186). Societies in pre-colonial contexts (and still today) recognize more than two genders, such as Native American communities (Gunn, 1992), or function without the construction of gender, such as the Yoruba (Oyěwùmí, 1997). These examples illustrate how sex and gender are not naturally occurring structures to organize society. Rather, the modern/colonial gender system, developed and strategically employed, enables the 'inferiorization' of gender, sexual, and racialized groups that fall outside the European white bourgeoise sex-gender binary construction embedded in heteronormativity (Lugones, 2007).

Further, Lugones (2007) ties the sex-gender binary that functions today as a process to instill this domination for those who are transgender, intersex, non-binary, or non-conforming. In other words, the modern/colonial gender system narrows the concept of gender to enable domination, thereby marginalizing household structures, genders, sexes, and sexualities that are not aligned with the modern/colonial gender system. We recognize that not all regions developed gender systems as a result of European colonization, as other mechanisms for social stratification have deemed some genders and sexes less than human. Nevertheless, Lugones' (2007) theorizing of the colonial/modern gender system illuminates for scholars and practitioners how, in many places, colonialism implemented strict gender and sexual systems; further, how this implementation attempted to move cultures toward the cis, heterosexual gender and sexual structures accepted in colonizing countries.

Researchers, practitioners, and global development organizations bring colonial ideologies to bear, especially in the case of gender and sexuality. We explore the Women's Empowerment in Agriculture Index (WEAI) as one example of how international development organizations and women's empowerment efforts fail to fully account for gender and sexual diversity in research and practice.

Box 2: A note on terminology.

Queer theory has largely been developed in the Western context (Hawley, 2001). As such, words like 'lesbian', 'gay', 'queer', 'non-binary', and 'gender non-conforming' have become terminology adopted by members of the queer community largely in English-speaking places in the United States and Europe to encapsulate gender and sexual identities. At the same time, terms like 'berdache', 'third gender', 'same gender loving', and others are useful among many populations (Driskill et al., 2011). The shifting acronym lesbian, gay, bisexual, transgender (LGBT), LGBTIA+ (with the inclusion of intersex and asexual), and 2SLGBTIA+ (with the inclusion of Two-spirit) illuminate the efforts to move away from cisgender and settler privilege in the queer community. We largely use 'queer' when discussing the literature on farmers, as this is the approach of these scholars. In other cases, we use gender or sexual minorities to capture the wide-ranging terminology of those who are not considered cisgender or heterosexual. However, minorities carry a certain mechanism for othering and dismissing that fails to fully demonstrate the presence and resistance to oppression embodied by these communities. Here, language fails us and likely excludes us from portraying a truly global description of queer farmers.

2.1 Case study: re-examining the women's empowerment in agriculture index

The project-level WEAI (referred to as pro-WEAI) offers practitioners and researchers a survey-based approach for 'measuring empowerment, agency, and inclusion of women in the agriculture sector' (IFPRI, 2022). Further, WEAI allows researchers and practitioners to 'diagnose areas of women's disempowerment, design strategies to address deficiencies, and monitor project outcomes' (IFPRI, 2022). The index was conceived as a collaboration between the International Food Policy Research Institute, the Oxford Policy and Human Development Initiative, and partner projects within the Gender, Agriculture, and Assets Phase 2. Programs have been applied pro-WEIA across multiple projects and geographies (Malapit and Quisimbing, 2015; Martinez et al., 2021; Seymour, 2017).

We first critique the survey's most recent version for embracing a primarily heteronormative vision of the family, namely a two-person monogamous household consisting of a biological male and female partnership. Although we recognize that this form of household may be the most dominant form of household organization in many geographical locations, we offer this critique to highlight how other gender and sexual minorities are (dis)empowered in agriculture but not featured in gender in agriculture research and practice. The aim of this critique is to assist in bringing equity to global agriculture. Although the WEAI includes a handful of questions related to polygamous[2] households, these questions only occur toward the end of the survey. By failing to measure the full range of household structures, genders, and sexualities, pro-WEAI fundamentally misses the opportunity to capture robust and appropriate

2 Polygamous references to men or women being married to more than one person at a time. In contrast, polygynous refers only to men being married to more than one person at a time. We use polygamy here to be more inclusive of varying household structures globally, not limited to those in patriarchal societies wherein only men have the ability to engage in this form of household structure.

potentials for empowerment. Finally, by confronting the internalized heteronormativity and the persistence of the modern/colonial gender system, we encourage future scholarship and research to revisit the pro-WEAI and other gender surveys to develop more effective means to include additional gender minorities and household models.

Pro-WEAI adopts the modern/colonial gender system by offering only the categories of 'male' and 'female' when asking questions about the sex of the respondent. By failing to offer additional categories, even 'other' or 'not listed', the structure of the instrument fundamentally fails to capture the full range of gender (dis)empowerment in agriculture. The equation of sex as gender occurs throughout the survey. For example, the instrument's directions for questions note, 'For Females only: Does respondent have a child under 5 years old?' Similarly, when inquiring about intimate partner violence, the survey includes 'husband' and 'wife' language, thereby failing to account for same-gender households involved in agricultural production. Although same-gender households and relations are illegal in some geographic contexts, the failure to include flexibility in these categories in pro-WEAI leads to further erasure as law and practice are often not congruent.

When asking about the groups in the community, pro-WEAI uses the 'gender' (which is actually biological sex) makeup of these groups to analyze empowerment potential. The instrument asks: '[I]s the group composed of all male or female or mixed-sex members?' The phrasing of this question erases the presence of any non-cisgender groups. In the context of industrialized nations, queer farmer networkers offer meaningful support (Wypler, 2019). Without including similar types of groups in the pro-WEAI context, this mechanism for empowerment among gender minorities cannot be fully measured and understood.

On page 21 of the 25-page survey, a set of questions measuring intrahousehold relationships includes a question about the 'most senior co-wife (the person who was in the household just before you, or, if you are the most senior wife, the one who married into the household after you).' This set of questions inquiries about respect, trust, and decision-making between the respondent and the senior co-wife (as well as other members of the household). As Bryan and Garner (2022) note, in polygynous households, the availability of land is impacted by the presence of co-wives or 'rival' wives and thereby can dramatically shape the questions that occur earlier in this survey, such as access to certain agricultural production. By failing to capture robust data, including non-heteronormative household models, pro-WEAI both reifies the colonial household model and fails to capture the most accurate information on gender minorities' empowerment.

These critiques of pro-WEAI should be expanded and brought to bear in all surveys seeking to capture women's involvement in global agricultural production. Although the efforts to improve women's empowerment in agriculture through pro-WEAI and other approaches offer continuously needed efforts to confront poverty, food insecurity, and gender-based violence, we urge researchers and practitioners to consider and include additional gender minorities such as transgender, non-binary, third gender, and others in this research. Doing so is crucial to eradicating gender oppression and inequities in all forms.

3 Queer farmers: existing research

In this section, we review the emerging debates and research in academia that begin to recognize the role of sexuality in agrifood systems. Agricultural practices are gendered and sexualized. Practices come with rules on whom is to produce, where, how, and what. At the same time, through agriculture itself, gender, sexual, and farming identities are constructed (Leslie et al., 2019; Pfammatter and Jongerden, 2023). Researchers have started to question the dominant agricultural model, constructed around heterosexuality and a gender binary, thereby unveiling the sexual nature of agrifood relations (Little, 2003; Rosenberg, 2016; Leslie et al., 2019). Applying a queer lens in agriculture has led to the recognition of how the cisheterosexual couple is the basis of the traditional family farm, and at the same time, the family farm upholds heterosexual norms and binary gender (Hoffelmeyer, 2020; Leslie et al., 2019). Many queer farmers perform gender and sexuality away from dominant binary and heteronormative framings, thus showing how agriculture is inherently 'relational', involving intimate (sexual) and working connections (Leslie et al., 2019). Further, queer farmers create new models by establishing intimate and working ties that question gender-sex-sexual norms (Pfammatter and Jongerden, 2023). In doing so, queer farmers demonstrate how agriculture can produce and allow more than two genders (Pfammatter and Jongerden, 2023).

Deviating from the heteropatriarchal norm, however, does not come without consequences. Queer farmers face effective discrimination from programs, policies, forms, and actions that require urgent addressing. For queer farmers, discrimination can manifest in various forms, from overt to more covert institutionalized oppressions, such as the promotion of the traditional (white, heterosexual, cisgender) family farm in and through policies, media, and educational programs (Hoffelmeyer, 2021; Hoffelmeyer et al., 2023 Leslie, 2019, 2023; Pfammatter and Jongerden, 2023). Heteropatriarchal dominance impacts farmers in numerous ways as queer farmers struggle to access land, financial assets, and agricultural knowledge and ensure farm viability (Hoffelmeyer et al., 2023; Leslie, 2019; Wypler, 2019). More than that, heteropatriarchal relations and structures have adverse effects on the psychological and physical well-being of queer farming communities. These communities also experience barriers to accessing adequate healthcare due to stigmatization and mistrust in the healthcare system (Wypler and Hoffelmeyer, 2020). Moreover, queer communities face disadvantages not only as food producers but also as consumers. For instance, Leslie et al. (2022) illustrate how food insecurity disproportionately affects LGBTQ+ groups in the New England region of the United States, highlighting the intersectional dimensions of heterosexism in the agrifood system.

Despite the negative outcomes of heteropatriarchal dominance in agriculture for queer farmers, these farmers do exist. More than that, on their farms, queer people manage to construct a space where gender is not performed and/or is performed differently. In doing so, queer farmers first uncover the performative nature of gender; second, queer farmers illuminate ways to farm away from binary gender and heterosexual relations and the strong hierarchies and inequalities that often accompany them (Pfammatter and Jongerden, 2023). By recognizing queer farmers as actors, learning from their adaptive capability, and including these farmers in decision-making processes, scholars and practitioners can learn from queer farmers to develop more resilient and sustainable rural areas and food systems (Hoffelmeyer, 2021; Leslie et al., 2019; Pakin-Albayrakoğlu, 2022; Pfammatter and Jongerden, 2023).

Queer farmers adopt various strategies to gain access to land, financial assets, and agricultural knowledge and ensure farm viability within a heteropatriarchal context. They form networks with other queer individuals, seek out queer-friendly working environments, and, at times, establish non-traditional households to build queer communities (Leslie, 2019; Wypler, 2019). Selective outing is another tactic employed by queer farmers to navigate heterosexism, revealing their identity only in safe and appropriate contexts (Hoffelmeyer, 2021). However, this approach has limitations, as not all queer farmers can easily pass as cisgender and heterosexual, particularly non-binary and transgender individuals who, at times, are more readily recognized as queer (Hoffelmeyer, 2021). Lesbian farmers, as a part of marginalized gender and sexuality, negotiate their way through heteropatriarchal farming environments by 'proving' themselves as farmers (Cramer, 2020; Pfammatter and Jongerden, 2023). This matches the literature on women in male-dominated professions, with the added layer in the case of lesbians: not only are they operating in a marginalized gender, but also operating in a marginalized sexuality. A lesbian farmer illustrates that establishing herself as a farmer is crucial not only for gaining community acceptance but also for validating her right to marry a woman and have children (Cramer, 2020). Being a hardworking contributor to the rural community thus justifies women's existence as rural producers, but also lesbians' existence as reproducers: whereas their marriage and childbearing practices had previously led to gossip and blame, through hard work, they compensate for their deviant sexuality and gain back their rights. To overcome the limitations of conventional productivist agriculture, queer farmers often turn to alternative agriculture (Hoffelmeyer, 2021; Leslie, 2017). This parallels the circumstances faced by women farmers in industrialized nations, where alternative agricultural approaches are more within reach due to gender inequities compared to the resource-intensive methods of industrialized farming (Sachs et al., 2016).

The literature reviewed primarily focuses on the industrialized countries, specifically the United States and Switzerland, and examines the challenges faced by queer individuals in farming. Even in these geographies, queer farmers are largely erased from censuses, surveys, and statistics (see Hoffelmeyer et al., 2023 for exception). Through complex coding methodology and access to non-publicly available data, Dentzman et al. (2021) find that same-sex married couples make up more than 1.2% of two-producer farms – nearly 23 000 farmers. Although this number is still likely a dramatic undercount of queer farmers in the United States. The case study of pro-WEIA and the lack of queer-inclusive data in agriculture in the United States and elsewhere calls for concrete action to eliminate heteronormative bias in forms, programs, policies, and outreach in all agriculture research.

Research in agriculture often fails to capture gender and sexuality or reifies the colonial/gender system. The handful of studies which engage with questions of queerness and heterosexism in the agrifood system demonstrate not only the need to examine heteronormativity in data instruments and collection but to see sexuality as a core organizational tool of oppression in agriculture – in need of analyzing and understanding – for efforts to equity to be advanced.

3.1 Case study: reexamining Switzerland's 'progress' toward queer inclusion

We discuss agricultural educational programs and farmland inheritance in Switzerland against the prevailing image of progressiveness associated with industrialized countries and their strides toward queer inclusivity. First, we illustrate how binary gender and compulsory heterosexuality are not only anchored in but factually codified through educational paths and socio-cultural ideals that offer little alternative to the cis-hetero family farm model. We then go on to show how policies sexualize and gender farmland acquisition patterns leading to queer farmers' exclusion. Furthermore, we use this case to challenge the oversimplified narrative that portrays certain regions as 'progressive' or 'backward' in their treatment of queer individuals. Through an examination of the effects of policy on queer farmers, this case highlights that strict regulations pertaining to gender roles and sexual relationships continue to persist.

Today, in Switzerland, farming practices remain largely heteropatriarchal. The dominant agricultural discourse and practice are implicitly yet manifestly organized around a cisgender heterosexual dyad (Contzen and Forney, 2017; Widmer, 2016), a cisgender man (the main farmer) and his wife (the farmwife, a cisgender woman). Traditional heterosexual dyads function according to particular 'social logics' composed of strong ideas of complementarity and family solidarity between husbands and wives, but also between generations (Amiotte-Suchet et al., 2017). These social logics are strongly gendered within and across generations. Land, farms, power, and the profession of the main farmer are handed down from man to man. Simultaneously, the unpaid reproductive work, precarity, and the 'helping role' from woman to woman. As a result, 93% of farms are owned and managed by men (Bundesamt für Landwirtschaft, 2022), while women's reproductive work is often re-defined as part of their natural unnegotiable responsibility. Women's productive work is constructed as 'helping'. The productive and reproductive work of women is accordingly left unpaid and unrecognized (Whatmore, 1991). The exclusion of gender minorities from positions of power in farming is also evident from the striking absence of women in leadership positions in agricultural organizations (Straub, 2019). In 150 years of history, neither the Federal Office of Agriculture (Moser, 2023) nor the Swiss Farmers' Union (Moser, 2023) have ever appointed a woman as director.

The heteropatriarchal dyad and the oppressive relations that often go with it are not invented on the farm; they are anchored in and simultaneously co-construct socio-cultural contexts and ideologies. These are exemplified by gendered and sexualized educational paths. In Switzerland, one can either study to become a farmer or a farmwife. The latter is constructed as a complementary yet subordinate profession to the former. This traditional view presumes that heterosexual male farmers will marry heterosexual female farmwives. As of 2019, no man had ever successfully completed the farmwife apprenticeship (Contzen, 2019). In 2021, the percentage of women who completed the farmers' school stood at 17%. Given the legal and practical acceptance of only two genders (women and men) and two sexes (female and male), no information is available outside of binary gender (The Federal Council, 2022). As such, no data is available on genderqueer persons in agricultural schools. Statistics on the sexualities of agricultural workers are not collected whatsoever.

Gender and heterosexuality in agriculture are not only educated but also legislated. The concept of a family farm is, in fact, anchored in the first article of Article 1 of the Federal Law on Peasant Land Law, the Bundesgesetz über das bäuerliche Bodenrecht (Art. 1 Abs. 1. lit. a, 1991) that emphasizes the importance of family farms for sustainable land use and a healthy farming community. The concept of the family farm lacks a formal and juristic definition; however, it is important to recognize that it is constructed in a practical and performative manner that effectively excludes queer individuals and communities. This exclusion is evident in various domains, including governmental policies, media representations, and the perceptions fostered by agricultural and environmental organizations. These influential institutions predominantly nearly exclusively portray family farms as the domain of white, heterosexual households with children, thereby reinforcing the notion of a normative, hegemonic family structure within the agricultural context. The same law package further regulates pre-emption rights. The agricultural pre-emption right is a legal right that gives family members the first option to buy agricultural land and businesses. This means that when owners of a farm and business want to sell, there is a legalized hierarchy that they are required to follow: first, the children may buy; then, under certain conditions, the siblings and the siblings' children of the transferring farm manager may purchase the farm or the land. The partner (husband or wife, yet mostly wives) is left empty handed. When, in 2022, the Federal Office of Agriculture tried to change the law to include the partner of the owner in the pre-emption rights, the Swiss Farmers' Union opposed the proposal because 'farms must be kept in the family' (Jäger, 2019). This has the effect of excluding the understanding of family that goes beyond blood ties. Here we are not only referring to the exclusion of the married partner (mostly cisgender heterosexual women) from inheritance paths but rather to the effective hindrance of queer inheritances.

First, 'coming out' is still often connected to cutting blood-family relations, whereby queer people might be rather excluded from farmland inheritance (Leslie, 2017, Pfammatter and Jongerden, 2023). Second, a queer perspective of family farms goes beyond blood and marriage ties, beyond the dyad, and beyond sexual relationships (Leslie, 2017, 2019; Leslie et al., 2019; Hoffelmeyer, 2021; Pfammatter and Jongerden, 2023). This raises the question of what a queer-inclusive inheritance policy might look like. Currently, some queer farmers that wish to keep their farms in queer hands are left with one fear and one hope when thinking about farm inheritance patterns. The fear that one of their traditional cis-heterosexual blood siblings, or their offspring, will develop an interest in farming. The hope is that the farm will be further led by queers and not fall back into traditional schemes (Pfammatter and Jongerden, 2023).

Despite queer exclusive policies and education, queer farmers in Switzerland manage to farm away from the heterosexual dyad, thus constructing alternative structures in and through which they challenge not only heteronormative expectations but also binary gender constructions and constrictions. Queer farmers who challenge traditional gender norms establish alternative queer norms on their farms. However, beyond their farm, gender becomes constrictive: constructed socially as a binary, naturalized, and then fixed as fact in policies and education. This leads to real-life consequences: advantages for those who adhere to the norms (such as heterosexual cisgender family farms), while those who experience gender binaries as restrictive and therefore challenge them – including queer farmers – face discrimination.

While queer farmers show us how gender can and is already effectively constructed otherwise, policies and educational paths are reluctant to follow. As the Swiss case illustrates, stringent regulation of gender and sexual relationships continues to occur in countries considered inclusive of 'queers' and globally marginalized groups. While countries like Switzerland receive praise, other countries, particularly in Africa and the Middle East, are analyzed, criticized, and labeled backward in their treatment of queer people. While many countries in, for example, Africa harshly penalize queerness (Human Rights Watch, 2022), countries like Switzerland maintain pernicious mechanisms for enforcing and disciplining gender and sexuality, especially in farming. To this end, sexuality research and queer-inclusive actions are desperately needed to understand and address heteropatriarchal policies and practices that underpin the agrifood systems in all geographies.

4 Future trends

In returning to the development of gender in agriculture research as compared to the much smaller and less mainstreamed sexuality in agriculture research, we offer several avenues for future research to connect these literatures and expand gender in agriculture research. Expansions and connections across these research areas are required to move toward more equitable and inclusive agrifood systems.

First, gender research among smallholders is necessarily sexuality research; however, without examining and documenting sexuality, gender research fails to address heterosexism and heteronormativity. Building on an understanding of colonization as implementing the colonial/modern gender system, we call for future research to elucidate how relations are not only gendered but also sexualized. For example, researchers and development practitioners attempted to bring women into economic reproduction, particularly to support empowerment (Chant, 2016). But oftentimes, their efforts failed to recognize that even when involved in economic production, women smallholders continued social reproductive labor, thereby increasing their labor responsibilities (Bain et al., 2018). The compulsory heterosexuality, in combination with gendered expectations, requires increased research to understand how the mechanisms of heterosexuality can be disentangled with patriarchal structures that disempower women and other gender(ed) minorities. By doing so, programs and policies will be better equipped to confront the production/reproduction, which, by and large, devalues women's labor in households and privileges economic reproduction. The role of sexuality in this is less clear because gender is considered the driving factor, but research and practitioners fail to see how gender is embedded in sexual relationships.

Further, we argue for the need to integrate the results of queer research, sexuality research, and gender performativity in all research, because gender construction and sexualities are not only relevant to those who bend, challenge, and deviate from the norm. We need to have a clear understanding of how gender is constructed, and how sexuality underpins agricultural practices throughout our research. We need to understand how a woman is constructed and how gender and sexual norms act in situated contexts. Sexuality and LGBTQ+-focused research often takes the position as a subgroup of women farmers' discussions, as if they were a special form of being a woman, thus reproducing the othering effect. Understanding sexuality and gender beyond binary framing does not mean including additional gender categories in one demographic question on gender and sexuality as often there are 'other(ing)' categories. Again, understanding gender and sexuality as constructed and performed, which implies giving more space to these categories to engage with them.

Second, many gender researchers and practitioners call for increased attention to intersectional oppression. Of course, considering sexuality another axis of intersectionality is one potential approach to deepening engagement with sexuality research in agriculture. Nevertheless, intersectionality is not the only approach to confronting the co-construction of oppression.

Intersectionality has become a cornerstone of women's studies' organizational programming and training (Nash, 2019). This widespread investment in studying, teaching, and signaling intersectionality is based on the idea that intersectionality is 'an unqualified ethical good' (Nash, 2019: 16), and more intersectionality will only provide greater ethical good. This institutionalization illustrates the process of conflating intersectionality with positive ethics.

While intersectionality has demonstrated theoretical and practical power in elucidating the relationship between power and oppression, we join a growing body of scholarship that offers some pause regarding the widespread calls for intersectionality (see Nash, 2019 for this discussion). We by no means discourage the use of intersectionality to better understand the experiences of smallholder women farmers and see rich opportunities to utilize intersectional approaches on this topic. In fact, leading Black Feminist scholars Nash (2019) and Hill Collins (2009) specifically call for intersectionality to be used globally among women of color. However, we highlight how a queer of color critique may, too, be useful in this area of study.

A queer of color critique examines how racialized (whiteness) and sexualized (heteropatriarchy) norms undergird gender, citizenship, and other social categories. Ferguson (2004) asserts that a queer of color critique can 'debunk the idea that race, class, gender, and sexuality are discrete formations, apparently insulated from one another' (4). A queer of color critique treats gender, race, class, and sexuality not as distinct social hierarchies but instead examines how normativity, particularly whiteness and heteronormativity, are infused into each category, thereby creating different forms of exclusion predicated on normativity. Intersectionality largely views oppression and privilege as static categories that can mitigate or enforce one another (Ferguson, 2004). However, queer of color critique examines how normativity is leveraged among those both privileged and oppressed. From this perspective, normativity and the ability to move toward normativity is a source of privilege (Ferguson, 2004).

Although Nash (2019) suggests a queer of color critique is 'sidestepping' (55) intersectionality by using some of the same ideas, namely the heterogeneity of racialized communities, we would argue that by examining the racialization – in particular whiteness – and heteronormativity infused into social categories, a queer of color critique expands intersectionality by calling attention to how racism/sexism/classism is not a result of different social locations but inherent in the very understanding of normativity and thus oppression. By privileging intersectionality as the core of Black Feminist works rather than a piece of this rich body of knowledge, this limits deep engagement with other tools, such as a queer of color critique. A queer of color critique and other approaches offer useful tools to interrogate heteronormativity further.

Finally, we also bring a perspective for which natural science research, in particular, may engage in sexuality research. Drawing from theorizing in Queer Ecology (Mortimer-Sandilands and Erikson, 2010), we offer questions about the gender binary, constructions of 'naturalness', and the human/more-than-human relationship. What if scholars and practitioners were to include animal and plant (projected) gender and sexuality in their research? How is the colonial/modern gender system used not only to oppress colonized human bodies, societies, countries, and cultures but also the more-than-human? This is a particularly interesting reflection on the context of agriculture, where binary gender and sexual normativity are projected and imposed, most notably on livestock, whereby the animal body is not only a negotiation site for gender relations but also a place where these can be challenged (Bull, 2016). How are scholars and practitioners leveraging the more-than-human by imposing the gender binary on farm animals, systematically killing gay bulls and intersex calves, and doing away with pollination on the farm by taking it to sterile standardized labs, where plant sexual reproduction is controlled and standardized (Bull, 2016)? Understanding how the modern/colonial gender model works on human and more-than-human bodies has the potential for an expansion of programs and policies beyond the concept of the patriarchal household structure. What can scholars and practitioners learn from those who bend, do away, and never took up normative colonizing concepts of gender, and sexuality?

5 Conclusion

As we have highlighted in the chapter, the majority of research engaging with queer farmers has occurred in the United States and Europe (see Leslie, 2017, 2019; Hoffelmeyer, 2020, 2021; Pfammatter and Jongerden, 2023; Wypler, 2019). It would be a mistake to use this research to suggest that these geographies are somehow 'more advanced' than other parts of the world where fewer academic discussions of LGBT+ people and heteronormativity are taking place (at least visibly) in agriculture. The painful irony remains that many of the places in which queerness is illegal and most harshly penalized (see map at Human Rights Watch, 2022) are places where colonialism sought to implement the modern/colonial gender system. The modern/colonial gender system, which directly ties gender and sexual subordination to colonial violence, means that any efforts to engage gender and sexuality minorities necessitate an anti-colonial approach. Part of this approach requires resistance to labeling cultures and communities as 'backward', 'ignorant', or 'primitive.'

Rather, expanding gender and sexuality studies in agriculture necessitates a deeper understanding of how colonialism and other structures of social stratification sought to marginalize gender and sexual minorities. Furthermore, how this marginalization shapes smallholders' ability to access land, food, knowledge, and social supports crucial for their livelihoods. We return to future research in the penultimate section. Nevertheless, we hope that by expressing this crucial link between colonialism and gender and sexual inferiorization, future research and theorizing may avoid/undo criticism of geographies that, at present, have not made socio-political shifts to support gender and sexual minorities.

The deeply rooted heteronormativity among individuals, organizations, and institutions fails to confront issues of cisheterosexism in the global agrifood systems. By first recognizing the legacy of colonialism in developing cisheterosexism in today's global society, then by viewing gendered inequalities as the products of sexual relationships, we have suggested future pathways for gender and agriculture research to bring sexuality to the forefront of this body of work. We suggest that scholars and practitioners have failed to recognize gender and sexually diverse participants as participants in smallholder or small-scale farming, thus effectively invisiblizing alternatives to cisgender heteronuclear farming practices. By understanding how cisheterosexism and heteronormativity are co-constructed with other inequalities, policies, programs, and research might better address inequities in agrifood systems.

6 Where to look for further information

- European Coordination Via Campesina. (2021). *Embracing rural diversity: Genders and sexualities in the peasant movement.*
- Hoffelmeyer, M., Sexsmith, K., & Glenna, L. (2024). Divergent approaches to the 'family farm': Celebrate, reform, or abolish? *Agriculture and Human Values*. https://doi.org/10.1007/s10460-024-10628-6
- Pfammatter, P. & Jongerden, J. (2023). Beyond farming women: Queering gender, work and family farms. *Agriculture and Human Values, 40*(4), 1639–1651. https://doi.org/10.1007/s10460-023-10449-z
- Raj, G. (2024). Selective, reciprocal and quiet: Lessons from rural queer empowerment in community-supported agriculture. *Agriculture and Human Values*. https://doi.org/10.1007/s10460-024-10552-9
- Gagliano, J. C. & Liebman, A. (2024). Trans*agrarian Marxisms?: On Unruly Matters, Rural Sexualities, and Problematic Campesinas. *TSQ: Transgender Studies Quarterly, 11*(2), 385–412. https://doi.org/10.1215/23289252-11215561
- Leslie, I. S., Wypler, J. & Bell, M. M. (2019). Relational Agriculture: Gender, Sexuality, and Sustainability in U.S. Farming. *Society & Natural Resources, 32*(8), 853–874. https://doi.org/10.1080/08941920.2019.1610626

7 References

Amiotte-Suchet, L., et al. (2017). *La petite entreprise au péril de la famille ? L'exemple de l'Arc jurassien franco-suisse*. Presses universitaires de Franche-Comté.

Art. 1 Abs. 1. lit. a. (1991). *Bundesgesetz vom 4. Oktober 1991* über das *bäuerliche Bodenrecht* (Bundesgesetz SR 211.412.11; BGBB). https://www.fedlex.admin.ch/eli/cc/1993/1410_1410_1410/de

Bain, C., Ransom, E., & Halimatusa'diyah, I. (2018). 'Weak winners' of Women's empowerment: The gendered effects of dairy livestock assets on time poverty in Uganda. *Journal of Rural Studies, 61*, 100–109. https://doi.org/10.1016/j.jrurstud.2018.03.004

Barker, M. (2007). Heteronormativity and the exclusion of bisexuality in psychology. In V. Clarke & E. Peel (Eds.), *Out in psychology: lesbian, gay, bisexual, trans and queer perspectives* (pp. 95–117). Chichester: John Wiley & Sons Ltd. https://doi.org/10.1002/9780470713099.ch6

Bryan, E., & Garner, E. (2022). Understanding the pathways to women's empowerment in Northern Ghana and the relationship with small-scale irrigation. *Agriculture and Human Values, 39*(3), 905-920. https://doi.org/10.1007/s10460-021-10291-1

Bull, J. (2016). Illdisciplined gender: nature/culture and transgressive encounters. In J. Bull & M. Fahlgren (Eds.), *Illdisciplined gender: engaging questions of nature/culture and transgressive encounters* (pp. 1–9). New York: Springer International Publishing. https://doi.org/10.1007/978-3-319-15272-1_1

Bundesamt für Landwirtschaft. (2022). *Agrarbericht*. Bern, Switzerland: Bundesamt für Landwirtschaft. https://www.agrarbericht.ch/de/betrieb/strukturen/betriebe

Chant, S. (2016). Women, girls, and world poverty: empowerment, equality or essentialism? International Development Planning Review, 38 (1). pp. 1-24. ISSN 1474-6743 DOI: 10.3828/idpr.2016.1

Contzen, S. (2019). Bäuerliche Familienbetriebe – eine Vielfalt an Arbeits- und Lebensformen. *Agrarwirtschaft und Agrarsoziologie*, *2019*, 19–43. https://doi.org/10.5281/ZENODO.3553655

Contzen, S. & Forney, J. (2017). Family farming and gendered division of labour on the move: a typology of farming-family configurations. *Agriculture and Human Values*, *34*(1), 27–40. https://doi.org/10.1007/s10460-016-9687-2

Cramer, S. (2020). Gaining ground: an exploration into the lives of Missouri's lesbian farmers. *Journal of Agriculture, Food Systems, and Community Development*, 1–16. https://doi.org/10.5304/jafscd.2020.093.017

Dentzman, K., et al. (2021). Queer Farmers in the 2017 US Census of Agriculture. *Society & Natural Resources*, *34*(2), 227–247. https://doi.org/10.1080/08941920.2020.1806421

Driskill, Q.-L., et al. (2011). *Queer indigenous studies: critical interventions in theory, politics, and literature*. Tucson: The University of Arizona Press.

Ferguson, R. A. (2004). *Aberrations in black: toward a queer of color critique*. Minneapolis: University of Minnesota Press.

Geneva, U. H. R. C. (39th sess : 2018 : (2018)). *United Nations Declaration on the Rights of Peasants and Other People Working in Rural Areas.* https://digitallibrary.un.org/record/1650694

Gunn, P. A. (1992). *The sacred hoop: recovering the feminine in American Indian traditions: with a new preface* (2nd ed.). Boston: Beacon Press.

Hawley, J. C. (Ed.). (2001). *Postcolonial, queer: Theoretical intersections*. Albany: SUNY.

Hill Collins, P. (2009). *Black feminist thought* (2nd ed.). New York: Routledge.

Hoffelmeyer, M. (2020). *Queer farmers*. New York: Routledge Handbooks Online. https://doi.org/10.4324/9780429199752-32

Hoffelmeyer, M. (2020). Queer farmers: sexuality on the farm. In C. Sachs, et al. (Eds.), *Gender and agriculture handbook* (pp. 348–359). New York: Routledge. https://doi.org/10.4324/9780429199752-32

Hoffelmeyer, M. (2021). "Out" on the farm: queer farmers manoeuvring heterosexism & visibility. *Rural Sociology*, ruso.12378. https://doi.org/10.1111/ruso.12378

Hoffelmeyer, M., et al. (2023). Surveying queer farmers: how heteropatriarchy affects farm viability and farmer wellbeing in U.S. agriculture. *Journal of Agriculture, Food Systems, and Community Development*, 1–15. https://doi.org/10.5304/jafscd.2023.123.005

Human Rights Watch (2022). *#OUTLAWED: "The love that dare not speak its name"*. https://features.hrw.org/features/features/lgbt_laws

International Food Policy Research Institute. (2022). weai.ifpri.info/files/2022/03/Pro-WEAI_main-module_01.03.2022_online.pdf.

Jäger, H. (2019, March 13). Kein Vorkaufsrecht für Partner. *Bauern Zeitung*. https://www.bauernzeitung.ch/artikel/kein-vorkaufsrecht-fuer-partner

Leslie, I. S. (2017). Queer farmers: sexuality and the transition to sustainable agriculture: queer farmers. *Rural Sociology*, *82*(4), 747–771. https://doi.org/10.1111/ruso.12153

Leslie, I. S. (2019). Queer farmland: land access strategies for small-scale agriculture. *Society & Natural Resources*, *32*(8), 928–946. https://doi.org/10.1080/08941920.2018.1561964

Leslie, I. S., et al. (2019). Relational agriculture: gender, sexuality, and sustainability in U.S. farming. *Society & Natural Resources*, *32*(8), 853–874. https://doi.org/10.1080/08941920.2019.1610626

Leslie, I. S., et al. (2022). LGBTQ+ food insufficiency in New England. *Agriculture and Human Values*. https://doi.org/10.1007/s10460-022-10403-5

Little, J. (2003). 'Riding the rural love train': heterosexuality and the rural community. *Sociologia Ruralis*, *43*(4), 401–417.

Lugones, M. (2007). Heterosexualism and the colonial / modern gender system. *Hypatia: A Journal of Feminist Philosophy*, *22*(1), 186–209. https://doi.org/10.2979/hyp.2007.22.1.186

Malapit, H. J. L. & Quisumbing, A. R. (2015). What dimensions of women's empowerment in agriculture matter for nutrition in Ghana? *Food Policy*, *52*, 54–63. https://doi.org/10.1016/j.foodpol.2015.02.003

Martinez, E. M., et al. (2021). The women's empowerment in agriculture index. In C. E. Sachs et al. (Eds.), *Routledge gender in agriculture handbook* (pp. 298–312). Abingdon: Routledge.

Mortimer-Sandilands, C. & Erickson, B. (2010). *Queer ecologies: sex, nature, politics, desire* (C. Mortimer-Sandilands & B. Erickson, Eds.). Bloomington: Indiana University Press.

Moser, P. (2023). *Schweizer Bauernverband, AfA110, Portal ARH People and institutions.* https://www.histoirerurale.ch/pers/personnes/Schweizer_Bauernverband,_AfA110.html

Nash, J. (2019). *Black feminism reimagined*. Durham: Duke University Press.

Oyěwùmí, O. (1997). *The invention of women: making an african sense of western gender discourses*. Minneapolis: University of Minnesota Press.

Pakin-Albayrakoğlu, E. (2022). Out and proud in the field: eco-queers for climate adaptation. *Peace Review*, *34*(1), 51–63. https://doi.org/10.1080/10402659.2022.2023429

Pfammatter, P. & Jongerden, J. (2023). Beyond farming women: queering gender, work and family farms. *Agriculture and Human Values*. https://doi.org/10.1007/s10460-023-10449-z

Rosenberg, G. (2016). *The 4–H -harvest: sexuality and the state in Rural America*. Philadelphia: University of Pennsylvania Press.

Sachs, C. E., et al. (2016). *The rise of women farmers and sustainable agriculture*. Iowa City: University of Iowa Press.

Seymour, G. (2017). Women's empowerment in agriculture: implications for technical efficiency in Rural Bangladesh. *Agricultural Economics*, *48*(4), 513–522. https://doi.org/10.1111/agec.12352

Straub, U. (2019). *Frauen in landwirtschaftlichen Organisationen*. Agroscope. https://2019.agrarbericht.ch/de/mensch/bauernfamilie/frauen-in-landwirtschaftlichen-organisationen

The Federal Council. (2022). *Einführung eines dritten Geschlechts oder Verzicht auf den Geschlechtseintrag im Personenstandsregister – Voraussetzungen und Auswirkungen auf die Rechtsordnung: Bericht des Bundesrates in Erfüllung der Postulate 17.4121 Arslan vom 13.12.2017 und 17.4185 Ruiz vom 14.12.2017* (p. 25) [Report of the Federal Council].

UNGA22. (2022). *Agriculture development, food security and nutrition: Resolution* (p. 15). New York: UN.

Whatmore, S. (1991). *Farming women: gender, work and family enterprise*. London: Macmillan.

Widmer, E. D. (2016). *Family configurations* (0 ed.). London: Routledge. https://doi.org/10.4324/9781315581903

Wypler, J. (2019). Lesbian and queer sustainable farmer networks in the midwest. *Society & Natural Resources*, *32*(8), 947–964. https://doi.org/10.1080/08941920.2019.1584834

Wypler, J. & Hoffelmeyer, M. (2020). LGBTQ+ farmer health in COVID-19. *Journal of Agromedicine*, *25*(4), 370–373. https://doi.org/10.1080/1059924X.2020.1814923

Chapter 10

Negotiating the intersection of gender and race in agriculture

Hannah Whitley[1], Independent Scholar, USA

1 Introduction

Scholars of food and agriculture have long documented the deep-rooted social, cultural, and political issues that plague global agricultural systems (Alkon and Agyeman 2011; Lobao and Meyer, 2001; Sachs et al., 2021). The experiences of women and smallholder farmers are not exempt from this systemic institutional and interpersonal discrimination. McKune and Touré (2021), Smith II (2019), and Tavenner and Crane (2019), for example, have all argued that an individual's (in) ability to meet agricultural production and profit goals is not merely a function of individual failures. Instead, the intersection of multiple levels of oppression manifests from a person's demographic characteristics, their self-identified expressions, and personal background. The shortcomings of agricultural support agencies and organizations to serve these individuals are evident in studies that document the effects of interpersonal and systematic inequities, as the agricultural operations struggling the most to establish and maintain sustainable practices are those led by members of socially disadvantaged

1 whitley.hannah93@gmail.com

http://dx.doi.org/10.19103/AS.2024.0148.12

groups[2] (Alkon and Agyeman, 2011; Doherty et al., 2023; Rodriguez et al., 2008). These findings parallel research in critical agri-food systems scholarship that has documented the differences in the productivity and profitability of smallholder farm operations in the Global South (Amekawa, 2009; Lowder et al., 2016; Vicol et al., 2019).

The representation and inclusion of women and people of color in global agri-food systems scholarship have historically been inadequate despite the well-documented disparities in production, profitability, program engagement, and operational sustainability. While multiple scholars have highlighted gender- (Henderson, 1995; Sachs et al., 2016; Sachs et al., 2021; Quisumbing et al., 2014) and race-related challenges (Bowens, 2015; Penniman, 2018; Reese, 2019; White, 2018) in agri-food systems, there is a noticeable lack of empirical research that explicitly examines the intersection of gender and race and its impact on individuals' lived experiences within agricultural spaces.

This chapter reviews the key issues and challenges to navigating the intersection of gender, race, and agriculture on a global scale. First, I describe the necessity of intersectional scholarship in contemporary agri-food systems research and argue that, for people who identify as members of a socially disadvantaged group, agricultural challenges are often amplified and compounded by multi-faceted cultural, historical, political, and socio-economic barriers. After providing a description of five major challenges related to the intersection of gender and race in agriculture identified within academic literature, I provide insight from a case study of women agriculturalists in Pittsburgh, Pennsylvania. Here, I demonstrate how agriculturalists' various socially constructed identities, including race, gender, and age, perpetuate barriers and opportunities for agriculture production and profitability. I conclude with a call for agricultural researchers, practitioners, and educators to reckon and engage with questions of intersectionality in relation to agri-food systems work and point toward additional scholarship and organizations whose work is at the forefront of these issues.

2 Contemporary agri-food systems research and the necessity of intersectional scholarship

Intersectionality has been identified as a theoretical and analytic concept that can help disentangle complex layers of identity and experiences that form the everyday lives of agricultural operators (Tavenner et al., 2022). Still, despite a contemporary interest in the paradigm shift toward intersectionality

2 The Consolidated Farm and Rural Development Act (1961) defines a "socially disadvantaged group" as one whose members "have been subject to racial, ethnic, or gender prejudice because of their identity as members of a group without regard to their individual qualities." USDA regulations further define socially disadvantaged farmers and ranchers as belonging to the following groups: American Indians or Alaskan Natives, Asians, Blacks or African Americans, Native Hawaiians or other Pacific Islanders, Hispanics, and women.

in domestic agricultural research, many studies of global agricultural systems ignore intersectional complications (Leder and Sachs, 2019; Smith II, 2019).

The roots of intersectionality and intersectional analyses lie in Black feminist scholarship and activism, as the concept considers how socially constructed identities are complexly related, overlapping, and produce different outcomes and experiences. Beverly Guy-Sheftall (1995) made one of the first arguments for intersectional analyses, writing that academics must reconceptualize social issues to include analyses of poverty, racism, imperialism, welfare, economic exploitation, etc. alongside concerns of gender discrimination. Black legal scholar, feminist theorist, and social activist Kimberlé Crenshaw created intersectionality theory to explain how a person's multiple identities intersect to create a whole experience that differs from human to human (1989). In other words, the overlapping of one's various socially constructed identities (e.g. race, ethnicity, gender, class, location, ability, and sexuality) is directly related to systems of power, privilege, societal oppression, domination, and discrimination. As Zambrana and Dill (2009) note, intersectionality deals with the "complexity rather than the singularity of human experience" (p. 2). It recognizes that individuals can occupy multiple social locations. In agri-food systems scholarship, a growing body of literature strives to recognize how intersecting identities complicate the barriers and opportunities for women and smallholder farmers. The intersection of one's gender and race are just one example of how intersecting socially constructed identities can shape a person's experiences in global agri-food systems.

Over the last decade, a growing number of peer-reviewed studies have applied intersectionality as a theoretical or analytic concept to empirical studies of global agri-food systems. Scholars have critiqued gender stereotypes and exploitation within crop production (Addinsall et al., 2023; Kimanthi et al., 2022; Luna, 2018) and extension and development programming (Tavenner et al., 2022; Witinok-Huber et al., 2021), led comparative analyses of intergenerational farm transfer dynamics (Akimowicz et al., 2022), and examined gender-species hierarchical arrangements in livestock-dependent cultures (Hovorka, 2012). Some have focused on the role of intersecting identities in the face of climate change and related adaptations to social-ecological change (Erwin et al., 2021; Mungai et al., 2017; Ravera et al., 2016). Many agri-food systems scholars have documented social (in)justices in urban agriculture spaces (Mackay, 2019; Motta, 2021; Reynolds, 2014; Young, 2019), within local food systems (Saville and Adams, 2021; Smith II, 2019) and amid complicated power dynamics (Meierotto et al., 2020; Tavenner and Crane, 2019). Others have taken a more focused look at the ways gender and race influence agricultural movements (White, 2011; White, 2018), experiences with discrimination (Nichols and Carter, 2023), and shape land ownership and farm operations (Horst and Marion, 2019; Pilgeram et al., 2022). These contemporary agri-food studies have raised awareness of the importance of intersectionality in identifying barriers and opportunities for farmers and growers.

3 Challenges for negotiating the intersection of gender and race in agriculture

As demonstrated above, there is a growing body of literature that strives to recognize how intersectional identities complicate the barriers and opportunities for agricultural production, profitability, and sustainability (Farhall and Rickards, 2021; McKune and Touré, 2021; Rybaczuk, 2023; Smith II, 2019; Tavenner et al., 2022; Tavenner and Crane, 2019). Globally, academic research has well-documented the benefits of participating in agricultural activities (Anderson et al., 2021; Ilieva et al., 2022) and the challenges unique to the experiences of women (Keller, 2014; Rissing, 2013; Trauger, 2004) and smallholder famers (Giordano et al., 2019; Von Loeper et al., 2016; Zerssa et al., 2021). Nevertheless, for agriculturalists who identify as members of a socially disadvantaged group, these challenges are often amplified and compounded by multi-faceted cultural, historical, political, and socio-economic barriers. My review of contemporary academic literature points to five primary challenges relating to the intersections of gender and race in agriculture. These include: (1) lack of access to and tenure on land, (2) limited access to and control over capital and resources, (3) limited agricultural background and restricted knowledge of technical and business skills, (4) lack of mentorship and feelings of isolation, and (5) the interplay between home, family, and the agricultural division of labor. The next five subsections detail each of these challenges.

3.1 Lack of access to and tenure on land

In order to farm, it is necessary to have access to land. Property, whether owned or rented, is essential to farming. However, property is not a commodity equally accessible to all. Globally, land access and tenure[3] is one of the most pressing challenges facing women and smallholder farmers. The challenge sits at a prime position at the intersection of gender and race in agriculture.

Though lack of access to and tenure on land for agricultural use is a challenge for all operators regardless of gender, women are differentially affected by patriarchal ownership norms and inheritance legacies. Globally, the majority of people who farm have not acquired their land through the market, but rather, through inheritance (Leonard et al., 2017). Inheritance, a closed system of intergenerational transfer of land within families, persists as land is passed on intact to one heir generation after generation (ibid.). Inheritance

3 Land tenure refers to 'the legal regime in which land is owned by an individual' (USAID 2018); who is said to "hold" the land. Land tenure determines who can use land, for how land, and under what conditions.

practices are shaped and negotiated by culturally accepted social norms, economic factors, and legal frameworks (Shortall, 1997, p. 112). Each of these aspects intersect to influence the ability of women and smallholder farmers to secure access to land. Shortall (1997) notes: "Economic considerations prompt impartible inheritance, that is, the passing on of the farm intact to one heir. Social norms dictate that the heir is male, and the legal system regulates the transfer of property" (p. 112). These inheritance patterns deny socially disadvantaged agriculturalists – women in particular – access to property and the ability to farm in their own right.

Along with inheritance practices, cultural traditions pose additional barriers that prevent women from inheriting and controlling land, crops, and animals on an equal basis with men (Adebisi and Monisola, 2012; Doss et al., 2011; Gamhewage et al., 2015; Hovorka et al., 2009; Maxwell et al., 1998). These culturally specific obstacles that prevent women from equitable control of land is complex and to a large extent context-specific. Still, barriers that prevent women's access to, control and use of land and other resources often include inadequate legal standards and/or ineffective implementation at national and local levels, as well as discriminatory socio-cultural attitudes and practices at both the institutional and community level (United Nations, 2013). These barriers have been observed with both rural and urban women farm operators (Fletcher and Kubik, 2016; Hovorka et al., 2009; Patil and Babus, 2008; Whitley, 2019).

Nevertheless, while systems of cultural tradition and inheritance complicate land access, so, too, does the legacy and practice of institutional discrimination against non-white farmers. This reality has been well-documented in the Global North and the US, in particular. In the US, decades of discrimination by the United States Department of Agriculture (USDA) (Graddy-Lovelace, 2016; Minkoff-Zern and Sloat, 2017; Van Sant et al., 2022), white supremacist violence (Billups, 2023; Williams, 2017; Woods, 1998), and legal exploitation of heir property (Baily and Thomson, 2022; Dyer and Bailey, 2009; Rivers, 2007) has led people of color to be almost entirely dispossessed of their land. In 1920, for example, 14% of all landowning US farmers were Black; today, approximately 1% of farms are operated by Black people, an estimated loss of over 14 million acres in 100 years (Penniman, 2018, p. 12). An array of farm loan legislation enacted over the years has contributed to the rapid decline in the number of Black farm acres and farm operators, Tyler and Moore (2013) argue.

Turning to institutional discrimination, Browne (2003) contended that the destruction of Black agriculture was in large part the work of "American governing institutions and the politics of institution building." For example, New Deal programs institutionalized commodity price support programs, which contributed to the concentration of land ownership and reduced opportunities for Black farmers. Fite (1984) maintains that these programs were run by and for relatively rich landowners and commercial farmers, most of whom were not Black farmers. Moreover, various scholars (Gilbert et al., 2002; Reid, 2007) contended that civil rights legislation failed to address Black farmers' needs as it did the needs of other farmers, mainly white farmers. By the late 1900s, legislation began to focus on righting the wrongs of the USDA discrimination practices toward minority farmers. The 2008 Farm Bill, for example, provided financial support to socially disadvantaged farmers and ranchers by mandating the creation of the USDA Office of Advocacy and Outreach to ensure equity in provision of funding and services (Tyler and Moore, 2013). Even so, the legacy and practice of discrimination within land access and tenure systems persist and add complications for navigating challenges related to gender and race in agriculture.

3.2 Limited access to and control over capital and resources

Women and smallholder farmers also struggle with restricted access to and control over capital and resources required to create and sustain profitable agricultural operations. One of the largest limitations to financial capital concerns inadequate access to farm credit and loans (Dorward et al., 2013; Gamhewage et al., 2015; Meinzen-Dick et al., 2014; Pattnaik et al., 2018; Sebata et al., 2014). Without adequate access to credit or loans, many women and smallholder farmers – especially farmers of color and those from the Global South – cannot purchase land, animals, seed, equipment, or other operational necessities, which restricts productivity and profitability, further deepening global agricultural inequities (Carter, 2019; Ransom and Stagner, 2019).

Access to and control over resources pose significant constraints on the operational capacity of women and smallholder farmers, shaping the foundation of their economic agency and productivity. This constraint has garnered substantial attention from global agri-food scholars. Wilbers (2003), for example, highlights two pivotal concerns: (1) the multifaceted challenges related to gaining access to and exercising control over productive resources and (2) the intricate task of managing the benefits derived from production. These concerns reflect the complex interplay of power dynamics, gender norms, and economic systems that shape the lives of women engaged in agricultural activities.

Maxwell et al. (1998, p. 2), moreover, further document challenges for smallholder farmers, describing how many women, frequently subject to economic dependence on their husbands, retain a portion of their earnings from produce sales in secrecy, thereby securing a semblance of financial autonomy and establishing a personal financial reserve. This covert action speaks to the agency exercised by women in response to their limited control over resources, underscoring their resourcefulness and determination to navigate within an often-constrained environment. The practice illuminated by Maxwell et al. (1998) resonates within a broader discourse on resource distribution and gender dynamics, highlighting the subtle negotiations and strategies that women employ to bolster their economic resilience in agricultural spaces (Sachs et al., 2016). This nuanced perspective sheds light on the intersection of economic forces, social norms, and individual agency and emphasizes the multifaceted nature of women's roles as economic actors within global agricultural spaces.

Other scholars have sought to clarify the intricate interplay of external factors – such as ideological, cultural, and economic influences – which frequently underlie the allocation of resources within households (Adesugba et al., 2019), as well as across governmental and non-governmental entities (Alkon and Agyeman, 2011; Baily and Thomson, 2022). As noted above, traditions rooted in patrilineal property inheritance traditions not only curtail women's land access but also reinforce male-centric perceptions that hinder their participation as proprietors and overseers of financial and built assets (Shortall, 1997). This, in turn, limits their access to vital financial support systems and lending services (Wilbers, 2003). The subsequent restricted access to and inadequate control over capital and resources substantially curtails the agricultural prowess of women, leading to constraints in terms of profitability, productivity, and the capacity to amass resources.

3.3 Limited agricultural background and restricted knowledge of technical and business skills

As demonstrated in the literature above, gendered and racialized assumptions that women farmers and farmers of color are less serious about agriculture or less dependent upon its profits have led to difficulties for farmers and growers in accessing land, financial capital, and needed equipment, along with technical assistance and business skills. One such example is demonstrated by Carter's (2021) account of gender-based discrimination by the USDA exposed in the 2000 *Love v Vilsack* case. Concurrently, cases of racial discrimination lodged by Latinx, Indigenous, and Black farmers against the USDA further compounded concerns. These collective grievances spurred the USDA and US Department of Justice to initiate a voluntary claims process that culminated in a $1.33 billion resolution in 2013 (as seen in Feder and Cowan, 2008). Still, despite financial compensation for some claimants, limited agricultural background and resulting constrained knowledge of technical and business skills continues to privilege white men in the global agricultural realm.

The challenges faced by women and farmers of color in the agri-food sector are further exacerbated by a lack of accessible extension services and technical support. These services are pivotal in providing valuable information, training, and guidance to farmers, allowing them to adopt improved agricultural practices and technologies that enhance productivity and sustainability (Carter, 2019; Sachs et al., 2016). Women, especially those from historically marginalized racial and ethnic backgrounds, may have limited access to agricultural education and training programs. This can hinder their ability to acquire the necessary skills and knowledge needed for participation in the agri-food sector, thus perpetuating the prevalence of primary farm operators who identify as white men. However, for people to fully benefit from extension services and other educational programming, they must be accurately identified as farmers or growers. Sadly, many institutions still fail to recognize women as active participants in agriculture, perpetuating a gender bias that obstructs women's access to crucial resources and opportunities (Farhall and Rickards, 2021). This omission not only undermines the potential of women as agents of change in the sector but also perpetuates gender disparities that hinder the overall progress of the agri-food domain.

Extension information is commonly provided through community meetings, community-based organizations, producer associations, and cooperatives (Barbercheck et al., 2009, p. 231). However, women and people of color may encounter significant barriers when attempting to join white, male-majority cooperatives, leading to their limited representation within these groups (Carter, 2019). Moreover, when extension services are delivered through such existing structures, women and people of color often face unequal treatment and reduced engagement opportunities (Manfre et al., 2013). Barbercheck et al. (2020) elaborates on how the persistence of unequal gender norms and relations contributes to a lower socio-economic status for women compared to men, further limiting their access and participation in formal groups (p. 231). These gendered dynamics exacerbate the challenges faced by historically disadvantaged farmers, particularly in terms of equitable access to valuable information and resources. The restricted ownership of land, control over credit, and access to vital information further exacerbates this disadvantage, constraining their ability to fully benefit from the information and services provided to farmer groups.

The inadequate attention to the intersectionality of gender and race in agricultural policies and programs can continue to perpetuate inequities. There is a pressing need for targeted interventions that comprehensively address the specific needs and challenges faced by women and women of color in the agri-food sector. Policies and programs that acknowledge and address the compounding effects of gender and racial discrimination can play a crucial role in dismantling systemic barriers and promoting inclusivity. For example, to navigate these challenges, some agriculturalists have turned to create alternative networks, including those formed by sustainable agricultural women (Sachs et al., 2016; Trauger, 2004), Black farmers (Taylor, 2018; White,

2018), and queer farmers (Hoffelmeyer, 2021; Leslie, 2017; Wypler, 2018). Carter (2019) describes how some organizations prioritize skill-based training to get socially disadvantaged farmers "up to speed" to be competitive within the industrialized food system and, it is implied, with white men (p. 277). Other support organizations are explicitly political in their prioritization of agricultural equity and systems change (White, 2018). For these organizations, gender- and race-based inequity in agriculture is not symptomatic of skills or knowledge that historically disadvantaged farmers and growers need to develop, but part of the larger systemic intersecting inequalities of racism, sexism, and classism maintaining the current agricultural system. The women and smallholder farmers who guide these organizations undertake their roles while confronting both historical and ongoing systemic invisibility (Sachs, 1983) and persistent vulnerability (Kruzic, 2017).

3.4 Lack of mentorship and feelings of isolation

Mentorship is pivotal in professional development across all sectors, providing guidance, support, and access to networks that facilitate career advancement. In agricultural spaces, however, mentorship opportunities for women and people of color remain disproportionately limited. This scarcity can be attributed to various factors, including historical gender and racial biases (Sachs et al., 2016), implicit stereotypes (Ramirez, 2015), and institutional structures that perpetuate social inequities (Tavenner et al., 2022). As a result, women agriculturalists, particularly those belonging to historically and systematically excluded demographic groups, often navigate their careers without mentors' guidance and advocacy.

Farming is often seen as an isolated industry, and increasingly so over the last two decades (Budge and Shortall, 2023). The absence of meaningful mentorship has significant implications for the sense of isolation experienced by women and smallholder farmers. Isolation can manifest in various ways, ranging from exclusion during professional gatherings to a lack of representation in leadership positions (Sachs et al., 2016, p. 57). Without mentors who can provide insights into navigating the complex dynamics of the industry, farmers and growers may struggle to identify pathways for career growth, further exacerbating feelings of isolation (Parent, 2012). The resultant sense of being "alone" in their professional journeys can hinder confidence, diminish job satisfaction, and even drive people away from agricultural endeavors (ibid.).

Feelings of isolation among farmers and growers were exacerbated by the COVID-19 pandemic. Rose and colleagues (2023) describe how, during the pandemic, agricultural shows were canceled and marts and sales of crops and livestock moved online, thus eliminating critical social hubs for farmers and growers. At the same time, household and caring responsibilities increased –

especially for women (ibid.) – as university students returned home and younger children were home-schooled. The impacts of COVID-19 on women and smallholder farmers' mental health and well-being became evident as the pandemic disrupted established routines, strained support networks, and accentuated the challenges of balancing agricultural work with domestic demands (Castellanos et al., 2022). The confluence of these factors underscores the urgency of addressing the structural barriers that contribute to isolation and its associated toll on the livelihoods of marginalized agricultural communities.

It is crucial to acknowledge that the experiences of isolation and the lack of mentorship are compounded for women of color in agriculture due to the intersectionality of their identities. These individuals face gender-based discrimination and racial biases that create unique barriers to their career progression (Whitley, 2019). Their experiences may include microaggressions, cultural misunderstandings, and limited access to networks that could facilitate mentorship relationships (Bowens, 2015; Penniman, 2018). As such, it is essential to recognize that addressing the issue of mentorship deficit requires a nuanced understanding of how multiple dimensions of identity interact to shape professional experiences.

3.5 Interplay between home, family, and the agricultural division of labor

Empirical studies examining household structures, family responsibilities, and their relationship to the gendered effects of agriculture are well-documented (Adesugba et al., 2019). While the majority of academic scholarship on household structures, "women's work," and the division of labor in agricultural activities have been primarily conducted in rural settings with predominantly white study participants, recent empirical studies have added a more racialized perspective to the discussion. For example, recent research by Rissing and colleagues (2021), Shisler and Sbicca (2018), and Smyth and colleagues (2019) has illuminated the profound impact of juggling childcare, caring responsibilities, and non-farming employment on the mental well-being of women in agriculture. The multifaceted nature of these roles places an overwhelming burden on women, magnifying the challenges they face and taking a toll on their mental health (Budge and Shortall, 2023; Gregoire, 2002; Yazd et al., 2019). As women navigate the intricate interplay of their roles as caregivers and agriculturalists, the accumulation of these demands amplifies stressors and underscores the imperative for targeted interventions that acknowledge and address the intersectional challenges that women face in agriculture confront.

Overwhelmingly, the division of labor within agricultural contexts has traditionally been influenced by gender norms, roles, and expectations (Flora,

2014; Pyburn and Kruijssen, 2019). Historically, such norms have confined women to domestic roles while assigning men primary responsibility for agricultural production. While these norms persist to varying degrees, shifts in societal perceptions of gender roles have prompted changes in the division of labor (Flora, 2014). Women's increasing participation in agriculture has led to evolving dynamics in households where women, specifically women of color, increasingly identify as farm operators (Whitley and Brasier, 2019).

Still, the labor of women and women of color within agricultural spaces persists in facing devaluation and underestimation, often being regarded as less significant when juxtaposed with operations led by men; here, white men, in particular, experience a distinct advantage (Pilgeram et al., 2022). Despite women's vital contributions to agricultural production, their efforts are frequently marginalized, and their expertise is undermined. This pattern of belittlement perpetuates gender disparities and emphasizes the need for concerted efforts to challenge deeply ingrained biases that hinder equitable recognition and representation within the agricultural domain.

4 Case study: women agriculturalists in Pittsburgh, Pennsylvania, USA

To illustrate the intersection of gender and race in agriculture and the resulting challenges for agricultural productivity, profitability, and sustainability, I offer a case study of women agriculturalists in Pittsburgh, Pennsylvania, USA.

4.1 Case study background

Located at the confluence of the Allegheny, Monongahela, and Ohio rivers in western Pennsylvania, Pittsburgh is the second-most populous city in Pennsylvania with a population of almost 303 000 as of the 2020 census (U.S. Census Bureau, 2020). Once a center for heavy industry, the post-World War II deterioration of traditional manufacturing led to the collapse of Pittsburgh's industrial core. Though Steel City continues to recover from industrial collapse and the subsequent Great Recession of 2008, Pittsburgh's local government and non-governmental organizations (NGOs) have prioritized local agricultural operations as a key component of citywide sustainability and community development initiatives. In the last 10 years, Pittsburgh has developed neighborhood-based food trust and food security programs, established food policy recommendation councils, and has witnessed the creation of multiple coalitions of urban farmers and gardeners who operate within the city. As of 2014, Pittsburgh's metropolitan statistical area boasted almost 8 000 farms, comprising over 908 000 acres of farmland (Rogus and Dimitri, 2014, p. 4). The city's commitment to sustainable agriculture and the improvement of community-based food systems is witnessed through policy and programming supported by NGOs like the Black Urban Growers and Farmers of Pittsburgh Cooperative, Grow Pittsburgh, Homegrown, Pittsburgh City Council, and the Pittsburgh Food Policy Council. With the support of its many collaborators, the City of Pittsburgh is considered by some scholars to be at the forefront of neighborhood revitalization and sustainable development efforts made via green space incorporation (Barreiro, 2012; Meenar et al., 2017) and has been recognized as one of the national leaders in urban agriculture policy (Haywood, 2017).

In 2019, I conducted six (6) months of ethnographic fieldwork with women farmers, growers, food activists, and advocates in Pittsburgh as part of a broader qualitative thesis that examines how dominant social hierarchies manifest within urban agriculture operations and the institutional, community, and regional organizations that support them (Whitley, 2019). During fieldwork, I collected qualitative data with 36 women urban agriculturalists who varied in age and racial and ethnic identities. The participant sample for this study includes three categories of urban agriculturalists: education and/or extension experts, urban agriculture NGO representatives, and Pittsburgh residents who identified as urban farmers, gardeners, and/or growers. Of the 36 participants, 17 people identified as Black or African American, 16 people identified as white, 2 people identified as Latina, and 1 person identified as Black and Indigenous. Participants ranged in age from 20 to 74 years, with the median age being 34–44 years.

4.2 Case study findings

Findings from this case study illustrate how intersectional the challenges related to agricultural production can be. First, the frequency of women working in Pittsburgh's agriculture spaces was a communal observation made by study participants. "I think the food systems world in Pittsburgh is dominated by women," explained one grower. "It's mostly white women who are working [in the space], which is problematic for many people who are deciding whether they want to get involved or not." Some participants attributed this high number to a correlation of women – especially white women – working in the 'female-dominated' non-profit field. Others believe that the high number of women working in urban agriculture is due to the 'natural correlation' between growing food and nurturing a family.

Because of the large number of women working in Pittsburgh's agriculture scene, many participants felt an absence of gender-specific discrimination, or, as one urban grower described, the "kind of gender-related professional barriers experienced in non-ag jobs." Black women working in for-profit urban agriculture, however, expressed feelings of isolation and marginalization in their food systems work. In Pittsburgh, for-profit urban agriculture operations are typically owned and operated by people who are "very white, a lot more male." Gendered and racialized differences in production spaces were consistently prevalent when comparing the experiences of growing food in non-profit versus for-profit spaces. One respondent clarified, "pretty much all the people working in production spaces that aren't working for a non-profit, are white men." Though many Black women are motivated to change the narrative of food production, they often stand with few individuals who have similar backgrounds and look like they do.

For many, it is difficult to identify how gender has influenced their experience with urban agriculture, as the majority of struggles they face seem to derive from issues related to race and class, not gender. One grower and non-profit representative clarified: "I ultimately don't even recognize [discrimination based upon my] gender because I see a lack in race and ethnicity. It's because I can't afford things that other people can. It's because I'm Black, not because I'm a woman." Some participants described how they felt "at the bottom" because white women are assumed to have better access to loans, grants, acreage, and resource donations. One urban agriculture non-profit representative believes that her privilege as a light-skinned, educated Black woman has allowed her to exist in spaces where she had greater access to financial and built capital resources.

A glaring lack in the representation of farmers of color in public policy settings has also hindered urban agriculture work, which speaks to the pervasiveness of unrecognized and uncompensated invisible labor that Black women often feel required to maintain their positions and navigate majority-white spaces. Many urban agriculture NGO representatives who work and grow in Pittsburgh's majority Black neighborhoods described how the "experience barrier" (i.e. not having lived, worked, or socialized in Pittsburgh's historically Black neighborhoods) is one of the main difficulties they experience when applying for resources or working on policy committees.

Challenges extended to perceptions of agricultural "legitimacy." Some participants described frustration with their agricultural involvement often labeled as a hobby rather than a legitimate for-profit operation that benefits themselves and their greater community. Many described how resource allocators and policy officials considered their limited land access and participation in land rental programs "unserious endeavors." "People think we have to actually buy a plot to be taken seriously," one grower and urban agriculture NGO representative clarified. Unable to afford the physical purchase of land for their operations, many women of color believe their credibility is challenged, and their reputation is at stake in urban agriculture spaces. For many, the purchase of land seemed necessary to be considered a "real" farmer or grower.

Moreover, Pittsburgh's industrial history has restricted the ease and accessibility of land use throughout the city, which further complicates the effects of food apartheid and structural inequities in the city's urban agriculture community. For example, there are very few land parcels in Pittsburgh's historically Black neighborhoods that have not been affected by urban blight, a process of deterioration characterized by deteriorating and abandoned homes and buildings, vacant lots with trash, high weeds and grass, and often abandoned or vandalized cars. The bureaucratic red tape navigation required of land purchase in combination with limited financial capital possessed by many Black women is only amplified by the city's pattern of residential segregation and food apartheid. "It's more than just a food desert; it's more than just a food-insecure community," one grower and urban agriculture non-profit representative explained. "Food apartheid in Pittsburgh is systematic racism that was done on purpose to oppress the people. The lack of food here and the lack of land was done to this area to further oppress people of color who live in this community." Connections between Pittsburgh's redlining zoning policies and impacts on the sustainability of urban agriculture spaces are troublesome.

Despite the increase in city-wide gentrification, Pittsburgh's agricultural spaces remain heavily segregated in terms of what land has been deemed suitable for food production and, demographically, by race. The segregated landscape results from long-standing policies that have isolated Pittsburgh's communities of color, resulting in limited land access and tenure. This segregated landscape thus influences a grower's capacity to produce and affects the conditions under which they grow. Although there are organizations that provide education, financial, and social support to Pittsburgh's agriculturalists, this support is not enough to overcome these institutional barriers. Currently, the majority of support services available to Pittsburgh's growers address the basic needs of agriculture production: access to and remediation for water and soil and access to land. These organizations, programs, and services support operational sustainability's environmental and financial tenants; however, social sustainability support is severely lacking. A few coalitions, such as the Black Urban Growers and Farmers of Pittsburgh Cooperative, strive to address institutional inequity through food and agriculture. However, Pittsburgh's current city-sponsored programs do not directly address the social, political, and economic system barriers tied to urban agriculture operations. Absent programs and policies that aim to acknowledge the historic institutional inequity endemic to food systems and increase the agricultural participation of marginalized growers, cities such as Pittsburgh will continue to overlook how US agriculture is rooted in white privilege and inequitable resource allocation.

Frustrations with land access and tenure are compounded by Black women's limited inclusion in resource allocation processes, often amplified by inconsistent, inaccessible agricultural education and training opportunities.

Though many of Pittsburgh's agriculturalists possess traditional agricultural knowledge that has been passed on from friends or family members, lacking a certificate or degree, they are commonly excluded from formal mentorship, education, or training positions. One grower and urban agriculture NGO representative described how individuals possessing traditional knowledge are often left out of these programs and training opportunities. This application of intersectionality to challenges for urban agriculture operations illustrates how various socially constructed identities, including race, gender, and socio-economic status, perpetuate challenges for urban agriculturalists. These dynamics document how opportunities and constraints for engagement, productivity, and profitability are not just determined by one's ability to secure tenured land, identify and navigate support programs and organizations, and access funding, but also by societal perceptions of legitimacy and the historical trauma and invisible labor held by many Black women in agriculture.

4.3 Case study take-aways: implications for understanding the intersection of gender and race in agriculture

Findings from this case study, like that of critical agri-food scholars Bowens (2015), Cohen and Reynolds (2014), Hanna and Pikai (2000), Haywood (2017), Hovorka et al., 2009), Levkoe (2006), Meenar and Hoover (2012), Ramirez (2015), Reynolds and Cohen (2016), and Rosan and Pearsall (2018), demonstrate how women agriculturalists in Pittsburgh, Pennsylvania, experience a variety of gender- and race-related inequities that inhibit the sustainability of their operations. These findings are consistent with the literature of intersectional theory and demonstrate that growers' various socially constructed identities, including race, gender, and location, perpetuate the barriers and opportunities for agriculture production and profitability. All research participants identified as women, yet their additional socially constructed identities all contribute to differing experiences in urban agriculture. For example, white women and women of color both expressed concerns regarding representation and legitimacy. Many struggled with the anxiety that they were not taken seriously and that they were unable to find mentorship within their profession. For white women, these concerns were typically attributed to their gender; race was never an expressed reason for worry. For women of color, particularly African American and Black growers, however, the concern was not only that there is a lack in women mentors or that women are not taken seriously as agriculturalists; the concern was that there is a specific lack of ethnic and racial diversity in agriculture and that they were not taken seriously because they are women of color.

This case study also found that for some growers of color, race was frequently the most important socially constructed identity that posed a barrier to their experiences in urban agriculture. This stands in contrast to white growers whose gender identity was typically their largest perceived social barrier. This finding raises an important question about intersecting identities in agriculture: are there times when one socially constructed identity overrides the others? For the majority of respondents of color, the answer appears to be "Yes."

5 Conclusion

As illustrated in the above case study, within global agricultural systems, institutional racism and sexism are illustrated in study participants' struggles to navigate institutions and support organizations, access funding and education, acquire relevant education and information, and secure tenured land. A person's set of identities is always simultaneously at play; their race overlaps with gender, their gender with urbanity, their urbanity with race, and so on with added demographic and social identities. However, although one's gender remains consistently active in the matrix of one's identities, it may not be the most relevant in each moment or context (Lazar 2014). For example, in the case study highlighted above, many women felt that their race was a more significant barrier to sustainability than their gender. Wodak (2008) has proposed that "there are many situations where other identities *override* gender identities" (p. 194; emphasis original). This was the case with my sample of women agriculturalists in Pittsburgh. For many Black and Latina women, racial identity was foregrounded within the context of barriers to operation sustainability. For white growers, on the other hand, race was not always the most salient social barrier they experienced, gender, rather, was seen as the biggest obstacle.

These findings are important for our understanding of how intersectionality modifies experiences within global agri-food systems: structural inequalities related to agriculturalists' intersecting identities influence potential for operational equity. However, there may be situations in which specific identities, such as race, override impacts of other identities, like gender. Implications for this finding demonstrate why it is not enough for agri-food scholars to recognize and ameliorate gender- or race-specific barriers; we must identify how identities overlap and stratify experiences in agriculture.

The concept of food apartheid further illuminates these structural inequities, particularly in urban settings like Pittsburgh. Unlike the term "food desert," which suggests a natural absence of resources, food apartheid explicitly addresses the deliberate, systemic inequalities created by racist policies and practices. As illustrated in the above case study, Pittsburgh's industrial history has restricted land use, complicating the effects of food apartheid and its impact on local food systems. In this case, historically Black neighborhoods face urban blight and bureaucratic challenges in land acquisition, exacerbating food insecurity and residents' ability to grow their own food. Adopting the language of "food apartheid" rather than "food desert" calls attention to institutional inequities (namely, racist redlining and zoning policies) that hinder sustainable urban agriculture and maintain racial inequities.

Addressing these various challenges requires concerted efforts to promote equity in global agri-food systems. This includes implementing policies and programs that address discriminatory practices, provide equal access to resources and services, encourage and promote women and women of color's leadership and voices, and support their capacity-building and empowerment in agricultural activities. Harnessing the potential of women and smallholder farmers in global agriculture is not only a matter of gender and racial equity, but for achieving sustainable development, food security, and social justice in this era of raising global temperatures, increasing extreme weather events, and the changing climate. Navigating the intersection of gender and race is incredibly challenging, but its navigation is indispensable for the creation of resilient, equitable, and thriving agri-food systems that can address our contemporary and future challenges.

6 Where to look for further information

The following articles provide an overview of intersectionality within agri-food systems:

- Farhall, K. & Rickards, L. 2021, 'The "gender agenda" in agriculture for development and its (lack of) alignment with feminist scholarship,' *Frontiers in Sustainable Food Systems*, vol. 5, no. 1, pp. 1–15.
- Pilgeram, R., Dentzman, K., & Lewin, P. 2022, 'Women, race and place in US Agriculture,' *Agriculture and Human Values*, vol. 39, no. 1, pp. 1341–1355.
- Smith II, B.J. 2019, 'Food justice, intersectional agriculture, and the triple food movement,' *Agriculture and Human Values*, vol. 36, no. 3, pp. 825–835.
- Tavenner, K. & Crane, T.A. 2019, 'Beyond 'women and youth': applying intersectionality in agricultural research for development,' *Outlook on Agriculture*, vol. 48, no. 4, pp. 316–325.
- Tavenner, K. & Crane, T.A. 2022, 'Intersectionality in gender and agriculture: toward an applied research design,' *Gender, Technology and Development*, vol. 26, no. 3, pp. 385–403.

Key research on the intersection of gender and race in agriculture can be found at the following organizations:

- Agriculture, Food, and Human Values Society: https://afhvs.wildapricot.org/

- Black Urban Growers: https://www.blackurbangrowers.org/
- Rural Sociological Society: https://www.ruralsociology.org/
- CGIAR GENDER (Consultative Group on International Agricultural Research, Generating Evidence and New Directions for Equitable Results): https://gender.cgiar.org/

7 References

Addinsall, C, et al. 2023, 'Applying an intersectional lens to addressing gender disparities and disadvantage in rural Melanesian agriculture,' *Journal of International Development*, vol. 35, no. 7, pp.1574–1594.

Adebisi, A & Afolabi Manisola, T 2012, 'Motivations for women involvement in urban agriculture in Nigeria,' *Asian Economic and Social Security*, vol. 2, no. 3, pp. 337–343.

Adesugba, M, et al. 2019, 'Farm household livelihood strategies,' in *Routledge Handbook of Gender and Agriculture* by CE Sachs, L Jensen, P Castellanos, & K Sexsmith. New York: Routledge.

Akimowicz, M, et al. 2022, 'Toward agricultural intersectionality? farm intergenerational transfer at the fringe. A comparative analysis of the urban-influenced Ontario's Greenbelt, Canada and Toulouse InterSCoT, France,' *Frontiers in Sustainable Food Systems*, vol. 5. https://www.frontiersin.org/journals/sustainable-food-systems/articles/10.3389/fsufs.2021.642787/full

Alkon, AH & Agyeman, J 2011, *Cultivating Food Justice: Race, Class, and Sustainability*. Cambridge, MA: MIT Press.

Amekawa, Y 2009, 'Reflections on the growing influence of good agricultural practices in the Global South,' *Journal of Agricultural and Environmental Ethics*, vol. 22, no. 1, pp. 531–557.

Anderson, CL, et al. 2021, 'Economic benefits of empowering women in agriculture: assumptions and evidence,' *The Journal of Development Studies*, vol. 57, no. 2, pp. 193–208.

Baily, C & Thomson, R 2022, 'Heirs property, critical race theory, and reparations,' *Rural Sociology*, vol. 87, no. 4, pp. 1219–1243.

Barbercheck, Mary, Kathryn J. Brasier, Nancy Ellen Kiernan, Carolyn Sachs, Amy Trauger, Jill Findeis, Ann Stone, & Lisa Stewart Moist. 2009. 'Meeting the Extension Needs of Women Farmers: A Perspective from Pennsylvania.' *Journal of Extension* 47(3): 2–11.

Barbercheck, M., et al. 2020.'Gender and agricultural extension.' *Routledge Handbook of Gender and agriculture*. Routledge. 225–238.

Barreiro, L 2012, 'Rooftop gardening in an urban setting: impacts and implications,' Master's Thesis, Duquesne University. https://dsc.duq.edu/etd/278/.

Billups, WR 2023, 'The cost of civil rights: White supremacist violence and economic resistance against Koinonia farm during the civil rights era,' *Journal of Southern History*, vol. 89, no. 1, pp. 89–130.

Budge, H & Shortall, S 2023, 'Agriculture, COVID-19 and mental health: does gender matter?' *Sociologia Ruralis*, vol. 63, no. 1, pp. 82–94.

Bowens, N 2015, *The Color of Food: Stories of Race, Resilience, and Farming*. Gabriola Island: New Society Publishers.

Browne, WP 2003, 'Benign public policies, malignant consequences, and the demise of African American agriculture,' in *African American Life in the Rural South, 1900–1950* by R Douglas Hurt (ed.). Columbia, MO: University of Missouri Press.

Carter, A. 2019. 'We Don't Equal Even Just One Man: Gender and Social Control in Conservation Adoption.' *Society & Natural Resources* 32(8): 893–910. DOI: 10.1080/08941920.2019.1584657

Carter, A. L., and Adam A.2020.'Soul food:[Re] framing the African-American farming crisis using the culture-centered approach.' *Frontiers in communication* 5: 5.

Carter, A 2021, 'Women's farm organizations in the United States,' in *Routledge Handbook of Gender and Agriculture* by CE Sachs, L Jensen, P Castellanos, & K Sexsmith. New York: Routledge.

Castellanos, P, et al. 2022, *Gender, Food, and COVID-19: Global Stories of Hope and Harm*. New York: Routledge.

Cohen, N & Reynolds, K 2014, 'Urban agriculture policy making in New York's 'new political spaces,'' *Journal of Planning, Education, and Research,* vol. 34, no. 2, pp. 221–234.

Consolidated Farm and Rural Development Act 1961, 18-128 U.S.C. § 294.

Crenshaw, K 1989, 'Demarginalizing the intersection of race and sex: a Black feminist critique of antidiscrimination doctrine, feminist theory, and antiracist politics,' *University of Chicago Legal Forum,* vol. 1989, no. 8, pp. 57–80.

Doherty, FC, et al. 2023, ''The farm has an insatiable appetite': a food justice approach to understanding beginning farmer stress,' *Journal of Agriculture, Food Systems, and Community Development*, vol. 12, no. 3, pp. 69–92.

Dorward et al., 2013. The Economics of Urban Farming. In *The Urban Farming Guidebook* (pp.37–44). Chapter: Six Publisher: EcoDesign Resource Society.

Doss, C, et al. 2011, 'The role of women in agriculture,' Food and Agriculture Organization, viewed 25 August 2023, http://www.fao.org/docrep/013/am307e/am307e00.pdf.

Dyer, JF & Bailey, C 2009, 'A place to call home: cultural understandings of heir property among rural African Americans,' *Rural Sociology*, vol. 73, no. 9, pp. 317–338.

Erwin, A, et al. 2021, 'Intersectionality shapes adaptation to social-ecological change,' *World Development*, vol. 138, no. 1, pp. 105282.

Farhall, K & Rickards, L 2021, ' The "gender agenda" in agriculture for development and its (lack of) alignment with feminist scholarship,' *Frontiers in Sustainable Food Systems*, vol. 5, no. 1, pp. 1–15.

Feder, J., and Cowan, T.2008.'The pigford case: USDA settlement of a discrimination suit by black farmers.' *Congressional Research Service*, Library of Congress.

Fite, GC 1984, *Cotton Fields No More: Southern Agriculture, 1865-1980*. Lexington, KY: University Press of Kentucky.

Fletcher, AJ & Kubik, W (eds.) 2016, *Women in Agriculture Worldwide: Key Issues and Practical Approaches*. London: Routledge.

Flora, C. 2014. 'Agricultural Labor: Gender Issues.' *Encyclopedia of Agriculture and Food Systems*. 123–130.

Gamhewage, MI, et al. 2015, 'Women participation in urban agriculture and its influence on family economy – Sri Lankan experience,' *The Journal of Agricultural Sciences*, vol. 10, no. 3, pp. 192–206.

Gilbert J, et al. 2002, 'The loss and persistence of Black-owned farms and farmland: a review of the research literature and its implications,' *Southern Rural Sociology*, vol. 18, no. 2, pp. 1–30.

Giordano, M, et al. 2019, 'Water scarcity and challenges for smallholder agriculture,' in *Sustainable Food and Agriculture: An Integrated Approach* by C Campanhola & S Pandley (eds.). Academic Press. 75–94.

Graddy-Lovelace, G 2016, 'The coloniality of US agricultural policy: articulating agrarian (in)justice,' *The Journal of Peasant Studies*, vol. 44, no. 1, pp. 78–99.

Gregoire, A 2002, 'The mental health of farmers,' *Occupational Medicine*, vol. 52, no. 8, pp. 471–476.

Guy-Sheftall, B 1995, *Words of Fire: An Anthology of African-American Feminist Thought*. New York: The New Press.

Hanna, AK & Pikai, O. 2000, 'Rethinking urban poverty: a look at community gardens,' *Bulletin of Science, Technology & Society,* vol. 20, no. 3, pp. 207–216.

Haywood, K 2017, 'Urban agriculture: a guide for municipalities,' viewed 25 August 2023. https://iop.pitt.edu/sites/default/files/Reports/StatusReports/Urban%20Agriculture%20-%20A%20Guide%20for%20Municipalities.pdf.

Hendersen, HK (ed.) 1995, *Gender and Agricultural Development: Surveying the Field*. Tucson, AZ: University of Arizona Press.

Hoffelmeyer, M 2021, '"Out" on the farm: queer farmers maneuvering heterosexism and visibility,' *Rural Sociology*, vol. 86, no. 4, pp. 752–776.

Horst, M & Marion, A 2019, 'Racial, ethnic and gender inequities in farmland ownership and farming in the U.S.,' *Agriculture and Human Values*, vol. 36, no. 1, pp. 1–16.

Hovorka, A 2012, 'Women/chickens vs. men/cattle: insights on gender–species intersectionality,' *Geoforum,* vol. 43, no. 4, pp. 875–884.

Hovorka, A, et al. 2009, 'Women feeding cities: Mainstreaming gender in urban agriculture and food security,' RUAF Global Partnership on Sustainable Urban Agriculture and Food Systems. https://cgspace.cgiar.org/handle/10568/81070.

Ilieva, RT, et al. 2022, 'The socio-cultural benefits of urban agriculture: a review of the literature,' *Land*, vol. 11, no. 5, pp. 1–21.

Keller, JC 2014, '"I wanna have my own damn dairy farm!": women farmers, legibility, and femininities in rural Wisconsin, U.S.,' *Journal of Rural Social Sciences*, vol. 29, no. 1, pp. 75–102.

Kimanthi, H, et al. 2022, 'Exploring gender and intersectionality from an assemblage perspective in food crop cultivation: a case of the Millennium Villages Project implementation site in western Kenya,' *World Development*, vol. 159, no. 1, pp. 106052.

Lazar, MM 2014, 'Feminist critical discourse analysis: relevance for current gender and language research' in *The Handbook of Language, Gender, and Sexuality* (2nd ed.) by S Ehrlich, M Meyerhoff, and J Holmes (eds). Malden, MA: Blackwell Publishing.

Leder, S & Sachs, CE 2019 'Intersectionality at the gender–agriculture nexus: relational life histories and additive sex-disaggregated indices' in *Gender, Agriculture and Agrarian Transformations*. New York: Routledge. https://www.taylorfrancis.com/chapters/edit/10.4324/9780429427381-5/intersectionality-gender%E2%80%93agriculture-nexus-stephanie-leder-carolyn-sachs

Leonard, B, et al. 2017, 'Policy drivers of farm succession and inheritance,' *Land Use Policy*, vol. 61, no. 1, pp. 147–159.

Leslie, IS 2017, 'Queer farmers: sexuality and the transition to sustainable agriculture,' *Rural Sociology*, vol. 82, no. 4, pp. 747–771.

Levkoe, CZ 2006, 'Learning democracy through food justice movements,' *Agriculture and Human Values,* vol. 23, no. 1, pp. 89–98.

Lobao, LM & Meyer, K 2001, 'The great agricultural transition: crisis, change, and social consequences of 20th century U.S. farming,' *Annual Review of Sociology,* vol. 27, no. 1, pp. 103–124.

Lowder, SK, et al. 2016, 'The number, size, and distribution of farms, smallholder farms, and family farms worldwide,' *World Development*, vol. 87, no. 1, pp 16–29.

Luna, JK 2018, 'The chain of exploitation: intersectional inequalities, capital accumulation, and resistance in Burkina Faso's cotton sector,' *The Journal of Peasant Studies*, vol. 46, no. 7, pp. 1413–1434.

Mackay, H 2019, 'Food sources and access strategies in Ugandan secondary cities: an intersectional analysis,' *Environment & Urbanization*, vol. 31, no. 2, pp. 375–396.

Manfre, C., et al. 2013. 'Reducing the gender gap in agricultural extension and advisory services.' Modernizing Extension and Advisory Services Discussion Paper United States Agency for International Development.

Maxwell, D, et al. 1998, 'Farming in the shadow of the city: changes in land rights and livelihoods in Peri-Urban Accra,' Food Consumption and Nutrition Division, International Food Policy Research Institute, Washington, DC, viewed 25 August 2023. https://idl-bncidrc.dspacedirect.org/bitstream/handle/10625/22606/108520.pdf?sequence=11.

McKune, S, et al. 2021, 'Gender and intersectional analysis of livestock vaccine value chains in Kaffrine, Senegal,' *PLOS ONE*, vol. 16, no. 7. https://pubmed.ncbi.nlm.nih.gov/34197467/

Meenar, M, et al. 2017, 'Regulatory practices of urban agriculture: a connection to planning and policy,' *Journal of the American Planning Association* vol. 83, no. 4, pp. 389–403.

Meenar, MR & Hoover, BM 2012, 'Community food security via urban agriculture: Understanding people, place, economy, and accessibility from a food justice perspective,' *Journal of Agriculture, Food Systems, and Community Development*, vol. 3, no. 1, pp. 143–160.

Meierotto, L, et al. 2020, 'Isolation and fear of deportation: intersectional barriers to well-being among Latina farmworkers in Southwestern Idaho,' *Culture, Agriculture, Food and Environment*, vol. 42, no. 2, pp. 93–102.

Minkoff-Zern, LA & Sloat, S 2017, 'A new era of civil rights? Latino immigrant farmers and exclusion at the United States Department of Agriculture,' *Agriculture and Human Values*, vol. 34, no. 1, pp. 631–643.

Motta, R 2021 'Social movements as agents of change: fighting intersectional food inequalities, building food as webs of life,' *The Sociological Review*, vol. 69, no. 3, pp. 603–625.

Mungai, C, et al. 2017, 'Uptake of climate-smart agriculture through a gendered intersectionality lens: experiences from western Kenya' in *Climate Change Adaptation in Africa,* WL Filho et al. (eds.). New York: Springer.

Nichols, C & Carter A 2023, 'It doesn't bother me: an intersectional analysis of discrimination among white women farmers in the US Corn Belt,' *Journal of Rural Studies*, vol. 101, no. 1, pp. 103054.

Parent, D 2012, 'Social isolation among young farmers in Quebec, Canada,' viewed 25 August 2023, https://ifsa.boku.ac.at/cms/fileadmin/Proceeding2012/IFSA2012_WS3.2_Parent.pdf.

Patil, B & Babus, VS 2008, 'Role of women in agriculture,' *International Journal of Applied Research*, vol. 4, no. 12, pp. 109–114.

Pattnaik, I., et al. 2018. 'The feminization of agriculture or the feminization of agrarian distress? Tracking the trajectory of women in agriculture in India.' *Journal of the Asia Pacific Economy* 23.1: 138–155.

Penniman, L 2018, *Farming While Black: Soul Fire Farm's Practical Guide to Liberation on the Land*. White River Junction, VT: Chelsea Green Publishing.

Pilgeram, R, et al. 2022, 'Women, race and place in US Agriculture,' *Agriculture and Human Values*, vol. 39, no. 1, pp. 1341–1355.

Pyburn, R & Kruijssen, F 2019, 'Gender dynamics in agricultural value chain development: foundations and gaps,' in *Routledge Handbook of Gender and Agriculture* by CE Sachs, L Jensen, P Castellanos, & K Sexsmith. New York: Routledge.

Quisumbing, AR, et al. 2014, *Gender in Agriculture: Closing the Knowledge Gap*. New York: Springer.

Ramirez, MM 2015, 'The elusive inclusive: Black food geographies and racialized food spaces,' *Antipode* vol. 47, no. 3, pp. 748–769.

Ransom, E & Stagner, F 2019, 'Gender and livestock production,' in *Routledge Handbook of Gender and Agriculture* by CE Sachs, L Jensen, P Castellanos, & K Sexsmith. New York: Routledge.

Ravera, F, et al. 2016, 'The diversity of gendered adaptation strategies to climate change of Indian farmers: a feminist intersectional approach,' *Ambio*, vol. 45, no. 1, pp. 335–351.

Reese, A 2019, *Black Food Geographies: Race, Self-Reliance, and Food Access in Washington, D.C.* Chapel Hill, NC: University of North Carolina Press.

Reid, DA 2007, *Reaping a Greater Harvest: African Americans, the Extension Service, and Rural Reform in Jim Crow Texas*. College Station, TX: Texas A&M University Press.

Reynolds, K 2014, 'Disparity despite diversity: social injustice in New York City's urban agriculture system,' *Antipode*, vol. 47, no. 1, pp. 240–259.

Reynolds, K & Cohen, N 2016, *Beyond the Kale: Urban Agriculture and Social Justice Activism in New York City*. Athens, GA: University of Georgia Press.

Rissing, A 2013, 'Iowan women farmers' perspectives on alternative agriculture and gender,' *Journal of Agriculture, Food Systems, and Community Development*, vol. 3, no. 2, pp. 127–136.

Rissing, A, et al. 2021, 'The invisible labor and multidimensional impacts of negotiating childcare on farms,' *Agriculture and Human Values* vol. 38, no. 1, pp. 431–447.

Rivers, F 2007, 'Inequity in equity: the tragedy of tenancy in common for heirs' property owners facing partition in equity,' *Temple Political & Civil Rights Law Review*, vol. 17, no. 1, pp. 1–83.

Rodriguez, JM, et al. 2008, 'Barriers to adoption of sustainable agriculture practices: change agent perspectives,' *Renewable Agriculture and Food Systems*, vol. 24, no. 1, pp. 60–71.

Rogus, S & Dimitri, C 2014, 'Agriculture in urban and peri-urban areas in the United States: highlights from the 2012 census of agriculture,' *Renewable Agriculture and Food Systems*, vol. 30, no. 1, pp. 739–750.

Rosan, C & Pearsall, H 2018, *Growing a Sustainable City?: The Question of Urban Agriculture*. Toronto: University of Toronto Press.

Rose, DC, et al. 2023, 'Farming wellbeing through and beyond COVID-19: Stressors, gender differences and landscapes of support,' *Sociologia Ruralis*, vol. 63, no. 1, pp. 3–10.

Rybaczuk, R 2023, 'Who's your farmer? how race, class, gender and sexuality intersect in contemporary agriculture,' viewed 25 August 2023, https://scholarworks.umass.edu/dissertations_2/2775/.

Sachs, C. 1983. The invisible farmers. *Women in agricultural production*.

Sachs, CE, et al. 2016, *The Rise of Women Farmers and Sustainable Agriculture*. Iowa City, IA: University of Iowa Press.

Sachs, CE, et al. 2021, *Routledge Handbook of Gender and Agriculture*. New York: Routledge.

Saville, A & Adams AE 2021, 'Environmental justice in the American south: an analysis of black women farmworkers in Apopka, Florida,' *Agriculture and Human Values*, vol. 38, no. 1, pp. 193–204.

Sebata, N et al. 2014. Does Urban Agriculture help improve women`s resilience to poverty? Evidence from low- income generating women in Bulawayo. *IOSR Journal of Humanities and Social Science*. 19. 128–136. 10.9790/0837-1943128136.

Shisler, RC & Sbicca, J 2018, 'Agriculture as carework: the contradictions of performing femininity in a male-dominated occupation,' *Society & Natural Resources*, vol. 32, no. 8, pp. 875–892.

Shortall, S 1997, 'Women, farming, and access to land,' in *Iris Socheolaiochta nah Eireann*, Symposium: Women and the Land. *Irish journal of sociology*, 7, 1, 111–118.

Smith II, BJ 2019, 'Food justice, intersectional agriculture, and the triple food movement,' *Agriculture and Human Values*, vol. 36, no. 3, pp. 825–835.

Smyth, JD, et al. 2019, 'Women's work? the relationship between farmwork and gender self-perception,' *Rural Sociology*, vol. 83, no. 3, pp. 654–676.

Tavenner, K & Crane, TA 2019, 'Beyond 'women and youth': applying intersectionality in agricultural research for development,' *Outlook on Agriculture*, vol. 48, no. 4, pp. 316–325.

Tavenner, K & Crane, TA 2022, 'Intersectionality in gender and agriculture: toward an applied research design,' *Gender, Technology and Development*, vol. 26, no. 3, pp. 385–403.

Taylor, DE 2018, 'Black farmers in the USA and Michigan: longevity, empowerment, and food sovereignty,' *Journal of African American Studies*, vol. 22, no. 1, pp. 49–76.

Trauger, A 2004, ''Because they can do the work': women farmers in sustainable agriculture in Pennsylvania, USA,' *Gender, Place & Culture*, vol. 11, no. 2, pp. 289–307.

Tyler, SS & Moore, EA 2013, 'Plight of Black farmers in the context of USDA farm loan programs: a research agenda for the future,' *Professional Agricultural Workers Journal*, vol. 1, no. 1, pp. 1–11.

United Nations Human Rights 2013, 'Realizing women's rights to land and other productive resources,' viewed 25 August 2023. https://www.ohchr.org/sites/default/files/Documents/Publications/RealizingWomensRightstoLand.pdf.

U.S. Census Bureau 2020, 'QuickFacts, Pittsburgh city, Pennsylvania,' viewed 25 August 2023. https://www.census.gov/quickfacts/fact/table/pittsburghcitypennsylvania/PST045222.

USAID (United States Agency for International Development) 2018, 'Securing land tenure and property rights for stability and prosperity,' viewed 25 August 2023. https://www.usaid.gov/land-tenure.

Van Sant, L, et al. 2022, 'A 'cultural transformation' at the US Department of Agriculture?: Examining racial (in)equality through federal farmland protection programs in Georgia,' *The Journal of Peasant Studies*, vol. 50, no. 4, pp. 1636–1660.

Vicol, M, et al. 2019, 'Global production networks, regional development trajectories and smallholder livelihoods in the Global South,' *Journal of Economic Geography*, vol. 19, no. 4, pp. 973–993.

Von Loeper, W, et al. 2016, 'Analysing challenges facing smallholder farmers and conservation agriculture in South Africa: a system dynamics approach,' *South African Journal of Economic and Management Sciences*, vol. 19, no. 5, pp. 747–773.

Young, L, 2019, 'Growing food in the city: urban agriculture in Quito, Ecuador, through a feminist lens,' *Journal of Agriculture, Food Systems, and Community Development*, vol. 9, no. 1, pp. 151–158.

White, MM 2011, 'Sisters of the soil: urban gardening as resistance in Detroit,' *Race/Ethnicity: Multidisciplinary Global Context*, vol. 5, no. 1, pp. 13–28.

White, MM 2018, *Freedom Farmers: Agricultural Resistance and the Black Freedom Movement*. Chapel Hill, NC: The University of North Carolina Press.

Whitley, H 2019, 'Power, privilege, and 'Playing in the Dirt': an intersectional exploration of women's agricultural experiences in Pittsburgh, Pennsylvania.' Master's thesis, The Pennsylvania State University, viewed 25 August 2023. https://etda.libraries.psu.edu/fles/fnal_submissions/20398.

Whitley, H & Brasier, K 2019, 'Women farmers and women farmer's identities,' in *Routledge Handbook of Gender and Agriculture* by CE Sachs, L Jensen, P Castellanos, & K Sexsmith (eds.). New York: Routledge.

Wilbers, J. 2003. "Urban Agriculture and Gender: some key issues." Resource Centre for Urban Agriculture & Forestry (RUAF), Leusden, the Netherlands. Retrieved from https://www.ruaf.org/sites/default/files/econf4_submittedpapers_wilbers.pdf

Williams, B 2017, 'Articulating Agrarian racism: statistics and plantationist emprics,' *Southeastern Geographer*, vol. 57, no. 1, pp. 12–29.

Witinok-Huber, R, et al. 2021, 'Gender, place, and agricultural extension: a mixed-methods approach to understand farmer needs in Liberia,' *The Journal of Agricultural Education and Extension*, vol. 27, no. 4, pp. 553–572.

Wodak, R 2008, 'Controversial issues in feminist critical discourse analysis' in *Gender and Language Research Methodologies* by K Harrington, L Litosseliti, H Saunston, and J Sunderland (eds.). Basingstoke: Palgrave Macmillan.

Woods, C 1998, *Development Arrested: The Blues and Plantation Power in the Mississippi Delta*. New York: Verso Books.

Wypler, J 2018, 'Lesbian and queer sustainable farmer networks in the Midwest,' *Society & Natural Resources*, vol. 32, no. 8, pp. 947–964.

Yazd, SD, et al. 2019, 'Key risk factors affecting farmers' mental health: a systematic review.' *International of Environmental Research and Public Health*, vol. 16, no. 23, p. 4849.

Zambrana, RE & Dill, BT 2009, 'Conclusion: Future directions in knowledge building and sustaining institutional change' in *Emerging intersections: Race, class, and gender in theory, policy, and practice* by BT Dill & RE Zambrana (eds.). New Brunswick, NJ: Rutgers University Press.

Zerssa, G, et al. 2021, 'Challenges of smallholder farming in Ethiopia and opportunities by adopting climate-smart agriculture,' *Agriculture*, vol. 11, no. 3, p. 192.

Chapter 11

Agriculture, power and gender-based violence in the Dedza and Salima districts of Malawi

Veronica Kaitano, World Relief, USA; and Vincent Achikulire Kaitano, University of Bolton, UK

1 Introduction

Agriculture is a major economic activity in Africa, contributing on average 15% of gross domestic product (GDP), although this varies widely across the continent, from 2.5% in South Africa to 56% in Sierra Leone (Palacios-Lopez et al., 2017). The sector employs about 65% of the labor force and is the main income-generating activity for many (Kamara et al., 2019). In Africa, more than 70% of farmers are small-scale or smallholder farmers, mainly engaged in subsistence agriculture. In SSA, 80% of agricultural production is by smallholder farmers based in rural areas, most of whom are women (https://www.farmafrica.org/what-we-do-1/women-in-the-field).

Smallholder farms, defined as 2 hectares or less, represent 80% of all farms in SSA and contribute up to 90% of production in some SSA countries (FAO, 2017). Although these farmers produce mainly to feed their families, they are crucial to food security in SSA (Nyambo et al., 2022). Women play a key role in agriculture, accounting for 43% of the agricultural labor force in developing countries (FAO, 2011), rising to more than 75% in SSA countries.

http://dx.doi.org/10.19103/AS.2024.0148.13

However, notwithstanding women's substantial role in agricultural production such as land clearing, cultivating, planting, weeding and pre- and post-harvest handling and value- addition activities like sheaving, drying, sorting, grading and processing, many barriers stand in their way because of their gender. Barriers include limited or no access to land, credit, markets, information and other productive resources, illiteracy and decision-making power apart from the burden of domestic duties which they are often solely responsible for. The agricultural sector is male-dominated, which results in women frequently lacking power and control in decision-making processes. This power imbalance often results in various forms of gender-based violence (GBV). Women who are economically dependent on their partners are particularly vulnerable to violence (Malawi Ministry of Gender, 2014). The resilience of smallholder women has been greatly affected by power dynamics within the gender roles at both community and household level. Power dynamics affect women smallholder agricultural production and gender inequality exacerbates GBV among women smallholder farmers. Innate gender bias within smallholder farming, such as inadequate access to land, finance, markets, agricultural training and education, could be a significant gender-specific hindrance that puts women at a critical disadvantage in making contributions toward sustainable development of agriculture and food production (Assan, 2022).

2 Smallholder women in agricultural production in sub-Saharan Africa

Although in SSA rural women make a significant contribution to the agricultural sector, they face several challenges that hinder their productivity and growth, resulting in weak rural economies and affecting local and global food systems. These challenges include limited access to education and employment, and most importantly, restricted access and ownership of productive resources like land and livestock, which creates disparities in farmland access, agricultural inputs, credit, extension services, autonomy, and decision-making regarding farming activities (Noshaba et al., 2021). Limited and restricted access to factors of production and other farm inputs results in lower productivity on land farmed by women (https://www. solidaridadnetwork.org/story/ celebrating-rural-women-contributions-in-sub-saharan-africa/) increasing their vulnerability and trapping them into the poverty. Some of the key challenges faced by SSA smallholder women farmers are expounded below.

2.1 Access to land

It has been acknowledged that in patriarchal, matriarchal and matrilocal land tenure systems in SSA, men wield decision-making power, leading to powerful inequalities that favor men (Chigbu, 2019). Consequently, women in SSA generally access land through male relations such as brothers, uncles, fathers or spouses rather than having direct or independent rights over the land. Even when they own land, women experience challenging land tenure systems, which are frequently influenced by biased customary norms and traditions. The majority of these women lack the necessary legal documents, which limits their ability to make long-term decisions regarding land use including activities such as forestry, tea and coffee farming, etc. (Phiri et al., 2022). Consequently, women smallholder farmers lack access to credit, such as bank loans, and technology and extension services due to their inability to secure collateral through land ownership.

2.2 Access to finance and agricultural inputs

Ugwu (2019) asserts that lack of collateral, high transaction costs, limited financial literacy and mobility, socio-cultural barriers, irregular employment account for women farmers' struggle to access credit. Furthermore, several authors argue that this situation is even more challenging in SSA because women are less likely than men to utilize credit (OECD/FAO, 2016; Dillon and Barrett, 2017). This is primarily due to their limited decision-making power, leading to reduced control over credit. The women farmers are therefore unable to access critical agricultural inputs like improved seeds, fertilizers, pesticides, machinery, etc., with the high market costs of these inputs also negatively impacting their productivity.

2.3 Limited access to markets

In SSA, women smallholder farmers face challenges in accessing markets and value chains, which limit their ability to sell their produce at fair prices and earn a sustainable income (Keane, 2018). Men have access to markets as they can travel long distances while women are time constrained as they have to remain home to cater for their families given their reproductive, productive and community-managing activities whereas men primarily undertake productive and community-politics activities (Torday-Gulden, 2019).

2.4 Limited participation in decision-making process

Owing to sociocultural norms coupled with low-income levels and lack of access to credit and/or land, smallholder women farmers have limited decision-making power within households and communities, which can affect their ability to make choices related to agricultural production and resource allocation (Ben-Ari, 2014). Women's low decision-making power is attributable to their age, education, land tenancy and wealth status. Subsequently, local decision-making and control over resources reside with male members of society which re-enforces inequality and propagating the suppression of individual rights.

3 Power dynamics in agricultural production in sub-Saharan Africa

Agricultural production in SSA is a male-dominated industry, which means that women often lack power and agency in decision-making processes. This power imbalance can lead to intimate partner violence (IPV) and other forms of GBV. Women who are economically dependent on their partners are particularly vulnerable to violence (Government of Malawi, 2014). The resilience of smallholder women has been greatly affected by power dynamics within the gender roles at both community/household level. Power dynamics affect women smallholder agricultural production and gender inequality exacerbates GBV among them. These power dynamics are also seen in market access, financial literacy and land ownership laws in patriarchal societies.

Women play a key role in agriculture and account for 43% of the agricultural labor force in developing countries. In Africa, they are considered responsible for producing up to 80% of the locally consumed food. Smallholder women play an important role in agricultural production in SSA (Mila & Minot, 2018). Despite this, the roles of the majority of women are limited to domestic chores while the men participate in commercial activities, resulting in unequal financial endowment between men and women. Societal gender norms often assign women as family nurturers and men as bread winners – an uneven playing field for the two sexes.

Gender roles and responsibilities in smallholder farming in SSA are dynamic. For instance, men are often thought to allocate disproportionately more of their time to (non-edible) cash crops, while women are believed to concentrate more on the production of staple and other food crops. Compared to the overall female share of agricultural labor, the female share allocated to non-edible crops is slightly lower in Malawi and Uganda, while it is slightly higher in Tanzania and a lot higher in southern Nigeria overall, while there is some variation in the female share of agricultural labor across crop categories within countries (especially Niger, Nigeria), these patterns are not generalizable across countries (Palacios-Lopez et al., 2017).

4 Gender-based violence in agricultural production in sub-Saharan Africa

GBV is a complex global social problem affecting millions of people. GBV is an umbrella term for any harmful act that is perpetrated against a person's will and that is based on socially ascribed (gender) differences between females and males (https://trainingcentre.unwomen.org/mod/glossary/view.php?id=36&mode&hook=ALL&sortkey&sortorder&fullsearch=0&page=-1). The United Nations (UN) defines it as 'any act that results in, or is likely to result in, physical, sexual, or mental harm or suffering to women, including threats of such acts, coercion or arbitrary deprivation of liberty, whether occurring in public or in private life'. The nature and extent of specific types of GBV vary across cultures, countries and regions. Examples include sexual violence (i.e. sexual exploitation/abuse and forced prostitution); domestic violence; trafficking; forced/early marriage; and harmful traditional practices such as female genital mutilation, honor killings and widow inheritance. It can be defined depending on the relationship between the perpetrator and the victim (IPV and non-IPV) or by type of the act of GBV, such as sexual, physical or emotional violence (Muluneh et al., 2020). Although it affects many individuals globally, women seem to experience the worst of it in terms of prevalence, severity and impact (Rogers and Ali, 2023). It is rooted in gender inequality and power imbalances, lop-sided power relations disproportionately targeting women. Patriarchal norms condone and perpetuate the dominance by men over women and often 'blame' women for failing to meet stereotypical gender norms and behaviors (Action Aid, 2012).

The World Health Organization estimates that about one in three women worldwide has been subjected to either physical or sexual (IPV) or to non-partner sexual violence in her lifetime (https://www.worldbank.org/en/topic/socialsustainability/brief/violence-against-women-and-girls). The numbers are staggering: 35% of women worldwide have experienced either physical and/or sexual IPV or non-partner sexual violence. Globally, 7% of women have been sexually assaulted by someone other than a partner. Globally, as many as 38% of murders of women are committed by an intimate partner, 200 million women have experienced female genital mutilation/cutting. Regions with the highest levels of violence against women ages 15–49 years are the Pacific (51%), South Asia (35%) and SSA (33%) (https://www.worldbank.org/en /topic/poverty/publication/what-works-to-narrow-gender-gaps-and-empower-women-in-sub-saharan-africa). GBV can have a significant impact on women's ability to participate in agricultural production and can exacerbate the gender gap in agricultural productivity (Muluneh et al., 2020).

4.1 Social and cultural norms affecting gender-based violence in agricultural production

Ahinkorah et al. (2018) posits that in SSA, a significant proportion of both men and women endorse a man's prerogative to physically discipline his wife with more women than men endorsing what they view as justified abuse, such as when a wife appears to neglect the children or argues with her husband.

The World Bank publication (https://www.worldbank.org/en/topic/poverty/publication/what-works-to-narrow-gender-gaps-and-empower-women-in-sub-saharan-africa) shows that 60% of women aged 15–49 believe that a husband is justified in hitting or beating his wife for at least one of five reasons. The lowest percentages of men and women who believe a man is always sometimes justified in beating his wife are observed in Cabo Verde, Eswatini, Madagascar, Malawi and Mauritius.

4.2 Food insecurity

Food insecurity has increased amidst the rapid population growth and global climate change challenges affecting agricultural production, while prevalent gender-based inequalities have also negatively affected agricultural productivity and compromised communities' climate resilience targets hence driving SSA further into severe food insecurity. Food insecurity also contributes to GBV due to the fact that women who are the sole producers of food for their families face constraints including low income, lack of access to factors of production, lack of training and low education levels, lack of decision-making power and negative cultural social norms. All these lead to low food production and adds additional pressure during times of food scarcity thus putting the women in vulnerable situations as they rely heavily on their husbands. This can lead to increased tension and violence within households. These gender inequalities therefore make smallholder women particularly vulnerable to GBV. Addressing the issue food security through increase in food production among the smallholder women farmers will reduce GBV; however, this requires a multi-faceted approach that includes improving women's access to resources, empowering them in decision-making processes, and addressing the root causes of gender-based power imbalances (Assan, 2022).

5 Case study: agricultural production and marketing, power and gender-based violence among smallholder farmers in Malawi – case of Dedza and Salima districts

Agriculture is the cornerstone of Malawi's economy, with more than 83% of its people living in the rural areas and engaged in agriculture and agriculture-related activities. Agriculture, forestry and fishing contribute 27.1% to GDP, comprising crop and animal production (18.7%) and forestry and logging (6.7%), and fishing and aquaculture (1.2%) (National Export Strategy II, 2021). The main agricultural commodities include (1) cash crops like tobacco, sugar, tea, coffee and cotton; (2) food crops like maize, rice, sorghum, groundnuts, bananas, potatoes, peas, beans, mangoes, tomatoes, among others; (3) livestock and poultry; and (4) fisheries. However, agriculture, being rain-fed, remains vulnerable to the vagaries of nature (drought, floods, cyclones) and to commodity price fluctuations, and external shocks, amongst other uncertainties.

Agriculture accounts for more than 80% of Malawi's export earnings, contributes 38% of GDP and supports 85% of the population (World Bank, 2021). Smallholder farmers, who contribute 75% of the agricultural production, cultivate maize, cotton, groundnuts, tobacco, groundnuts, soybeans, rice, sweet and Irish potato. Smallholder women in agriculture tend to have marginalized gender roles and suffer GBV apart from having limited access to credit, training and necessary technology – inequalities that help create a gender gap in productivity between women and men on farms of equal size. Gender inequalities especially among female smallholder farmers stagnates social progress in Malawi. There are gender differences in agricultural productivity because of women's limited access to and control of inputs such as seed and fertilizers. Access to markets and credit is also another huge challenge for female farmers apart from access to weather forecasts and other agricultural related information. It is estimated that up to 34% of women in Malawi have reported to have experienced physical violence since age 15, and 20% of them have experienced sexual violence, and 42% of ever-married women have experienced spousal violence (NSO, 2017).

5.1 Introduction

The case study looked at power dynamics (who has power, who makes decision regarding inputs, production, marketing and income) and violence related to agricultural production amongst smallholder women farmers in two districts of Malawi. Specifically asking what types of GBV women experience and why, before exploring available mitigation measures GBV. It used focus group discussions (FGDs), key informant interviews (KIIs) and household surveys[1] with randomly selected women smallholder farmers in Dedza and Salima for data collection. FGDs assisted in unpacking answers given in the KIIs. In the spirit of inclusivity, participants' age, disability, education, ethnic, religion and socioeconomic status were inconsequential. Importantly, participants were selected based on their informed verbal consent and willingness to participant in FGDs and interviews. The face-to-face KIIs and FGDs took place in private locations. Permission for the research was granted through district Ministry of Gender staff. Data analysis was done using Excel and Ona.

1 Salima (20 KIIs +2 FGDs); Dedza (20KIIs + 2FGDs)

Dedza and Salima districts partly follow the matriarchal system where women are accorded some roles in the social governance structure. Subsistence agriculture is the mainstay of these districts, with female women farmers being central contributors.

Although both men and women are involved in agricultural production, the degrees of involvement are different significantly. Agricultural production begins with land preparation, planting, fertilizer/manure application, weeding, harvesting and post-harvest handling, and value addition and marketing. This study discusses the involvement of men and women in the agriculture value chain.

5.2 Smallholder women in agricultural production

5.2.1 Roles and involvement during the various stages from production to harvesting at the farmer level by sex

In Dedza and Salima districts, the agricultural season begins by land preparation stretching from August to the onset of rains in November/December. Table 1 shows who is mostly involved in land preparations, weeding and harvesting.

i. Land preparation

What can be deduced from above is that women are very involved in land preparation (85%). This indicates a significant proportion of women taking an active role in land preparation activities. Men have a lower percentage (50%) of being very involved in land preparation compared to women. This was confirmed in FGDs where it was mentioned that land preparation is mainly done by women as men say that *'this work is for women'*. According to women in Dedza this was *'because there are weeds and thorns which prick our legs and so the men do not want to do this work.'* This was particularly true for previously cultivated fields. Men tend to be more involved in land preparation if virgin fields are considered as these would involve felling trees.

Table 1 Engagement in land preparation, weeding and harvesting

	Land preparation						Weeding						Harvesting					
	N	W	*N*	M	*N*	Y	*N*	W	*N*	M	*N*	Y	*N*	W	*N*	M	*N*	Y
Very	34	8	20	50	7	17.5	32	80	28	70	8	20	35	88	27	68	7	17.5
Little	6	15	17	42.5	14	35	20	20	10	25	10	25	5	13	10	25	11	28
None	0	0	3	7.5	16	40	0	0	2	5	20	50	0	0	3	7.5	20	50
Other	0	0	0	0	3	7.5	0	0	0	0	0	0	0	0	0	0	2	5

Source: Household Survey Interviews. Legend: W = Women; M = Men; Y = Youth.

Land preparation also involves application of homemade organic manure (combo of dried maize stocks/goat/pig droppings) made by women smallholder farmers in the wake of rising fertilizer costs on the market.

ii. Weeding

As can be noted in Table 1, overall, women (80%) are very involved in weeding than men (70%) or youth. In addition to that, 20% of women are little involved in weeding, compared to 25% of men. In both districts, youth are more likely not involved in weeding.

iii. Harvesting

Table 2 shows how respondents perceived harvesting of their crops. In general, there are some differences in the way that harvesting is perceived and practiced by women, men and youth.

Women are more likely than men to be involved in harvesting, with 88% of women saying they are very involved compared to 68% of men. Youth are also more likely than men and women not to be involved in harvesting, with 50% of youth who were said they were not involved compared to 7.5% of men and 0% of women.

5.2.2 Roles and involvement during the various stages in value addition and marketing at the farmer level by sex

In terms of marketing and value addition, it is evident that there is little value addition being done in the communities while marketing is mainly a man's domain as can be deduced from Table 2.

i. Value addition

Overall, the majority of respondents fall into the 'None' category for value addition, accounting for 50% of women, 60% of men, and 78% of youth as depicted in Table 3. This suggests that the practice of value addition is not

Table 2 Engagement in value addition and marketing

	Value addition						Marketing					
	N	W	*N*	M	*N*	Y	*N*	W	*N*	M	*N*	Y
Very	12	30	0	0	1	25	22	55	32	80	5	12.5
Little	8	20	16	40	7	18	14	35	5	12.5	5	12.5
None	20	50	24	60	31	78	4	10	3	7.5	28	70
Other	0	0	0	0	1	2.5	0	0	0	0	2	5

Source: Household Survey Interviews.
Legend: W = Women; M = Men; Y = Youth.

Table 3 Understanding of the term 'gender-based violence'

Theme	Percentage
Power dynamics Abusive behaviors (verbal, physical, psychological, property confiscation) of men toward women and children. Violation of women's rights and mistreating women. There is no peace in a household that experiences GBV. Women persist in marriage even if they face GBV for fear of separation.	45
Involvement in decision-making Men do not respect women's opinions, e.g., when your views about what and when to buy something after selling produce are ignored. Spouse refusing to support you on what you want and unequal sharing of market proceeds. Failure to reach fair conclusions on key decisions, e.g., not given a chance to say your voice on money. Not being involved in decisions, e.g., not supporting your spouse's ideas and the failure of my husband to accommodate my opinion.	38
Unequal and unfair treatment of women Mistreatment in marriage (physical, verbal abuse) Violating one's rights in marriage (not given chance to air out one's views) Illtreatment of the other gender (men) such as being abandoned or neglected by your spouse Husband marrying another wife, husband neglecting his responsibility and substance abuse (alcohol) and lack of child care Unequal sharing of farm inputs Male dominance over financial resources	28

Source: Household Survey Interviews.

widely prevalent among the respondents in both districts. A small proportion of respondents fall into the 'Very' category for value addition, with 30% of women and 2.5% of youth. Interestingly, there were no men who reported being very involved in value addition. The 'Little' category shows moderate involvement in value addition, with 20% of women, 40% of men and 18% of youth reporting as such.

ii. Marketing

As depicted in Table 2, overall, 55% of women, 80% of men and 12.5% of youth were 'very' involved in marketing, 35% of women, 12.5% of men and 13% of you were 'little' involved and 10% of women, 7.5% of men and 70% of you had 'none' involvement. This was echoed in FGDs where the women pointed out that '*men take the harvest to sell and disappear for some days and come back and report that they either lost the money or paid back a loan they took and if you ask for more details, you get beaten*'. It can be deduced from this that men dominate marketing of the crops than women in both districts.

Men dominating sales of produce is as a result of lack of decision-making power by women, as echoed by a woman in Dedza: *'When doing all our farming activities we do not have any problems with our husbands but when it comes to marketing then there are a lot of problems. The men usually say they are the heads of the house and so we should not question them and so they take the produce and go to town to sell it but then we do not see the money from those sales. It is like preparing a meal, everything is alright up until the meal is ready but when we are about to eat, we are left with nothing. So, we never see the profit of our hard labor.'*

A woman in Salima said, *'Most of the farming work is done by the women in the fields but the money and the profit is not seen by us as we have no authority over the money we make out of farming.'*

In both districts, most women are subjected to physical abuse when they confront their spouses regarding their marketing errands: *'We are beaten if we refuse to allow him to sell the produce. Even when he comes back with nothing, we do not have the authority to question his decision because doing so will end up in a quarrel leading to a beating'* and *'most of the time we endure this type of behavior in order to keep peace in our homes. Sometime he will even sell the seed subsidy given by government and when asked about it, the reaction is the same ... you are verbally abused or beaten into submission.'* Dedza and Salima FGDs.

5.2.3 Challenges facing women smallholder farmers from production to marketing

Table 4 depicts the challenges that smallholder women face at various levels of the agricultural cycle.

On challenges faced by women during production (from land preparation to value addition), most women responded that they are 'involved as men yet they had to do other household chores' with 55% in land preparation, 80% during weeding and 75% during harvesting while value addition is viewed as a woman's job (35%). So, although both men and women are involved in smallholder farming, women have the extra burden of reproductive and community managing responsibilities on their shoulders.

As regards marketing, 52.5% of women have little involvement during marketing. This echoes the findings earlier on that most women labor during different stages of farming except for marketing. In FGDs, women were very candid in describing the challenges they face with marketing. There was general consensus that women do not have any say on how the money accrued from sales of crops is used: *'It is the husbands that go to town to sell the crops. When it comes to the money that we get from the produce after selling what we have produced, most of us women do not have a say over the money. Since it is the*

Table 4 Challenges facing women smallholder farmers from production to marketing.

Land preparation	*N*	%
Spend time more than men	16	40
Involved as men yet have to do other household chores	22	55
Other	0	5
Weeding		
Only women are involved	5	12.5
Involved as men yet have to do household chores	32	80
Other	3	7.5
Harvesting		
A woman's job	6	15
Involved as men yet have to do household chores	30	75
Other	4	10
Value addition		
A woman's job	14	35
Involved as men yet have to do household chores	0	0
Other	26	65
Marketing		
Not involved	3	7.5
Little involvement	21	52.5
Other	16	40

Source: Household Survey Interviews.

man that does the marketing, and they decide how to spend the money.' One woman said, *'After coming back from selling the produce, I asked my husband what he had brought of us; how much he had sold.' Angrily he said, 'I am the head of this house; you have no right to question me on anything and what I say goes. What is it to you if I came back with nothing!' So instead of being accused of insubordination I just agree with what he says. Because if I resist, he can beat me. So that is why I let him go to the market. And accept whatever he brings back'.* Another woman said, *'we are beaten. If we do not agree to allow the man to go and sell the produce. Most of the time, we endure this. There was a time that he even sold the seed subsidy that we had received. So, we did not grow much that year due to insufficient seed.'*

5.2.4 Involvement in decision-making regarding the use of income generated from sales of farm produce/products

Figure 1 indicates that decision-making power regarding the use of income from farm produce sales is skewed heavily toward men.

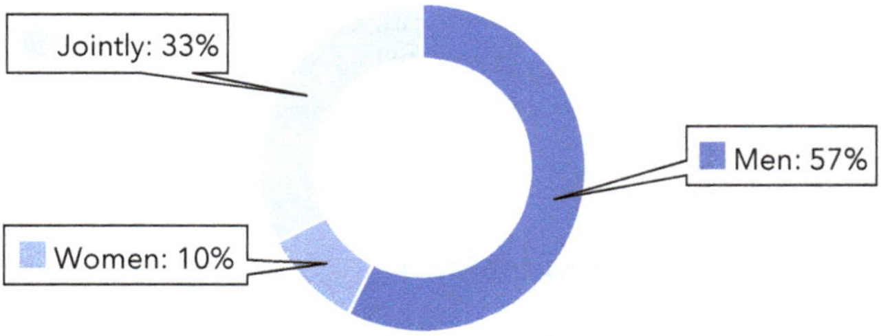

Figure 1 Decision-making in sales of farm produce.Source: Household Survey Interviews

The majority of respondents reported that men decide how to use income from sale of produce in both districts, while only 10% mentioned that women are involved in decision-making. Additionally, 33.5% said that decisions are made jointly.

This distribution of decision-making authority suggests an unequal and unfair decision-making practices, with men having more control over how the income from agricultural sales is used. Resultantly, women's economic empowerment and ability to invest in their own priorities, e.g., education, health is denied. This dependence on their male counterparts perpetuates inequality, making women more susceptible to GBV.

5.3 Gender-based violence

5.3.1 What does gender-based violence mean to women smallholder women farmers

Table 4 demonstrates the multi-faceted nature of GBV and its connection to broader issues of power, resource distribution and gender norms within society.

i. Unequal power dynamics

Recognizing GBV for what it is, is important for it to be combatted. Results indicate that about half (45%) of respondents said GBV is related to unequal power dynamics between women and men. Respondents associate GBV with abusive behaviors including verbal, physical, psychological abuse and property confiscation of men toward women and children, mistreatment of women and violations of women's rights. Respondents viewed GBV as rooted in unequal power relationships, where men exert control and dominance over women, leading to various forms of abuse. For example, in FGDs, respondents said GBV was *'a man leaving all the work to the woman; and if the woman complains, the man says he is the head of the house and he must be obeyed so the woman is obliged to submit. The woman ends up just doing whatever the man says because that's what culture has taught us. If you want to challenge this more often than not, nobody supports you, even your own relatives so women just endure GBV."* This indicates that GBV is often rooted in unequal power relationships, where men exert control and dominance over women, leading to various forms of abuse.

ii. Involvement in decision-making

Another issue is that almost two in five (38%) mentioned was lack of women's involvement in decision-making at the household level. As such, GBV is associated with women's opinions and views being disregarded or ignored; e.g. where the woman suggests how money from agricultural produce should be used and the man says he is head of the house and she must listen to him and not the other way round. In addition, women are not given a chance to contribute on how the money should be used hence not involved in making decisions in the home and the man acts as a lone ranger and informs the woman what he has decided and his word must be obeyed without any question to avoid repercussions. These unequal power relations often are the basis for violence and abuse within relationships as can be deduced from an FGD respondent in Salima who said, *'after all the hard work, my husband sold all the produce and when I asked how much money he had come back with, he said he had given it all back to creditors he owed money. It was the first time I was hearing about it and when I probed further, I was told it was none of my business and that if I continued to "interrogate" him, I would be beaten, so I stopped asking any further questions because I knew I would be beaten as I was in the past. All he brought back home was a packet of salt and a nsalu (cotton cloth wrapper commonly worn by women). To be honest, I was deeply hurt because I had worked so hard but what could I do? In fact, this is what most of us endure year after year.'* Most of the women in the FGD nodded in agreement. In KIIs, results indicated that 10% of women were responsible for making decisions as regards the income generated from the sales of farm produce and 57% of them said that men lead decision-making while 32.5% indicated that they made decisions jointly with their spouses. The results indicate that decision-making regarding the use of income from farm produce sales is skewed heavily toward men. The majority of respondents reported that men are the ones who make these decisions, while only a smaller percentage mentioned that women are involved in decision-making. Male-dominated decision-making perpetuates poverty as the money accrued from sale of produce is rarely used for household development leaving the women to take the sole burden of taking care of the children's needs such as clothes, school fees, as one woman said, *'I have to take care of the children myself … but I just put up with it.'* And another woman said: *'the result is poverty, instead of going forward, you go back, for example, all those ten bags of groundnuts (sold) no money on the ground, instead of buying something to change the home. Every time you speak, they say be careful, they just buy a paper (for roofing).'* Another woman said, *'children fail to go to school, the man has spent all the money, he has used the money, and like that development cannot happen!'*

iii. Unequal and unfair treatment of women by men

Almost one in three (28%) associate GBV with discrimination and unequal and unfair treatment of women by men within marriages and relationships. The unfair treatment included being mistreated in marriage, abandonment and lack of support. In addition, GBV is associated with unequal sharing of resources such as farming inputs, i.e. seed and fertilizers and financial control and male dominance of financial resources. In FGDs, this was a major issue and there was consensus in both districts that there is physical abuse if a woman asks for the money which the man is supposed to bring after selling produce, the couple together with their children have moved hard for: *'yes, they will beat us, they just tell you to be careful and keep your mouth shut'* and *'we can tell you that there are several cases of beatings, beatings and running away to sleep at your friend or neighbor's house, running away from the house. These bodies are often beaten and if our children want to intervene they are likely to be beaten as well' – FGDs.*

Others link GBV to lack of provisions and support in relationships, e.g., where husbands neglect their responsibilities, engage in substance abuse or marry other wives, leading to neglect and abandonment of their spouses and children. Respondents in Dedza FGD the women said, *'There is one of us whose husband always leaves after the harvest, yet it's the woman who does most of the farming activities. He usually goes off drinking and to marry another woman after each harvest. He usually leaves the family without food. He only comes back after squandering the money.'*

One woman in Salima summed GBV as follows, *'What happens in most of our homes is GBV! When men deny us our rights and deprive us of any authority that is GBV, when we do most of the work and men sell the produce on their own and use the money however they like, that is GBV! When I have to run away to find peace in someone else's home- that is GBV, when I don't have any say over what happens to the produce I have worked so hard for – that is GBV.'*

5.3.2 Who is affected by gender-based violence

Nearly all (97.5%) respondents identified women as the primary group affected by GBV as per Fig. 2, highlighting the widespread recognition of women's vulnerability to various forms of abuse in the study communities.

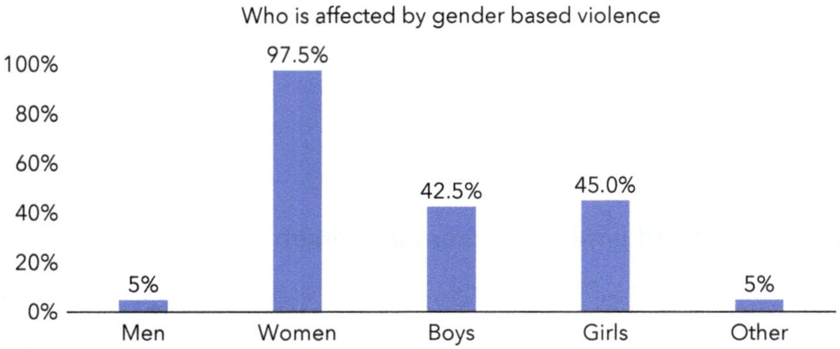

Figure 2 Proportion of those affected by gender-based violence. Source: Household Survey Interviews.

Additionally, 45% and 42.5% stated that girls and boys, respectively, are impacted by GBV, revealing violence against youth. Only 5% said men were affected by GBV. This was echoed in FGDs where there was unanimity that *'women are the ones that are mostly affected by gender-based violence in this community because they do not have power and often cannot stand on their own.'*

5.3.3 How are woman smallholder farmers affected by gender-based violence

Findings of household interviews reveal several areas of how women smallholder farmers are affected by GBV. The findings provide valuable insights into the various ways GBV impacts women's lives in the context of smallholder farming. The most prevalent theme reported (62.5%) is 'financial sabotage,' where women face challenges related to financial control and decision-making within the household. Examples include women not seeing the profit from their farm labor, men using money without consulting their wives, withholding resources or profits from women, women not developing to their full potential and the funds from sales of produce being used by women for unproductive activities such as womanizing. All this results in self-inflicted poverty as one woman during FGDs said, *'development does not happen at home. Because men squander the farming produce, there is hunger at home, we can't build good houses, we sleep on reed mats because we cannot afford beds and mattresses'.* This hinders women's economic empowerment and development.

Another significant theme reported by 40% of respondents is 'support and care.' This theme encompasses the lack of support, care and provision from husbands, affecting women's well-being and their ability to provide for their families adequately. Some women endure abusive situations for the sake of their children, while others expressed frustration at the husband's failure to contribute to household responsibilities, leading to food scarcity and limited access to farming inputs like fertilizer. During FGDs, the women said *'children fail to go to school, the man has spent all the money recklessly for example on drinking, premier betting or with prostitutes…, instead of buying iron sheets they are buying (plastic) paper, that means no development at home, and like that development cannot happen'*. This results in women becoming the primary providers in the home since the men abrogate their responsibility to provide for the family. GBV has detrimental effects on women smallholder households and overall community resultantly.

Yet another prominent theme reported by 30% of respondents is 'low self-esteem.' GBV erodes women's confidence and self-worth, with women feeling powerless because they are not involved in decision-making at any level. Women's ideas may be disregarded or dismissed, and men often exert disproportionate control over family decisions, perpetuating traditional gender norms that limit women's agency and participation.

Lastly, a smaller group of respondents (2.5%) said they were *'not really affected because I am on good terms with my husband.'* This indicates that some women have positive relationships with their spouses and have not experienced GBV in their contexts.

5.3.4 Type(s) of gender-based violence experienced by women smallholder farmers

Women smallholder farmers experience a wide range of GBV as illustrated in Fig. 3. The most common types of violence they experience are verbal (77.5%), psychological and socioeconomic (65%) and physical (52.5%). This was echoed

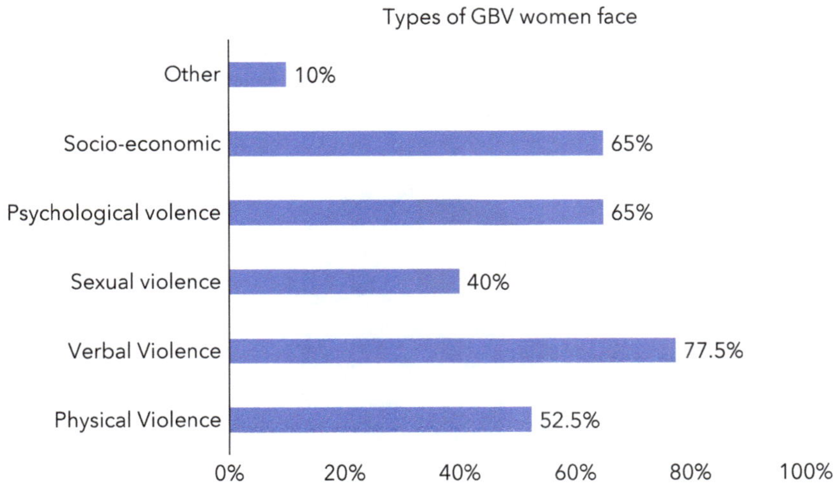

Figure 3 Types of gender-based violence experienced by women. Source: Household Survey Interviews.

by a woman in FGDs who said *'we face all types of GBV including physical, verbal, emotional and psychological violence. Physical violence often comes when the man says I will teach you a lesson for being argumentative with me.'*

The high rates of violence experienced by women smallholder farmers are a serious problem that can have a devastating impact on their physical and mental health, their economic security and their ability to participate in agriculture. This was echoed in FGDs by one woman who said that *'another challenge is that sometimes the man will take the woman's national identity document and use it to collect seed and fertilizer subsidy but instead of taking the inputs home he sells them and squanders the money.'* Another woman said that she was told when she went to collect her subsidized fertilizer and seed that her husband had already collected the inputs and when she asked him he said she had been told a lie she should go back and ask the distributors but when she went back they checked in the system and told her that her inputs had already been collected by somebody who came with her ID and when she confronted her husband again asking him to accompany her to the distributors he refused to go. She concluded that he had taken the inputs and had most likely sold them and subsequently squandered the money.

5.3.5 Prevalence of gender-based violence during any part of the agricultural season

With reference to Figs 3 and 4, the findings indicate that a significant proportion of respondents (72.5%) have experienced GBV during various stages of the agricultural season from production to marketing.

Among those who reported experiencing GBV, the most commonly mentioned types of violence were verbal (72.4%), socioeconomic (69%) and physical (34.5%). Additionally, a considerable number of respondents reported experiencing sexual violence (41.4%) and psychological violence (37.9%). A smaller percentage (13.8%) mentioned other forms of GBV.

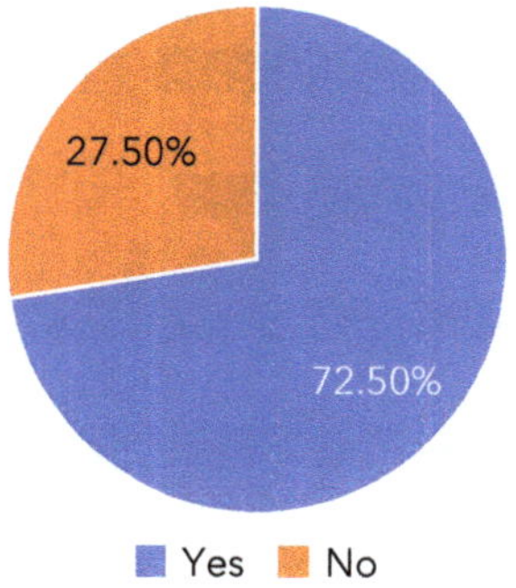

Figure 4 Proportion of women who responded 'yes' to gender-based violence. *Source: Household Survey Interviews*

5.3.6 Causes of gender-based violence

The findings from the household survey on the causes of GBV experienced by women smallholder farmers, overall, the data indicate that GBV among women smallholder farmers is influenced by complex interactions of trust or the lack thereof, poor communication, power dynamics and financial challenges.

The most common causes include 'lack of love, unfaithfulness, and lack of trust which leads to misunderstanding and disagreements,' which accounts for 86% of responses. Participants mentioned issues related to loss of trust, disagreements, misunderstandings, communication breakdown and infidelity as factors contributing to conflicts and escalation of domestic violence within the household.

'Poor resource utilization' is the next source of GBV (31%) which includes financial problems in the family as the man abrogates his responsibility and spends money recklessly, poor leadership by the husband as they do not have vision for the family and are unconcerned about their wives' and children's welfare. This results in a vicious cycle of poverty and pushes the burden of child care on women. '*Lack of love is one of the causes of gender-based violence. Also, men don't listen. They don't want to take advice. They say an outsider cannot tell me what to do. They don't take responsibility for their families. Headship, the question of headship always comes in where money is concerned*' (FGD participant).

'Power dynamics' was cited by 13% of respondents, highlighting the perception of men as the head of the household, leading to abusive behaviors and decisions imposed on women. The men always tell the women that they are heads of the houses and should the woman want to argue, as one woman said during FGDs '*... yes, they will beat us, they just tell you to be careful and keep your mouth shut.*' Some women have been abandoned for asking or arguing with their husbands, as one woman said '*and so soon as they have planned that today I want to leave, they climb up on the roof and remove the iron sheets from the family house. The house they built together but when the men leave, they take with them the iron sheets, so the other thing that I also see is that you can't speak, so you have nothing to say.*' Another woman pointed out that '*... the men do not listen to the women and so couples are not working together. They accuse women of insubordination if the woman wants to ask a question*' (FGD participant).

5.3.7 Consequences of gender-based violence

The two most significant consequences of GBV are 'economic growth and production and death and injury' (50%) according to household interviews. In terms of impact on economic growth and productivity, GBV negatively impacts agricultural production, leading to low yields and poor participation in community activities. It further affects household development and contributes to poverty. Additionally, 'hunger since you are forced not to eat' was mentioned as a consequence, indicating that GBV can result in deprivation of food and basic needs. Another FGD participant said, *'children are weak, there is not enough food in the house'* while another said *'seeing a 4-year-old child going to school without food but you already have another baby, so we are in trouble because there are many drunkards here who are not working during the planting season but become hungry for money during the harvest'* and another said *'when we cook food, we wait for each other so that we all eat together at 12 o'clock .... .maybe just to eat in the afternoon because hunger hurts the most in the afternoon and in the evening, we just go to sleep, the days go by, we have just gotten used to poverty ... but the men are there.'* Another woman summed it very appropriately when she said, *'Some of the consequences of GBV is that there is no development in the area and in the family apart from hunger. Families cannot build better housing because the money realized from farming is used for beer and womanizing. This is because there is no cooperation between the couple. And so there are no developmental programs for the family. Also, there is poor nutrition for the children and generally poor meals for the family. No proper clothing. No mattresses and so forth.'* GBV inevitably leads to a life of poverty with the circle repeated generation after generation.

In terms of 'death and injury,' this includes incidents of physical harm leading to injuries or even death. This form of GBV has been on the rise and men believe that beating a wife is their right and that a woman needs to be beaten in order for her to be disciplined.

'Broken (family) relationships' (49%) includes strained relationships within the family, leading to disagreements and ultimately divorce or separation. Participants also highlighted how GBV can promote infidelity and emotional distress. Once divorce takes place, the woman is left with the children and becomes the sole bread winner which exacerbates poverty. *'There are incidences of separation of couples and the man goes away with their resources including iron sheets and leaves the woman and the children destitute; the women have no say for fear of repeated violence.'* Another woman said, *'some of the women neglect their own fields and go and do piece work to supplement for their family because the man is not there to provide for his family.'*

'Poor well-being' was mentioned as a consequence (30%). It involves the negative effects of GBV on child development and school absenteeism *'... this leads to poorly nourished children which leads to stunted growth and high levels of absenteeism from school.'* And, *'there are no school fees for children so they drop out of school and they remain poor year after year'* (FGD participants).

5.3.8 Main perpetrators of gender-based violence

When asked the question *'who were the perpetrators of gender-based violence,'* the majority (80%) mentioned that their spouses were the main perpetrators of GBV as depicted in Fig. 5. Other perpetrators include family members, in-laws, parents and others with a representation of 23%, 33%, 25% and 13%, respectively.

Spouses of women smallholder farmers were said to be the perpetuators of GBV while the women said none of them were perpetuators because they are not the house heads.

Within each district, Dedza has the highest representation of spouses as the main perpetrators (95%) and parents and in-laws having following (45% and 40% respectively) as compared to Salima (65%, 5% and 25% respectively).

5.3.9 Training received on the prevention of gender-based violence

From the Fig. 6, only 40% of the respondents had ever received any training on GBV.

In FGDs, the women said that organizations such as World Relief put them into groups and trained them on village savings and loans and this has been very helpful because, as one woman put it *'this is how I see it, because if you save enough after a year, you will come back with some money'.* In addition, participants mentioned that World Relief also taught them about having respectful and healthy families through an intervention called Banja la moyo -Chichewa for Families for Life, a couple strengthening program. One women said that this program has saved her marriage because after participating

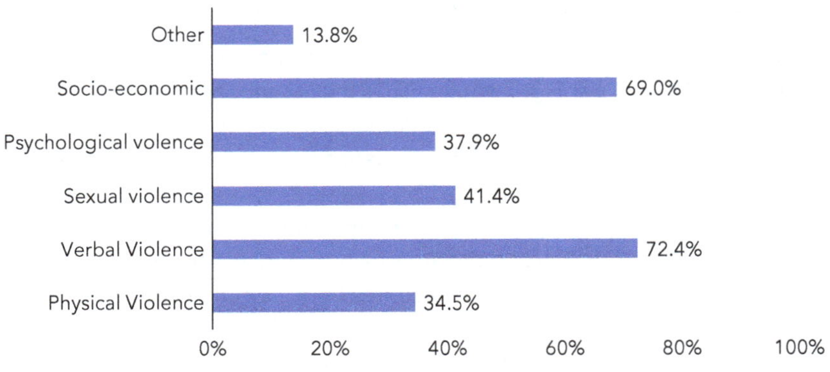

Figure 5 Types of gender-based violence experienced by those who had answered yes. Source: Household Survey Interviews.

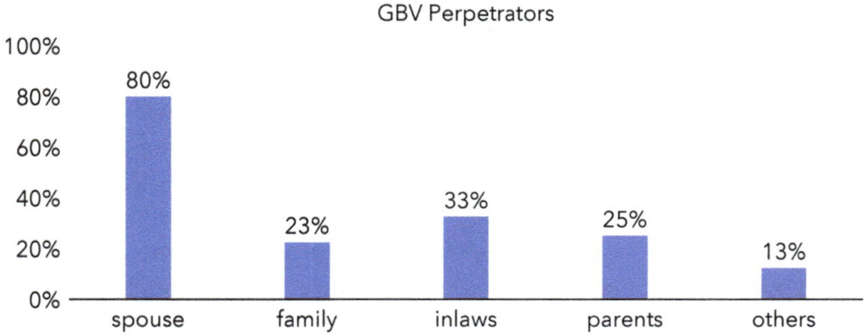

Figure 6 Main perpetrators of gender-based violence.

in the program *'...he (the husband) changed because he used to drink alcohol at the time, but when we went through the families for Life program he stopped drinking, so now, he tries to make a budget to help the family, so we can eat but since there was none before, I saw the change because things can't change all at once and that's why I'm testifying'.* Another woman also shared how her husband has been transformed after they became members of the Families for Life program. She said that before joining the program, her husband would not leave her any money for the house *'but now he brings money home. And the family is much better.'* Another woman said, *'my husband used to drink but now he has stopped and we now work together and make decisions together. So, this is one program that can be done also to help stop gender-based violence in our community.'*

In addition, Malawi police trained community members on how to combat GBV and as one woman pointed out in FGDs, *'The police victim support unit came and taught us about gender-based violence. They told us where to report in case of abuse.'*

5.3.10 Seeking help for gender-based violence

Overall, 78% of the respondents responded 'yes' they would seek help if they experienced GBV as per Fig. 7 and in FGDs they mentioned they would report to the police and a local NGO called Primary Justice. Some FGD participants said often many women choose to keep silent so as to preserve their marriages and for the sake of the children (Fig. 8).

From FGDs, findings revealed that women mostly choose to remain silent for fear of reprisals if they report the abusive husband and he gets arrested, *'we want to report GBV cases but when he comes back, he can beat you up until you faint -this happened to someone here and your in-laws will scorn you and call you all sorts of negative names, condemning your behavior'.* Many women in

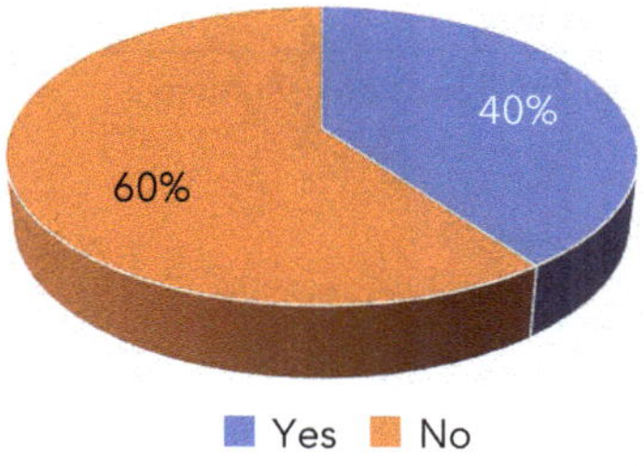

Figure 7 Training received on gender-based violence.

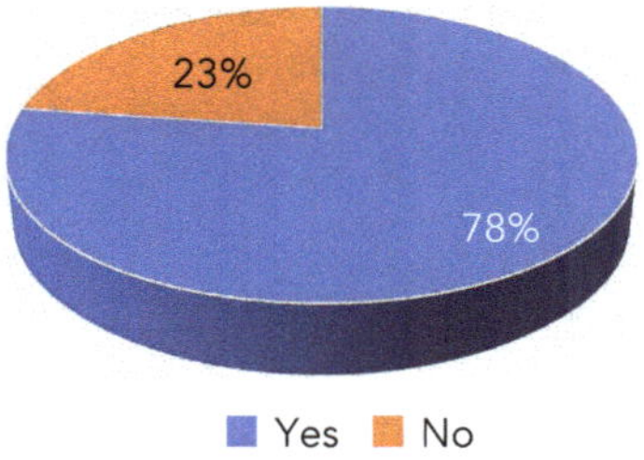

Figure 8 Would you look for help if you face gender-based violence?

the FGDs said women were pressured by relatives to withdraw reported cases to the police against their husbands as a result of fear of reprisal.

5.3.11 Where to go for help

The results show that 71% of the responses fall into Group 1 (Seeking Help and Support from individuals or Institutions), indicating that the majority of respondents would seek help and support from various sources as illustrated in Table 5. The most common places they go to seek help include marriage counselors, police victim support unit, NGOs and churches. Community policing which *'constitutes community members who round up anyone causing any kind of trouble within the village and it deals with all GBV in our community'* was highly commended in FGDs especially in Dedza.

Table 5 indicates that women can seek assistance from diverse sources when facing challenges or conflicts.

On the other hand, only 3% of the responses fall into Group 2 (Uncertain or No Help). This suggests that there is a smaller proportion of respondents who expressed uncertainty about where to seek help or feel there are no trusted organizations available to them. The response mentions keeping things in their heart, possibly indicating a lack of accessible resources or support systems for such individuals.

Table 5 Where would you go for help?

Seeking help and support from individuals or institutions

- Church (4)
- Marriage counselor (11)
- Relatives
- Victim Support (5)
- Elders (2)
- Non-Governmental Organizations (6)
- WOLREC
- Police (includes police and community police)
- Local leaders (includes chiefs, community leaders and village chief)
- People that have been married before for some time, community police
- Government, legal courts
- Counsellor
- Community police
- Parents
- Court
- YONECO, justice and peace
- Community leader
- DCs
- Husband's relatives

Source: Household Survey Interviews.

Table 6 What is done after seeking help

Seeking help and support (33%)
To help dialogue/To ask for counseling/To help in managing the situation/To get help/To seek help that I can trust
For moral support
They help you resolve your issues or separate ways
They help to equip people to keep marriages
Nothing happened when I sought support until WR came into my life to provide moral and spiritual support through our conservation agriculture group

Reconciliation and conflict resolution (26%)
Reconciliation
Mediating/they mediate and bring reconciliation or otherwise
Perpetrators are brought to justice/Perpetrators are arrested/perpetrators are punished/perpetrators pay if found guiltyperpetrators are called for a hearing/some perpetrators stop but others do even more

Community police and reporting (10%)
The community police escalate the issue to other authorities
The community police demand K5000 as a service fee, as a result, choose not to report to avoid incurring extra costs
Not much is done as perpetrators are part of the system

Uncertain or no help (6%)
Not sure
I have never sought help because there is nowhere to go

Source: Household Survey Interviews.

5.3.12 What is done for them after seeking assistance

Table 6 indicates that all responses in the 'Seeking Help and Support' (32%) involves various actions taken to seek help, such as seeking counseling, moral support, help in managing situations and trusting sources of support.

In the 'Reconciliation and Conflict Resolution' category (26%) of the responses indicate actions taken to promote reconciliation, mediate conflicts or address the perpetrators through justice systems or hearings.

The 'Community Police and Reporting' category (10%) highlights the role of community police in escalating issues to higher authorities. However, there were concerns about service fees and the involvement of perpetrators within the system.

Lastly, 6% of the responses fell into the 'Uncertain or No Help' category, where some respondents' express uncertainty about seeking help, while others stated that they have not sought help due to a perceived lack of available resources or options.

5.3.13 Barriers to seeking help

Table 7 show that the most significant barrier (52%) to seeking help is related to general challenges faced by the respondents, such as lack of support from spouses, lack of awareness about where to get support, fear of reprisal from in-laws, financial limitations to access factors of production, the fear that once reported the spouse becomes even more abusive and the fact that even if reported, nothing will change. In an FDG, a woman pointed out that when the

Table 7 Barriers to seeking help

General barriers to seeking help (52%)
- Non-compliance of the husband
- Threats by the other party
- Lack of awareness about where to get support and fear of reprisal
- Marital conflict is exacerbated when you report
- Lack of financial resources like land and money
- Sometime you can't go to report because nothing changes

Fear and guilt (20%)
- Fear of losing husband
- Fear that the perpetrator can be more angry
- Fear of reprisal
- Feeling of guilt for those who reported. This can lead to the withdrawal of the case.

None or failed to mention (19.4%)
- No
- None (2)
- None mentioned
- Failed to answer
- Some perpetrators are part of the system

Source: Household Survey Interviews.

husband came out from the police station, he beat her so hard until she fainted and since then she stopped reporting him. In another instance, the woman said that the husband's relatives castigated and accused her of being evil after the police arrested him for beating her. Others faced divorce after the spouse came from the police station.

The second most common barrier revolves around fear and guilt (20%). Fear of reprisals, losing the husband or facing anger from the perpetrator, as well as feelings of guilt, can deter individuals from seeking help.

Another 19.4% of respondents either mentioned no barriers or failed to specify any. This indicates that a portion of the respondents might not perceive any particular obstacles or chose not to disclose them.

5.3.14 What community structures that exist to support women who suffer gender-based violence

Table 8 indicates that community structures and support such as community police, local courts and chiefs and victim support entities such as the police are the main community structures that women who suffer GBV seek assistance from. The FGDs in Dedza pointed out that community policing was very active in keeping order in the community including dealing with violence against women.

Table 8 Community structures that support women who face gender-based violence

	%
Victim support	**55**
• Police victim support unit (40%)	
• Other NGOs	
• Mother groups	
• Primary justice	
Community structures and support	**85**
• Community	
• structures are there as trained by WOLREC	
• Churches	
• Teen mother groups	
• Community police	
• but the issues are handled in a way that it doesn't really satisfy the subject	
• Social welfare	
• Courts	
• Chiefs (traditional leaders)	
Unspecified community structures or support	**9**
• Community structures are there but failed to recall them	

Table 9 Measures to prevent gender-based violence

	%
Advocacy	**47.5**
Dialogue	
Discussing issues about GBV openly	
Putting in place groups that will help to deal with such cases	
Joining action groups	
Heading to good counsel	
Awareness	
Reporting issue	
Relationship-building	**30**
Love and trust among spouses	
Develop a mutual understanding as spouses to resolve your conflicts amicably	
Initiate dialogue – discussing with my husband on family affairs and make decision together	
Should join women's development groups where we discuss marital issues	
Should learn good behavior, working together for family well-being	
Planning together	
Self-control and faithfulness	
Being faithful to my husband	
Respecting my husband	
Self-empowerment	**23**
Awareness among women for those who do not know GBV	
Join social networks where we receive social skills and empowerment	
Working hard to make money to improve their livelihood, invest in small-scale business	
Self-reliance	
Promotion of counseling among spouses	
Separation with the husband	
Legal and social support	**15**
Families should be counseled on GBV	
Building strong relationships with other families will lead to doing things together	
Avoid bad company	
Others	**7.5**
Obedience	
Divorce	
It's difficult	
Just accept that some things can't change. Not reacting to any abuse done to you. This way they stop	
Stay quiet when a man misuses resources	

5.3.15 Measures to prevent gender-based violence

The results according to Table 9 show that the most common approach mentioned to prevent GBV is 'Advocacy' which had the highest percentage at 47.5%. This indicates that a significant number of respondents believe that advocacy efforts such as groups that discuss GBV issues, reporting issues and adhering to counsel given are crucial in preventing violence. In FGDs, the women pointed out that police victim support unit coming into the area and speaking to both men and women has been very helpful as they highlight the dangers of GBV and what happens to perpetuators of GBV. In addition, it was pointed out that Families for Life program by World Relief advocates for strengthening mutual trust and communication between couples by raising awareness and educating families through community sensitization and training in partnership with local churches.

The second most mentioned approach is 'Relationship-building' (30%). This includes strategies like promoting love, trust, understanding, dialogue and joint decision-making among spouses. In FGDs, women pointed out that participating in Families for Life program by World Relief was a very important one for building and strengthening relationships between husband and wife. As one woman who is a trainer in the program said that the program teaches them that '*fathers and mothers should live peacefully together and how a family should live, and this helps to end violence. Some men are practicing this and are teaching their friends.*' Another woman said, '*...because World Relief's Families for Life brings prayers and teaches against GBV some men are changing for the better*'. Strong couple relationships were cited as an antidote against GBV.

'Self-empowerment' (22%) is the third category. To support this, a woman in a FGDs said, '*I think the men can change if we women are empowered, if we are able to stand on our own and don't have to depend on them and they will begin to respect us and in that way they will not be abusing us*'. Strategies under this category include raising awareness, joining social networks for empowerment, self-reliance and seeking counseling among spouses. The women in FGDs emphasized that women having access to factors of production such as land and inputs (certified seeds, fertilizers) is key to empowering them. '*Giving us access to agricultural inputs would help us as women to be financially stable. Because what is purely ours usually our spouses have no right over it. When they have put a little bit of money in it, then they claim total control over the produce' – Salima FGD.*

Therefore, access to finance through village and savings loans (VSLs) is a major way to empower women and reduce dependence on men. '*The ones who engage in business are mostly women who rely on themselves, they do not depend on men...' – Salima FGD.*

'Women must own income generating activities – they are their own activities which men have no control over; the women say that when men have put money in their business that gives them the authority to control the women and tell them what they should do with their money but when the women start their own income generating activities without the men's involvement, they has no authority to dictate what we women should do with the money' – Salima FGD

A teenage mother said that World Relief brought a program to support teen mothers which '*... taught us about plowing and they started giving us fertilizer and when they gave us fertilizer that year, we had a bumper harvest and we were very proud of the program ... they (men) didn't interfere (with our harvest) because it was ours*'. This was emphasized by another woman who said '*... the women should have their own field – for example a field of groundnuts that they can have absolute control over so that a man cannot claim the produce thereof.*'

Women smallholder farmers in both districts commended organizations like WOLREC, Primary Justice, World Relief, YONECO and CUMO Finance and others for starting VSLs among women which they said had empowered women to become independent, secure loans, make savings which at the end of the year they are able to procure their own farming inputs. This has assisted in reducing dependance on men. '*Women should work together because one of the causes of gender-based violence is poverty*'. As one woman pointed out '*when women belong to clubs such as the village savings loan clubs, they are able to have access to loans and those loans help us to come up with small income generating activities which helps us to be independent. And usually when we are independent men are ashamed to touch what is ours. As opposed to when we depend upon them for everything. Village savings loans have greatly helped us' – FGDs*

'Legal and social support' is the fourth category (15%). This includes counseling for families, strong relationships with other families and avoiding bad company. In FGDs, participants pointed out that police are very crucial to curb GBV through litigation of perpetrators. And social support networks such as Families for Life couple groups provide safe spaces for couples to discuss issues affecting them resulting in strengthened couple relations which is the bedrock for strong families.

Finally, 'Other' strategies, such as obedience, divorce and being fatalistic that things will never change and becoming resigned to one's condition of abuse, account for 7.5% of the responses. A case in point is an FGD participant who said, '*Iam staying in the marriage for the sake of my children and also because I have nowhere to go.*'

6 Conclusion and future trends

Key challenges facing women smallholder farmers results in lowered productivity in 'their' farms leading to low household income, greater food insecurity and lowered well-being of their families and wider communities. Thus, the fact that there is gender discrimination in access to resources means that total output is lower than it would be if women had greater access to these resources. The broad evidence from the literature and the case study on women's roles in agriculture shows that gender inequality slows development. At the same time, there is broad-based agreement that women and men farmers do not generally face the same production conditions and, as a result, do not necessarily make the same production choices, with implications for output and incomes. Understanding the constraints on women farmers and the forces that drive gender gaps in agricultural productivity is critical for policy and practice. It is argued that increasing women's empowerment by extending women's abilities to make decisions and take advantage of opportunities is essential for broad-based agricultural development in low-income countries. Decreasing gender inequalities is fundamental for reducing poverty and increasing growth.

This paper has shown that men have power and/or authority in agricultural production because in SSA in general and Malawi specifically, men control most of the land and their production is higher than women farmers due to the gender gaps in agricultural production. Women have limited access to resources such as land, credit and other production inputs, which affects their productivity and therefore their income. Women's work in high-value cash crops does not guarantee their control over the generated income. In fact, women's increasing role as economic producers does not necessarily increase their decision-making power in the household and community. In addition, women do not have control over the income that is accrued from sales of produce which exacerbates poverty. Agricultural extension services throughout the region have traditionally focused on men and their production needs, despite women constituting about 60–80% of the agricultural workforce in some countries leaving women out of the fold. From participation in decision-making to access and control of assets and resources in agricultural value chain, social norms dictate and hold a woman's place in her society. Simply put, women cannot empower themselves in agriculture unless social norms are addressed. With the above challenges, women continue to have low productivity therefore low incomes which keeps the communities in a vicious cycle of poverty. They rely on their spouses and therefore being vulnerable to GBV. This situation exacerbates the already negative cultural norms which give men power over women. Poor yields also mean inadequate food for family consumption let alone to sell which puts them in a situation of food insecurity.

The broad evidence from the literature and the case study on women's roles in agriculture shows that gender inequality slows development and perpetuates poverty and GBV. Policy-makers and international organizations cannot ignore the interests of women agriculturalists if they are to have an impact on household and national food security. Women are ten times more likely than men to invest their income in the health, education and nutrition of their children with long-term effects on human capital formation and economic security. In analyzing gender roles and their impacts on agriculture, the specific context is of great importance. Focusing on support for men's contributions to agricultural production will clearly leave major sources of agricultural output flowing from women's labor untouched. Reducing gender inequality is a critical element in promoting agricultural development in low-income countries.

The abuses of power and gender inequality are the underlying causes of GBV. In the case study, violence, exploitation and abuse occur because power is misused by those closest to women farmers as they lack decision-making powers in their marital relationships. Inadequate education, information and access to services, economic inequalities and control over resources, as well as inappropriate or inexistent policies, laws and institutions are factors that contribute to women's vulnerability to GBV. In practice, however, various causes and contributing factors to GBV often converge. Improvement of women's security and access to time saving equipment and services is needed for these women farmers.

The case study has shown that Couple-Focused Interventions (CFIs) such as World Relief's Families for Life program are very effective in curbing GBV. Families for Life is a gender-transformative program aimed at changing power dynamics within couple relationships, transforming masculinities, promoting couple communication, shared decision-making and challenging and altering harmful social-cultural norms and practices. Studies have in fact shown that CFIs can increase gender-equitable norms (Speizer et al., 2018), joint decision-making within the couple (Doyle et al., 2018) and reduce intimate partner violence (Doyle et al., 2018). This chapter further agrees with Farnworth et al. (2013) who argue that there is a causal relation between more equal gender relations in the household and in the community and better agricultural outcomes: the one underpins the other.

Transforming masculinities is essential in curbing GBV due to widespread patriarchal practices in Malawi. Concerted efforts of the Government of Malawi, UN agencies, local and International non-governmental organizations, the Malawi Irish consortium on GBV, faith and traditional leaders and communities among many actors are required to address the challenges faced by women smallholder farmers. Malawi Government's recent male engagement strategy could not have come at a better time.

7 Where to look for further information

- Malawi Government's male engagement strategy.
- Malawi Irish consortium on gender-based violence.
- Malawi Ministry of Gender, Community Development and Social Welfare website.

- Oxfam Malawi.
- NGO GCN.
- UNDP's Spotlight Initiative.
- Malawi Human Rights Commission.
- MEGEN.

7.1 Key Journals/Conferences

- Malawi Irish Consortium on GBV Conference which brings together all actors combating GBV in Malawi.
- National Gender conference organized by Ministry of Gender and UN agencies.
- UNDP's Spotlight Initiative which is a nationwide programme addressing GBV.

8 References

Action Aid. (2012). The policy brief on advancing the rights of women smallholder farmers with lessons from Covid-19, accessed on https://actionaid.org.au/wp-content/uploads/2021/07/Advancing_the_rights_of_women_farmers.pdf, February 2022

Ahinkorah, B. O., et al. (2018) Women decision-making capacity and intimate partner violence among women in sub-Saharan Africa. *Archives of Public Health* 76(5). doi: 10.1186/s13690-018-0253-9.

Assan, N. (2022). It's time for reimagining the future of food security in sub- Saharan Africa: gender-smallholder agriculture-climate change nexus. *Trends Journal of Sciences Research* 1, 504. doi: 10.31586/ujfs.2022.504.

Aziz, N., et al. (2021). Women's empowerment in agriculture and household food insecurity: evidence from Azad Jammu & Kashmir (AJK), Pakistan. *Land Use Policy* 102, 105249. doi: 10.1016/j.landusepol.2020.105249, https://www.sciencedirect.com/science/article/pii/S0264837720325874.

Ben-Ari, N. (2014). Gendering agriculture. https://www.un.org/africarenewal/magazine/special-edition-agriculture-2014/gendering-agriculture, Accessed 8 September 2023.

Chigbu, U. E. (2019). Anatomy of women's landlessness in the patrilineal customary land tenure systems of sub-Saharan Africa and a policy pathway. *Land Use Policy* 86, 126–135. doi: 10.1016/j.landusepol.2019.04.041, https://www.sciencedirect.com/science/article/pii/S0264837718319975.

Dillon, Brian & Barrett, Christopher B., 2017. Agricultural factor markets in Sub-Saharan Africa: An updated view with formal tests for market failure, *Food Policy*, Elsevier, vol. 67(C), 64–77.

Doyle K, et al. (2018). Gender-transformative Bandebereho couples' intervention to promote male engagement in reproductive and maternal health and violence prevention in Rwanda: Findings from a randomized controlled trial. *PloS One* 13(4), e0192756. https://pubmed.ncbi.nlm.nih.gov/29617375/.

FAO. 2017a. The State of Food and Agriculture: Leveraging Food Systems for Inclusive Rural Transformation. http://www.fao.org/3/a-i7658e.pdf

Farnworth, C., et al. (2013). *Transforming Gender Relations in Agriculture in Sub-Saharan Africa*. Swedish International Agricultural Network Initiative (SIANI).

Food Agriculture Organization of the United Nations. (2011). Documenting the gender gap in agriculture women in agriculture closing the gender gap (the state of food and agriculture 2010-2011). Rome. https://www.fao.org/4/i2050e/i2050e.pdf

Government of Malawi. (2014) National plan of action to combat gender-based violence in Malawi (2014-2020). Lilongwe: Ministry of Gender, Children, Disability and Social Welfare. www.togetherforgirls.org/wp-content/uploads/2017/10/National-Plan-of-Action-to-CombatGender-Based-Violence-in-Malawi-2014-2020.pdf.

Kamara, A., et al. (2019). The relevance of smallholder farming to African agricultural growth and development. *African Journal of Food, Agriculture, Nutrition and Development* 19(1), 14043–14065.

Keane, R. (2018). Identifying challenges and constraints experienced by women smallholder farmers in northern Ghana. https://a4nh.cgiar.org/2018/09/05/identifying-challenges-and-constraints-experienced-by-women-smallholder-farmers-in-northern-ghana/, *Accessed 9 September 2023*.

Malawi Investment and Trade Centre. (2021). *National Export Strategy II 2021-2026*. https://mitc.mw/trade/images/NES_II.pdf, Accessed 11 September 2023.

Mila, S., and Minot, N. (2018). What factors explain women's empowerment? Decision-making among small-scale farmers in Uganda. *Women's Studies International Forum*, 71, 46–55. doi: 10.1016/j.wsif.2018.09.005.

Muluneh, M. D., et al. (2020) Gender based violence against women in Sub-Saharan Africa: a systematic review and meta-analysis of cross-sectional studies. *International Journal of Environmental Research and Public Health* 17(3), 903. doi: 10.3390/ijerph17030903.

National Statistical Office (NSO) [Malawi] and ICF. (2017). *Malawi Demographic and Health Survey 2015-16*. Zomba, Malawi, and Rockville, MD: NSO and ICF.

Nyambo, P., et al. (2022). Sub-Saharan Africa smallholder farmers agricultural productivity: risks and challenges. In: Mupambwa, H. A. et al. (eds) *Food Security for African Smallholder Farmers. Sustainability Sciences in Asia and Africa*. Singapore: Springer. https://doi.org/10.1007/978-981-16-6771-8_3, https://link.springer.com/chapter/10.1007/978-981-16-6771-8_3

OECD/FAO (2016), OECD-FAO Agricultural Outlook 2016-2025, OECD Publishing, Paris, https://doi.org/10.1787/agr_outlook-2016-en.

Palacios-Lopez, A., et al. (2017). How much of the labor in African agriculture is provided by women? *Food Policy* 67, 52–63. doi: 10.1016/j.foodpol.2016.09.017, https://www.sciencedirect.com/science/article/pii/S0306919216303852.

Phiri, A. T., et al. (2022). A review of gender inclusivity in agriculture and natural resources management under the changing climate in sub-Saharan Africa. *Cogent Social Sciences* 8(1). doi: 10.1080/23311886.2021.2024674

Rogers, M. and Ali, P. (2023). Understanding gender-based violence. In *Gender-Based Violence: A Comprehensive Guide*. Springer International Publishing. doi: 10.1007/978-3-031-05640-6_1

Solidaridad. (2022). Celebrating rural women contributions in sub-saharan Africa. https://www.solidaridadnetwork.org/story/celebrating-rural-women-contributions-in-sub-saharan-africa/, accessed 11 September 2023.

Speizer, I. S., et al. (2018). Changing sex risk behaviors, gender norms, and relationship dynamics among couples in Cape Town, South Africa: efficacy of an intervention on the dyad. *Social Science & Medicine* 209, 95–103. https://pubmed.ncbi.nlm.nih.gov/29843074/

Tantoh, H. B., et al. (2021). Gender roles, implications for water, land, and food security in a changing climate: a systematic review. *Frontiers in Sustainable Food Systems* 5. https://www.frontiersin.org/articles/10.3389/fsufs.2021.707835 DOI=10.3389/fsufs.2021.707835 ISSN=2571-581X

Torday Gulden, K. (2019). International Women's Day: closing the gender gap among smallholders. Empowering women is a key factor in Resilience and InnovAfrica, two projects working towards improving the agricultural productivity and adaptive capacity of smallholders in India and the Sub-Saharan Africa, accessed on https://www.nibio.no/en/news/international-womens-day-closing-the-gender-gap-among-smallholders

Ugwu, (2019). Women in agriculture : challenges facing women in Africa farming, accessed on https://www.researchgate.net/publication/332053861

UN Women. (2019). *The gender gap in agricultural productivity in sub-saharan Africa: causes, costs and solutions*. https://www.unwomen.org/sites/default/files/Headquarters/Attachments/Sections/Library/Publications/2019/UN-Women-Policy-brief-11-The-gender-gap-in-agricultural-productivity-in-sub-Saharan-Africa-en.pdf

World Bank. 2021. Malawi Economic Monitor, December 2021: Addressing Macro and Gender Imbalances. © World Bank. http://hdl.handle.net/10986/36747 License: CC BY 3.0 IGO

Chapter 12

The impact of migration on women smallholders in Ghana

Jemima Nomunume Baada, University of British Columbia, Canada

1 Introduction

A significant proportion of the world's smallholder farmers are women (HLPE, 2013; Othman et al., 2021). According to a 2023 report by the Food and Agriculture Organisation of the United Nations (UN-FAO), 36% of all women employed globally are based in the agricultural sector (FAO, 2023). This figure is even greater for sub-Saharan Africa (SSA) – like much of the Global South – where 66% of women are employed in agrifood systems, compared to 60% of men (FAO, 2023). In addition, studies show that women, particularly in rural and low-income regions and communities, contribute a large proportion of smallholder farm labour (Geleta et al., 2018; Nchanji et al., 2020). Despite this high involvement, many women remain unrecognised as standalone farmers across different agricultural spheres. Even in cases where women smallholders receive recognition, their contributions to farming tend to be underreported, undervalued and underpaid (Doss, 2011; FAO, 2023). Moreover, prevailing gender norms in rural areas imply that women still remain largely responsible for household and caregiving responsibilities, thereby causing them to be

http://dx.doi.org/10.19103/AS.2024.0148.15

overburdened with their multiple responsibilities at home and on the farm (Aregu et al., 2019; Ganguly et al., 2019). This situation is more pronounced in climate-affected, low-income settings where women are reported to experience more time poverty due to needing to source household fuel, water and other domestic resources for family upkeep (Sultana, 2018; USAID, 2020).

Climate change threatens agrarian farming livelihoods. For instance, several studies show that SSA is already experiencing extreme climatic stressors such as rising temperatures, reduced and erratic rainfall, poor soil fertility, droughts and declining agricultural production (Ayanlade et al., 2022; IPCC, 2022). Given that, about 60% of people in SSA are smallholders (Ayanlade et al., 2022), and around 95% of farming in the region is rainfed (Hlophe-Ginindza and Mpandeli, 2021), these climate-related agricultural changes have dire effects on farming options, food security and livelihoods. Furthermore, as mentioned earlier, studies show that women and girls experience vulnerability to climate change effects in different ways than men and boys. For example, women and girls have been found to be more likely to experience immobility and suffer morbidity and mortality, during environmental hazards – including climate-induced ones such as hurricanes, flooding and drought. This is partly due to sociocultural norms around women's mobility and access to resources (Ayeb-Karlsson, 2020; Zahan Tanny and Wakilur Rahman, 2017). Studies also show that in many parts of SSA, including in countries such as Ghana, Kenya and Uganda, more girls and women must forego school attendance and other obligations to source water for household needs (Makina and Moyo, 2016; Nunbogu et al., 2023; USAID, 2020).

These inequalities in climate change vulnerabilities extend to other aspects of climate adaptation and responses as well. To illustrate, research shows that even with the recognition that climate-resilient farming technologies and methods are crucial to mitigating climate-related deteriorating agricultural production, women across many countries in the Global South still have limited access to these innovations (Badstue et al., 2020; Shibata et al., 2020). Again, this is largely due to gendered social and cultural norms that tend to inhibit women's access to resources, and/or their participation in agricultural innovations and services (Badstue et al., 2020; Tsige, 2019). Moreover, in response to the impacts of climatic stressors on agrarian livelihoods, many smallholders dependent on farming for their food and upkeep resort to alternative coping strategies, paramount among which is migration (Kuuire et al., 2016; Luginaah et al., 2009). Many communities, particularly rural ones in SSA, are therefore witnessing high volumes of internal out-migration to areas with relatively better climatic and farming conditions (ibid). However, despite the importance of migration as a livelihood adaptation strategy for many smallholders across SSA, decision-making regarding migration and access to migration opportunities, are very gendered.

For instance, Langill et al. (2023) in their study on Burkina Faso found that sociocultural norms still tend to privilege the migration of men over that of women. Furthermore, studies such as Gaurtala et al. (2010), Lawson et al. (2019) and Tacoli and Mabala (2010) show that the growing emigration of men in agrarian communities is leading to the feminisation of farm labour – referring to the increasing representation of women in farming activities. Despite this agricultural feminisation, in migration origins, the absence of structural changes targeted at reforming social and cultural barriers that hinder women's access to land, inputs, credits and technologies implies that women end up being overburdened with work (Aregu et al., 2019; Tsige, 2019). This is because women must take up roles and perform the labour that male household members were responsible for prior to migrating, and without the privilege of being able to access vital agricultural resources that some of these men did.

Similarly, in migration destinations, Baada et al. (2019, 2021) in their study on Ghanaian women smallholders' experiences of farming, post-migration, revealed that women continue to face structural challenges – rooted in climate change and patriarchal social norms – in accessing and using agricultural resources. This has concomitant effects on their livelihoods and wellbeing. Considering the exacerbated challenges that women face in accessing agricultural resources and opportunities, it is therefore not surprising that many women-led farming households, including in migration origin and destination areas, have been shown to experience higher rates of poverty and food insecurity (Atuoye et al., 2019; Felker-Kantor and Wood, 2012; Kansanga et al., 2020).

To this end, given the impacts of climate change on smallholder livelihoods and the resulting need to migrate for better farming opportunities – as well as the fact that agricultural, migration and other vulnerabilities and adaptation options tend to be gendered – it is crucial to explore the effects of migration on women smallholder farmers in both origin and destination areas. This is timely and urgent as, first, many studies that look at the effects of migration on smallholders often focus on men (Abdul-Korah, 2008; Kuuire et al., 2016; Luginaah et al., 2009). Second, the few that examine migration and women smallholders tend to focus on either the experiences of people in the migration origin/sending society or those in the destination/receiving society (Lawson et al., 2019), but never both simultaneously. It is therefore important to examine the experiences of women smallholders in both origins and destinations given the growing involvement of women smallholders in (non)migration patterns (Atuoye et al., 2021; Baada et al., 2019). It is also essential to recognise that while some women engage in migration of their own volition, others may do so due to gendered obligations rooted in sociocultural norms. Likewise, although some women may be unable to migrate due to gender and sociocultural norms, others may choose to engage in voluntary non-migration (Wiegel et al.,

2019; Zickgraf, 2021). Hence, the need to explore the experiences of women smallholder migrants, non-migrants and return-migrants regarding farming.

Thus, drawing from 30 and 12 in-depth interviews (IDIs) with smallholders and key informants (working with governmental/non-governmental organisations), respectively, as well as five focus group discussions (FGDs) with women and men migrant, non-migrant and return-migrant smallholders, this chapter explores the effects of migration on women farmers in both rural migration origins and destinations of Ghana. This exercise is critical, as worsening climatic conditions in SSA and globally will result in more migrations. Moreover, as indicated, many of the world's smallholder farmers are women; yet, they continue to be unrecognised, undervalued and underpaid across many agricultural circles. These study findings contribute towards understanding the gendered impacts of migration on women smallholders in Ghana and SSA more broadly.

2 Rural migration in Ghana

Stark geographical, economic, climatic and gender inequalities exist in Ghana, and have contributed to long histories of migration between northern and southern Ghana, both voluntary and involuntary (Abdulai and Hulme, 2015; Songsore, 1979; Songsore et al., 1995; USAID, 2020). Specifically, the northern sector of the country, including the Upper West Region (UWR), has high levels of economic and infrastructural deprivation relative to the south. These structural inequalities are rooted in colonial and neo-colonial legacies that intentionally left northern Ghana underdeveloped in comparison to the southern sector (Songsore, 1979; Songsore et al., 1995). During colonial rule, British colonists settled in southern Ghana due to its proximity to their trade route (Gulf of Guinea) and the trade goods they were interested in. Roads, railways, hospitals and other infrastructure were subsequently established in southern Ghana to facilitate colonial trading (Garrett, 1980). Northern Ghana was excluded from these infrastructural developments. Instead, young people (predominantly men) were often recruited, sometimes forcibly, as labour for the mines and plantations in the south. After independence in 1957, subsequent Ghanaian governments continued these uneven development patterns as a way of encouraging 'voluntary' migrations of people from northern to southern Ghana, to keep the now established industries in the south in operation (Songsore, 1979; Songsore et al., 1995). In the early 1980s, Ghana underwent the Structural Adjustment Programme (SAP) at the recommendation of the Bretton Woods twins. As part of this, the government had to cut back on national spending, including removing agricultural subsidies (Konadu-Agyemang, 2000). Smallholders in northern Ghana were the most affected by the SAPs, as the majority were resource-poor and depended on government subsidies to undertake their agricultural activities. These north-south economic and geographical inequities prevail (Abdulai and Hulme, 2015).

Rural-urban geographical inequalities are also prevalent in the country. In 2018, Ghana's multidimensional poverty stood at 0.156, with deprivation in rural areas contributing about 72% to this index – although this figure was as high as 93% in the UWR (FAO and Ecowas, 2018). This is both understandable and concerning, as around 84% of the UWR's population is rural, compared to the national average of 49% (Ghana Statistical Service, 2013). Given the bleak economic outlook in the UWR, with time, a culture of migration emerged, where young adults migrated from northern to southern Ghana for various reasons including to find wage work, as a rite of passage, for exploration, and to accumulate resources to start their own families (Abdul-Korah, 2007, 2008). These migrations were predominantly among men due to sociocultural norms that discouraged the migration of women at the time (Grier, 1992; Koenig, 2005). From the 1970s, however, these migration trends began to change mainly due to worsening climate change effects.

Apart from the geographic and socioeconomic inequalities described above, there are also climate change-related inequities in Ghana. The UWR, like most of northern Ghana, is particularly susceptible to climate change effects due to its vulnerable positioning in the Guinea Savannah Vegetation Zone (Luginaah et al., 2009; van der Geest, 2011). Although middle and southern areas of Ghana are increasingly witnessing climate change effects as well (Adu et al., 2018), these areas still fare relatively better as they are located in the Semi-Deciduous and Deciduous Ecological Zones. As a result, many in northern Ghana migrate to rural areas of the middle belt to undertake farming and generate food and remittances for family upkeep (Kuuire et al., 2016; Luginaah et al., 2009). Rural middle belt areas are the preferred destination due to comparatively better farming conditions here, as well as the area's relative geographical proximity to UWR, affordability of farmlands, and existing migrant networks that have developed over years of in-migration (Abdul-Korah, 2008; Kuuire et al., 2016). In recent times, these rural-rural migration trends have witnessed the involvement of many women farmers. Despite this, women migrant farmers continue to face gendered marginalisation rooted in patriarchal structural, communal and household policies and traditions.

As of 2020, Ghana ranked 133 out of 189 countries in the Human Development Report's (HDR) Gender Inequality Index (GII), which uses sociocultural indicators such as education, health and economic disposition to assess women's empowerment (USAID, 2020). According to an Ecowas and FAO report on Ghana, gender inequalities are prevalent within the country, mainly due to sociocultural norms that privilege men as household heads and women as 'contributing farm workers' (FAO and Ecowas, 2018). Consequently, many women in Ghana have been found to have limited access to agricultural resources such as land, inputs and technologies. Women in this context also have limited power in household and communal decision-making, including ones that involve agricultural production (FAO and Ecowas, 2018; USAID, 2020). The report places women's and men's employment in agriculture at 38% and 46%, respectively, although anecdotal evidence and studies from Ghana and other contexts point to the fact that this figure on women's participation might be underreported due to cultural understandings of who a farmer is (Badstue et al., 2020; Tsige, 2019). Given these gendered inequalities, it is important to examine smallholder women's experiences of migration and farming in rural contexts.

3 Studying the impact of migration on women smallholders

This study employs a qualitative methodology involving IDIs and FGDs to explore the impacts of migration on women smallholders. The data used in this chapter form part of a larger mixed methods study that examined the combined effects of climate change, domestic and foreign investment and rural migration in Ghana. In total, 30 IDIs were conducted among women and men migrants, non-migrants and return migrants in the UWR (origin) and middle belt (destination). These involved 15 IDIs each in the origin and destination communities. Of the 30 IDIs, 17 and 13 were women and men participants, respectively. Twelve IDIs were also conducted with key informants working in governmental and non-governmental organisations in the areas of agriculture, rural development, gender equality, among others. Four key informants were women and eight were men. Lastly, five FGDs were conducted among smallholders in UWR and the middle belt – with three and two in the origin and destination areas, respectively. Participants in FGDs numbered 55 in total, with 22 of them being women and 33, men. The average membership for FGDs was 11. All interviews lasted between 15 min and 2 h. Interview times and locations were scheduled based on participants' preferences, with many taking place in participants' homes and open, private spaces. Participants were advised of their rights and roles in the study, including the right to confidentiality and to withdraw from interviews or the entire study at any point without consequences. Written or verbal consent was obtained before any study activities began.

Data were gathered between October 2019 and August 2021. Participants were recruited from eight districts in both sites. These include the Nandom, Lawra, and Wa Districts in the UWR, the Dormaa East, Dormaa Central, Sunyani and Wenchi Districts in the middle belt, and the Accra Municipal District, which is Ghana's national capital. Only key informant interviews were conducted in Accra (the national capital). Interviews were conducted in English, Dagaare and Twi, the major languages spoken by participants in the study locations.

Semi-structured interview guides were used to facilitate IDIs and FGDs (Berg and Lune, 2017; Hesse-Biber, 2007). Interview questions asked smallholder participants about their experiences of climate change in both origin and destination areas, their motivations for choosing to migrate or not migrate, their experiences of settling in destination societies and/or resettling in the origin, and their experiences of farming in their respective communities. Key informant interview questions revolved around their experiences of working with smallholders in rural areas on issues of climate change, farming and gender equity. Interviews were audio recorded with permission from participants.

Interviews were transcribed directly from the local languages into English, using both literal and contextual translations. Following this, open coding was done to get a sense of the data and identify initial themes, after which detailed line-by-line coding was done (Crang, 2005) with the help of NVivo – a qualitative data management tool. Specifically, each transcript was read thoroughly, and emerging ideas were coded under initially created themes and later themes identified during this follow-up coding process. Data were analysed using an inductive, explanation-building approach. However, only the findings related to the impact of migration on women smallholders are presented in this chapter. Ethics approval for this study was obtained from Western University's Research Ethics Board before any study activities commenced.

4 Findings: migration motives and patterns

Study findings are organised under the major themes related to gender, migration and smallholder farming that emerged during interviews. First, an overview of migration motives and patterns in both origin and destination communities is presented. Subsequently, the impacts of migration on women smallholders in the migration origin (UWR) are outlined, followed by those from the destination (middle belt). Participants' names have been replaced with pseudonyms to protect their identities. Direct quotations from IDIs and FGDs are contextualised with participants' pseudonyms, sex, district of residence, age (where applicable) and interview method. Quotations from women participants are prioritised in this chapter.

To understand how migration impacts women smallholders, it is important to first grasp UWR-rural middle belt migration patterns and motives. In the migration origin, study findings revealed a variety of migration motives and im/mobilities. Overall, several communities in UWR are witnessing high volumes of outmigration to the middle belt for farming purposes. This is mainly due to the combined effects of climate change and economic deprivation in the region. Migration decisions are made by both individuals and households. At the individual level, several participants indicated that migration is largely a self-selection process. This entails individuals making the decision to migrate based on their economic and social needs, as well as the resources and support available to undertake this migration.

Despite the self-selected nature of individually-decided migrations, some general patterns stood out. First, most migrants are youth and adults aged 18–45, although persons outside this age range migrate as well. Second, adults between the ages of 28-45 often prefer to migrate to rural areas for farming purposes, whereas those 27 or younger favour migration to urban areas to undertake wage work. In some instances, younger adults migrated to rural areas as well, but often for jobs outside of farming – for example, artisanal small-scale mining or commercial agriculture. According to participants, the preference of younger migrants for urban areas is due to the desire to enjoy the benefits of urban life – such as social amenities like electricity and technology, and entertainment options like bars and sports. For unmarried migrants, urban areas also provided more options for finding a potential partner, compared to rural areas that tend to be remote and secluded. Third and relatedly, many migrants who move to rural middle belt areas often do so as a family, due to the need for informal labour to undertake their farming in the destination. Fourth, while more women are participating in migrations, again, younger and/or unmarried women often prefer urban areas for the reasons listed above, whereas older and/or married women are more likely to move with family and settle in rural locations for farming.

Regarding household migration decisions, these were predominantly undertaken by the extended family and still tended to favour adults between the ages of 18–45. In addition to encouraging the emigration of younger adults, households also prioritise the migration of members considered to be of good health, as persons within this age and health group are considered stronger, and would therefore be more 'productive' in the middle belt. Furthermore, household decisions favour male migration; and in cases where these men had nuclear families of their own, they might move with the family or engage in seasonal migration. This preference for male migration is rooted in sociocultural norms and expectations of men as breadwinners and household heads. However, another reason why younger men are preferred migration candidates is because, traditionally, the oldest male sibling is expected to remain in the origin to keep the homestead going. In cases where older siblings migrate, sometimes, the youngest male sibling is expected to take over this role. In addition, households prioritise migration for farming purposes, as they found this to be more reliable for meeting food needs, compared to wage activities such as artisanal mining or construction work, which fluctuated and were unpredictable. Moreover, participants expressed that even if household members made some money from jobs outside farming, it might not be enough to meet household food needs after other expenses were deducted. Lastly, households often choose members regarded as having good work, financial and family ethics. A participant quote from an IDI captures this really well:

> If you send someone who is lazy, they will go and won't work. If you send someone who doesn't respect money, they can work but will end up using all the money to drink. And if you send someone who is hardworking and can keep money well, but who doesn't think about or have the family at heart, they will forget about you once they get there... So that is why we choose those who have all these qualities (Man, Nandom, FGD 5-Mixed).

Migration types also encompassed diverse forms of mobilities. The majority of migrations from UWR are considered temporary, although interviews with participants in destination areas reveal otherwise. Specifically, many migrants had been residents in the middle belt for over 10 years, some as long as 40, but still considered themselves temporary migrants. Within these temporary forms of migration, many engaged in cyclical/circular movements, migrating once or twice a year during the rainy season and returning once the farming season was over. Others engaged in longer forms of temporary migration, moving for between 2 and 5 years and sending regular food and monetary remittances during this period. Some migrants had (re)settled in more than one destination community. A few acknowledged being permanent migrants; and some engaged in what scholars term pendulum migration, where migrants set up permanent homes in origin and destination societies, and travel between both for almost equal amounts of time.

A noteworthy finding is that many in UWR are also immobile. Older adults, women, and people with disabilities/chronic illnesses form the majority within this group. However, there are within-group variations, as some participants' non-migration was voluntary, while others would have preferred to migrate but either lacked the physical and economic resources to do so or were passed over during household migration decision-making processes. For people who migrated and returned to their origin, many cited ageing and related health decline, although some younger adults also returned due to deteriorating health. Lastly, some non-and-return migrants indicated that their stay in their origin was due to the need to perform caretaking and caregiving responsibilities ascribed to them. The impacts of these migration motives and patterns on women smallholders in origin and destination areas are discussed in the following sections.

5 Findings: impact of migration on women smallholders in the Upper West Region

5.1 Loss of household farm and domestic labour

Most participants in the migration origin acknowledged that emigrations to the middle belt are helpful for meeting household food and income needs, however, they added that these migrations come at a cost. For the majority of women smallholders, a downside was the loss of household labour. As indicated, many in the UWR are resource-poor and therefore cannot afford to hire wage labour for farming; hence, they depend on household members (partners, children) to farm. As such, the migration of family implies that labour gaps are created, as the participant below discusses.

> Well, we don't have money to hire people to do by-day [wage work]. So it's the young ones that we usually rely on to do the farming. But now all of them are running down there because they say there's nothing here for them. That really affects us, because even if you get the land, you don't have anyone to help you farm (Woman, Nandom, FGD 5-Mixed).

Although both women and men smallholders are affected by this loss of labour, the effects are particularly pronounced for women, given the male-oriented migration dynamics discussed prior. A participant in an IDI speaks to this.

> Okay, both women and men migrate. However, more times it's the men, because they say they are stronger, and they also want to be able to take care of their families. So it's the men who go, leaving the women. But you think about it, if they say it's the men who are strong, but it's also the men who leave, how are the women supposed to farm now?... But still we have to farm (Mercy, Woman, Lawra, 47, IDI).

Similar to the sentiments expressed in the above participant's response, conversations with other female participants revealed a feeling of a sense of unfairness that women had to farm under the same climatic and agricultural conditions that many men were unable to farm in, hence their migration.

> I don't think the men migrate because they intentionally want to punish us. But the truth too is, it's the women who end up suffering. We all agree that our land isn't good for farming, which is why many run down south to get something. Yet, those who have travelled still expect that those of us who remain here do something small too. If they come back and you haven't done anything, they'll call you a lazy person (Nomu, Woman, Lawra, 49, IDI).

As shown in the quote from Nomu, despite the widespread recognition that farming in the UWR is becoming more challenging, with associated declining agricultural productivity, many in the origin are still expected to 'avoid idling' during the farming season. The fact that men are encouraged to migrate while women stay in the origin indirectly means that women smallholders are expected to farm in the absence of male migrants, even under the stressful agricultural conditions. Although many participants talked about the impacts of male outmigration on farming, others discussed the effects of female outmigration on women smallholders as well.

> Growing up, I was expected to be by my mothers' side at all times – helping them in cooking, fetching water, weeding, and taking food to farmers on the field. But now our daughters, as soon as they come of age, they tell you, 'mama, I want to go and wash chop bar [local restaurant] bowls in Techiman'. If you say no, the next day, she'll pack her bag and run away. So now if our daughters aren't here to help us in the housework, it means we have do everything by ourselves: fetch the water, gather firewood, forage for shea nuts, prepare the shea butter, and still farm (Puvila, Woman, Lawra, 55, IDI).

Puvila's sentiments succinctly capture the increased workloads that women smallholders in the origin have to perform as a result of the outmigration of younger people. In this case, the emigration of female children has adverse effects on women as, in addition to stepping into men's previous roles in farming, they also have to take up labour roles that younger women previously performed or were responsible for. Many participants added that the impact of this household labour loss is more detrimental to women smallholders due to their inability to access resources to support farming in the origin.

5.2 Lack of economic, social and cultural resources

In heavily patriarchal contexts such as rural Ghana, societies, roles and responsibilities are highly gendered. For example, men are considered household heads and must meet breadwinning responsibilities, whereas women are considered 'helpers' and must perform caregiving responsibilities. Due to these expectations of men as household heads, many agricultural services and groups are often male-biased, whether intentionally or unintentionally, inevitably limiting women's access to some of these farming resources and groups. Even in cases where group membership is open to women, some female participants expressed discomfort in participating in these activities for reasons of cultural propriety. Consequently, for women of the origin whose male household members have migrated, accessing agricultural resources and services becomes challenging.

> Sometimes they come and share fertilisers, seeds and cutlasses at the farmers' meetings. I don't attend the meetings because I'm a woman. I don't want them to say to say I want to wear the same trousers as the men. So I don't attend. But if I don't go to the meetings, then I can get some of the things that they're sharing, have you seen? (Woman, Nandom, FGD 3-Women Only).

Thus, some participants noted that prior to migrating, men within their households attended these meetings and benefitted from such initiatives. However, ever since they left, the women at home, despite still farming, could not access the same spaces and resources, as elaborated below.

> When my sons were here, they used to attend the meetings. And because of that, if some help came to the community, they would hear about it and also go for some. Sometime ago they even gave them small loans. But since they [sons] left, I haven't heard or received anything (Kyaapuorey, Woman, Nandom, 61, IDI).

Given the high rates of economic deprivation in the UWR, the ability to access subsidised agricultural tools, inputs, loans and services makes a substantial difference in women smallholders' farming outcomes. Apart from the exclusion from communal resources due to sociocultural reasons, many women also expressed that farming in the origin could be even more challenging if emigrant children did not provide financial support. The need to find better livelihoods and transfer food and monetary remittances – to support household food requirements and expenses in the origin – is the major reason for encouraging migrations at the individual and household levels. Nonetheless, participants emphasised that experiences of farming in the origin varied greatly depending on who had migrated and the support that they provided for those in the origin.

> I can't speak for everyone, but, the fingers are not the same [proverb]. Maybe for some people, migration is helping them. But for some of us, we don't reap any benefits. Some children usually send their families food, money, clothing, fertilisers and many more. For my sons, if they even pick up my phone call, I'll be happy. So obviously, if you have someone to help you, you can make progress with the farming here. But if you don't, like me, you are doomed (Mwinsom, Woman, Lawra, 72, IDI).

A sense of abandonment by her migrant sons is embedded in Mwinson's narrative. Speaking to this theme of abandonment, a key informant noted that in his work with rural, smallholder communities, he has often witnessed situations of men who claimed to be migrating temporarily to gather resources for their family in UWR, only to never return.

> You see it and you hear about it many times. Some of the men will tell their wives that they're going to farm and return with food, and that's end. They disappear and go and start new families in the south. It's very sad really (Key Informant, Man, Community Development Alliance (CDA), Upper West).

In addition to highlighting the importance of economic, social and cultural support/resources for women's farming in the origin, these findings also underscore the differentiated nature of the experiences of women smallholders in the UWR. I elaborate on these differentiated and intersectional vulnerabilities in the next section.

5.3 Intersectional experiences of smallholder women in the migration origin

Although, generally, all women smallholders in UWR experience challenges in farming and their daily lives as a result of migration, these vulnerabilities are felt differently based on individual and household dispositions. For example, the findings reveal that older women, women with disabilities or chronic illnesses, women caring for disabled or sick household members, and widows, experience heightened marginalisation in smallholder farming due to migration. Regarding age, many older women talked extensively about how the emigration of younger adults, both male and female, leaves them with limited options for farming and caring for themselves.

> You see the dirty clothes piled over there, I need to wash them, but I can't; my body aches and pains. My fingers feel cramped and my shoulder is constantly hurting... How then can I hold a hoe to farm? I can't any more. I used to work hard back in the day, but now if my sons don't come and help me, it means I won't eat. As for the girls, they're all focused on their husbands and children down there (Faa-Mmenga, Woman, Lawra, 80, IDI).

In addition to the feelings of neglect reflected in this quotation by an older participant, the response also illuminates how older smallholder women whose household members have all migrated experience heightened marginalisation, with exacerbated impacts on their food security. Another group that experienced particularly negative effects of migration on their farming abilities were women who either had disabilities or chronic illnesses themselves, or were caring for a disabled or sick relative.

> I used to migrate, but since Junior's father [participant's son] came and left him here with me, I can no longer do anything. He's like a baby; I have to feed him, bathe him, everything. Where will I get the time to farm? I can't put him on my back and farm. At least, occasionally his father sends us a little food. If not, I don't know how we would survive (Kyaapuorey, Woman, Nandom, 61, IDI).

Thus, due to the need to care for her grandson with cerebral palsy, Kyaapuorey is unable to make time for other household activities, let alone farming. This is made worse by the fact that all other relatives within her household have migrated, consequently implying that Kyaapuorey must forego her own migration aspirations to care for Junior. Widowed women smallholders were another group who faced challenges within UWR, as their marital status limited their access to land, a crucial farming resource, due to customary traditions around land ownership and access/use.

> How can four widows live alone? ... After we all lost our husbands, they just came and left us together in this house. Women don't own land. We don't have land to farm. Even if we did, we can't afford hoes and cutlasses, let alone fertilisers. So we're just here, begging to eat... All the men have either died or scattered [migrated] (Kaunsob, Woman, Nandom, 61, IDI).

Similar but distinct sentiments were expressed by women who had never given birth to male children. According to them, male sons came with benefits such as being able to secure favourable land leasing options and receiving remittances from these migrant male children.

> My husband and I tried for a long time, but no boys. All six children are girls. There are benefits to having girls, but there are some things they just cannot do. So when our neighbours' sons are migrating and sending them food and money, we cannot enjoy those benefits (Woman, Nandom, FGD 5-Mixed).

Despite these challenges and multiple burdens that women smallholders must perform regarding farming and household sustenance, many participants acknowledged that women in the origin still have limited access to food harvests and decision-making regarding agricultural activities, as a male participant in an FGD recounts.

> Let us not lie, the truth is that, when the maize is on the farm, it belongs to both partners. When it's time to weed around the groundnuts, the wife and husband are both farmers. When it's time to harvest the millet, the farm belongs to both. But once the food makes it to the barn, the wife no longer has a say. She ceases to be a farmer. That is the truth... So the women suffer more (Man, Nandom FGD 4-Men Only)

Unsurprisingly, the marginalisation faced by women smallholders in the migration origin leads to higher rates of poverty among these groups, and especially in predominantly female or female-led households. Yet, women in UWR are still expected to meet gendered/sociocultural obligations as food providers. Given the challenges that even families with migrant sons experience, one can only imagine the hardships that women-led households undergo, even ones that have supportive migrant relatives.

> Poverty is killing us here. My sister remarried and migrated, but her children remained. They are mouths to feed. School fees must be paid. Where do I get the money? Our farming isn't going well. Even though my sister tries to help, she is still settling. They just went this year. So it will take time before she can send us food and money. It's really tough (Zinni, Woman, Nandom, 30, IDI).

This quotation illuminates the fact that even if benefits will be derived from smallholder migration, this takes time, as migrants must first settle in the destination before they can begin sending remittances. In such cases, women in the origin must figure out how to sustain the household until such a time. Nevertheless, some participants acknowledged that if farming is productive for household members in destination areas, and these household members are responsible and family-oriented, migration does provide immense positive benefits for people in UWR.

6 Findings: impact of migration on women smallholders in the Middle Belt Region

6.1 Access to farming resources and decision making

The findings in the migration destination reveal that women smallholders also face challenges in farming and benefiting from their farming livelihoods. These challenges are similarly rooted in gendered/sociocultural norms reproduced postmigration, that affect their access to resources such as land, inputs, credits and others. However, these challenges in the destination play out in distinct ways. Postmigration, men are still regarded as household heads and breadwinners, whereas women are considered 'helpers'. Some women highlighted the contradictions in these assumptions for migrant farming in the middle belt, as shown below.

> Hmm, my daughter, you think about it. I was at home [UWR] and you [husband] told me to come and join you so we could farm together. If I wasn't a farmer, would you bring me here? If I wasn't strong enough, would you have asked me to come? So that's how it is. When it suits them, we are farmers, and when it doesn't, we aren't farmers (Tiere-bio, Woman, Wenchi, 50, IDI).

Participants added that the traditional conceptualisations of men as farmers and women as helpers have detrimental effects on women's ability to access agricultural resources like land, and/or undertake independent farming.

> When we go to ask for land, the owners ask, 'where is your husband?' If you say you're by yourself, they'll ask why you didn't come with your husband. 'Is it because you're trying to hide the land from him... only difficult women do that.' We don't lease to difficult women (Felicity, Woman, Wenchi, 37, IDI).

The perception of women smallholders with interests in undertaking independent farming as 'improper' or 'difficult' discourages many migrant women from venturing down the path of individual farming. In cases where women needed their own farms, these had to be smaller (typically regarded as gardens) and were often obtained with permission from their partners. Women smallholders added that even in cases where landowners were willing to lease to them in the absence of their husbands, the lease terms tended to be exploitatory – as many landlords perceived the women as desperate and therefore figured they would be willing to pay more for the lands. Also, landowners often assumed women knew nothing about issues related to land (acquisition), hence, they could get away with charging exorbitant amounts. Lastly, given the restrictive customs around women's land acquisition, many landlords felt that women smallholders' options for securing lands elsewhere were limited. Many women participants opined that these dynamics had detrimental effects on them, especially in cases where they were not benefitting from household farming and needed their own source of food/income even more:

> It's not all the time that you must rely on your husband, especially if he doesn't allow you to touch the farm produce without his permission... if you need to buy salt, it must be your husband. Even underwear that you must buy, it has to come from him. So it's good to have something on the side. But it is very difficult to do so here (Sung, Woman, Sunyani Municipal, 49, IDI).

Or where household income/livelihoods needed to be supplemented, as the participant below describes.

> It's good to farm with your husband. But it's not every year that the maize and cassava will do well. Sometimes, as a woman, you need to also have something on the side. A vegetable farm, so that if your husband says one day, 'I don't have money for food', you can also go into your farm and harvest your produce to help out (Kuu-ima, Woman, Dormaa Central, 43, IDI).

The ability to own one's farm is especially necessary for women smallholders in instances where migrant men, who are ascribed headship and breadwinning roles, are struggling to meet their obligations. However, some women noted that interestingly, even when it was obvious their partners were facing difficulties in farming, they still did not take kindly to women taking up major roles in agricultural decision-making and activities. For instance, a participant notes how her husband (who was struggling with alcohol dependency) resorted to violent behaviour when he found out that she had replaced yam seedlings with cassava ones when the yams perished.

> Two years ago, he insisted on planting yams, even though I told him cassava would be better suited to the land. He didn't listen and went ahead. Okay, the yams died and I realised we might sleep hungry that year. So I quickly planted cassava in the hopes that we would at least get something. When he found out, he got violent and physical with me. He told me to never forget that he is the man of the house (Zenebia, Woman, Dormaa East, 38, IDI).

Participants like Zenebia whose partners were not supportive of their contributions to farming added that this reduces their (women smallholders) productivity and prevents them from living up to their full capability as farmers, as captured below.

> I acknowledge he is the man of the house; I will never even think of competing with him for that position. But now, he holds me back. You can ask anyone that knows me; I don't fear hard work. I can farm like a man. So even if he's not interested in farming – and it's obvious he isn't – he should at least let me farm so the children won't sleep

> hungry. He won't take farming seriously -- always talking ambitiously about growing teak for export, which isn't realistic. But he won't let me farm too (Zenebia, Woman, Dormaa East, 38, IDI).

Women smallholders added that while their marginalisation in deciding on and leading farming is not solely due to migrating, migration does introduce new dimensions to these challenges. For instance, in Zenebia's case where her partner is struggling with alcohol dependency, if they were back in the UWR, extended family members could have intervened to support them with both drinking and farming. Similarly, another participant notes that in the migration origin, the extended family served as mediators during conflicts, including farming-related ones. As a result, women could more easily express their displeasure with some farming decisions knowing that support from other household members was available.

> They say we lack farming knowledge and are not farmers. But the truth is, we can tell you how the rainy season will go and what crops will do well that year. Back home, you can sometimes voice these opinions. If other household members consider what you've said and think you're right, they will call your husband and tell him he has no case, and he'll backtrack. But over here, if you joke, it'll be a big fight with no referee (Antuna, Woman, Wenchi, 48, IDI).

The above findings underscore the role of extended family in supporting women's farming decisions, and how migration affects this support system. Like their counterparts in the migration origin, women smallholders in destination areas also experience agricultural vulnerabilities in unique ways based on their social and identity categories.

6.2 Intersectional experiences of women smallholders in the destination

In destination areas, women smallholders' background as migrants/outsiders affects their options for buying and leasing lands, as well as their ability to access some communal-based farming resources. Thus, many indicated that getting fertile lands at an affordable cost is challenging, as better-yielding lands are more expensive. Given that many migrants are already resource-poor, coming up with the extra amounts needed to purchase these lands proves difficult. Likewise, many migrant women mentioned that due to their status as outsiders, they are not always informed of, or and involved in, communal activities within their settlement sites. Consequently, even when agricultural support comes to their communities, they often hear about them after the fact.

> They don't see us as one people, so when any help comes to this village, they never tell you anything. You only hear whispers of it, and if you follow up, they'll tell you

> that they brought the help for community members, but you are from north (Christy, Woman, Dormaa East, 25, IDI).

Although the exclusion of migrant farmers from agriculture-based communal support is across-the-board, again, women are disproportionately affected; because in the few instances where migrants are invited to participate in activities or share in resources, these invitations are typically extended to male migrants, further marginalising women smallholders. Apart from these effects of migration status on farming, participants added that women smallholders who are widowed also experience more barriers to farming in the destination society mainly due to the difficulties of securing land, affording farm labour and inputs, and receiving fair prices for the sale of farm produce. A widowed woman speaks to her experiences.

> When we first came, things were good. My husband was a hard worker so we used to get plenty food. Everything changed when he died... The owner came for his land because we couldn't renew the rent... said we were wasting his land. Like I said, well, the children try their best, but they're still too young and lack the strength to farm like their father (Lebkaa, Woman, Wenchi, 48, IDI).

As highlighted in the above quote by Lebkaa, the ability to provide labour for farms in the middle belt makes a significant difference in farming productivity, especially as many migrants cannot afford external labour. Also, given that many rely on family labour for their farms, having an adult family who can undertake this farming is crucial; and for migrant households that lack this, agricultural output is greatly affected. Still, regarding differentiated vulnerabilities and smallholder migrant labour, some participants shared that despite their hopes of making the most of their stay in the middle belt, slow-onset or sudden illness had thrown them off course, as described below.

> I was very strong when we first came here, so I managed to get something from farming. Over time, my health started to decline. They say I have hypertension. I also have terrible back pain. I can no longer work the way I used to. This affects my farming (Kanyiri, Woman, Sunyani Municipal, 48, IDI).

These findings illustrate that migrating to the middle belt has the potential to help improve agricultural productivity for migrant women smallholders from UWR – with the associated benefits of meeting food and income needs for both migrants and their families in the origin. However, the ability to do so is dependent on the availability of economic, social and cultural resources/support to undertake this farming.

7 Summary and conclusion: how migration affects women smallholders who stay and those who migrate

This chapter explored the impacts of migration on women smallholders in rural origin and destination communities of Ghana. The findings show that women face gendered, economic and sociocultural marginalisation in undertaking farming and accessing/using agricultural resources and produce. Although women in both the origin and destination face these challenges, there are distinct differences among women in each site.

In the UWR (origin), high outmigration leads to the loss of household farm and domestic labour. Given that women are ascribed caregiving responsibilities while still required to farm, they end up overburdened with work. These findings support those of studies such as Langill et al. (2023) and Tacoli and Mabala (2010) who observe that in areas of high outmigration, women who remain in their communities end up with a triple labour burden, as they must continue to perform their own dual responsibilities of domestic work and farming, while also filling the farming roles that male relatives performed prior to migrating. The findings further reveal that in places like UWR where significant numbers of (younger) women are also migrating, women smallholders are left with even more demands on their time, since the labour support that these younger women could have provided – for example in household chores – is lost.

In addition, the economic deprivation of many in UWR implies that women smallholders have limited access to farming tools and inputs that are crucial for farming. Again, this is consistent with studies such as Tsige (2019) and Nchanji et al. (2020) who found that in countries like Ethiopia and Tanzania, women smallholders continue to face barriers to obtaining agricultural resources, which affects their productivity. For some women in this study, these challenges of accessing structural-level farming support in the origin are worsened by the fact that they also receive minimal support from migrant relatives in the destination, while still needing to shoulder the responsibilities that these relatives left behind. Thus, some women smallholders acknowledged that in situations where they received financial and food support from migrant relatives, migration provided immense benefits. This finding supports prior research by Kuuire et al. (2016) and Luginaah et al. (2009) who found that migration has positive outcomes for food and income security among households in UWR. However, it also contrasts the aforementioned studies, as it highlights the fact that not every household in the origin experiences these benefits, hence the importance of examining within-group differences.

Study findings also demonstrate that migration affects women smallholders differently based on individual and social factors. To illustrate, older women, widows, women without male relatives, women with disabilities/chronic illnesses, and those caring for sick/disabled relatives, experienced more farming-related vulnerabilities. In the case of sick and older women smallholders, they lack the physical strength to undertake farming, particularly given the climatic conditions that make farming in UWR challenging. Women caring for sick/disabled relatives on the other hand struggle to make time to farm, given the intensive nature of caregiving in the absence of structural and household support. For widowed smallholder women, their access to land and other crucial agricultural resources such as inputs are affected due to gendered social norms that restrict women's land and resource rights. Likewise, the absence of a male relative to attend farming meetings and advocate on households' behalf, affects women smallholders' ability to benefit from agricultural resources and services that come to their communities. These findings underscore the importance of improving rights/access to land and agricultural resources for women in Ghana, like much of the Global South (Baruah, 2009; Aregu et al., 2019). They also emphasise the need to pay attention to how women smallholders' individual and social identities shape their agricultural activities and outcomes (Badstue et al., 2020).

In the middle belt (migration destination), the findings show that women smallholders similarly face gendered, economic and sociocultural challenges in accessing farming resources and undertaking productive farming. Although some of these barriers stem from patriarchal norms reproduced in the destination area, there are differences. For instance, women in the destination area also face difficulty renting land due to both gendered customs and their status as migrants or outsiders. Hence, women explained that it was difficult to lease/buy lands by themselves, and in situations where they did, these were often on exploitatory lease agreements. This finding is buttressed by earlier studies that found similar dynamics of inequitable land negotiations among migrants and/or women smallholders in destination communities in southern Ghana (Baada et al., 2019; Kansanga et al., 2020).

Another distinctive finding in the destination was that, apart from experiencing restricted access to farm produce, post-harvest, migrant women smallholders also experience limitations in the extent to which they can undertake farming-related decision-making. For many, this stems from the fear that their input could anger their partners, and without extended family to serve as mediators, household disputes could get out of hand. Consequently, even when migrant men were struggling to fulfil ascribed obligations as breadwinners, they still resented the idea of having their female partners lead farming decisions and activities. Subsequently, taking farming initiatives without a partner's knowledge/consent opened women up to gender-based violence (GBV). This is consistent with other studies in SSA that show that household negotiations around agriculture may leave women vulnerable to GBV (Jakobsen, 2014; Ragetlie et al., 2022). This finding also speaks to the ways in which rapidly changing economic and social contexts that destabilise gender roles – particularly for low-income men – may be seen as a threat to masculinities and male privileges (Kelbert and Hossain, 2014), through their revelations of the limits of patriarchy. As a solution, affected men may resort to violent behaviour as a way of reasserting their masculinity.

Lastly, the findings reveal that women smallholders in the destination also face intersectional challenges in undertaking farming in their settlement communities. Specifically, women's outsider status excluded them from many communal activities and support, similar to what Baada et al. (2021) found in their study on migrant farming in rural communities of Ghana. Moreover, widowed women and those with disabilities/chronic illnesses were also exposed to more vulnerability, due to their inability to afford land, labour and inputs to undertake productive farming.

Based on these study findings, and interviews with key informants, some policy interventions at the structural, local, household and individual levels would help to promote better agricultural outcomes among women smallholders in rural areas of Ghana. At the structural level, more attention must be paid to implementing equitable, climate-resilient/adaptive farming strategies in both migration origin and destination communities. These could include the provision of improved seed, and water access during the dry season (e.g. irrigation systems adapted to low-income, rural Ghanaian contexts). This adaptive farming could be supplemented with subsidised agricultural equipment and inputs that prioritise women and low-income individuals. Further, given that some people (older adults, people with disabilities, caregivers for disabled family) are unable to undertake farming due to their individual circumstances, it is important that the government of Ghana consider social protection nets that guarantee consistent, monthly welfare stipends. This would ensure that persons living in vulnerability are able to meet their basic needs. While a version of this (the Livelihood Empowerment Against Poverty Programme-LEAP) exists, interviews with participants and key informants revealed that the current amounts are insufficient (maximum of GHS 212 or USD 22.7 per month for a four-member eligible household) and disbursement of funds is irregular.

At the local level, given the high outmigrations in the country and lack of targeted interventions for migrant groups, establishing migrant desks/representatives within government administrative offices at the regional, district and communal levels would provide (prospective) migrant women smallholders with avenues to seek and access migration information/resources, and file grievances. Also, providing women in climate-affected rural areas with off-farm employment opportunities such as informal businesses, trades and crafts would help to ensure that they have livelihood options outside of farming. This may also help to reduce outmigration for people who would prefer to remain in UWR, with resulting benefits for mitigating the challenges that accompany high volumes of outmigration. Lastly, at the communal and individual levels, undertaking sensitisation campaigns rooted in culturally sensitive and gender transformative approaches – among women, men, migrants and host populations in origin and destination areas – may help to reduce the discrimination that women smallholders face due to the intersections of their gender, migrant and low-income status.

In conclusion, this study explored the impact of migration on women smallholders in both rural origin and destination areas of Ghana. In doing so, it has illuminated the unique ways that the migration of household members, and/or women smallholders themselves, shapes their experiences of farming. It also emphasised the gendered and intersectional differences faced by diverse women smallholders in carrying out their livelihoods. These findings are useful for local and national policymakers in Ghana, as they inform our understanding of women smallholders' experiences of farming within a migration context. They could further help in the design of locally relevant policies that mitigate gendered economic and sociocultural barriers to farming in migration origin and destination communities, to ensure that women smallholders can thrive.

8 References

Abdul-Korah GB (2007) 'Where is not home?': Dagaaba migrants in the Brong Ahafo Region, 1980 to the present. *African Affairs* 106(422): 71–94. DOI: 10.1093/afraf/adl023.

Abdul-Korah GB (2008) Ka Biɛ Ba Yor: labor migration among the Dagaaba of the Upper West Region of Ghana, 1936– 1957. *Nordic Journal of African Studies* 17(1): 1–19.

Abdulai A-G and Hulme D (2015) The politics of regional inequality in Ghana: state elites, donors and PRSPS. *Development Policy Review* 33(5): 529–553. DOI: 10.1111/dpr.12124.

Adu DT, et al. (2018) Application of livelihood vulnerability index in assessing smallholder maize farming households' vulnerability to climate change in Brong-Ahafo region of Ghana. *Kasetsart Journal of Social Dciences*, 39(1): 22–32.

Aregu L, et al. (2019) Implications of agricultural innovation on gender norms. In: Sachs CE (ed.) *Gender, Agriculture and Agrarian Transformations: Changing Relations in Africa, Latin America and Asia*. New York: Routledge, pp. 162–179. DOI: 10.4324/9780429427381-10.

Atuoye KN, et al. (2019) Household income diversification and food insecurity in the upper west region of ghana. *Social Indicators Research* 1–22. DOI: 10.1007/s11205-019-02062-7.

Atuoye KN, et al. (2021) Who are the losers? Gendered-migration, climate change, and the impact of large scale land acquisitions on food security in coastal Tanzania. *Land Use Policy* 101: 105154. DOI: 10.1016/j.landusepol.2020.105154.

Ayanlade A, et al. (2022) Extreme climate events in sub-Saharan Africa: a call for improving agricultural technology transfer to enhance adaptive capacity. *Climate Services* 27: 100311. DOI: 10.1016/j.cliser.2022.100311.

Ayeb-Karlsson S. (2020) When the disaster strikes: Gendered (im) mobility in Bangladesh. *Climate Risk Management* 29: 100237.

Baada JN, et al. (2019) 'What we were running from is what we're facing again': examining the paradox of migration as a livelihood improvement strategy among migrant women farmers in the Brong-Ahafo Region of Ghana. *Migration and Development* 1–24. DOI: 10.1080/21632324.2019.1573564.

Baada JN, et al. (2021) Looming crisis – changing climatic conditions in Ghana's breadbasket: the experiences of agrarian migrants. *Development in Practice* 31(4): 432–445. DOI: 10.1080/09614524.2020.1854184.

Badstue L, et al. (2020) Women farmers and agricultural innovation: marital status and normative expectations in rural ethiopia. *Sustainability* 12(23): 9847. DOI: 10.3390/su12239847.

Baruah B (2009). Monitoring progress towards gender-equitable poverty alleviation. *Progress in Development Studies* 9(3): 171–186.

Berg LB and Lune H (2017) A dramaturgical look at interviewing. In LB Berg and H Lune *Qualitative Research Methods for the Social Sciences* (9th edn). New York, NY: Pearson, pp. 65–93

Crang M (2005) Qualitative methods: there is nothing outside the text? *Progress in human geography* 29(2): 225–233. DOI: 10.1191/0309132505ph541pr.

Doss C (2010) *The role of women in agriculture Prepared by the SOFA Team and The Role of Women in Agriculture 1 Prepared by the SOFA Team*. ESA Working Paper No. 11-02. The Food and Agriculture Organization of the United Nations. https://openknowledge.fao.org/server/api/core/bitstreams/a2537cd3-2aec-4146-8670-521bd77e06e7/content

FAO (2023*) THE STATUS OF WOMEN IN AGRIFOOD SYSTEMS*. Rome. https://doi.org/10.4060/cc5343en

FAO and Ecowas (2018) *National Gender Profile of Agriculture and Rural Livelihoods – Ghana*. Country Gender Assessment Series. Accra: FAO and ECOWAS Commission.

Felker-Kantor E and Wood CH (2012) Female-headed households and food insecurity in Brazil. *Food Security* 4(4): 607–617. DOI: 10.1007/s12571-012-0215-y.

Ganguly S, et al. (2019) Changes in participation of women in rice value chains: Implications for control over decision making. In *Gender, agriculture and agrarian transformations*: Routledge, pp. 239-253

Garrett LM (1980) The comprador on the Gold Coast: A study of merchant entrepreneurs during the seventeenth and eighteenth centuries (Doctoral dissertation, Centre of West African Studies, University of Birmingham).

Gaurtala HN, et al. (2010) Feminisation of agriculture as an effect of male out-migration: unexpected outcomes from Jhapa District, Eastern Nepal. *International Journal of Interdisciplinary Social Sciences* 5(2): 407–412.

Geleta EB, et al. (2018) "The pluses of pulses": haricot beans and women's empowerment in Ethiopia. *Development in Practice* 28(2): 311–317. DOI: 10.1080/09614524.2018.1417353.

Ghana Statistical Service GSS (2013) The 2010 population and housing census. *National Analytical Report*. Government of Ghana, Accra.

Grier B (1992) Pawns, porters, and petty traders: women in the transition to cash crop agriculture in colonial ghana. *Signs: Journal of Women in Culture and Society* 17(2): 304–328. DOI: 10.1086/494732.

Hesse-Biber SN (2007) The Practice of Feminist In-Depth Interviewing. In: *Feminist Research Practice*. Sage, pp. 110–148. DOI: 10.4135/9781412984270.n5.

Hlophe-Ginindza and Mpandeli (2021) The role of small-scale farmers in ensuring food security in africa. In: *Food Security in Africa*. pp. 1–14. IntechOpen.doi:10.5772/intechopen.91694

HLPE (2013) *Investing in smallholder agriculture for food security.* A report by the High Level Panel of Experts on Food Security and Nutrition of the Committee on World Food Security. High Level Panel of Experts on Food Security and Nutrition (HLPE), Rome.

Intergovernmental Panel on Climate Change (IPCC) (2022) Technical summary. In: *The Ocean and Cryosphere in a Changing Climate*. Cambridge University Press, pp. 39–70. DOI: 10.1017/9781009157964.002.

Jakobsen H (2014) What's gendered about gender-based violence? An empirically grounded theoretical exploration from Tanzania. *Gender & Society* 28(4): 537–561.

Kansanga MM, et al. (2020) Agricultural mechanization, environmental degradation, and gendered livelihood implications in northern Ghana. *Land Degradation & Development* 31(11): 1422–1440. DOI: 10.1002/ldr.3490.

Kelbert A and Hossain N (2014) Poor man's patriarchy: gender roles and global crises. *IDS Bulletin* 45(1): 20–28. DOI: 10.1111/1759-5436.12064.

Koenig D (2005) Multilocality and social stratification in Kita, Mali. In: Trager L (ed.) *Migration and Economy: Global and Local Dynamics*, 77–102. Oxford: Alta Mira Press.

Konadu-Agyemang K (2000) The best of times and the worst of times: structural adjustment programs and uneven development in africa: the case of ghana. *The Professional Geographer* 52(3): 469–483. DOI: 10.1111/0033-0124.00239.

Kuuire VZ, et al. (2016) Abandoning land in search of farms: challenges of subsistence migrant farming in Ghana. *Agriculture and Human Values* 33(2): 475–488. DOI: 10.1007/s10460-015-9612-0.

Langill JC, et al. (2023) Gendered effects of migration on social organization and smallholder production in Oubritenga Province, Burkina Faso. *Journal of Rural Studies* 98: 19–33.

Lawson ET, et al. (2019) Dealing with climate change in semi-arid Ghana: understanding intersectional perceptions and adaptation strategies of women farmers. *GeoJournal* 1–14. DOI: 10.1007/s10708-019-09974-4.

Luginaah I, et al. (2009) Environment, migration and food security in the upper west region of ghana. In: Luginaah IN and Yanful EK (eds.) *Environment and Health in Sub-Saharan Africa: Managing an Emerging Crisis*. Dordrecht: Springer, pp. 25–38. DOI: 10.1007/978-1-4020-9382-1_2.

Makina A and Moyo T (2016) Mind the gap: institutional considerations for gender-inclusive climate change policy in Sub-Saharan Africa. *Local environment* 21(10): 1185–1197. DOI: 10.1080/13549839.2016.1189407.

Nchanji EB, et al. (2020) What does gender yield gap tell us about smallholder farming in developing countries? *Sustainability* 13(1): 77. DOI: 10.3390/su13010077.

Nunbogu AM, et al. (2023) I feel the pains of our past water struggles anytime I turn on the tap: diaspora perceptions and experiences of water, sanitation, and hygiene

(WaSH) gendered violence in Ghana. *Social Science & Medicine* 317: 115621. DOI: 10.1016/j.socscimed.2022.115621.

Othman MS, et al. (2021) Farming groups and empowerment of women smallholder farmers. *Development in Practice* 31(5): 676–689. DOI: 10.1080/09614524.2021.1911947.

Ragetlie R, et al. (2022) Association between poor food production and intimate partner violence among smallholder farmers in northwestern Benin. *Global Public Health*, 17(11): 2737-2751.

Shibata R, et al. (2020) Gendered intra-household decision-making dynamics in agricultural innovation processes: assets, norms and bargaining power. *Journal of International Development* 32(7): 1101–1125. DOI: 10.1002/jid.3497.

Songsore J (1979) Structural crisis, dependent capitalist development and regional inequality in Ghana, I.S.S. Occasional papers 71. Institute of Social Studies, The Hague.

Denkabe A, Trondheim UI,

Sultana F (2018) Gender and water in a changing climate: challenges and opportunities. In: Fröhlich C, Gioli G, Cremades R, et al. (eds.) *Water Security across the Gender Divide*. Water security in a new world. Cham: Springer International Publishing, pp. 17–33. DOI: 10.1007/978-3-319-64046-4_2.

Tacoli C and Mabala R (2010) Exploring mobility and migration in the context of rural–urban linkages: why gender and generation matter. *Environment and Urbanization* 22(2): 389–395. DOI: 10.1177/0956247810379935.

Tsige M (2019) Who benefits from production outcomes? Gendered production relations among climate-smart agriculture technology users in rural Ethiopia. *Rural sociology* 84(4): 799–825. DOI: 10.1111/ruso.12263.

USAID (2020) *USAID/GHANA GENDER ANALYSIS REPORT APRIL 2020*. Washington, DC: USAID.

van der Geest K (2011) North-South migration in Ghana: what role for the environment? *International Migration* 49: e69–e94. DOI: 10.1111/j.1468-2435.2010.00645.x.

Wiegel H, et al. (2019) A mobilities perspective on migration in the context of environmental change. *WIREs Climate Change* 10(6). DOI: 10.1002/wcc.610.

Zahan Tanny N and Wakilur Rahman M (2017) Climate change vulnerabilities of woman in bangladesh. *The Agriculturists* 14(2): 113. DOI: 10.3329/agric.v14i2.31355.

Zickgraf C (2021) Theorizing (im)mobility in the face of environmental change. *Regional Environmental Change* 21(4): 126. DOI: 10.1007/s10113-021-01839-2.

Part 4

Facilitating access to resources and technology

Chapter 13

Addressing gender inequities in accessing agricultural extension and advisory services

Kathleen Earl Colverson[1], Arati Joshi, and Fallon Y Riaño J, University of Florida, USA; and Steven Franzel, Independent Consultant, USA

1 Introduction

Women are extensively involved in agriculture and rural development and contribute to food security in their families and communities. Many smallholders are family farms, where livelihoods depend on the joint efforts of all family members. Family farming is the most prevalent form of agriculture today, with over 500 million family farms producing 80% of the world's food supply (Lowder et al., 2014; FAO, 2014). Women participate actively in all aspects of family farming and represent, on average, 43% of the agricultural labor force. Yet, in many countries, their productive role is that of unpaid family labor, and in many cases, their work is excluded from statistics (Byram, 2001). Although the majority of small-scale farmers are women, they have lower productivity levels compared with men due to structural barriers, and thus their contributions are often marginalized (Byram, 2001). For women to increase their productivity, there must be substantial changes in gender relations and ways of engaging men more successfully, particularly since the vast majority of extension agents are men (Farnworth and Colverson, 2015).

Despite some evidence of increased attention by governments, NGOs, and bilateral and multilateral agencies in securing access for women to extension

1 Animal Sciences Department, University of Florida, kcolverson@ufl.edu.

http://dx.doi.org/10.19103/AS.2024.0148.17

and advisory services (EAS) (Manfre et al., 2013), women's access to extension services remains low. The role of the male as 'farmer' remains the norm; however, outmoded this may be, even though when it comes to the actual relative numbers of women and men in farming, women's numbers are higher (Farnworth, 2010). A study in Ethiopia found that women receive fewer extension services through various channels than their male counterparts (Ragasa et al., 2013). Women farmers also have less access to land and agricultural inputs which can also explain lower technology adoption by women households. Farnworth and Colverson (2015) posit that there needs to be a complete change in the way that extension approaches norms and power structures that create and reproduce gender inequalities. This new system would challenge existing structures and emphasize not only the creation of knowledge products for dissemination to end users but also the process of creating knowledge with both men and women.

This chapter will address the gendered inequities that women face when accessing EAS. It will also explore examples of projects or programs that are tackling women's access through the application of gender-transformative approaches. It will not address the quality or content of extension programming as that is an extensive topic on its own. The first part of the chapter will examine the current state of gender integration in extension. The authors will use the Gender Continuum and its application to extension programs to describe examples of extension programs or projects that fit the gender-accommodating and gender-transformative categories, emphasizing lessons learned via specific case studies of gender integration in extension programs. The chapter will conclude with a summary of future trends and what is needed to successfully address gender inequities in EAS.

Transforming gender relations – part of which involves building on existing collaborative gender norms and practices at the household and community level – will help to make smallholder agriculture and associated agricultural development more effective and efficient, with benefits for a variety of development outcomes. To do this, Extension Services (or Rural Advisory Services (RAS)) must move from ignoring or minimizing gender relations in farming households to understanding and actively addressing these inequities. The Gender Integration Continuum (https://www.unfpa.org/sites/default/files/pub-pdf/UNFPA_GTA-2023.pdf) is a method of examining gender integration from a series of steps that gradually increase gender integration across a spectrum. At the far end of the spectrum is the *gender-blind* perspective, which avoids or ignores gender issues in project or program planning either intentionally or without conscious knowledge of its impacts. As gender becomes more apparent and significant in regard to its impact, the continuum moves from *gender sensitive* to *gender responsive* to finally *gender transformative,* which seeks to address underlying gender norms and constraints and redistribute power and resources.

2 The Gender Integration Continuum and extension services

Figure 1 shows the Gender Integration Continuum framework adapted to demonstrate women's limitations when accessing extension services. Extension programs may frame their efforts with or without a gender lens. As a result, they may fall under the gender-blind category, the gender-transformative one, or somewhere in between. This framework does not represent a linear process or imply that extension programs projects fit neatly into a single category or move in one direction. This is because extension program approaches may have unintended outcomes, or their gender approaches may simply reach or benefit women instead of empowering them, by transforming and removing harmful gender norms that limit women's access to extension services.

2.1 Gender-blind approach

Historically, women farmers have faced challenges with agricultural extension services, which often ignore or undermine their contributions. Extension services for women frequently focused on home economics and reproductive tasks such as childcare, perceiving them as subsistence producers rather than commercial farmers (Buchy and Basaznew, 2005). Additionally, the emphasis on cash crop production for export, rather than local food production, further limited women's involvement in agricultural extension, despite their significant contributions to both cash and food crop farming (Berger et al., 1984). Lack of female extension agents in gender segregated societies and absence of gender sensitizing training further creates a barrier for women's access to extension services (World Bank, 2010).

Examples of "gender-blind extension" approaches include the Progressive Farmers, a program implemented in African countries that focused on well-educated, land-owning, affluent, politically influential farmers. Women often did not meet these criteria and thus had limited access to these services

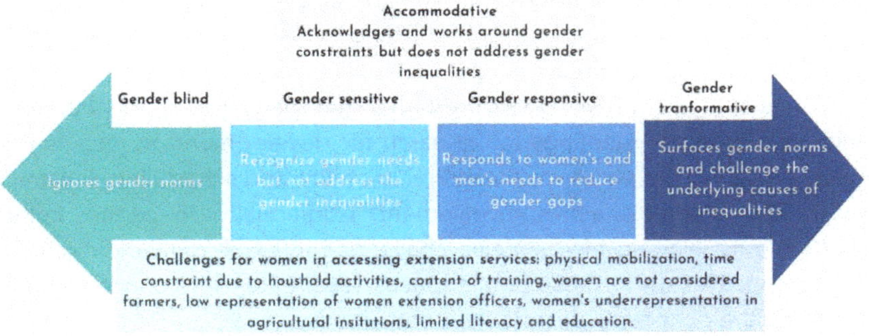

Figure 1 Gender Integration Continuum.

(Berger et al., 1984). Extension programs focusing on 'progressive farmers' used the model farmer approach as a delivery method which used a gender-blind approach as women often encountered difficulties in being elected as model farmers due to their limited political influence. Even if women were elected as model farmers, the requirement to attend frequent meetings at the local administrator's office posed additional challenges, such as travel constraints and the burden of managing their domestic responsibilities (Berger et al., 1984). Similarly, Buchy and Basaznew (2005) identified that gender-blind extension practices undertaken by the Awasa Bureau of Agriculture in Southern Ethiopia as extension programs did not account for the seasonal and daily calendars of women, and since the extension services were provided during women's busiest periods their participation was limited.

The "transfer of technology" approach disseminates knowledge from privileged universities and research stations to extension agents and ultimately to farmers, often overlooking the gendered roles of farmers and assuming they are passive recipients of knowledge. This approach is techno-centric, focusing services on 'innovators' (Buchy and Basaznew, 2005) and fails to account for the specific technological needs and social realities of women farmers in farm households, thereby excluding them from accessing appropriate support (Koutsouris, 2017; International Assessment of Agricultural Knowledge, Science and Technology for Development, 2014).

The "training and visit" approach exemplifies another gender-blind extension approach that was developed using a top-down model, neglecting the role of women farmers as active participants in agricultural production. This approach faced criticism for its failure to effectively engage women farmers, small-scale producers, and specific ethnic communities (Jafry et al., 2014).

Farnworth and Colverson (2015) used the term 'conceptual lock-in' to highlight the pervasive issue of extension programs being developed with a deep-rooted patriarchal mindset. These programs often overlook gender roles and inadvertently or purposefully cater primarily to male farmers, thus exhibiting a gender-blind approach. Even the relatively newer systems thinking extension approaches (Cook et al., 2021) tend to be implemented without due consideration of gender roles.

Swaans et al. (2013) discovered that the participation of women in innovation platform meetings for research for development projects is often lower, indicating the presence of cultural, mobility, and scheduling constraints that hinder women's active involvement. Additionally, in certain cultural environments, women face specific challenges in expressing their opinions and perspectives within these public spaces, leading to a lack of accountability regarding their concerns despite their participation.

2.2 Gender-accommodative approach

Numerous programs and development projects have fallen under the scope of the gender-accommodative approach, which acknowledges and works around gender constraints but does not address gender inequalities which may limit women's participation and access to extension services (Interagency Gender Working Group, 2017). Under the accommodative approach, there are two levels in which projects may address gender inequalities, *gender-sensitive* and *gender-responsive*. *Gender-sensitive* projects recognize gender needs but do not challenge the underlying structural barriers contributing to gender inequality, while *gender-responsive* projects directly respond to women's and men's needs to reduce inequalities without confronting the root causes (UNICEF, 2020; UNFPA, 2023).

Physical mobility is one frequent constraint that limits women's access to agricultural extension services (FAO and ECOWAS, 2018; Petrics et al., 2022), as noticed by extension agents in the Egiye Jai and Nijera Gori Projects in Bangladesh (Lee et al., 2022). The program implemented food production extension services for men and women farmers and were aware of women's limitations in accessing advisory services. To address these constraints, they carried out a cluster-level training approach to increase women's participation by implementing multiple small training groups close to the villages. This project addressed the visible barriers demonstrating a *gender-sensitive approach,* resulting in 90% women's participation. Conversely, men's participation was weak because men did not feel the training and schedule benefited them. Although the project improved women's participation, it did not directly tackle or uncover the social norms or power dynamics (root causes) that limited women's access to extension services.

In another example of the *gender-sensitive* approach, extension officers and researchers in Zambia implemented a project that integrated men and women of the fisher value chain in an experiment to improve productive fish technologies. Project staff arranged the meeting times to accommodate women's schedules to guarantee active participation (Cole et al., 2020). After finishing the project, researchers found that this strategy and others that acknowledged the gender roles and responsibilities of participants made it possible to conduct the project. However, there were no significant changes in attitudes or other social norms compared to incorporating a more gender-transformative approach (Cole et al., 2020).

In the next level of the gender-accommodative approach, the Rural Capacity Building Project (RCBP) shows a model of an extension *gender-responsive* approach. The government of Ethiopia implemented the RCBP to strengthen the country's agricultural extension services by offering a more responsive system that fulfills the needs of men and women farmers (Bueheren et al., 2019). One primary target was to increase woman's participation. To do so, some of the program strategies were to mainstream the gender lens at each level of the extension system and hire women extension agents. Ultimately, the program made it possible for women farmers (heads of households or living in male-headed households) to increase access to extension services and benefit from the information on improved agricultural practices. However, the program's positive impact was not sustained. The literature does not show evidence that the project addressed the structural barriers contributing to the gender gap. Considering the communities' social norms would possibly have contributed to ensuring a long-lasting positive result.

Citing the limitations of the technology transfer model, Chambers and Jiggins (1987) proposed the farmer-first-and-last (FFL) approach, which aimed to improve upon traditional approaches by prioritizing the needs and priorities of farmers. The FFL approach actively involved farmers in the identification of research topics and methodologies, making it a participatory approach that recognized the complex and diverse needs and conditions of resource-poor farmers. Implementing participatory approaches has effectively identified and addressed the challenges women farmers face. Notably, the Participatory Research and Gender Analysis Program conducted by the International Center for Tropical Agriculture demonstrated success in enhancing women's access to productive resources in Syria and Colombia (Biermayr-Jenzano, 2011).

The Farmer's Field School (FFS) initiative is another participatory approach that originally focused on integrated pest management but over time has expanded to include various agricultural practices tailored to specific contexts and has been implemented in more than 90 countries worldwide (Waddington et al., 2014). Initially, the participation of women in FFS was a subject of debate due to factors such as the selection process, time constraints, land ownership, religious beliefs, and cultural barriers, all of which restricted women's involvement (Bello-Bravo et al., 2011). However, FFS has since been adapted to specifically target and empower women farmers, addressing their time and mobility constraints and designing curricula relevant to their specific needs. These efforts have resulted in improved household decision-making for women (DANIDA, 2011), change in gender roles (Friis-Hansen et al., 2012), and increased empowerment (Westendorp, 2012).

Haku Wiñay is a rural development program implemented in Peru by the government (Hernández Asensio et al., 2022). The program uses the farmer-to-farmer methodology to reach a wide range of community needs, including family farming, health, rural businesses, and financial literacy. This program does not have a written policy that mandates a gender-sensitive/responsive approach. Nevertheless, the extension agents and other program staff have fostered strategies to boost women's participation and break gender stereotypes. In this program, most women are not formally registered, but their husbands are. However, women are considered legitimate clients, so training activities and meetings have been adapted to cope with their time and mobility constraints and education level. Additionally, extension agents have facilitated women representing and voicing their interests. At this level, the role of the extension agent has been instrumental in making the program more gender-sensitive and responsive to women's and men's needs, although there is no formal institutional mechanism that supports this gendered approach.

Extension programs with an accommodative approach attempt to cope with visible barriers, such as time and mobility. However, hidden constraints deeply rooted in time-honored cultural norms also limit women's access to extension services and could not be perceptible by rural advisory programs. For instance, researchers in Tajikistan found that elderly women like mothers-in-law tend to be a significant obstacle that prevents younger women from participating in RAS limiting their involvement (Wood et al., 2018). Mudege et al. (2015) emphasize that changes in attitudes, roles, and relations at different societal levels will promote gender-transformative outcomes. These previous examples demonstrate that extension services with accommodative approaches have made it possible to develop and implement strategies that are aware and sensitive to social and cultural norms. Additionally, the projects have transitioned to action levels responding to specific needs that benefit women by engaging them in a meaningful way and enabling the environment to achieve transformation. However, this approach has yet to achieve remarkable changes in gender equality or women's empowerment since it did not tackle the underlying structural barriers framed by social-cultural beliefs and power dynamics in specific contexts.

2.3 Gender-transformative approach

The prevailing gender norms in society place an undue burden on women, assigning them the roles of productive, reproductive, and community management spheres (Moser, 1989). Cultural barriers and these triple roles limit women's access to and engagement with agricultural extension services. As previously mentioned, the traditional exclusion of women's contributions in farming and denying their status as farmers have prompted various approaches aimed at challenging these social norms and promoting gender equality. One such approach is the use of gender-transformative approaches, which strive to equally distribute power, resources, and services among all genders in society (UNFPA, 2023). The gender-transformative approach also challenges underlying power structures and gender norms (UNFPA, 2023).

Enhancing women's access to extension services does not guarantee the automatic adoption and implementation of best practices. According to Farnworth and Colverson (2015), in order for women to become empowered and unhindered decision-makers, it is crucial to address and transform gender relations. This necessitates EAS to collaborate with both male and female farmers to address prevailing gender norms and inequalities. By shifting extension approaches from a mere 'service' dissemination of knowledge to a 'facilitation system' creating knowledge with the participants can effectively facilitate the process of empowering women and promoting gender equality within agricultural practices. This involves empowering women ('agency'), addressing the existing cultural, social, and economic structures and the underlying assumptions, values and ideologies ('structure'), and expanding women's network ('relations') (Hillenbrand et al., 2015) with an understanding of micro-, meso-, and macrolevels of society that influence gender norms.

Within the field of extension services, there is an increasing utilization of gender-transformative approaches. For instance, the gender model family approach was implemented in Sierra Leone to promote nutrition education, agricultural support, water, sanitation, and hygiene promotion, and microfinance. This approach sought to mobilize husbands by encouraging their involvement in household chores, promoting joint decision-making on family and agribusiness matters, and providing equal opportunities to boys and girls. The success of this approach was evident in changing male behaviors, such as increased contributions to household chores like cooking and feeding, improved joint decision-making, and a shift in attitude toward equitable distribution of meals to wives and children (FAO, IFAD and WFP, 2020).

Another extension approach, known as the Gender Action Learning Systems (GALS), was implemented in Zimbabwe for the Livelihoods and Food Security Program (LFSP) targeting rural farming households. The objective of GALS was to enhance gender equity by improving household decision-making processes and ensuring the equal distribution of benefits. The LFSP program in Zimbabwe reported transformative outcomes, including changed gender relations, promotion of collaborative decision-making, shifts in gender division of labor within productive and reproductive domains, and numerous other benefits (LFSP, 2020).

Similar positive outcomes were observed through the implementation of the Hanns R. Neumann Stiftung Gender Household Approach in Uganda, Tanzania, Ethiopia, and Côte d'Ivoire for smallholder coffee producers. In this approach, selected smallholder couples who are members of producer organizations received training and support from extension agents to promote gender equity in household and farm activities. Extension agents visited each couple's household to assist them in developing their household visions and plans. The intervention had a positive influence on spousal cooperation, household welfare, household income, household food security, adoption of agronomic practices, and the use of improved seed varieties for smallholder coffee producers in Uganda and Tanzania (Lecoutere and Van Campenhout, 2023).

Similarly, in Guatemala, Equimundo and Global Communities implemented a similar approach called Journeys of Transformation (JoT) to enhance women's economic empowerment within women empowered (WE) groups. JoT involved a comprehensive 7-month training program that actively engaged couples, providing them with a non-judgmental space to reflect on prevailing gender norms and practice gender equality with same sex-peers and their partners. The program worked in collaboration with local community leaders to recruit participants and promote their active involvement. The outcomes of the program were positive in changing gender norms that hindered women's participation in economic activities. Furthermore, the program fostered increased participation of men in household tasks. Couples also experienced improved communication regarding financial planning and engaged in joint decision-making regarding income sources and expenditures (Equimundo and Global Communities, 2022).

3 Case studies of gender integration in extension programs

3.1 Case study 1 – Increasing the number of female farmer trainers: Lessons from Africa

Farmer-to-farmer extension (FTFE) programs involve the provision of training and information by farmers to other farmers through a structure of farmer trainers, also called lead farmers, farmer promoters, or other similar names (Franzel et al., 2018). FTFE is one of the most widely used extension approaches and is common throughout Latin America, Africa, and Asia. Until recently, most FTFE programs were gender blind, resulting in very low proportions of, or no, female farmer trainers. But in recent years, many public extension programs, NGOs, and projects have made significant efforts to increase the proportion of female farmer trainers, as well as sensitizing male farmer trainers to serve female farmers as well as men.

In most programs, community leaders or members choose the farmer trainers which involves efforts to sensitize leaders and members about the importance of including women as farmer trainers. For example, the Development Fund of Norway has helped the Ministry of Agriculture, Malawi and NGOs it works with to provide gender sensitization training to Village Development Committees, who select lead farmers. The program has trained 3000 lead farmers, 50% of whom are women (NORAD, 2016). In Kenya, Rwanda and Tanzania, Technoserve, a US-based NGO, implemented the Coffee

Initiative, 2008–13. They established a quota of 30% female farmer trainers and attracted women in several ways: advertising through women's associations, using trainings rather than interviews to identify potential candidates and offering childcare during training. By the end of the program, 40-% of the farmer trainers were female (Technoserve, 2013).

The increases in the proportions of farmer trainers who are female has had several impacts. First, in many cases, it has shown that women train more women than men do. For example, in a survey of farmer trainers working in 20 programs in Cameroon, researchers found that male and female farmer trainers trained similar numbers of farmers. But women comprised 74% of farmers trained by female farmer trainers, compared to 41% of those trained by male farmer trainers (Tsafack et al., 2015). In Mozambique, women's adoption of improved farm techniques increased by five percentage points in communities with female farmer trainers as compared to villages with only male farmer trainers (Kondylis et al., 2016). Finally, it was noteworthy that many programs were able to achieve higher proportions of women among their farmer trainers than among their field staff. For example, in the Ministry of Agriculture, Malawi, in 2013, 21% of the extension staff were female whereas 40% of the Ministry's 12 000 lead farmers were female (Franzel et al., 2018).

3.2 Case study 2 – Are women's membership in self-help groups and improved access to information a recipe for improved agricultural outcomes? A case study from India

Women's groups are a widely recognized platform for improving women's livelihoods throughout the tropics (Petrulla et al., 2021). In India, Raghunathan et al. (2019) assessed the degree to which women's self-help groups (SHGs) increased women's access to agricultural information, decision-making, and use of improved agricultural practices. The SHGs are community savings and loans groups comprised of 10–20 women and were supported by a national NGO, Professional Assistance for Development Action, which provided agricultural extension services. The analysis involved comparing SHG members with non-members who had similar characteristics. The sample of 977 women was spread across five states: Odisha, Madhya Pradesh, Jharkhand, Chhattisgarh, and West Bengal.

Participation in SHG's was found to increase women's access to agricultural information, significantly increasing the probability of women receiving eight of ten types of information, such as on improved seeds, crop selection, and rotation. However, SHG membership had little effect on use of improved agricultural practices or other outcomes, such as crop diversification, possibly due to other constraints such as lack of finance, social traditions, and women's domestic responsibilities. The findings demonstrate that membership in groups and access to agricultural information alone may not be sufficient to improve women's agricultural outcomes and needs to be accompanied by other types of support. These may include accessing capital and changing social norms around women's participation in agriculture.

3.3 Case study 3 - Can community health workers provide training in nutrition and parenting to rural couples that enhances gender equality and empowers women? Experiences from Tanzania

Much of the progress in advancing gender equality and women's empowerment in rural areas of the tropics is being led by health and nutrition NGOs. In Tanzania, for example, Global Communities (formerly Project Concern International) supports paraprofessional government community health workers, with minimal levels of education and health training, to train rural couples in nutrition and parenting. To assess the impact of the training, researchers compared four treatments: nutrition training for mothers, nutrition and parenting training for mothers, nutrition training for fathers, and nutrition and parenting training for fathers (Galvin et al., 2023). Training was highly participatory and included critical reflection and discussion, problem solving, practical skill building, and commitment setting. Sessions examined gender norms and roles, positive masculinity and healthy communication, and relationships. The training was spread over 24 biweekly sessions.

The results showed considerable impact on attitudes and behaviors. Engaging couples, as compared to training only mothers, resulted in significantly increased paternal and maternal gender-equitable attitudes, paternal time spent on domestic chores, and maternal decision-making power. Bundling nutrition and parenting training, as compared to providing only nutrition training, increased maternal leisure time, reduced intimate partner violence, and increased women's dietary diversity. The combination of engaging couples and bundling nutrition and parenting improved paternal gender attitudes, couples' communication frequency, and women's dietary diversity. That the training was conducted by community health workers, and not skilled professionals, suggests that it can be scaled up more quickly and at lower cost than otherwise would be the case.

3.4 Case study 4 - Can information and communication technologies such as videos improve women's access to extension services and empowerment?

Digital information and communication technologies, such as the internet, cell phones, video, and radio have great potential for improving farmers' access to advice and information. However, women are at a disadvantage because of their limited access to many of these technologies. For example, whereas 32% of men in least developed countries use the internet, the corresponding percentage for women is only 21% (ITU, 2022). Moreover, women face further constraints such as lower literacy and technical skills (BMGF, undated).

Impact assessments of phone and internet-based extension tools tend to show that they are more widely used by men than by women. For example, women farmers in India were less likely than men to use the Direct2Farm mobile agri-advisory service (Kansiime et al., 2019). In Kenya, over 70% of the viewers providing phone and SMS feedback to Shamba Shape-up, an 'edutainment' TV show about farming, were men (Kiptot et al., 2017).

Nevertheless, there is evidence that women with access to ICTs find them effective in improving their agricultural performance. In Uganda, Lecoutere et al. (2023) compared the effects of showing women agricultural extension videos about improved farming practices to showing them only to men or to couples. They found that giving women direct and exclusive access to agricultural extension videos advanced their knowledge, decision-making, and use of the practices. Providing information through videos to couples, rather than to individuals, resulted in increases in joint decisions to use improved practices. The researchers also varied the actors in the videos to see if women responded differently when viewing a video with female rather than male actors. Viewing female actors had no effect on women's overall uptake of improved practices.

Whereas ICTs such as videos can help women access information, they may not have a direct impact on empowering women. Lecoutere et al. (2023) note that if the aim is to empower women, most gains can be made by re-designing advisory services to target information exclusively to the female co-head within the household.

4 Conclusion and next steps

The Gender Continuum provides a unique way to assess EAS from a gender equity perspective. It encourages those who deliver extension services to examine their outreach approach to ensure that both men and women are able to access assistance. The evolution of EAS along the continuum provides recommendations for how to more fully engage all genders. Key takeaways include the following:

- Include both men and women in training sessions to increase participation and equitably share information, particularly in topics considered 'male or female'. (Petrics, et.al. 2022; Aregu, et.al. 2016)

- Conduct a gender analysis to determine women's and men's roles in agricultural as well as domestic responsibilities and use this information to address inequitable gender constraints in projects or programs.
- Use a Systems Approach to understand the role of gender norms. (Bevitt, 2017)
- Recognize women's triple roles to accommodate their responsibilities, particularly the time and location of training.
- Create women's self-help groups to address the lack of micro-credit and loans for women farmers.
- Engage men and boys in what is perceived as traditionally female training that address household responsibilities such as nutrition and reproductive health. (Poulsen, 2018.)
- Utilize participatory methods and tools (such as INGENAES) to engage both women and men in examining and addressing gender norms and constraints.
- Work with boys and girls in school settings to engage in discussion and address issues around gender inequities.
- Provide training using both male and female facilitators.
- Increase the integration of rigorous monitoring and impact evaluation using both quantitative and qualitative methods for assessing progress toward gender equity in extension projects and programs.
- Design programs addressing the needs of particular categories of women such as youth, spouses, and single heads of households and determining the effects of project interventions on the different categories.

Providing extension and RAS to women has numerous challenges to overcome, particularly as it relates to gender norms. RAS agents tend to approach male farmers more often than female farmers because of the common beliefs that women's role in farming is less important and that advice will eventually 'trickle down' from the male household head to all other household members (Bello-Bravo et al., 2011; Farnworth and Colverson, 2015). Numerous gender-responsive strategies exist to counteract these challenges, including: strategies that specifically target female household members, creating and strengthening self-help groups and women's associations, adopting affirmative action in user group associations, and promoting the political awareness, leadership, and advocacy abilities of women. Targeting service providers such as recruiting and training women extension agents and designing, implementing, and monitoring projects in a gender-sensitive manner; and targeting public administration, elected representatives, and political parties, such as reserving seats for women representatives on local councils or committees and providing gender-sensitive training for staff also contribute to addressing challenges with existing gender norms .

Moving forward in addressing gender inequities in EAS will take continued efforts in not only tackling technical issues related to production and processing in agriculture but also a comprehensive tackling of underlying causes of the issues that maintain gender norms. Gender-transformative approaches offer a way forward to EAS in creating a brighter future for both women and men farmers.

5 Where to look for further information

- Consultative Group for International Agricultural Research (CGIAR) Gender Impact Platform. https://gender.cgiar.org/gender-research.
- FAO (2020). Gender transformative approaches for food security, improved nutrition, and sustainable agriculture – A compendium of fifteen good practices. https://www.fao.org/documents/card/en/c/cb1331en.
- Global Forum for Rural Advisory Services (GFRAS). https://www.g-fras.org/en/.
- Integrating Gender and Nutrition into Agricultural Extension Systems (INGENAES). https://ingenaes.illinois.edu/ifta/.
- FAO (2018). The gender and rural advisory services assessment tool. https://www.fao.org/3/CA2693EN/ca2693en.pdf.

6 References

Aregu, L., Darnhofer, I., Tegegne, A., Hoekstra, D. and Wurzinger, M. (2016). The impact of gender-blindness on social-ecological resilience: the case of a communal pasture in the highlands of Ethiopia. *Ambio* 45(Suppl. 3), 287. https://doi.org/10.1007/s13280-016-0846-x.

Bello-Bravo, J., Seufferheld, F. and Agunbiade, T. A. (2011). Gender and farmer field schools in agricultural production systems in West Africa. *The International Journal of Science in Society* 2(4), 13. https://doi.org/10.18848/1836-6236/CGP/v02i04/51280.

Berger, M., Delancey, V. and Mellencamp, A. (1984). *Bridging the Gender Gap in Agricultural Extension*. [online], pp. 1, 87. Available at: https://www.icrw.org/wp-content/uploads/2016/10/Bridging-the-Gender-Gap-in-Agricultural-Extension.pdf.

Bevitt, K. (2017). *Video Stories to Help Families Overcome Gender-Based Challenges in Zambia*. WorldFish Center. Available at: http://blog.worldfishcenter.org/2017/03/video-stories-tohelp-families-overcome-gender-based-challenges-in-zambia/.

Biermayr-Jenzano, P. (2011). *The Use of Participatory Methodologies to Increase Women Farmers' Access to Productive Resources*. [online] UN Women, pp. 1–9. Available at: https://www.un.org/womenwatch/daw/csw/csw56/egm/Biermayr-EP-10-EGM-RW-Sep-2011.pdf [Accessed 20 June 2023].

Bill & Melinda Gates Foundation (n.d.). *Gender and Agricultural Advisory Services Issue Brief*. [online] Gender Equality Tool Box, pp. 1, 15 Available at: https://www.gatesge

nderequalitytoolbox.org/wp-content/uploads/BMGF_AG-Advisory-Services-Brief_web.pdf.

Buchy, M. and Basaznew, F. (2005). Gender-blind organizations deliver gender-biased services: the case of Awasa bureau of agriculture in Southern Ethiopia. *Gender, Technology and Development* 9(2), 235. https://doi.org/10.1177/097185240500900204.

Buehren, N., Goldstein, M., Molina, E. and Vaillant, J. (2019). The impact of strengthening agricultural extension services on women farmers: evidence from Ethiopia. *Agricultural Economics* 50(4), 407. https://doi.org/10.1111/agec.12499.

Byram, M. (2001). *From Farmer to Planner & Back*. [online] Available at: www.fao.org. https://www.fao.org/3/y0354e/y0354e00.htm#TopOfPage.

Chambers, R. and Jiggins, J. (1987). Agricultural research for resource-poor farmers part II: a parsimonious paradigm. *Agricultural Administration and Extension* 27(2), 109. https://doi.org/10.1016/0269-7475(87)90021-3.

Cole, S. M., Kaminski, A. M., McDougall, C., Kefi, A. S., Marinda, P. A., Maliko, M. and Mtonga, J. (2020). Gender accommodative versus transformative approaches: a comparative assessment within a post-harvest fish loss reduction intervention. *Gender, Technology and Development* 24(1), 48. https://doi.org/10.1080/09718524.2020.1729480.

Cook, B. R., Satizábal, P. and Curnow, J. (2021). Humanising agricultural extension: a review. *World Development* 140, 105337. https://doi.org/10.1016/j.worlddev.2020.105337.

DANIDA (2011). *Evaluation of the Farmer Field School Approach in the Agriculture Sector Programme Support Phase II, Bangladesh*. [online] UM-ENEN. Available at: https://um.dk/en/danida/results/eval/eval_reports/evaluation-of-the-farmer-field-school-approach-in-the-agriculture-sector-programme-support-phase-ii [Accessed 21 June 2023].

Equimundo and Global Communities (2022). *Journeys of Transformation Adaptation Program to Engage Men as Allies in Women's Economic Empowerment*. Washington, DC: Equimundo.

FAO (2014). *Family Farmers*. [online] Food and Agriculture Organization of the United Nations. Available at: https://www.fao.org/3/mj760e/mj760e.pdf.

FAO and ECOWAS (2018). *Ghana, Country Gender Assessment Series. National Gender Profile of Agriculture and Rural Livelihoods*, pp. 1, 60 Available at: https://www.fao.org/3/i8639en/I8639EN.pdf.

FAO, IFAD and WFP (2020). *Gender Transformative Approaches for Food security, Improved Nutrition and Sustainable Agriculture – A Compendium of Fifteen Good Practices*. [online] Rome, Italy: FAO, IFAD and WFP. Available at: https://www.fao.org/documents/card/en/c/cb1331en.

Farnworth, C. R. (2010). *Gender Aware Approaches in Agricultural Programmes A Study of Sida-supported Agricultural Programmes*. [online] Sida, pp. 2, 104 Available at: https://cdn.sida.se/publications/files/sida61292en-gender-aware-approaches-in-agricultural-programmes.pdf.

Farnworth, C. R. and Colverson, K. E. (2015). Building a gender-transformative extension and advisory facilitation system in sub-Saharan Africa. *Journal of Gender Agriculture and Food Security* 1(1), 20. Available at: http://agrigender.net/views/Building-gender-transformative-extension-JGAFS-112015-2.php.

Franzel, S., Degrande, A., Kiptot, E., Kirui, J., Kugonza, J., Preissing, J. and Simpson, B. (2018). Farmer-to-farmer extension. In: *What Works in Rural Advisory Services?*

Global Good Practice Notes. [online] Switzerland: Global Forum for Rural Advisory Services, GFRAS, pp. 1, 137. Available at: https://www.g-fras.org/en/gfras/1040-what-works-in-rural-advisory-services.html.

Franzel, S., Kiptot, E. and Degrande, A. (2018). Farmer-to-farmer extension: a low-cost approach for promoting climate-smart agriculture. In: Rosenstock, T., Nowak, A. and Girvetz, E. (Eds.), *The Climate-Smart Agriculture Papers*. [online] Cham: Springer, pp. 277–288. https://doi.org/10.1007/978-3-319-92798-5_24.

Friis-Hansen, E., Duveskog, D., Taylor, W. and E. (2012). Less noise in the household: the impact of farmer field schools on gender relations. *Journal of Research in Peace, Gender and Development* 2(2), 44. Available at: https://www.academia.edu/6494928/Less_noise_in_the_household_the_impact_of_Farmer_Field_Schools_on_Gender_Relations [Accessed 21 June 2023].

Galvin, L., Verissimo, C. K., Ambikapathi, R., Gunaratna, N. S., Rudnicka, P., Sunseri, A., Jeong, J., O'Malley, S. F., Yousafzai, A. K., Sando, M. M., Mosha, D., Kumalija, E., Connolly, H., PrayGod, G., Endyke-Doran, C. and Kieffer, M. P. (2023). Effects of engaging fathers and bundling nutrition and parenting interventions on household gender equality and women's empowerment in rural Tanzania: results from effects, a five-arm cluster-randomized controlled trial. *Social Science and Medicine* 324, 115869. https://doi.org/10.1016/j.socscimed.2023.115869.

Hernández Asensio, R., Burneo de la Rocha, M. L., Loaiza Díaz, P. A., Nova Arismendi, R., Fernández, J., Díaz, V., Cortínez, V. and Petrics, H. (2022). The Power of Farmer-to-farmer advisors: a case study of Haku Wiñay of Foncodes/MIDIS in Peru. In: Petrics, H., Barale, K., Kaaria, S. K. and David, S. (Eds.), *Good Practices for Promoting Gender Equality through Rural Advisory Services Case Studies from Ethiopia, India and Peru*. [online] Rome, Italy: Food and Agriculture Organization, pp. 1, 96. Available at: https://www.fao.org/publications/card/en/c/CC3539EN.

Hillenbrand, E., Karim, N., Mohanraj, P. and Wu, D. (2015). *Measuring Gender-Transformative Change: A Review of Literature and Promising Practices*. [online] WorldFish and the CGIAR Research Program on Aquatic Agricultural Systems. Available at: https://digitalarchive.worldfishcenter.org/bitstream/handle/20.500.12348/248/3997_AAS-Working-Paper-Measuring-Gender-Transformative-Change.pdf?sequence=1&isAllowed=y [Accessed 23 June 2023].

Interagency Gender Working Group (2017). *The Gender Integration Continuum Training Session User's Guide*. [online] Interagency Gender Working Group (IGWG), pp. 1, 28. Available at: https://www.igwg.org/wp-content/uploads/2017/12/17-418-GenderContTraining-2017-12-12-1633_FINAL.pdf.

International Assessment of Agricultural Knowledge, Science and Technology for Development (2014). *Agriculture at a Crossroads - Global Report*. [online] Unep.org. International Assessment of Agricultural Knowledge, Science and Technology for Development. Available at: https://wedocs.unep.org/20.500.11822/8590.

ITU (2022). *Measuring Digital Development: Facts and Figures 2022*. [online] International Telecommunication Union. Available at: https://www.itu.int/en/ITU-D/Statistics/Pages/facts/default.aspx#:~:text=Latest%20figures%20show%20that%20an.

Jafry, T., Moyo, B. and Mandaloma, L. (2014). *MEAS Evaluation Series Assessment of Extension and Advisory Methods and Approaches to Reach Rural Women -Examples from Malawi*. [online] United States Agency for International Development (USAID) Project 'Modernizing Extension and Advisory Services' (MEAS). Available at: www

.meas-extension.org. https://meas.illinois.edu/wp-content/uploads/2017/02/MEAS-EVAL-2014-EAS-Reaching-Rural-Women-Report-Malawi-Jafry-March-2014.pdf [Accessed 20 June 2023].

Kansiime, M. K., Alawy, A., Allen, C., Subharwal, M., Jadhav, A. and Parr, M. (2019). Effectiveness of mobile agri-advisory service extension model: evidence from Direct2Farm program in India. *World Development Perspectives* 13, 25. https://doi.org/10.1016/j.wdp.2019.02.007.

Kondylis, F., Mueller, V., Sheriff, G. and Zhu, S. (2016). Do female instructors reduce gender bias in diffusion of sustainable land management techniques? Experimental evidence from Mozambique. *World Development* 78, 436. https://doi.org/10.1016/j.worlddev.2015.10.036.

Koutsouris, A. (2017). Role of extension in agricultural technology transfer: a critical review. In: *Innovation Technology and Knowledge Management*. [online] Springer Link, pp. 337–359. https://doi.org/10.1007/978-3-319-67958-7_16.

Lecoutere, E. and Van Campenhout, B. (2023). Joint forces: the impact of intrahousehold cooperation on welfare in East African agricultural households. *Feminist Economics* 29(1), 266. https://doi.org/10.1080/13545701.2022.2120206.

Lecoutere, E., Spielman, D. J. and Van Campenhout, B. (2023). Empowering Women through targeting information or role models: evidence from an experiment in agricultural extension in Uganda. *World Development* 167, 106240. https://doi.org/10.1016/j.worlddev.2023.106240.

Lee, H. B., McNamara, P. E. and Bhattacharyya, K. (2022). Does linking women farmers to markets improve food security? Evidence from rural Bangladesh. *Agriculture and Food Security* 11(1). https://doi.org/10.1186/s40066-022-00373-6.

LFSP (2020). *Impact and Lessons Learned Through the Gender Action Learning Systems (Gals) Methodology in the Lfsp-Apn Project in Zimbabwe*. [online] Available at: https://lfspzwcom.files.wordpress.com/2020/06/gals-review-report.pdf.

Lowder, S. K., Skoet, J. and Singh, S. (2014). *What Do We Really Know About the Number and Distribution of Farms and Family Farms Worldwide? Background Paper for the State of Food and Agriculture 2014*. [online] Rome: Food and Agriculture Organization of the United Nations FAO, pp. 1, 39. Available at: https://www.fao.org/documents/card/en/c/34559307-55db-53ad-ae1b-cec4bef37deb/, 1, 39.

Manfre, C., Rubin, D., Allen, A., Summerfield, G., Colverson, K. and Akeredolu, M. (2013). *Reducing the Gender Gap in Agricultural Extension and Advisory Services How to Find the Best Fit for Men and Women Farmers*, pp. 1, 28 [online] Available at: https://meas.illinois.edu/wp-content/uploads/2015/04/Manfre-et-al-2013-Gender-and-Extension-MEAS-Discussion-Paper.pdf.

Moser, C. O. N. (1989). Gender planning in the third world: meeting practical and strategic gender needs. *World Development* 17(11), 1799. https://doi.org/10.1016/0305-750X(89)90201-5.

NORAD (2016). *Mid-term Review Malawi Livelihood Program*. Lilongwe, Malawi: Norwegian Agency for Development Cooperation.

Petrics, H., Barale, K., Kaaria, S. K. and David, S. (Eds.). (2022). Good practices for promoting gender equality through rural advisory services case studies from Ethiopia, India and Peru. In: *FAODocuments*. [online] Rome, Italy: Food and Agriculture Organization, pp. 1–96 Available at: https://www.fao.org/publications/card/fr/c/CC3539EN.

Petrics, H., Blum, M., Kaaria, S., Tamma, P. and Barale, K. (2015). Enhancing the potential of family farming for poverty reduction and food security through gender-sensitive rural advisory services. In: *FAOdocuments*. Rome, Italy: Food and Agriculture Organization, pp. 1, 76. Available at: https://www.fao.org/documents/card/en/c/7c2c9631-c91b-4a6c-9cfd-5b571e39c0d6/.

Petrulla, N., Cheney, K., Guy, A. and Hohenberger, E. (2021). *Evidence Scan: Increasing Women's Roles in Agricultural Decision-Making*. [online] Rockville, MD: United States Agency for International Development under the Advancing Women's Empowerment Program. Available at: https://agrilinks.org/sites/default/files/media/file/AWE%20CO7_Evidence%20Scan%20Report%20and%20Summary%20Deck.pdf.

Poulsen, L. (2018). Implementing gender transformative approaches (GTAs) in agricultural initiatives. In: *INGENAES Feed the Future*. [online] Integrating Gender and Nutrition within Agricultural Extension Services. Available at: http://ingenaes.illinois.edu/wp-content/uploads/ING-DP-2018_06-Gender-Transformative-Approaches-in-Agricultural-Initiatives-Poulsen.pdf [Accessed 21 June 2023].

Ragasa, C., Berhane, G., Tadesse, F. and Taffesse, A. S. (2013). Gender differences in access to extension services and agricultural productivity. *The Journal of Agricultural Education and Extension* 19(5), 437. https://doi.org/10.1080/1389224X.2013.817343.

Raghunathan, K., Kannan, S. and Quisumbing, A. R. (2019). Can women's self-help groups improve access to information, decision-making, and agricultural practices? the Indian case. *Agricultural Economics* 50(5), 567. https://doi.org/10.1111/agec.12510.

Swaans, K., Cullen, B., Rooyen, A. van, Adekunle, A., Ngwenya, H., Lema, Z. and Nederlof, S. (2013). Dealing with critical challenges in African innovation platforms: lessons for facilitation. *Knowledge Management for Development Journal* 9(3), 116. Available at: http://journal.km4dev.org/.

Technoserve (2013). *Lessons Learned the Coffee Initiative 2008 to 2011*, pp. 1, 41. Available at: https://www.technoserve.org/wp-content/uploads/2013/04/coffee-initiative-lessons-learned.pdf.

Tsafack, S. A. M., Degrande, A., Franzel, S. and Simpson, B. (2015). *Farmer-to-farmer Extension: A Survey of Lead Farmers in Cameroon*. [online] Nairobi, Kenya: World Agroforestry Centre, pp. 1, 52. Available at: https://apps.worldagroforestry.org/downloads/Publications/PDFS/WP15009.pdf.

UNFPA (2023). *Gender Transformative Approaches to Achieve Gender Equality and Sexual and Reproductive Health and Rights*. [online] United Nations Population Fund, pp. 1, 43. Available at: https://www.unfpa.org/publications/gender-transformative-approaches-achieve-gender-equality-and-sexual-and-reproductive.

UNICEF (2020). *Gender-Responsive Age-Sensitive Social Protection: A Conceptual Framework*. [online] UNICEF Office of Research–Innocenti, pp. 1, 36 Available at: https://www.unicef-irc.org/publications/1116-gender-responsive-age-sensitive-social-protection-a-conceptual-framework.html.

Waddington, H., Snilstveit, B., Hombrados, J., Vojtkova, M., Phillips, D., Davies, P. and White, H. (2014). Farmer field schools for improving farming practices and farmer outcomes: a systematic review. *Campbell Systematic Reviews* 10(1). https://doi.org/10.4073/CSR.2014.6.

Westendorp, A. B. (2012). *The contribution of farmer field schools to rural development in Nepal*. [PhD Thesis] Wageningen University. Wageningen School of Social Sciences. Available at: https://edepot.wur.nl/238928 [Accessed 21 June 2023].

Wood, E. A., McNamara, K., Kowalewska, A. and Ludgate, N. (2018). Household decision-making around food in rural Tajikistan: a cross-sectional study to help extension workers in the field. *Food Nutr Res*. 62. doi: 10.29219/fnr.v62.1330.

World Bank (2010). *Gender and Governance in Rural Services: Insights from India, Ghana, and Ethiopia*. [online] Washington, DC: World Bank Publications. Available at: https://documents1.worldbank.org/curated/en/557271468331168780/pdf/528600PUB0Rura101Official0Use0only1.pdf [Accessed 23 June 2023].

Chapter 14

Women smallholders and technology: facilitating access to improve farm productivity and profitability

Girma Gezimu Gebre, Leibniz-Centre for Agricultural Landscape Research (ZALF), Germany; and Stefan Sieber, Leibniz-Centre for Agricultural Landscape Research (ZALF), Germany and Humboldt Universität zu Berlin, Germany

1 Introduction

Women are often the farmers who cultivate food crops and produce commercial crops alongside the men in their households as a source of income. Women are major contributors to the agricultural workforce and food security in developing countries, either as family members or in their own right as women heading households. However, the productivity of women in agriculture is often constrained by their limited access to productive resources (e.g. land, labor, fertilizer, improved seed, and machinery or capital) and services, which could have serious implications for food security of their family and women themselves (Meinzen-Dick et al., 2011; Agarwal, 2015; Nijuki et al., 2016; Aryal et al., 2018). Moreover, a mixture of economic constraints, cultural norms, and practices limit womens' contribution to household food security and, to a lesser extent, inhibit the commercialization of the agriculture sector. Gender roles and relationships influence the division of work, the use of resources, and the sharing of the benefits of production between women and men. In particular, the introduction of new technologies and practices, underpinned by improved service provision, often disregards the gendered-consequences of market-oriented growth and many benefits bypass women.

http://dx.doi.org/10.19103/AS.2024.0148.18

2 Women and agricultural technology adoption

An extensive body of literature has examined the determinants of agricultural technology adoption (e.g. Feder et al., 1985; Feder and Umali, 1993; Doss et al., 2003; Doss, 2006; Uaiene et al., 2009; Beshir and Wagary, 2014; Ogada et al., 2014; Zeng et al., 2018). The factors identified with technology adoption are most commonly found to be lack or limited access to credit, information and inputs, inadequate infrastructure, social networks, and social learning behaviors. However, these studies do not address gender issues in technology adoption. On the other hand, a substantial number of empirical studies have examined gender differences in agricultural technology adoption by using the sex of the household head as the gender indicator (e.g. Kassa et al., 2013; Diiro et al., 2015; Gaya et al., 2017, Gebre et al., 2022). Using headship as a gender indicator is not necessarily indicative of who actually makes decisions for agricultural production (Peterman et al., 2010); hence, such studies end up dismissing the decision-making role of women within male-headed households (MHHs) and males within female-headed households (FHHs) (Doss, 2015). Women in MHHs make production decisions often jointly with their spouses (Deere et al., 2009), while men (e.g. adult sons) in FHHs also make such decisions (Bourdillon et al., 2002; Gebre et al., 2019). Therefore, conceptualizing headship as a gender indicator does not tell us everything about the gender roles, gendered division of labor, and gender politics in agricultural technology adoption.

Some studies such as Tanellari et al. (2014) or Addison et al. (2018) go beyond headship and look at the difference in technology adoption at the plot-manager level using the sex of individual farmers as the gender indicator. However, this approach fails to consider gender difference in agricultural settings where plots are managed jointly by male and female farm household members and also the role of decision-making in technology adoption. Thus it may lead to an incorrect policy prescription. Male and female family members face different access to inputs and information, and thus, they make choices about technology adoption decisions differently. These choices may be made jointly or separately depending on the individual, household, or other conditions such as social norms and cultural dictates (Doss, 2015). Moreover, gender roles in technology adoption vary across countries and regions (Manfre et al., 2018). For example, in southern Ethiopia, the majority of plots are collectively farmed by households, while a few are farmed individually by males or females with little cooperation from family members.

Recently, several studies emerged that consider technology adoption decisions jointly made by men and women in the same household (e.g. Ndiritu et al., 2014; Marenya et al., 2015; Lambrecht et al., 2016; Diiro et al., 2018; Muriithi et al., 2018). They provide mixed evidence of gender differences in agricultural technology adoption, making it more challenging to design policies to address gender-related inequalities in agriculture. Some studies assert that there is no gender difference in technology adoption (e.g. Muriithi et al., 2018), while others claim that joint management positively influences technology adoption (e.g. Marenya et al., 2015; Lambrecht et al., 2016) or heterogeneous effects (Diiro et al., 2018). Thus, agricultural technology adoption in relation to gender is context-specific, and common conclusions are challenging to reach.

Many previous studies note that women are less likely to adopt improved technologies compared to men (e.g. Doss and Morris, 2001; Tiruneh et al., 2001; Chirwa, 2005; Ragasa et al., 2012; Kassa et al., 2013; Diiro et al., 2015; 2018; Gaya et al., 2017). These studies found that there is no difference between men and women in their adoption of agricultural technologies when controlling for other factors such as complementary inputs. Yet as Doss (2015) argued, in reality, men and women do not have equal access to the inputs for which the model controls, such as capital, credit, land, seed, fertilizer, and extension services. Hence, it is important for policy planners on technology adoption to consider the differential levels of resource accessibility by gender. Manfre et al. (2018) also note that agricultural technologies are not gender-neutral and their adoption could be considerably influenced by social and cultural forces that affect existing gender roles and relations. This implies that gendered adoption of agricultural technologies varies widely based on context and culture (Gebre et al., 2019). For example, in most parts of Ethiopia, plowing with oxen and planting are primarily considered as men's task, while women share other agricultural activities with men, such as weeding, harvesting, and collecting. The roles and decision-making power over agricultural activities of women vary across households, locations, and cultures/ethnicities in the country. Belay and Olijira (2016) note that women plow with oxen in 'Awra Amba' in south Gonder of northern Ethiopia, while it is culturally forbidden for women to plow with oxen in central and southern Ethiopia.

One of the most seminal studies on the relationship between gender and agricultural technology adoption is research by Doss and Morris (2001). It focused on the gender of a farmer and the gender of the household head in Ghana. However, they did not provide information on *de jure* FHHs and *de facto* FHHs; their analysis was premised that all farmers regardless of their marital status and all married female farmers lived in a MHH. This approach may mis-categorize married females with missing husbands as if they were living together in an MHH. Hence, it is important to note *de facto* FHHs and *de jure* FHHs are not equally constrained to adopt agricultural technology. Indeed, FAO (2011a) notes that, while the *de facto* FHHs who receive remittance from their husbands can mitigate the effects of the absence of male agricultural laborers by increasing investment in farm tools and improved inputs, remittance does not procure labor itself. Moreover, Doss and Morris (2001) analysis focused on the rate of technology adoption not on the intensity of technology used.[1] These assumptions made it difficult for the study to address gender inequalities regarding the adoption of agricultural technologies. Another seminal study on gender and technology adoption is Seymour et al. (2016). This study focused on the adoption of improved maize varieties (IMVs) in Ethiopia, Kenya, and Tanzania. In the case of Ethiopia, however, the maize technologies provided to farmers in a package are only complementary to each other. Therefore, a decision to adopt one technology necessarily entails the decision whether to adopt another technology, hence varied outcomes are incremental at a societal level. Thus, research on the adoption of IMVs alone is not sufficient to formulate policy that can enhance improved productivity.

3 Women and agricultural productivity sub-Saharan Africa

Agricultural productivity, in general, is low in many sub-Saharan African countries where most farmers are smallholders. It is even lower for female farmers, who comprise 50% of the agricultural labor force in the region (FAO, 2011b). A report by the International Fund for Agricultural Development cited in FAO (2011b) showed that the percentages of FHHs in rural eastern and southern Africa range from 25% to 60%. FHHs are not homogenous. They can largely be divided into two categories: (a) households headed by women who are not married, divorced, or widowed (*de jure* FHHs) and (b) women whose spouses are away from home because of work or other reasons (*de facto* FHHs). The main reasons for the increase in the number of FHHs are the migration of men away from rural areas to seek jobs elsewhere, widowhood, divorce, and other family disruptions (FAO, 2011b; Kassie et al., 2014).

The extent of agricultural productivity differences between male and female farmers varies across and within countries in sub-Saharan Africa. Empirical evidence shows that the gender differences in agricultural productivity across sub-Saharan African countries are generally around 20–30%, with an average of 25% (FAO, 2011b; Aguilar et al., 2014; Kilic et al., 2015; Mukasa and Salami 2015). The productivity difference between MHHs and FHHs from northern to southern Ethiopia ranges from 30% to 65%, respectively (Aguilar et al., 2014; Challa and Mahendran, 2015; Gebre et al., 2021). One of the key reasons women farmers have lower productivity is the difference in the use of inputs, such as improved seed, fertilizer, and labor, and access to other resources influencing productivity such as education, extension, and credit.

1 The rate adoption is the relative speed with which farmers adopt an innovation, it time as one of its elements. Intensity of adoption refers to the level of usage of a given technology in any time period (Bonabana-Wabbi, 2002).

The vast majority of the recent literature and a review of early studies undertaken by Quisumbing (1996) confirm that estimates of the gender productivity gap become insignificant taking into account the difference in use of inputs, access to productive resources, and characteristics of individual farmers. Following recent empirical reviews, FAO (2011b) takes this further: 'If women farmers used the same level of the resource as men on the land they farm, their land productivity could increase by 20-30%. This could raise the total agricultural output in developing countries between 2.5 and 4%, which could, in turn, reduce the number of undernourished people in the world by 12–17%' (p. 5). However, a study by the World Bank (2014) in six sub-Sharan Africa countries revealed that equal access to productive resources such as fertilizer, farm labor, and training does not always translate into equal returns for women farmers. In most households, the agricultural production system is based on social norms and practices that determine the gender division of labor (Smith et al., 2003; FAO, 2011b; ADB, 2013; Nijuki et al., 2016). Women from every region face gender-specific constraints related to sociocultural forces that serve to reduce their agricultural productivity and limit their ability to ensure production. For example, women in Ethiopia face a serious gender gap in access to productive resources, as farming there is culturally considered a man's task (Aguilar et al., 2014; Gebre et al., 2021).

Previous studies on gender and agriculture tend to focus on understanding efficiency and productivity differences between MHHs and FHHs (e.g. Njuki et al., 2006; Thapa, 2008; Ragasa et al., 2012; Challa and Mahendran, 2015). There has been little or no attention paid to gender differences between *de facto* FHHs and *de jure* FHHs.

4 Measuring gender differences in agricultural productivity

A substantial body of the existing studies indicates that gender inequality in access to productive resources such as land, improved varieties, fertilizers, farm equipment, labor, training, and information leads to the difference in agricultural productivity between male and female farm households. However, the extent of the differences and the relative importance of their potential drivers depend on the country or region, the sample size, the type of crop, the unit of measurement, or the method of analysis (Table 1). To measure the magnitude of agricultural productivity differentials and unpack their potential sources, the commonly used approaches by existing studies are production function estimates (e.g. Saito et al., 1994; Udry et al., 1995; Njuki et al., 2006; Ragasa et al., 2012; Challa and Mahendran, 2015) and Blinder-Oaxaca decomposition methods (e.g. Aguilar et al., 2015; Kilic et al., 2015; Oseni et al., 2015; Mukasa and Salami, 2015; Slavchevska, 2015; Ali et al., 2016).

Table 1 Summary of literature review on gender differences in agricultural productivity

Author/s and year	Country studied	Crop	Sample	Dependent variable	Gender indicator	Productivity gap	Main reasons for the gap
Saito et al.(1994)	Kenya	Main crops	750 households	Value of output/ha	Head dummy	8.4%	Difference in input use
Udry et al. (1995)	Burkina Faso	All crops, sorghum, and vegetables	150 households farming 4655 plots in 4 years	Yield/ha	Farmer dummy	18% for all crops, 40% for sorghum, and 20% for vegetables	Lower use of productive inputs by women
Tiruneh et al. (2001)	Ethiopia	All crops	180 households	Value of output/ha	Head dummy	35%	Women's lower levels of input and limited access to extension services
Njuki et al. (2006)	Kenya	Crops, tree, and cattle	40 farmers	Value of output/ha	Plot manager dummy	No significant difference	Age, female labor, and land size have a significant effect
Goldstein and Udry (2008)	Ghana	Maize and Cassava	60 married couples with each four villages	Profit	Farmer dummy	Wives achieve lower profit than husband	Women's land tenure insecurity leads to less investment in land fertility
Akresh (2008)	Burkina Faso	All crops	2406 households	Value of output/ha	Farmer dummy	32% (ranges from 32% to 50%)	Differences in farm size and vulnerability to rainfall shocks
Alene et al. (2008)	Kenya	Maize	800 households	Yield/ha	Head dummy	23%	Women's lower access to land and education
Kinkingninhoun-Meˆdagbe et al. (2010)	Benin	Rice	145 farmers	Yield/ha	Farmer dummy	27%	Women lack timely access to services for farming and transportation
Peterman et al. (2011)	Nigeria and Uganda	All major crops	Nigeria: 3706 households; Uganda: 3625 plots	Value of output/ha	Head and plot manager dummy	Nigeria: 75% Uganda: 28.6%	Differences in socioeconomic characteristics, farm input, and crop choice

Ragasa et al. (2012)	Ethiopia	All major crops	7530 households and 31 450 plots	Value of output/ha	Head and plot manager dummy	Women achieve lower productivity	Women's lower access to quality extension services, price information, and other inputs
Aguilar et al. (2014)	Ethiopia	All major crops	1518 households	Value of output/ha	Plot manager dummy	Average: 23.4% (Tigray: 33.1%, South: 61.4%)	Differences in access to productive resources and their returns
Kilic et al. (2015)	Malawi	All major crops	16 372 plots	Value of output/ha	Plot manager dummy	25%	Differences in productive resources/endowments
Challa and Mahendran (2015)	Ethiopia	Coffee, maize, and bean	150 households	Value of output/ha	Head dummy	66.76%	Women's lower access to productive resources/inputs
Backiny-Yetna and McGee (2015)	Niger	All crops	3968 households	Value of output/ha	Plot manager dummy	Average: 19% but ranges up to 66%	Shortage of male labor, fertilizer, and land owned
Mukasa and Salami (2015)	Nigeria, Tanzania, and Uganda	All major crops	Nigeria: 4017,Tanzania: 2530, and Uganda: 2029 plots	Value of output/acre	Plot manager dummy	Nigeria: 18.6%, Tanzania: 27.4%, and Uganda: 30.6%	Differences in endowment and structural disadvantages of women's land size, land quality, labor inputs, and household characteristics
Oseni et al. (2015)	Nigeria (North and South)	All major crops	2431 households farming 4240 plots	Value of output/ha	Sex of plot manager	North: 28% South: no significant difference	Structural disadvantages of women in the northern parts
Slavchevska, (2015)	Tanzania	All major crops	6945 plots from 2182 households	Value of output/ha	Sex of plot manager	Weak gender difference	Differences in size of plot area and family labor
Ali et al. (2016)	Uganda	All major crops	6999 plots	Value of output/acre	Sex of plot manager	17.5%	Differences in child dependency ratio, access to transportation, improved seed, and pesticide use

Source: Compilation of authors from various sources.

Production function estimates measure the differentials by coefficients while the decomposition methods measure them by covariates and its coefficients. Moreover, some studies used switching regression in a counterfactual manner to measure production efficiency (e.g. Solís et al., 2007; Ahmed, 2012, Gebre et al., 2021) and test whether the difference is driven by the differences in observable characteristics or return to these characteristics.

Concerning the measurement of agricultural productivity using the sex dummy, the commonly used approaches by existing studies consider the sex of the household head, the sex of the farmer, and the sex of the plot manager or decision-maker (Table 1). The headship is the most widely used approach by many studies; however, it does not take into account the contribution of women in a MHH and men in a FHH (Doss, 2018). The approach to using the sex of the farmer (or the plot manager) may be adapted to measure the productivity difference between men and women within the same household; however, this approach is also problematic if farmers are misidentified in the household.[2] In some cultural contexts, women are considered helpers for men working on the farm but not as farmers because they spend more time working in the homestead. This situation holds particularly true in Ethiopia (Gebre et al., 2021). Indeed, Ragasa et al. (2012) noted that extension services in Ethiopia do not always consider women to be farmers. This kind of study is usually completed in western Africa, where men and women often manage separate plots; meanwhile, this practice is much less common in eastern Africa. Even studies based on the sex of the plot managers in the east African region show a significant overlap along gender lines. For instance, in their study, de la O Campos et al. (2016) used nationally representative data from Uganda and found that 92% of female-held plots and 77% of female-managed plots belong to FHHs.

Regarding the unit of productivity measurement, the yield (quantity of output per unit area) and the value of output per unit of land are the most commonly used approaches in the existing studies (Table 1). Yield measurement is simple and works best when farmers grow a single crop on a plot; however, it is less straightforward when farmers grow multiple crops on a single plot of land at the same time (Saito et al., 1994). When one is interested in evaluating the productivity of a single crop across different periods, using the yield method is problematic for the plots where multiple crops are grown across different seasons. Even for a single production season, it is difficult to measure the yield of crops that are continuously harvested (Doss, 2018).

Table 1 provides a summary of extant studies on gender productivity differences conducted in sub-Saharan African countries.

2 According to Doss (2018), 'farmer' is defined as the owner, the manager, or the person providing the day-to-day labor for the plot or crop.

The existing studies used different units to gauge the contribution of different factors impacting gender differences in agricultural productivity. Among those factors, the most difficult to measure in the context of developing countries is labor, which is the most important input in the production process. Because smallholder farms in developing countries typically employ family labor, there is no wage income or written records of labor time to estimate the family labor input (Arthi et al., 2016). In most studies, farmers were asked to recall the amount of labor used for their plots for the previous farming season (Doss, 2018). This leads to biases in reporting on labor time and affects the quality of data collected in the developing country (Arthi et al., 2016). The farming activities are not as regular as office work, and many of these activities are carried out jointly with other household activities (Doss, 2018). The seminal work by Arthi et al. (2016) points out that the biases in farm labor data are derived from reports on the weeks and days worked not from the hours worked per day. This is because some farmers or family members will work longer hours per day than some other farmers or family members. As a result, a day laborer could have a different marginal contribution to farm output (Doss, 2018). Since male and female labor is not substitutive in agriculture (Doss, 1999), most of the existing studies have separately estimated the labor inputs of men and women. Some of these studies (e.g. Saito et al. 1994; Tiruneh et al., 2001; Njuki et al., 2006; Challa and Mahendran, 2015) estimated the marginal productivity of family labor using shadow wage rates, which represent the opportunity cost of the family labor time (Sharma, 2013).

5 Conclusion

There are numerous studies on the interlinkage between gender and agricultural productivity and gender and agricultural technology adoption. However, most of these studies use the sex of the household head as a gender indicator, which ends up dismissing the contributions of males in FHHs and females in MHHs. In addition, they fail to capture the actual gender power relationship over the relative positions of men and women in the same household in agricultural decision-making. Moreover, most of the studies that consider the individual plot manager as a gender indicator are based on the premise that men and women in the same household manage separate plots; they fail to examine the prevalent case of farm households where men and women manage the same plot of land jointly. In other words, they fail to consider the level of joint contributions by male and female family members to the production. To address these shortcomings of extant studies, some studies used gender of decision-maker (joint vs individual) as a gender indicator.

Gender dimensions influence agricultural activities of the household in various ways across different regions, countries, or cultural norms. For instance, in some regions of the world, the head of the household (either male or female) solely owns and manages the total plots of the household and its associated output because of their cultural norm. In some regions, men and women in the same household manage separate plots. In other regions, plots are managed jointly by male and female family members. These gendered ownerships and management of the plots and associated outputs have a direct influence on agricultural technology adoptions, level of farm productivity, marketing decisions, and food security of the agrarian households. Sociocultural norms and economic power structures assign different roles to men and women involved agricultural activities within the household. Gender norms and power structures influence labor divisions between men and women within the household. In such context, it is a challenge to reach a common consensus about men's and women's roles in the household regarding agricultural activities. Thus, agricultural production and technology adoption in relation to gender is context-specific and common conclusions are challenging to reach.

6 References

Addison, M., et al. (2018). Gender effect on adoption of selected improved rice technologies in Ghana. *Journal of Agricultural Science*, 10(7).

Agarwal, B. (2015). Food Security, productivity, and gender inequality. The Oxford handbook of food, politics and society, ed. by Ronald J. Herring. Oxford University Press.

Aguilar, A., et al. (2014). *Decomposition of gender differentials in agricultural productivity in Ethiopia*. The World Bank, Africa Region, Poverty Reduction and Economic Management Unit, Policy Research Working Paper 6764.

Aguilar, A., et al. (2015). Decomposition of gender differentials in agricultural productivity in Ethiopia. *Agricultural Economics*, 46(3), 311–334.

Ahmed, S. (2012). *Agricultural land tenancy in rural Bangladesh: productivity impact and the process of contract choice*. PhD dissertation, University of Adelaide.

Akresh, R. (2008). *Efficiency in intra-household allocations*. University of Illinois at Urbana Champaign working paper.

Alene, A. D., et al. (2008). Economic efficiency and supply response of women as farm managers: comparative evidence from Western Kenya. *World Development,* 36(7), 1247–1260.

Ali, D., et al. (2016). Investigating the gender gap in agricultural productivity: evidence from Uganda. *World Development,* 87, 152–170.

Arthi, V., et al. (2016). *Not your average job: measuring farm labor in Tanzania*. Development Economics, Development Data Group, WPS7773.

Aryal, J. P., et al., (2018). Untangling gender differentiated food security gaps in Bhutan: An application of exogenous switching treatment regression. *Rev Dev Econ*. 00: 1–21.

Asian Development Bank (ADB) (2013). Gender equality and food security: women's empowerment as a tool against hunger. Mandaluyong City, Philippines.

Backiny-Yetna, P. and McGee, K. (2015). *Gender differentials and agricultural productivity in Niger*. World Bank Group, Development Research Group, Poverty and Inequality Team, Policy Research Working Paper 7199.

Belay, F. and Olijira, A. (2016). Gender role in agricultural activities in Ethiopia: country review. *Journal of Culture, Society and Development,* 22, 1–7.

Beshir, B. and Wagary D. (2014). Determinants of smallholder farmers' hybrid maize adoption in the drought prone central rift valley of Ethiopia. *African Journal of Agricultural Research,* 9(17), 1334–1343. doi:10.5897/AJAR2013.8336.

Bonabana-Wabbi, J. (2002). *Assessing factors affecting adoption of agricultural technologies: the case of Integrated Pest Management (IPM) in Kumi District, Eastern Uganda*.

Bourdillon, M., et al. (2002). *Assessing the impact of HYV maize in resettlement areas of Zimbabwe*. Summary report, International Food Policy Research Institute, Washington, DC.

Challa, M. T. and Mahendran, Dr. A. (2015). Gender difference and its impact on agricultural productivity: the case of Sheko District in Bench Maji Zone of SNNP, Ethiopia. *International Journal of Current Research*, 7(11), 22938–22942.

Chirwa, E. W. (2005). Adoption of fertiliser and hybrid seeds by smallholder maize farmers in Southern Malawi. *Development Southern Africa*, *22*(1), 1–12.

de la O Campos, A. P., et al. (2016). How does the choice of the gender indicator affect the analysis of gender differences in agricultural productivity? Evidence from Uganda. *World Development,* 77(C), 17–33.

Deere, D., et al. (2009). *Poverty, headship and gender inequality in asset ownership in Latin America*. Gainesville, FL: Working Paper, University of Florida, Center for Latin American Studies.

Diiro G., et al. (2015).The role of gender in fertilizer adoption in Uganda. *African Journal of Agricultural and Resource Economics,* 10(2), 117–130.

Diiro, G., et al. (2018) Women's empowerment in agriculture and agricultural productivity: evidence from rural maize farmer households in western Kenya. *PLoS ONE*, 13(5), e0197995.

Doss, C. (1999). Twenty-five years of research on women farmers in Africa: Lessons and implications for agricultural research institutions; with an annotated bibliography. CIMMYT economics program paper No. 99-02. Mexico D.F.: CIMMYT

Doss C. (2006). Analyzing technology adoption using microstudies: limitations, challenges, and opportunities for improvement. *Agricultural Economics, 34*(3), 207–219.

Doss C. (2015). *Women and agricultural productivity: what does the evidence tell us?* Economic Growth Center Discussion Paper No. 1051.

Doss. C. (2018). Women and agricultural productivity: reframing the Issues. *Development Policy Review*, 36, 35–50.

Doss C. and Morris M. (2001). How does gender affect the adoption of agricultural innovations? The case of improved maize technology in Ghana. *Agricultural Economics,* 25(1), 27–39.

Doss, C., et al. (2003). *Adoption of maize and wheat technologies in Eastern Africa: a synthesis of the findings of 22 case studies*. CIMMYT Economics Working Paper 03-06. Mexico: CIMMYT.

Feder, G. and Umali, D. (1993). The adoption of agricultural innovations: a review. *Technological Forecasting and Social Change,* 43, 215–239.

Feder, G. et al. (1985). Adoption of agricultural innovations in developing countries: a survey. *Economic Development and Cultural Change,* 33. doi:10.1086/451461.

FAO (Food and Agriculture Organization). (2011a). The role of women in agriculture. ESA Working Paper No. 11-02.http://www.fao.org/docrep/013/am307e/am307e00.pdf

FAO (Food and Agriculture Organization). (2011b). The state of food and agriculture. In *Women in agriculture: closing the gender gap for development*. Washington, DC. https://www.fao.org/4/i2050e/i2050e01.pdf

Gaya, H., et al. (2017). Gender differentials and adoption of drought tolerant maize varieties among farmers in Northern Nigeria. *European Journal of Business and Management,* 9(5), 81-87.

Gebre, G. G., et al. (2019). Gender differences in the adoption of agricultural technology: the case of improved maize varieties in southern Ethiopia. *Women's Studies International Forum*, 76, 102264.

Gebre, G.G., et al. (2021). Gender differences in agricultural productivity: evidence from maize farm households in southern Ethiopia. *GeoJournal,* 86, 843–864.

Gebre, G. G., et al. (2022). Improved maize technology adoption and its intensity among male- and female-headed households in Dawuro zone, Southwestern Ethiopia, *Gender, Technology and Development,* 27(1), 42–59.

Goldstein, M. and Udry, C. (2008). The profits of power: land rights and agricultural investment in Ghana. *Journal of Political Economy,* 116(6), 981–1022.

Kassa Y., et al. (2013). Determinants of adoption of improved maize varieties for male headed and female-headed households in West Harerghe zone, Ethiopia. *International Journal of Economic Behavior and Organization*, 1(4), 33–38. doi:10.11648/j.ijebo.20130104.11.

Kassie, M., et al. (2014). What determines gender inequality in household food security in Kenya? Application of exogenous switching treatment regression. *World Development,* 56, 153–171.

Kilic, T., et al. (2015). Caught in a productivity trap: a distributional perspective on gender differences in Malawian agriculture. *World Development* Vol.70, pp.416–463.

Kinkingninhoun-Meˆdagbe, F. M., et al. (2010). Gender discrimination and its impact on income, productivity, and technical efficiency: evidence from Benin. *Agriculture and Human Values*, 27, 57–69. doi:10.1007/s10460-008-9170-9.

Lambrecht, I., et al. (2016). Agricultural extension in Eastern Democratic Republic of Congo: does gender matter? *European Review of Agricultural Economics*, 43(5), 841–874.

Manfre C., et al. (2018). *Assessing how agricultural technologies can change gender dynamics and food security outcomes*.https://ingenaes.illinois.edu/wp-content/uploads/Part-One-Learn-Final-10_17.pdf

Marenya P., et al. (2015). Fertilizer use on individually and jointly managed crop plots in Mozambique. *Journal of Gender, Agriculture and Food Security (Agri-Gender),* 1(2), 62–83.

Meinzen-Dick, R., et al. (2011). *Engendering Agricultural Research, Development, and Extension: Priority Setting, Research &Development, Extension, Adoption, Evaluation*. Washington: IFPRI.

Mukasa, A. N. and Salami, A. O. (2015). *Gender productivity differentials among smallholder farmers in Africa: a cross-country comparison*. Working Paper Series N° 231, African Development Bank, Abidjan, Côte d'Ivoire.

Mukasa, A. N. and Salami, A. O. (2016). Gender equality in agriculture: what are really the benefits for sub-Saharan Africa? *Chief Economic complex, African Economic Brief,* 7(3).

Muriithi B., et al. (2018). *Does gender matter in the adoption of sustainable agricultural technologies? A case of push-pull technology in Kenya*. Structural Transformation of African Agriculture and Rural Spaces, working paper.

Ndiritu W., et al. (2014). Are there systematic gender differences in the adoption of sustainable agricultural intensification practices? Evidence from Kenya. *Journal of Food Policy,* 49, 117–127. doi:10.1016/j.foodpol.2014.06.010

Njuki, M., et al. (2006). *Productivity differences between male and female managed farms in the Eastern and Central Highlands of Kenya*. Contributed paper prepared for presentation at the International Association of Agricultural Economists Conference, Gold Coast, Australia, August 12–18.

Njuki, J., et al,. (2016). Gender, agriculture, and food security: where are we? In Njuki, J., Parkins, J. R., and Kaler, A. (Eds.), *Transforming gender and food security in the global south*. New York: Routledge.1–18.

Ogada M., et al. (2014). Farm technology adoption in Kenya: a simultaneous estimation of inorganic fertilizer and improved maize variety adoption decisions. *Agricultural and Food Economics,* 2, 12.

Oseni, G., et al. (2015). Explaining gender differentials in agricultural production in Nigeria. *Agricultural Economics*, 46(3), 285–310.

Peterman A., et al. (2010). *A review of empirical evidence on gender differences in non-land agricultural inputs, technology, and services in developing countries.* International Food Policy Research Institute.

Peterman, A., et al. (2011). Understanding the complexities surrounding gender differences in agricultural productivity in Nigeria and Uganda. *The Journal of Development Studies*, 47(10), 1482–1509.

Quisumbing, A. R. 1996. Male-female differences in agricultural productivity: methodological issues and empirical evidence. *World Development,* 24(10), 1579–1595.

Ragasa C., et al. (2012). *Gender differntials in access to extension services and agricultural productivity.* ETHIOPIA Strategy Support Program II, working paper.

Saito, K. A., et al. (1994). *Raising the productivity of women farmers in Sub-Saharan Africa*. Discussion Paper Series 230, Africa Technical Department, the World Bank.

Seymour G., et al., (2016). Women's empowerment and the adoption of improved maize varieties: Evidence from Ethiopia, Kenya, and Tanzania. Paper prepared for presentation at the Agricultural & Applied Economics Association's Annual Meeting, Boston, MA, July 31 - August 2, 2016.

Sharma, P. B. (2013). Measuring the price of labor in agricultural economics: the shadow wage rate. *Economic Journal of Development Issues*, 15&16(1–2) combined issue.

Slavchevska, V. (2015). Gender differences in agricultural productivity: the case of Tanzania. *Agricultural Economics,* 46(3), 335–355.

Smith, Lisa C., et al,. (2003). *The Importance of Women's Status for Child Nutrition in Developing Countries*. IFPRI Research Report 131. Washington, DC: IFPRI.

Solis, D., et al,. (2007). Soil Conservation and Technical Efficiency among Hillside Farmers in Central America: A Switching Regression Model. *Australian Journal of Agricultural and Resource Economics,* 51(4), 491–510.

Tanellari, E., et al. (2014). Gender impacts on adoption of new technologies: the case of improved groundnut varieties in Uganda. *African Journal of Agricultural and Resource Economics,* 9(4), 300–308.

Thapa S., (2008). Gender differentials in agricultural productivity: Evidence from Nepalese household data. Working Paper. Trento, Italy: University of Trento.

Tiruneh, A., et al. (2001). *Gender differentials in agricultural production and decision-making among smallholders in Ada, Lume, and Gimbichu Woredas of the Central Highlands of Ethiopia.* Mexico: International Maize and Wheat Improvement Center (CIMMYT) and Ethiopian Agricultural Research Organization (EARO).

Uaiene, R. N., et al. (2009). *Determinants of agricultural technology adoption in Mozambique*. Discussion papers, No. 67E. National Directorate of Studies and Policy Analysis, Ministry of Planning and Development, Republic of Mozambique.

Udry, C., et al. (1995). Gender differentials in farm productivity: implications for household efficiency and agricultural policy. *Food Policy,* 20(5), 407–423.

World Bank. (2014). *Levelling the Playing Field: Improving Opportunities for Women Farmers in Africa.* Washington, DC: World Bank Group.

Zeng, D., et al. (2018). Land ownership and technology adoption revisited: improved maize varieties in Ethiopia. *Land Use Policy*, 72, 270–279.

Chapter 15

Advancing a feminist intersectional lens in gender transformative work with women smallholder farmers

Diana E. Lopez, KIT Royal Tropical Institute, The Netherlands; Ekaterina Gualoto, SDW Hamburg e.V, Germany; and Rhiannon Pyburn, KIT Royal Tropical Institute, The Netherlands

1 Introduction

Women smallholder farmers embody different dimensions of social identity and face related discrimination and hardship that creates barriers to full, fruitful and meaningful engagement in agrifood systems. This has been acknowledged in different ways since the 1970s with Esther Boserup's examination of African women's invisible contributions to the agricultural economy (Boserup, 1970). Global thinking on gender dynamics in development has evolved over the past 50+ years in a non-linear, overlapping, and uneven manner. Analysis from the first wave of women in development (WID) to a later iteration of gender and development (GAD) exposed and focused on different critical aspects of gender inequality. WID focused on 'integrating' women into prevailing development processes, highlighting their invisible roles and disadvantaged positions. A next wave – women and development (WAD) shifted focus from women's relationship to development to the relationship between patriarchy and capitalism (Rathgeber, 1990) and on identifying structural impediments

http://dx.doi.org/10.19103/AS.2024.0148.19

to women's equal and equitable participation.[1] The 1985 United Nation's Conference on Women in Beijing put women's rights as central, recognizing that policy-making and institutions reproduced gender inequalities and thus also needed to address them, kicking off gender mainstreaming efforts worldwide (Van Eerdewijk, 2916). GAD[2] reconsidered the focus on *women*(-only) and gender roles to context-specific and dynamic *gender relations*. Over the past 15 years, *Gender Transformative Approaches (GTAs)* have become more prominent, drawing on the advances made in previous waves of thinking and specifically digging into the systemic and structural roots of gender inequality (Wong et al. 2019). Most recently, an *intersectional feminist lens* has been brought into gender and development work to make explicit the diversity and heterogeneity among women and among men. This allows an understanding of how intersecting social markers (gender alongside age, ethnicity, education, social standing, and so on) and inequalities affect the most marginalized.

GTAs are increasingly being taken up in development research and practice as a way to transform constraining gender norms that limit smallholder women's engagement in agrifood systems and their enjoyment of its related benefits. It is widely recognized that GTAs are key for addressing gender inequality and structural barriers to women's empowerment in agrifood systems. Importantly, the literature on GTAs underscores the need to consider the diversity of identities and experiences and to let that diversity guide the design of effective gender transformative (GT) interventions. Nevertheless, to date, only fragmented efforts in agrifood systems embed an intersectional lens into their GT work, resulting in limited operationalization.[3] In this chapter, we make the case for a more robust embedding of intersectional considerations into GTAs for the benefit of smallholder women and men. Intersectionality provides a powerful framework for understanding the intricate connections between distinct dimensions of social identity and their influence on individual women's experiences of discrimination and privilege – including differentiated access to agricultural resources and technology. When applied to GTAs in agrifood systems, intersectionality becomes crucial for unraveling inequalities related to gender and intersecting axes of social differentiation and exclusion. An intersectional lens also helps to reveal how existing inequalities are exacerbated by shocks and long-term stressors such as malnutrition, political-system fragility, market failure, conflict, and climate change. The limited use of an intersectional approach in current GT work in agrifood systems translates into a lack of empirical evidence which is fundamental to understand the

1 Often referred to as *Women* AND *Development (WAD).*

2 GAD conceptualizes empowerment – women's individual and collective agency – as a way of addressing subordination. It digs into the power dynamics between and among men and women. The explicit inclusion of non-binary and intersex people is a more recent and important development.

3 The literature reviewed for this chapter confirms this: out of 94 sources reviewed on GTAs only 56 mention the need for intersectional considerations, with very few empirical studies.

effectiveness of GT approaches and their outcomes in achieving gender equality. A comprehensive understanding of the complex interactions between gender and other aspects of social differentiation can serve as a catalyst, influencing the social dynamics of agrifood systems and shaping the outcomes of GAD initiatives.

The foundation for this chapter is a critical literature review, based on a selection of sources from 2010 to 2023, to identify the state-of-the-art of GTAs in agrifood systems. The guiding question across the literature reviewed is how, within these works, intersectionality is considered. A critical literature review draws from 'a fair selection of works' to 'tell a story to advance our understanding of what is already known' in an 'original, perceptive and analytical' manner (Jesson and Lacey, 2006, pp. 139, 142). It is therefore neither an exhaustive nor a descriptive study rather it identifies key topics discussed to unravel issues that are not necessarily well-articulated in the mainstream literature (Jesson and Lacey, 2006). We used Google Scholar as the main search engine. The search included academic journals, books and book chapters, reports, and other scholarly publications. The search involved combinations of relevant keywords including 'gender transformative approaches,' 'gender transformative methods/methodologies,' 'agriculture,' 'food systems,' and 'transformative change.' In total, 94 sources were identified. These were assessed according to a set of parameters including number of citations, relevance in the field, and mention of intersectional considerations. Finally, 56 sources were selected for full review. Among them, 29 were research articles and 27 gray literature.[4] The analysis of gray literature included 19 working papers, reports, and 8 books and book chapters from organizations such as the CGIAR, the Food and Agriculture Organization of the United Nations (FAO), the International Food Policy Research Institute, Canada's International Development Research Center, and the International Fund for Agricultural Development (IFAD).

2 What the literature tells us about gender transformative approaches

Our literature review underscores the significance of feminist scholarship in agrifood systems. Feminist theories and perspectives critically analyze power dynamics, patriarchal systems, and structures of oppression, debunking the notion of a linear relationship between increased productivity and gender equality. To fundamentally challenge and transform gender norms and power dynamics, GTAs must be anchored in feminist scholarship.

Our analysis of the literature reveals intriguing patterns and shared perspectives on GTAs. Critiques directed at mainstream development approaches are

4 See full list in References marked with '*'.

prevalent, highlighting their inadequacy in effectively tackling gender inequalities within agriculture and agrifood systems. Mainstream approaches have primarily focused on enhancing women's agricultural productivity, profitability, and market access, assuming that these improvements automatically lead to gender equality. However, they fail to address structural barriers or challenge prevailing social constructs related to agricultural production and labor, thereby risking the reinforcement of existing gender disparities (Hillenbrand et al., 2023; McArthur et al., 2022; Timu and Kramer, 2023; Njuki et al., 2023; Farnworth and Colverson, 2015; Kantor, 2015; Verma, 2014). Such narrow approaches run the risk of perpetuating existing gender stereotypes and exacerbating disparities. In contrast, GTAs go beyond mere participation and prioritize challenging gender norms and fostering critical awareness (Freitas Lopez et al., 2020; Cole et al., 2020; Stacey et al., 2019; Wong et al., 2019). Research articles widely recognize that GTAs offer a fresh perspective, highlighting the limitations of gender-focused interventions that solely target women's participation.

The literature consistently acknowledges the influential role of gender norms and power relations in shaping gender inequalities within agricultural and agrifood systems. These deeply ingrained social and cultural norms contribute to the perpetuation of discriminatory practices, limiting women's agency, and decision-making power. Understanding and recognizing these norms and power dynamics becomes pivotal in dismantling the structural barriers imposed by harmful social norms, attitudes, behaviors, and systems that sustain gender inequality. Addressing these issues is deemed essential for transformative changes in the agricultural sector (Timu and Kramer, 2023; Tavenner et al., 2022; Lopez et al., 2022; Farhall and Rickards, 2021).

Furthermore, the importance of considering socio-cultural and political contexts and integrating gender perspectives into agricultural policies, programs, and projects is emphasized. Such integration ensures that the unique challenges faced by women – including smallholder women – in agriculture are properly acknowledged and addressed (Andersson et al., 2022; Berretta et al., 2023; Mulema et al., 2020; Padmaja et al., 2020; Bilfield et al. 2020; Kantor et al., 2015).

Our literature review underscores the significance of feminist scholarship in agrifood systems. Feminist theories and perspectives critically analyze power dynamics, patriarchal systems, and structures of oppression, debunking the notion of a linear relationship between increased productivity and gender equality. To fundamentally challenge and transform gender norms and power dynamics, GTAs must be anchored in feminist scholarship. By incorporating feminist perspectives, GT interventions can benefit from a comprehensive understanding of gender and social justice (Haque, 2023; MacArthur et al., 2022; Farhall and Rickards, 2021; McOmbre et al., 2021; Cole et al., 2020; Stacey et al., 2019; Aguiari et al., 2017).

Valuable insights for examining gender norms and power relations include recent publications by Lecoutere et al. (2023a, 2023b), Bryan et al. (2024), and Puskar et al. (2023), emphasizing the need for transformative changes in governance, policies, and institutions to eliminate structural constraints that hinder women's access to resources, technology, and services, addressing discriminatory practices and harmful social norms. Other studies such as Wong et al. (2019), Kleiber et al. (2017), and Kruijssen et al. (2022) offer guidance for development of GTAs within development agencies, in small-scale fisheries, and value-chain analysis. Pyburn and van Eerdewjik (2021) shift the focus from improving women's productivity to gender equality and women's empowerment, while Lopez and Ludwig (2021) stress the importance of transdisciplinary perspectives and strategic gender research for GT change. Sachs et al. (2021) call for recognition of queer farmers, and Colfer et al. (2018) highlight the importance of intersectionality. Finally, Sarpura Escobar et al. (2017) emphasize the need to address values and cultural aspects related to gender roles and relations.

GT interventions presented in the literature cover a wide range of focus areas, including women's economic empowerment and financial inclusion (McOmber et al., 2021; Njuki et al., 2019), farmer organizations and governance (Farnworth and Colverson, 2015), communication practices (McOmber, 2021; Mulema, et al., 2020), climate-change resilience (Haque, 2023;), agroecology through participatory plant-breeding initiatives (Benítez, 2020), nutrition (Berretta et al., 2023), irrigation systems and sanitation (Padmaja et al., 2020), livelihood improvements (Cole et al., 2020; Stacey et al., 2019), agricultural insurance (Timu and Kramer, 2023), and value-chains (Bilfield et al., 2020).

While GTAs work at multiple levels (see Table 1), to date, most GTA interventions and research have focused on farmer household and community levels with much less attention to institutional and organizational levels (c.f. Pyburn and Eerdewijk, 2021) or lead agency. This oversight is being addressed by recent initiatives that aim to move beyond the individual, household, and community to the more confronting (for development organizations, research institutes, and governments) organizational, institutional, and lead agency levels. Most proponents of GTAs posture that these 'higher' levels have explicitly or implicitly been a part of the conceptualization of GT change and GTAs from the start, despite the greater attention given to the others.

Research, advocacy, and practice related to GTAs is ongoing and in a period of expansion and deepening. For example, since 2016, the Rome-based agencies[5] have come together as the Joint Programme on GTAs for food security and nutrition[6] to tackle underlying social and behavioral causes

5 Namely, the Food and Agriculture Organization of the United Nations (FAO), the World Food Program (WFP), and the International Fund for Agricultural Development (IFAD).
6 https://www.fao.org/joint-programme-gender-transformative-approaches/overview/en

Table 1 Examples of entry points for different gender transformative approaches

Entry point	GTA
Couples and their households	• Gender Action Learning System (GALS) *(Oxfam Novib, IFAD, Hivos, Twin and Twin Trading)* • Gender Household Approach *(Hanns R. Neumann Stiftung)* • Gender Model Family *(SEND)* • Individual Household Mentoring *(IFAD)* • Journeys of Transformation *(Promundo)* • Nurturing Connections© *(HKI)* • Joint Programme to Accelerate Progress towards the Economic Empowerment of Rural Women (JP RWEE) *(FAO, IFAD, WFP and UN Women)*
Groups and communities	• Community Conversations *(WFP)* • Dimitra Clubs (FAO) • Farmers' Field and Business Schools (FFBS) (CARE) • Farmer Field and Life School (FFLS) and Junior FFLS (JFFLS) (FAO) • Gender Action Learning System (GALS) (Oxfam Novib, IFAD, Hivos, Twin and Twin Trading) • Gender Mainstreaming in Member-based Organizations (Trias) • Social Analysis and Action (SAA) (CARE) • Joint Programme to Accelerate Progress towards the Economic Empowerment of Rural Women (JP RWEE) (FAO, IFAD, WFP and UN Women)
Organizations/ formal institutions/ private sector	• Gender Action Learning System (GALS) (Oxfam Novib, IFAD, Hivos, Twin and Twin Trading) • Models to Empower Women in Outgrower Schemes (AgDevCo) • Joint Programme to Accelerate Progress towards the Economic Empowerment of Rural Women (JP RWEE) (FAO, IFAD, WFP and UN Women)
Lead agency/ project staff	• Journeys of Transformation (Promundo) • Social Analysis and Action (SAA) (CARE)

Source: FAO, IFAD and WFP (2020).

of gender inequality, and the CGIAR[7] is developing a gender norms index to inform GTAs and explore (these less understood) institutional and systemic dimensions.

While the literature review identified common understandings of GTAs in agriculture, some key insights and challenges, it also revealed key areas that need to be urgently developed. We highlight three. (1) Many of the reviewed interventions remain pilot projects or small-scale initiatives, which signals the lack of capacity and resources for implementing and scaling up GTAs in meaningful and sustainable ways. Scaling up GT interventions requires sustained financial investments, institutional support, and strong partnerships between government agencies, civil society organizations, and research

7 See ONE CGIAR initiative *HER+: Harnessing gender and social equality for resilience in agrifood systems.*

institutions. (2) Measuring and assessing GT change is challenging. Only seven research papers mentioned clear indicators, making it difficult to compare and synthetize findings or evaluate methodologies and the actual impact of GTAs (Njuki et al., 2023; FAO, IFAD, and WFP, 2020). (3) The main gap in the literature, however, is the limited attention given to how gender intersects with other social identities within GTAs. Hence, we have decided to make this concern the main focus of this chapter. Overall, the review shows there is no shortage of research avenues to further explore how GTAs can be more effective, better assessed, scaled, and generally strengthened in benefit of those who are most vulnerable.

3 What intersectionality brings to gender transformative interventions

Intersectionality provides a powerful framework for understanding the intricate connections between various dimensions of social identity and their influence on individuals' experiences of discrimination and privilege (Colfer et al., 2018, p. 2). When applied to GTAs in agriculture, the lens of intersectionality becomes crucial in unraveling the constraints and inequalities related to gender and intersecting axes of social differentiation and exclusion within agricultural production and agrifood systems (Pyburn and van Eerdewijk, 2021, p. 25; Hillenbrand et al., 2015, p. 6, 14; Cole et al., 2015, p. 11, 46).

Scholars such as Clancy et al. (2019, p. 2) argue that a comprehensive understanding of the interplay between gender and other social identities, including race, age, and sexual orientation, lies at the core of GTAs. Similarly, Tavenner et al. (2022, p. 386) emphasize how intersectionality sheds light on the complex interactions between gender and other aspects of social differentiation, such as age, assets base, marital status, race/ethnicity, and caste/class. This understanding serves as a catalyst, influencing the social dynamics of agricultural systems and shaping the outcomes of GAD initiatives (Njuki et al., 2023, p. 219, 224).

Some of the empirical studies analyzed related to agrifood systems highlight the significance of considering intersectional axes of differentiation rather than gender alone (Tavenner et al., 2022; Berretta et al., 2023; Singh et al., 2022). Other sources such as Colfer et al. (2018), Fischer et al. (2019), and Tavenner and Crane (2019) offer useful guidance on applying an intersectionality lens in forest and farming system, while Hillenbrand et al. (2015) propose participatory tools like community maps and social network analysis to reveal intersectional differences and promote equitable change. Importantly, McDougall et al. (2021), present a step-by-step manual for the integration of intersectionality in food systems research for development.

However, many challenges remain. Lecoutere et al. (2023a, 2023b, pp. 1, 2) identify persistent structural barriers in various domains and at multiple scales as the underlying causes of gender and intersecting differentiation gaps, highlighting the need for a holistic and inclusive approach to dismantle these barriers and achieve gender-transformative change in agrifood systems. Furthermore, the authors emphasize that these inequalities are further exacerbated by shocks and long-term stressors, including malnutrition, coronavirus disease 2019, political-system fragility, conflict, and climate change (p. 6). Likewise, the literature review reveals that men 'as a gendered category' has not been effectively incorporated into development programs (Singh et al., 2022; Farhall and Rickards, 2021): only 8 out of 28 research articles acknowledge the importance of and/or incorporate men in programs and interventions. Similarly, the review identified only four sources that addressed the perspectives of non-binary gender identities (Farhall and Rickards, 2021; MacArthur et al., 2022; Sachs et al., 2021; McDougall et al., 2021).

Thus, our analysis shows the need and potential for integrating intersectionality into GTAs in agrifood systems. By centering the experiences of individuals from diverse backgrounds, GT interventions can transcend the limitations of singular perspectives and encompass the multifaceted challenges faced by marginalized groups (Farhall and Rickards, 2021; Tavenner et al., 2022; Colfer et al., 2018). Furthermore, through a critical examination of power structures and the dismantling of oppressive systems, GTAs grounded in intersectionality can pave the way for social justice, inclusivity, and empowerment within agricultural systems (Rietveld et al., 2022, p. 9; Colfer et al., 2018, pp. 11, 30).

Nevertheless, while several of the sources reviewed underscore the need to consider the diversity of women's experiences and identities with some providing useful guidance and understanding for the design of GT interventions, only a fraction of the analyzed literature (14 out of 28 research articles and 19 out of 27 sources of gray literature) emphasize the importance of systematically adopting an intersectional approach in GT interventions. To illustrate this point, we highlight one of the oldest works reviewed, a 2014 study on communication practices in aquaculture systems, which stated the urgency to acknowledge intersectionality (Underwood and Hendrickson, 2014, pp. 7, 30), yet, a decade later, Rietveld et al. (2022, p. 9) note that GT interventions continue to fall short in effectively adopting an intersectional approach.

While the importance of using intersectional perspectives to strengthen GT work in agrifood systems has long been acknowledged, its limited application in GTAs not only translates into a lack of empirical evidence and case studies on intersectionality in food, land, and water systems but also stresses the need for better evidence to understand the effectiveness of GT approaches for different women smallholder farmers and men and how GT outcomes are supporting lasting progress in gender equality.

4 Deepening the roots of intersectionality in gender transformative approaches

Intersectionality is a major feature of feminist scholarship that is particularly useful for exploring and making sense of complex phenomena (Salem, 2018). Diverse conceptualizations and methodologies endeavor to understand and make use of analytical categories to explore the complexity of intersectionality in social life. Although no universally agreed definition of intersectionality can be drawn from the literature, the two that follow illustrate the nuanced and varied ways in which it can be approached. Leslie McCall defines intersectionality as 'the relationships among multiple dimensions and modalities of social relations and subject formations…' (2005 p. 1771). Olena Hankivsky brings power, privilege, and oppression into the definition, stating the following:

> Intersectionality promotes an understanding of human beings as shaped by the interaction of different social locations (e.g. 'race'/ethnicity, indigeneity, gender, class, sexuality, geography, age, disability/ability, migration status, religion). These interactions occur within a context of connected systems and structures of power (e.g. laws, policies, state governments and other political and economic unions, religious institutions, media). Through such processes, interdependent forms of privilege and oppression shaped by colonialism, imperialism, racism, homophobia, ableism and patriarchy are created. (Hankivsky, 2014, p. 2)

As a methodology, intersectionality is likewise highly debated with different views on how analytical categories should be used to explore the complexity of social life. To date, the most widely used intersectional approach is inspired by the works of Leslie McCall (2005; 2000). McCall's categorical approach seeks to explicate relationships of inequality among social groups through the use of provisional categories as 'anchor points' while recognizing that such categories are dynamic:

> The concern is with the nature of the relationships among social groups and, importantly, how they are changing, rather than with the definition or representation of such groups per se. (The categorical approach) goes further in exploring whether meaningful inequalities among groups even exist in the first place. Perhaps inequalities were once large but now they are small, or in one place they are large but in another they are small. This perspective leaves open the possibility that broad social groupings more or less reflect the empirical realities of more detailed social groupings, thus minimizing the extent of complexity . . . McCall. (2005, p. 1785)

However, the broad adaptation of this (categorical) approach by gender and other researchers working in development (and other fields) has led to a critique of intersectionality *as a moment of resistance* to intersectionality *as a neoliberal approach that erases inequality* and that *negates complexity* (see Salem, 2018). To partially respond to the critique about the dangers of category-based research and its homogenization and simplification of women's complex situations, McCall has attempted to explicate what the categorical approach does (and what it does not do):

> (The categorical approach) focuses on the complexity of relationships among multiple social groups within and across analytical categories and not on complexities within single social groups, single categories, or both ... the subject is multigroup, and the method is systematically comparative. ... Overall, complexity is managed in comparative multigroup studies by what at first appears to be a reductionist process – reducing the analysis to one or two between-group relationships at a time–but what in the end is a synthetic and holistic process that brings the various pieces of the analysis together. (McCall, 2005, p. 1786)

Importantly, and while McCall and others advocate for the use of an 'anchor point' as main unit of analysis, in particular that of 'women,' this can also be misleading as anchor points are often treated as static categories focused on *who* they represent (e.g. rural women farmers) rather than treated as relational categories that evidence the complex interrelationships among social groups (e.g. relationships between different types of rural women farmers and rural men farmers) and *how* those interrelationships are changing (e.g. young rural women farmers increased access to agricultural knowledge and inputs compared to a decade ago and in contrast to, e.g. young rural men farmers). In agrifood systems work, this reductionist treatment of anchor points leads to a primary category of a particular type/group (e.g. women, girls, rural women, women farmers, etc) with all other social categories perceived to be relevant in a given project or context[8] (e.g. age, class, or caste) analyzed only superficially and as means to identify or explain certain inequities/inequalities of the primary category instead on focusing on how the categories relate to one another and contribute (or not) to GT change. While this has evident strengths (see full discussion in McCall 2005 and overview in Table 2), such an approach can limit sustained structural change toward greater gender equality (see full discussion in Pyburn & van Eerdewijk, 2021). A more useful anchor point for GT change would be 'gender', understood as a social power relation between different peoples and scales; or 'gender norms', focused on how power shapes the choices, voices, and mindsets of distinct peoples in a particular context or sector (McDougall et al., 2021; Wong et al., 2019).

8 Or following specific donor requirements such as recent emphasis on 'gender and youth.'

Table 2 Intersectional approaches and its potential applications in gender transformative work

Author	Intersectionality approach	Potential analysis to strengthen GT work
Crenshaw (1989)	**Three dimensions:** structural, political, and representational	In-depth case studies of social groups at the intersection of different categories, and exploration of how they experience marginality due to social relations and social movements in which they are situated.
McCall (2005)	**Three approaches:** 1. *Anti*categorical 2. *Intra*categorical 3. *Inter*categorical or categorical	Individual life histories to demonstrate how individuals' experiences are mediated by different social relations, life-cycle processes and personal agency. Deconstructs categories. In-depth case studies of social groups at the intersections of different categories. Well suited for macro, quantitative or mixed methods analysis. Categories act as anchor points for analysis. The goal is to empirically chart the changing relationships among multiple social groups.
Hankivsky (2014)	**Eight principles:** (1) intersecting categories, (2) multi-level analysis, (3) power, (4) reflexivity, (5) time and space, (6) diverse knowledges, (7) social justice and equality, (8) resistance and resilience	Useful for (feminist) political economy/ecology analyses where the contextual, situational, fluid and constantly renegotiated and contested aspects of intersectionality are highlighted.
Kabeer (2016)	**Three types of intersecting inequalities:** vertical, horizontal and spatial	Well suited to highlight whose (socio)economic disadvantages are intensified by the discriminations faced by certain groups on the basis of their identity, as well as by their greater concentration in the least favourable locations in their society.
Colfer et al. (2018)	**Five lenses:** 1. Cognitive (individual understanding of the world) 2. Emotional (how marginalization affects an individual) 3. Social (narratives and norms) 4. Economic (access, employment, inheritance) 5. Political	As a framework to understand and organize intersectional doings at the micro, meso, and macro levels (linked to McCall's 'intracategorical' approach).

(Continued)

Table 2 (*Continued*)

Author	Intersectionality approach	Potential analysis to strengthen GT work
McDougall et al. (2021)*	**Six guiding principles:** 1. Avoid being gender blind and go beyond binary 2. Integration during entire project cycle with involvement of whole team 3. Aim for transformative change 4. Expand the frame 5. Be clear about intended gender outcomes 6. Recognize the research process as both empowering and disempowering	Focus on unraveling power issues/dynamics across different people and scales. Integration of intersectional considerations during project cycle – including integration at each project phase (i.e. design, implementation, PMEL, analysis, communications, and recommendations).

Source: authors based on works cited in first column.

*Recently, another resource to integrate intersectional considerations into the project cycle was developed. Notably, it distinguishes a 'segmented' vs an 'intersectional' approach. See: Barraza, A. (2023). FAO guidelines for the incorporation of an intersectional approach.

Therefore, the gender researcher needs to think critically and contextually to be able to determine whether the categorical approach to intersectionality is indeed the best suited to inform the particular GTA being pursued. As noted by McDougall et al. (2021, p. 20), integrating an intersectional lens in a project cannot be done prescriptively: 'It relies on social science and gender expertise and adaptation to the specific R4D [research for development] issue.' Some key considerations they note include the following:

1 Reflect the needs and preferences of different priority social-gender groups of people.
2 Surface potential risks or negatives and how they are distributed.
3 Make explicit which gender outcomes and aspects of equity will be investigated, such as in decision-making processes and in the distribution of benefits, negatives/costs or trade-offs between more and less powerful actors.
4 Ask 'why' (causes) and 'so what' (implications for well-being including nutrition and food security and sustainability).
5 As appropriate, be explicit about asking questions consistently in relation to the most marginalized or vulnerable groups. (Ibid.)

Based on the literature reviewed as well as on key feminist works on intersectionality, in Table 2 we distinguish six prominent ways to think about intersectionality.[9] Each approach results in specific ways to conceive of and implement GT work and will derive particular limitations, opportunities, and strengths that allow more realistic gender goals and ambitions to be set.

As Table 1 illustrates, GT change requires actions at distinct levels as the same subject embodies multiple layers of complexity – see Fig. 1 as an example. However, not all GTAs are equipped to entice change at the levels required for sustained structural change, e.g. replacing the dominant patriarchal system with one based on gender equality. As most GTAs still focus on the individual, couple, household, and community levels. One of the most successful GTAs to date, the Gender Action Learning System (GALS) methodology arguably follows a (categorical) intersectional approach as it frequently targets the poorest and most food insecure women and men including smallholder farmers, young people (particularly unemployed youth), small-scale and micro-entrepreneurs and people living with disabilities (FAO, IFAD, and WFP, 2020). Yet, it still primarily focuses on change at the household and community levels. Emphasis is placed on behavioral change (of women smallholder farmers, extension workers, entrepreneurs, etc.) with limited reflection (and space for action) on how certain processes (e.g. systematic

9 This list is not exhaustive as other approaches exist.

Figure 1 Example of different intersecting elements as highlighted in Table 2 in a single person. Source: Content developed by authors, photo publicly available from Storyblocks .com

exclusion of women farmers in land tenure regimes), actors (e.g. government, companies, etc.), or other structural elements (e.g. neoliberal economy) intersect with and affect the ability for behavioral change of the target group. One way to strengthen the GALS methodology is through the adoption of an intersectional lens specifically designed to recognize and address these issues such as the one advanced by Colfer et al. (2018), Hankivsky (2014), or McDougall et al. (2021).

5 Conclusion

In reviewing the literature on GTAs and intersectionality in agrifood systems research and practice, some key conclusions surface and a number of ongoing challenges can be identified.

To begin, intersectionality is a powerful yet under-utilized tool for unraveling inequalities related to gender and intersecting axes of social differentiation and exclusion in agrifood systems. The dearth of empirical evidence limits imagination as to what is possible, but conceptually there are rich, complex, and well-developed frameworks and categorizations to draw from and build on. As a researcher, an intentional decision needs to be made as to the specific objectives and relevant units of analysis for the problem being studied. On that basis, different aspects of intersectional approaches can be brought in. Likewise, within a project intervention, intentional decisions are needed to pull out the intersectional elements relevant to a particular GTA, and in working with smallholder women embedded in specific power dynamics.

Bringing intersectional considerations more prominently into GT research and interventions in agrifood systems will allow us to better identify what works best for whom (micro-level) including a better understanding and positioning of sociocultural and political (macro-level) factors that affect women farmers' opportunities including access to resources and technologies. One challenge moving forward in integrating an intersectional lens into GTAs will be to elegantly reconcile the value of recognizing individual identities that intersectionality brings, with the value of moving beyond an individual level focus – namely community, institutional, and systemic levels – that GT research and interventions are now delving into more deeply. This means, for instance, linking the smallholder woman farmer with the systemic and institutional processes and factors that shape her (local) opportunity structure while being more nuanced in examining the alliances and contradictions of her multi-layered identities and the power dynamics she is embedded in. Importantly, it means to openly identify and acknowledge the multiple power structures that influence her opportunities. As noted by McDougall et al. (2015) and Wong et al. (2019), gender and social relations operate at multiple, intersecting scales. As such, GT change informed by intersectional considerations requires a concerted, sustained effort across multiple actors. Indeed, projects that focus and operate at only one scale (e.g. household or community) may fail to recognize the influence of powerful drivers at other scales (e.g. policy, market, or research) limiting GT change. For researchers working in agricultural development, applying a feminist intersectional lens in GTAs means acknowledging the role of agricultural research – and of researchers – for bringing about change. As noted by McDougall et al. (2021, p. 15), research organizations and researchers often are 'powerholders relative to the food system actors who AR4D is intended to benefit' such as smallholder women farmers. This manifests in who is recognized as possessing expert knowledge (related to gender equality or other matters), who controls the research agenda or intervention, or who determines what will be assessed. Even the anchor point of analysis may be based on powerful actors' own worldviews. Acknowledging these power dynamics as well as the researcher's own positionality is key to 'leveling the playing field in terms of who controls and shapes, and ultimately gains from innovations and the processes themselves' (p. 15) (see full discussion in McDougall et al., 2021). GT research methodologies are therefore useful for gender researchers committed to transformative and socially just change processes as they are conceived to help the design and application of GTAs. Principles underpinning GT research include (1) conducting deep, intersectional analyses to understand the context and the multiple dimensions and layers of inequality and power within a particular group or area; (2) meaningfully engaging with and (if needed) building capacities of diverse actors to drive and sustain locally meaningful change processes; (3) recognizing and equally valuing distinct knowledges and approaches toward sustained GT change; (4) combining action to influence desired change with (transdisciplinary) research on how change happens; and (5) a focus on how norm and structural change can happen across scales (Lopez et al., 2023).

With the belated but fast-evolving moves toward the decolonization of social institutions, especially in relation to development – including of agrifood systems in low-and-middle-income countries – intersectionality, at least in some of its definitions (e.g. Hankivsky, 2014 and McDougall et al., 2021), offers insights into a way forward. An intersectional lens can support the deconstruction or unraveling of power related to colonialism and oppression and how these forces play out in individual (overlapping) identities. Sitting on the cusp of change toward the post-colonial, feminist conceptual developments, including intersectional analysis, offer important opportunities for development thinking and practice. These developments underscore that gender is not *just* about women, and certainly not about a non-existent *universal woman*. GTAs hold many ingredients that complement and pave the way for a more intersectional approach. Already by working with both men and women, and seeking out their common ground – common identities – as well as acknowledging their diversity, GTAs such as GALS break down stereotypes and offer new visions. However, they will only be strengthened by a more conscious and comprehensive embrace of what intersectionality has to offer, for instance, to unravel policy, market and research agendas and priorities in agrifood systems. As noted by Tavenner et al (2022, p. 396): 'Interventions informed by intersectionality provide empirics that lay out the underlying and intersecting inequalities in agricultural systems, building the evidence base for identifying and challenging harmful gender norms.'

6 Where to look for further information

The following articles provide a good overview of the subject:

- Tavenner, K., et al. (2022) Intersectionality in gender and agriculture: toward an applied research design, *Gender, Technology and Development*, *26*(3), 385–403.
- Colfer, C. J. P., et al. (2018). *Making sense of 'intersectionality': A manual for lovers of people and forests* (Vol. 184). CIFOR.

Key research in this area can be found at the following organizations:

- Joint Programme on Gender Transformative Approaches for Food Security and Nutrition available at: https://www.fao.org/joint-programme-gender-transformative-approaches/overview/gender-transformative-approaches/en.

- Community of practice to accelerate CGIAR learning on the use of gender-transformative research methodologies for systemic change available at: https://gender.cgiar.org/news/community-practice-accelerate-cgiar-learning-use-gender-transformative-research-methodologies.

7 References

Aguiari, S., et al. (2017). Beyond the double gender gap in science and agriculture: is there space for feminist-inspired transformation within AWARD?. *Journal of Gender, Agriculture and Food Security (Agri-Gender)*, *2*(302-2017-1465), 106–124.

Andersson, K., et al. (2022). Window dressing inequalities and constructing women farmers as problematic: gender in Rwanda's agriculture policy. *Agriculture and Human Values*, *39*(4), 1245–1261.

Benítez, B., et al. (2020). Empowering women and building sustainable food systems: a case study of cuba's local agricultural Innovation Project. *Frontiers in Sustainable Food Systems*, *4*, 554414.

Berretta, M., et al. (2023). Rapid evidence assessment on women's empowerment interventions within the food system: a meta-analysis. *Agriculture & Food Security*, *12*(1), 1–52.

Bilfield, A., et al. (2020). Brewing a more balanced cup: supply chain perspectives on gender transformative change within the coffee value chain. *International Journal on Food System Dynamics*, *11*(1), 26–38.

Boserup, E. (1970). *Woman's role in economic development*. London: George Allen and Unwin.

Bryan, E., et al. (2024). Addressing gender inequalities and strengthening women's agency to create more climate-resilient and sustainable food systems. *Global Food Security*, *40*, 100735.

Clancy, K., et al. (2019). *Transforming gender relations-insights from IDRC research*. Ottawa, Canada: International Development Research Center.

Cole, S. et al. (2020). Gender accommodative versus transformative approaches: a comparative assessment within a post-harvest fish loss reduction intervention. *Gender, Technology and Development*, *24*(1), 48–65.

Coles, A., et al. (Eds.). (2015). *The Routledge handbook of gender and development*. London: Routledge.

Colfer, C. J. P., et al. (2018). *Making sense of 'intersectionality': a manual for lovers of people and forests* (Vol. 184). Bogor, Indonesia: Center for International Forestry Research (CIFOR), CGIAR Research Program on Aquatic Agricultural Systems – Program Report AAS-2014-36, Penang, Malaysia.

Crenshaw, K. (1989). Demarginalizing the intersection of race and sex: a Black feminist critique of antidiscrimination doctrine, feminist theory and antiracist politics. *University of Chicago Legal Forum*, *1*, 8.

FAO, IFAD, and WFP. (2020). *Gender transformative approaches for food security, improved nutrition and sustainable agriculture: a compendium of fifteen good practices*. Rome. https://doi.org/10.4060/cb1331en.

Farhall, K. & Rickards, L. (2021). The "gender agenda" in agriculture for development and its (lack of) alignment with feminist scholarship. *Frontiers in Sustainable Food Systems*, *5*, 573424.

Farnworth, C. & Colverson, K. (2015). Building a gender-transformative extension and advisory facilitation system in sub-Saharan Africa. *Journal of Gender, Agriculture and Food Security, 1*(1), 20–39.

Fischer, G., et al. (2019). *Gender analysis in farming systems and action research: a training manual*. Ibadan: IITA.

Freitas Lopez, D. B., et al. (2020). *Gender integration in social and behavior change*.

Freitas Lopez, Debora B., Sanyukta Mathur, Haley Brightman, Kathryn Berryman, and Kamden Hoffmann. 2020. "Gender integration in social and behavior change," Breakthrough RESEARCH Scoping Report. Washington, DC: Population Council. 17 pages https://knowledgecommons.popcouncil.org/departments_sbsr-pgy/1253/

Hankivsky, O. (2014). Race and gender: Intersectionality theory. In *The Wiley Blackwell encyclopedia of health, illness, behavior, and society* (pp. 1–4). Chichester, UK: John Wiley & Sons, Ltd.

Haque, A. S., et al. (2023). Gendered perceptions of climate change and agricultural adaptation practices: a systematic review. *Climate and Development*, *15*, 885–902.

Hillenbrand, E., et al. (2015). *Measuring gender-transformative change: a review of literature and promising practices*. CARE USA, WorldFish, CGIAR Research Program on Aquatic Agricultural Systems, October 2015.

Hillenbrand, E., et al. (2023). "There is still something missing": comparing a gender-sensitive and gender-transformative approach in Burundi. *Development in Practice, 33*(4), 451–462.

Jesson, J. K. & Lacey, F. M. (2006). How to do (or not to do) a critical literature review. *Pharmacy Education*, *6*(2), 139–148.

Kabeer, N. (2016). *"Leaving no one behind": the challenge of intersecting inequalities*. ISSC, IDS and UNESCO, Challenging Inequalities: Pathways to a Just World, World Social Science Report, 55-8.

This report should be cited as follows: ISSC, IDS and UNESCO (2016), World Social Science Report 2016, Challenging Inequalities: Pathways to a Just World, UNESCO Publishing, Paris.

Kansanga, M. M., et al. (2019). A feminist political ecology of agricultural mechanization and evolving gendered on-farm labor dynamics in northern Ghana. *Gender, Technology and Development*, *23*(3), 207–233.

Kantor, P., et al. (2015). Amplifying outcomes by addressing inequality: the role of gender-transformative approaches in agricultural research for development. *Gender, Technology and Development*, *19*(3), 292–319.

Kleiber, D., et al. (2017). Promoting gender equity and equality through the small-scale fisheries guidelines: experiences from multiple case studies. In *The small-scale fisheries guidelines: Global implementation* (pp. 737–759).

Kruijssen, F., et al. (2022). *Gendered aquaculture value chain analysis and development: An analytical framework*.

Kruijssen F, Danielsen K, Newton J and Braaten Y. 2021. Gendered aquaculture value chain analysis and development: An analytical framework. Penang, Malaysia: CGIAR Research Program on Fish Agri-Food Systems. Manual: FISH-2021-27.

Lecoutere, E., et al. (2023a). Fostering an Enabling Environment for Equality and Empowerment in Agri-food Systems. *CGIAR GENDER Impact Platform Working Paper*.

Lecoutere, E., et al. (2023b). Equality and Empowerment by Gender and Intersecting Social Differentiation in Agri-food Systems: Setting the Stage. *CGIAR GENDER Impact Platform Working Paper*.

Leder, S., et al. (2019). Transformative engagements with gender relations in agriculture and water governance. *New Angle: Nepal Journal of Social Science and Public Policy*.

Lewis, D. (2018). Gender, feminism and food studies: a critical review. In *Gender, peace and security in Africa*, pp. 54–69.

Lopez, D. E. & Ludwig, D. (2021). A transdisciplinary perspective on gender mainstreaming in international development: the case of the CGIAR. In D. Ludwig et al. (Eds.), *The politics of knowledge in inclusive development and innovation* (p. 48). Routledge, London, UK.

Lopez, D. E., et al. (2022). Towards gender-inclusive innovation: assessing local conditions for agricultural targeting. *PLoS One*, *17*(3), e0263771.

Lopez, D. E., et al. (2023). Designing for change through "reflecting and doing": the CGIAR Community of Practice on Gender-Transformative Research Methodologies. *Frontiers in Sustainable Food Systems, 7*.

MacArthur, J., et al. (2022). Gender-transformative approaches in international development: a brief history and five uniting principles. *Women's Studies International Forum, 95*, 102635.

Manlosa, A. O., et al. (2019). Leverage points for improving gender equality and human well-being in a smallholder farming context. *Sustainability Science*, *14*, 529–541.

McCall, L. (2000). Explaining levels of within-group wage inequality in US labor markets. *Demography*, *37*(4), 415–430.

McCall, L. (2005). The complexity of intersectionality. *Signs: Journal of Women in Culture and Society*, *30*(3), 1771–1800.

McDougall, C., et al. (2015). Implementing a gender transformative research approach: early lessons. In *Research in development: Learning from the CGIAR research program on aquatic agricultural systems* (pp. 41–56).

McDougall, C., et al. (2021). *Gender integration and intersectionality in food systems research for development: a guidance note*. Penang: CGIAR Research Program on Fish Agri-Food Systems. Manual: FISH-2021-26.

McOmber, C., et al. (2021). Investigating the conceptual plurality of empowerment through community concept drawing: case studies from Senegal, Kenya, and Nepal. *Sustainability*, *13*(6), 3166.

Mudege, N. N., et al. (2020). Gender topics on potato research and development. In *The potato crop: Its agricultural, nutritional and social contribution to humankind* (pp. 475–506).

Mulema, A. A., et al. (2020). Clapping with two hands: Transforming gender relations and zoonotic disease risks through community conversations in rural Ethiopia. *Human Ecology*, *48*(6), 651–663.

Njuki, J., et al. (Eds.). (2016). *Transforming gender and food security in the global south*. New York: Routledge.

Njuki, J., et al. (2019). Beyond Access: gender transformative financial inclusion in agriculture and entrepreneurship. In A. R. Quisumbing et al. (Eds.), *ATOR: Gender equality in rural Africa: from commitments to outcomes*. IFPRI, Washington: ReSAKSS Trends and Outlook Report.

Njuki, J., et al. (2023). Meeting the challenge of gender inequality through gender transformative research: lessons from research in Africa, Asia, and Latin America.

Canadian Journal of Development Studies/Revue canadienne d'études du développement, *44*(2), 206–228.

Padmaja, R., et al. (2020). Gender transformative impacts from watershed interventions: insights from a mixed-methods study in the Bundelkhand region of India. *Transactions of the ASABE*, *63*(1), 153–163.

Pyburn, R. & van Eerdewijk, A. (2021). *Advancing gender equality through agricultural and environmental research: past, present, and future*. Washington, DC: International Food Policy Research Institute (IFPRI).

Puskar, R., et al. (2023). Closing gender gaps in productivity to advance gender equality and women's empowerment. CGIAR GENDER Impact Platform. Working Paper 014. Nairobi, Kenya.

Quisumbing, A., et al. (2023). Measuring women's empowerment in agriculture: innovations and evidence. *Global Food Security*, *38*, 100707.

Rao, N., et al. (2017). Gender justice and food security in India.

Rathgeber, E. (1990). WID, WAD, GAD: Trends in research and practice. *Journal of Developing Areas*, *24*(4), 489–502.

Rietveld, A. M., et al. (2022). A community of practice for gender-transformative research methodologies. CGIAR GENDER Impact Platform Working Paper.

Rohe, J., et al. (2018). A gender lens on women's harvesting activities and interactions with local marine governance in a South Pacific fishing community. *Maritime Studies*, *17*, 155–162.

Ros-Tonen, M. A., et al. (2019). Conceptualizing inclusiveness of smallholder value chain integration. *Current Opinion in Environmental Sustainability*, *41*, 10–17.

Sachs, C. et al (2021). Routeledge Handbook of Gender and Agriculture. Taylor and Francis. London and New York.

Salem, S. (2018). Intersectionality and its discontents: intersectionality as traveling theory. *European Journal of Women's Studies*, *25*(4), 403–418.

Sarapura Escobar, S., et al. (2017). Gender transformaive approaches in agricultural innovation: the case of the Papa Andina Initiative in Peru. In I. Oborn et al. (Eds.), *Sustainable Institificaiton in samllholer asgitvulutre – an integrated systems research approach Earthscan Food and Agriculture* (pp. 304–316). New York: Routledge.

Singh, S., et al. (2022). Protocol: gender transformative approaches in agriculture for women's empowerment: a sysemic review. *Campbell Systemic Reviews*, *18*, e1265.

Stacey, N., et al. (2019). Enhancing coastal livelihoods in Indonesia: an evaluation of recent initiatives on gender, women and sustainable livelihoods in small-scale fisheries. *Maritime Studies*, *18*, 359–371.

Tarjem, I. A., et al. (2021). *Tools and methods on gendered design, deployment and evaluation of agricultural technologies*. CGIAR GENDER Platform Working Paper.

Tavenner, K. & Crane, T. A. (2019). Beyond "women and youth": applying intersectionality in agricultural research for development. *Outlook on Agriculture*, *48*(4), 316–325.

Tavenner, K. & Crane, T. A. (2022). Hitting the target and missing the point? on the risks of measuring women's empowerment in agricultural development. *Agriculture and Human Values*, *39*(3), 849–857.

Tavenner, K., et al. (2022). Intersectionality in gender and agriculture: toward an applied research design. *Gender, Technology and Development*, *26*(3), 385–403.

Timu, A. G. & Kramer, B. (2023). Gender-inclusive, -responsive, and -transformative agricultural insurance: a literature review. *Global Food Security 36*, 100672.

Underwood, C. R. & Hendrickson, Z. (2014). *Communication interventions for gender equality and social equity in aquatic agricultural systems: A review of the literature*.

Van Eerdewijk, A. (2016). Gender mainstreaming: views of a post-beijing feminist. In W. Harcourt (Ed.), *Palgrave handbook on gender and development: critical engagements in feminist theory and practice* (pp. 117–131). New York: Palgrave MacMillan.

Verma, R. (2014). Business as unusual: The potential for gender transformative change in development and mountain contexts. *Mountain Research and Development*, *34*(3), 188–196.

Wong, F., et al. (2019). *Implementing Gender Transformative Approaches in Agriculture*. A Discussion Paper for the European Union. CGIAR Collaborative Platform for Gender Research.

Zhang, W., et al. (2021). Soil health and gender: why and how to identify the linkages. *International Journal of Agricultural Sustainability*, *19*(3–4), 269–287.

Chapter 16

Financial support for women smallholders: experience in Bangladesh

M. Jahangir Alam Chowdhury, University of Dhaka, Bangladesh

1 Introduction

Agriculture plays a significant role in economic growth in many countries. It is a particularly important sector for a developing country as it can serve as an "engine of growth" for the whole economy. In developing countries, the desired level of industrialization cannot be achieved without proper support from the agriculture sector. Overall economic growth can thus be promoted through supporting a country's agriculture sector (Schultz, 1964; Khan et al., 1986; Abdulai and Rieder, 1996; Gollin et al., 2002; Awokuse and Xie, 2014).

Johnston and Mellor (1961) argue that there are five channels through which agriculture contributes to economic growth. These five channels are:

1. supplying surplus labor to the industrial sector;
2. ensuring food security through producing enough food at an affordable price for the population;
3. supplying an economic surplus to allow investment in the industrial sector;
4. creating markets for the outputs of the industrial sector; and
5. earning foreign exchange to allow importing raw materials for industrial production.

http://dx.doi.org/10.19103/AS.2024.0148.20

However, the impact of the agriculture sector is not linear across countries as it varies from country to country (Awokuse and Xie, 2014).

The agriculture sector has been one of the main contributors to the total Gross Domestic Product (GDP) of Bangladesh. In 1987, it contributed 40% of total GDP (BBS, 1992). This reduced to 12% of GDP in 2022 (BBS, 2022b). Though this contribution has declined over the years, it is still one of the most important sources of employment in the country. In 2003, the agriculture sector created 52% of the total employment opportunities in the country (BBS, 2013a). This contribution reduced to a still significant 45% in 2022 (BBS, 2022a). Agriculture also plays a key role in food security. Bangladesh has achieved self-sufficiency in rice production, though not in other cereals such as wheat, where production needs to increase (due to a dietary shift away from rice towards other cereals over the past 30 years) (BPC, 2020).

The availability of land for farm production is important for the agriculture sector to play its desired role in growing GDP, employment creation, and improving food security in the country. However, the total area of land available for agriculture is declining gradually at a rate of 0.73% per annum in Bangladesh (BPC, 2020). This has resulted in a decline in per-capita land availability. In 1984, the area of land ownership per household was 1.67 acres. It declined by 61% to 0.65 acres per household by 2019 (BBS, 2022b). Hoque (1988) argues that the most efficient farm size is between 7 and 12 acres. Inequities in land ownership are even higher than income inequality in Bangladesh (Griffin et al., 2002). One of the main characteristics of a poor rural household in Bangladesh is landlessness or near landlessness (Khan et al., 1986).

More than 53% of households in Bangladesh are functionally landless. These households are either totally landless or own a small piece of homestead land of less than 5 decimals (i.e. around 240 square yards/200 square meters, where a decimal represents one hundredth of an acre). The majority of farm households are smallholders and headed by males. Only 5% of firm households are headed by females. In Bangladesh, a smallholder household is defined as a household that has land ownership of 5 to 250 decimals. Smallholders represent 44% of all households nationally and 92% of all farm households in the country (BBS, 2022a). Since agriculture remains one of the main contributors to GDP and employment, and the government has a policy of increasing agricultural production and exports, it is essential to promote smallholder agriculture to enhance overall agricultural productivity (Lele and Mellor, 1988).

Access to credit is important for the sustainable development of the agricultural sector in the country as it helps "production, crop diversification, processing, value addition, and marketing" (BPC, 2020). It is also important for promoting agriculture of smallholder firms as it helps to regulate consumption and to increase income through income-generating activities (Khandker and Faruqee, 2003). Smallholders have less access to formal sector financial institutions as they are less likely to have sufficient acceptable assets to satisfy the formal sector financial institutions as collateral. In addition, they are also excluded through bureaucratic and lengthy loan sanction procedures, preference for high-income clients, and credit rationing. As a result, they tend to rely on informal financial sources. These, however, are not beneficial for smallholder firms as they can be exploitative in nature and harm borrowers through monopolistic power, excessively high interest rates, and undervaluation of collateral (Bhaduri, 1977; Rao, 1980; Bardhan, 1980; Ghosh, 1986; Ghate, 1992; Foltz, 2004; Petrick, 2005).

The limitations of both formal and informal financial sectors in providing credit to households that lack the required collateral have led to the evolution of microcredit programs. The main objective of these programs is to offer households or individuals credit without collateral to help them start income-generating activities and reduce poverty. In microcredit systems, group commitment, strict discipline in credit provision and repayment collection, and supervision of borrowers have replaced the need for collateral (Yunus, 1999; Chowdhury et al., 2005).

Access to credit has a gender dimension (Chowdhury et al., 2016). Women are less likely to receive loans from formal sector financial institutions, as loans are heavily skewed to male borrowers. Bearing in mind the prevailing discrimination in the financial sector against women and the importance of promoting smallholder firms through ensuring access to credit , this chapter intends to explore women smallholder firms' access to credit in Bangladesh. There is a gap in the literature on the issue in this country, so the chapter aims to help bridge the gap in the literature on financial support for women smallholders in Bangladesh.

This chapter has five sections. Following this short introduction, Section 2 presents the available literature, Section 3 provides estimation strategy and data, after which the results are presented in Section 4. Concluding remarks are made in Section 5.

2 Access to credit for women smallholders

Credit markets in developing countries often function inefficiently, ultimately resulting in credit rationing of potential borrowers (Stiglitz and Weiss, 1981; Inderst and Mueller, 2007). The conditions that create credit market imperfections include (Carter, 1988; Bell et al., 1997; Foltz, 2004):

- government-imposed interest rate ceilings;
- monopolistic power of some credit suppliers, especially moneylenders;
- high transaction costs of borrowers; and
- borrowers' moral hazard (a term describing the willingness of individuals to take excessive risks if they consider the consequences to be limited).

Due to the problem of moral hazard on the part of borrowers, financial institutions request borrowers to provide them with collateral with a market value higher than the value of the loan. The ensuing credit rationing mostly affects poor people who lack assets, especially land, to provide financial institutions with collateral. Land ownership increases household access to credit from formal financial sources, especially commercial banks, and reduces dependence on informal financial sources (Piza and de Moura, 2015; Guirkinger and Boucher, 2008; Chandio and Jiang, 2018; Fafchamps, 2014). Access to credit is crucial for firm growth as it increases investment (Banerjee and Duflo, 2014; de Mel et al., 2008; Dupas and Robinson, 2013). It is also important as credit constraints hamper the flow of funds to the poor, which results in the reduced allocative efficiency of capital and increases income inequality (Aterido et al., 2013).

The credit constraint also has a gender characteristic (Arenius and Minniti, 2005; Klapper and Parker, 2011). This gender-based financial bias has negative impacts on the overall economy (Aterido et al., 2013). Women are more likely to be constrained than men in terms of accessing capital for starting new businesses (Fletschner, 2008), possibly because the financial sector is dominated by men. As a result, there is a bias toward female borrowers in formal financial institutions. Women's loan applications are more likely to be rejected (Aterido et al., 2013). The legal, social, cultural, and economic conditions that prevail in society affect women's demand for and supply of capital. The perception by financial institutions of women as insignificant and inexperienced borrowers constrains the supply of funds for women (Lycette and White, 1989). Due to the lack of required collateral and greater discrimination in the formal financial sector, women rely more on internal or informal financial sources (Richardson et al., 2004). However, Bruhn (2009) did not find any gender gap in access to credit in Latin America.

In an environment of adverse selection, the lack of knowledge about the contribution made by women contributes to the capital constraints for women in the formal financial sector. Having reviewed the available literature, Fletschner (2008) identified various contributing factors toward credit constraints for women. These factors are the requirement for collateral, difficulties in finding a guarantor, the requirement for authorization by the woman's husband or a male relative, preference for activities that are operated by men, lack of information regarding the availability of funds, high transaction costs of receiving a loan, women's risk-averse nature, the necessity for household work, low literacy rate, and lack of confidence when applying for loans. In addition, Aterido et al. (2013) identified that women's lesser involvement in the formal sector is another reason for the credit constraints they face. The traditional gender-based division of labor in the household due to prevailing social norms is one of the main reasons behind women's lower involvement in the formal sector, as they are left with insufficient time and opportunities to get involved.

Using data from a survey applied to 210 couples in rural Paraguay, Fletschner (2008) claims that intra-household dynamics are more important for women than men in getting a loan. Women are more likely to be constrained if they belong to a wealthier household and less likely to be constrained if the household owns titled land. Moreover, women are less likely to be constrained if they have control over a larger share of assets, which demonstrates their bargaining power in the household.

3 Investigating availability of credit for women smallholders in Bangladesh

3.1 Data used

This chapter uses data from the Household Income and Expenditure Survey (HIES) 2010 (HIES, 2010), generated by the Bangladesh Bureau of Statistics (BBS), a department of the Government of the People's Republic of Bangladesh (BBS, 2011). Khandker et al. (2013) use data from several rounds of the same survey (for example HIES 2000, HIES 2005, HIES 2010) to assess the effects of access to credit on microenterprise growth in Bangladesh. Chowdhury (2011) also uses HIES 2005 to establish the determinants of entrepreneurship in a conflict region in Bangladesh

HIES 2010 is a nationally representative dataset covering all parts of Bangladesh. A two-stage random sampling technique was applied to identify respondent households. In the first stage, the primary sampling units (PSUs) were randomly selected. A PSU is a cluster of households. In the second stage, 20 households were randomly selected from each of the PSUs. HIES 2010 covers data from 12 240 randomly selected households (N = 12 240) from all 64 districts and 384 sub-districts across seven divisions in the country. However, some households have been dropped from the analysis due to missing data. This chapter found data from 12 075 households usable.

Besides household demographic information, the dataset contains detailed information on the occupational status of household members and income, consumption, assets, and liabilities of households. It also contains detailed information on land ownership and loans of households from different sources. These sources include microfinance institutions (MFIs), commercial banks (CBs), cooperatives, moneylenders, friends, and family members. Moreover, this dataset also includes detailed information on other socio-economic and demographic aspects of the household. Data for three district-level variables – per capita agriculture value addition, labor force participation rate, and per capita availability of commercial bank loans – come from the Statistical Year Book of the Government of Bangladesh (BBS, 2013b).

The status of a household as a smallholder is defined on the basis of the size of the land ownership of that household. It varies from country to country. In India, households that have land ownership under 5 acres (500 decimals) are considered as smallholders. Households with land ownership of less than 12.5 acres (1250 decimals) are defined as smallholders in the province of Punjab of Pakistan (Hussain and Thapa, 2012). In Bangladesh, smallholder farms are identified on the basis of the definition used by the Bangladesh Bureau of Statistics, a government organization. Using land ownership data, all households in the country are divided into four categories – non-farm holdings, small farm holdings, medium farm holdings, and large farms. Non-farm holdings are those households that own no land or less than 5 decimals. Small farm holdings are households with land ownership of 5 to 249 decimals. The land ownership of medium farms is 250 to 749 decimals, whilst large farms are those households that own more than 749 decimals of land. According to the 2008 Agriculture Census, non-farm, small, medium, and large farms accounted for 47.09%, 44.65%, 7.44%, and 0.82%, respectively (BBS, 2022), whilst in HIES 2010, the distribution of non-farm, small, medium, and large farms was 33.01%, 60.81%, 5.62%, and 0.56%, respectively.

Bangladesh is predominantly a patriarchal society (Jaim, 2021; Murshid and Zippay, 2016; Chowdhury, 2009) in which the senior-most active male adult becomes the head of the household and is the main socio-economic decision-maker in the household. The senior-most female adult member becomes head of the household only if the senior-most male adult dies or is sick or disabled, or if there is no income-earning adult male member. In such a case, the senior-most female member of the household becomes the head of the household and the main socio-economic decision-maker. According to HIES 2010, the percentage of male and female households is 85.67% and 14.33% respectively. In this chapter, women smallholders are defined as those households that are headed by a female and have land ownership between 5 and 249 decimals. The number of women smallholders is 1000, which equates to 8.28% of all households. Table 1 shows that the total number of smallholders is 8343. Therefore, women smallholders represent 12% of all smallholders in the country.

3.2 Model

Following Chowdhury (2014), I specify the following reduced form equation for the financial support for a woman smallholder *j* conditional on individual, household, and district level characteristics:

$$C^{*}_{j} = \alpha_j X_j + u_j \tag{1}$$

Table 1 Descriptive statistics of variables used in the model

Variable	Definition	Mean	SD
Dependent variables			
Access to credit	=1 if the household has access to any credit sources; 0 otherwise	0.173	0.378
Access to CB credit	=1 if the household has access to credit from commercial and development banks; 0 otherwise	0.018	0.133
Access to MFI credit	=1 if the household has access to credit from microfinance institutions (MFIs); 0 otherwise	0.107	0.309
Access to informal credit	=1 if the household has access to credit from informal sources; 0 otherwise	0.049	0.216
Total credit (log)	Total amount of credit from all sources in Taka	1.682	3.723
Total CB credit (log)	Total amount of credit from all formal sector commercial and development banks in Taka	0.190	1.422
Total MFI credit (log)	Total amount of credit from all microfinance institutions in Taka	0.982	2.852
Total informal credit (log)	Total amount of credit from all informal sources in Taka	0.519	2.312
Household head's characteristics			
Age	Age of the household (HH) head in years	45.9	14.4
Age^2	Square of Age		
Education	HH head's education level in years	3.1	3.99
Married	=1 if the household head is married; 0 otherwise	0.49	0.50
Household characteristics			
HH members 0 to 5	Total household members in the age category of 0 to 5	0.42	0.66
HH members 6 to 25	Total household members in the age category of 6 to 25	1.94	1.35
HH members 26 to 60	Total household members in the age category of 26 to 60	1.10	0.76
HH members 60a	Total household members in the age category of 60 above	0.27	0.47
Minority	=1 if the HH belongs to the minority group; 0 otherwise	0.10	0.29
Land	Land ownership of the household in decimal	51.5	99.47
Financial assets (log)	Total value of all financial assets of the household in Taka	4.47	4.87
Business ownership	=1 if the household owns a business; 0 otherwise	0.04	.18
Other assets index	An index of all assets other than land and financial assets	0.37	2.05

(Continued)

Table 1 (*Continued*)

Variable	Definition	Mean	SD
District level variables			
District CB loans (log)	Per capita loans and advances from commercial and development banks in the district in Taka	5.075	5.595
District Agri value addition (log)	Per capita agricultural value addition in the district in Taka	4.014	3.776
District labor participation (log)	District labor force participation rate in percentage	1.537	0.896

In equation 1, C_j represents the dependent variables, which include different types of access to credit and loan amounts from difference sources.

Four different variables have been identified to determine a household's access to credit. These variables are:

- access to credit from any financial sources;
- access to credit from formal sector commercial and development banks;
- access to credit from microfinance institutions; and
- access to credit from informal sources.

Access to credit is binary in nature. The probit regression model is used for empirical estimations. Access to credit of a woman smallholder *j* (C^*_j) is a latent variable that is unobservable. In equation 1, $\alpha_j X_j$ is an index function that helps us to estimate the probability of a woman smallholder's access to credit:

$$C_j=1 \text{ if } C_j^*>0 \text{ or0 if otherwise} \quad (2)$$

Aside from the above-mentioned four dependent variables representing access to credit, there are four additional dependent variables that represent the total amount of credit from different sources, which include the total amount of credit from all sources, total amount of credit from formal sector commercial and development banks, total amount of credit from microfinance institutions, and finally, total amount of credit from informal sources. As these dependent variables are ratio label variables, the ordinary least squares (OLS) regression technique has been applied.

$$C_j=\eta_{ij}Ind_{ij}+\lambda_j HH_j+\Phi_k Dis+\upsilon_j \quad (3)$$

In equation 1, the independent variable X_j includes individual (ind_{ij}), household (HH_j), and district ($Dis_{k)}$ level variables that are considered determinants of access to credit and total amount of credit. In equation 3, a detailed version of equation 1, Ind_{ij} is a vector of individual characteristics; HH_j represents a vector of household characteristics; Dis_k specifies a vector of

district-level characteristics, and finally, u_j is the error that captures effects of other unobserved factors influencing access to microcredit that vary by smallholders and has an expected conditional mean of zero. As above, the parameters, η_{ij}, λ_j, and Φ_k are to be estimated. The vector ind_{ij} represents variables representing the age, education, and marital status of the household head. The vector HH_j represents the number of household members in four different age categories of 0–5, 6–25, 26–60, and 60 and above. It also includes minority status, total land ownership, ownership of total financial assets, business ownership, and ownership of other assets. The vector Dis_k represents district-level variables. These variables are the availability of commercial bank loans per capita, per capita agricultural value addition, and female labor force participation rate in the district.

Table 1 shows the descriptive statistics of the independent variables in the model. The average age of the women smallholders was around 46 years, which indicates that household heads were not that much older. The education level of the women smallholders was low as the average education level was just 3.1 years, meaning the majority of these women didn't even complete five years of primary education. It also confirms that these women came from households that could not afford to send their daughters to school. Among women smallholder households, 49% of the household heads were married, meaning 51% of female heads of households were either divorced or widowed. It also reveals that married women smallholders might have economically inactive husbands due to sickness or disability.

The number of household members in the age categories of 0–5, 6–25, 26–60, and 60 above were 0.49, 042, 1.94, 1.10, and 0.60 (means), respectively. This illustrates that these households had more members in school and college and in economically active age categories. The minority group accounted for 10% of the smallholders, which belonged to Hindu, Christian, and Buddhist religions. This number is representative nationally. The average household land ownership was around 52 decimals (0.52 acres) with a high variability. The average value of financial assets of women smallholders was Taka 29 593 (log value of 4.47) (Taka – the currency of Bangladesh; approximately 154 Taka to £1 sterling; Taka 29 593 = approximately £192). Only 4% of these households owned a business. The average index of assets other than land and financial assets was 0.37. This index was constructed using the principal component analysis (PCA) technique. The construction of this index took into account ownership of 25 household consumer durables, including radios, televisions, refrigerators, washing machines, fans, furniture, etc., as these assets reflect the economic status of a household. The average district level per capita loans and advances and agriculture value-addition were around Taka 119 000 (log value of 5.08) and Taka 10 320 (log value of 4.01) respectively. The average district-level labor force participation rate, including both male and female, was around 36% (log value of 1.54).

4 Availability of credit for women smallholders in Bangladesh: results

Table 2 shows the land ownership of smallholders, indicating that the average land ownership of smallholders is 57 decimals, whilst that of women smallholders is 34 decimals. The land ownership of women smallholders is 40% lower than that of all smallholders. This difference is statistically significant, indicating that the firm size of women smallholders is relatively small.

Access to credit by women smallholders is not widespread – Table 3 indicates that just 173 households (17% of all smallholders) had access to credit. The results also illustrate that some households had more than one loan. Among these households, 22 women smallholders had two loans and 14 women smallholders had three loans. These results indicate that the majority of women smallholders remain outside of the financial system in terms of getting access to credit. This might be because women typically fail to meet the requirements of the financial sector (particularly the formal financial sector) when applying for a loan. Moreover, the financial sector is biased towards male borrowers. Various other factors discourage women from the financial sector, including women's tendency to be more risk-averse, a lack of time outside of household chores, low literacy levels, and lack of confidence. Land is considered one of the preferred types of collateral, and so ownership of small plots of land does not lead to access to credit. Another reason might be that not having a male household head limits access to credit, as the formal financial sector requires a male guarantor to get a loan.

Table 4 shows the average size of loans from different sources. Loans have been divided into three categories – commercial and development banks,

Table 2 Land ownership of women smallholders

Variable	Obs	Mean	SD	Min	Max
All smallholders	7343	56.68	60.33	5.0	249.0
Women smallholders	1000	34.15	44.54	5.0	240.0

Table 3 Access to credit of all women smallholders

Access to credit	No. of women smallholders	Percent
No	827	82.70
Yes	173	17.30
Total	1000	100.00

Table 4 Amount of loans of borrowing women smallholders

Variables	Obs	Mean	SD	Min	Max
All loans	173	57 783	238 779	500	3 000 000
Commercial bank loans	18	200 556	699 705	3000	3 000 000
MFI loans	106	13 221	15 046	500	100 000
Other informal loans	49	99 700	127 596	2000	600 000

microfinance institutions (MFIs), and other informal sources. The development banks include two specialized formal sector banks – Bangladesh Krishi Bank and Rajshahi Krishi Unnoyan Bank – established to promote the country's agriculture sector. The commercial and development banks are formal sector financial institutions and are regulated by the country's Central Bank. MFIs provide small collateral-free loans to households. Other informal financial sources include moneylenders, local co-operatives and associations, input suppliers, friends, and family members.

The average size of all loans was approximately Taka 58 000. Only 18 women smallholders (2% of all women smallholders) had loans from commercial and development banks, the average size of which was around Taka 200 000. The size of these loans varied much more than the loans from the other two sectors. Around 11% of smallholders had loans from MFIs, the average size of which was around Taka 13 300. This loan size is much smaller than that of the formal sector commercial and development banks, and other informal financial sources. A total of 50 women smallholders (5% of all women smallholders) had loans from informal financial sources. The average size of these loans was approximately Taka 100 000. These descriptive statistics show that women smallholders are highly credit-constrained with minimal access to formal sector financial institutions. Due to their limited access to formal sector financial institutions and the small size of loans from microfinance institutions, women smallholders are dependent on informal sources for loans that are exploitative in nature.

4.1 Determinants of access to credit

Table 5 presents determinants of access to credit from different sources. Four different models have been estimated for this purpose – column (1) for access to credit from all sources, column (2) for access to credit from commercial banks and development banks, column (3) for access to credit from MFIs, and column (4) for access to credit from informal financial sources. The dependent variable is binary and displays 1 if the household has access to credit and 0 otherwise. Therefore, the probit regression technique has been applied when undertaking estimations.

Table 5 Determinants of access to credit

	(1)	(2)	(3)	(4)
	All loans	CB loans	MFI loans	Informal loans
Age	0.0300	0.0817	0.0491	−0.0250
	(1.15)	(1.28)	(1.51)	(−0.77)
Age^2	−0.000396	−0.000739	−0.000605	0.000184
	(−1.48)	(−1.14)	(−1.80)	(0.57)
Married	−0.166	−0.0708	−0.284*	0.133
	(−1.32)	(−0.26)	(−1.97)	(0.71)
Education	0.0165	0.0433	0.0188	−0.0132
	(1.10)	(1.45)	(1.07)	(−0.59)
HH members 0 to 5	0.175*	0.254	0.115	0.117
	(2.26)	(1.75)	(1.25)	(1.09)
HH members 6 to 25	0.0885*	0.105	0.0921*	0.0140
	(2.36)	(1.45)	(2.16)	(0.25)
HH members 26 to 60	0.0957	0.0265	0.110	0.0158
	(1.35)	(0.19)	(1.39)	(0.14)
HH members 60a	−0.0551	0.132	−0.00830	−0.0925
Minority	(−0.41)	(0.50)	(−0.05)	(−0.47)
	0.539***	−0.285	0.696***	−0.0422
	(3.58)	(−0.68)	(4.37)	(−0.17)
Land (log)	0.0177	0.108*	−0.0110	0.00957
	(0.76)	(2.13)	(−0.40)	(0.29)
Financial assets (log)	−0.00376	−0.0198	0.000150	−0.00105
	(−0.35)	(−0.89)	(0.01)	(−0.07)
Business ownership	0.528*	0.976**	0.532*	0
	(2.23)	(2.73)	(2.06)	(.)
Other assets index	−0.126***	−0.00389	−0.148***	−0.0693
	(−3.89)	(−0.06)	(−3.65)	(−1.49)
District CB loans (log)	−0.00286	−0.0586	−0.00981	0.0312
	(−0.09)	(−0.62)	(−0.25)	(0.70)
District agri value addition (log)	−0.00749	0.120	−0.0226	0.0312
	(−0.08)	(0.53)	(−0.22)	(0.25)
District labor participation (log)	0.429	0.147	0.555*	0.155
	(1.80)	(0.30)	(1.98)	(0.45)
Constant	−3.254*	−6.023*	−4.187**	−2.165
	(−2.45)	(−2.03)	(−2.64)	(−1.18)
N	1000	1000	1000	1000
pseudo R^2	0.076	0.128	0.114	0.029

t statistics in parentheses.
* $p < 0.05$, ** $p < 0.01$, *** $p < 0.001$.

The results (column (1)) indicate that the number of household members aged 0–5 and 6–25, the minority status of the household, business ownership status, and the ownership of other assets are significant determinants of access to credit from all sources. The number of family members aged 0–5 and 6–25 are significant positive determinants of access to credit. These results imply that women smallholders who have more household members in these two age categories are relatively young as they have young children. They are therefore more likely to be more energetic and enthusiastic than other women smallholders in terms of getting access to credit. They are also more likely to be economically active and thus more likely to demand and receive loans. Women smallholders who belong to the minority community are significantly more likely to receive a loan. The reason behind this might be to do with religion. Minority households belong to either Hinduism, Buddhism, or Christianity, which means that these women smallholders are non-Muslim. In Islam, interest is prohibited, and some Muslims follow this strictly. In non-Muslim households, there are no religious restrictions in terms of taking on interest-based loans, which might be why minority women smallholders are more likely to have access to loans from more financial sources. The business ownership of women smallholders is also a significant positive determinant of access to credit from financial sources. Women smallholders who own businesses are likely to have higher demand for loans to run their businesses and are more likely to have more assets to use as collateral for loans. Moreover, the formal sector commercial and development banks consider these women smallholders to be more creditworthy than other women smallholders, as they have a regular income source. The ownership of other assets, which are consumer durables, is a significant negative determinant of access to credit from any source. Women smallholders who have more consumer durables are more likely to be wealthier and have a higher level of income. As their income and wealth levels are higher, smallholders with more assets are less likely to require a loan for farming and running a business.

To access credit from commercial and development banks, there are only two significant determinants – land ownership and business ownership (column (2)). The size of the land ownership is a significant positive determinant of access to credit from commercial and development banks. Women smallholders with higher levels of land ownership are more likely to receive a loan from commercial and development banks. This finding is consistent, as we know that borrowers need to provide commercial and development banks with assets as collateral in order to receive a loan, and land is one of the preferred assets. A higher level of land ownership, therefore increases the probability of access to credit from commercial and development banks for a woman smallholder. The ownership of a business is also a positive determinant of access to credit from commercial and development banks. Women smallholders who own businesses are more likely to receive credit from commercial and development banks in order to run their businesses. Commercial and development banks likely consider these women smallholders more creditworthy as they have a regular income source to repay loans, so the probability of getting a loan from commercial and development banks increases.

There are six significant determinants of access to credit from MFIs (column (3)). These are marital status, the number of household members aged 6–25, minority status, business ownership, ownership of other assets, and the district-level labor participation rate. Married women smallholders are significantly less likely to receive a loan from MFIs. Unmarried smallholders are either widowed or divorced and are more likely to be destitute. Statistics show that married women smallholders own significantly more financial assets and consumer durables than those who are unmarried. The average value of financial assets of married women smallholders was approximately Taka 39 000, compared to Taka 21 100 for unmarried households. Similarly, married women in smallholder households own significantly more consumer durables than those who are unmarried. The average number of consumer durables of married and unmarried women smallholders was 7.3 and 5.7 items, respectively. As MFIs tend to give poorer households, collateral-free small loans and married women smallholders are more likely to be wealthy, a married woman smallholder is, therefore, less likely to receive a loan from MFIs.

Women smallholders with a greater number of family members aged 6–25 are more likely to receive a loan from MFIs, as they are typically relatively younger and economically active. In the microfinance system, women members are required to attend weekly meetings away from their homes, and younger women smallholders are more likely to be able to attend these meetings. This is likely why smallholders with more family members aged 6–25 are more likely to receive loans from MFIs. Moreover, women smallholders with a higher number of family members aged 6–25 have more school- and college-going children and thus require greater income to finance their education, which is likely to increase demand for credit from MFIs.

The minority status of women smallholders is a significant positive determinant of access to credit from MFIs. As minority women smallholders are non-Muslim, they will not be deterred from taking interest-based loans from MFIs, as interest is prohibited in Islam. The business ownership of women smallholders significantly increases the likelihood of access to credit from MFIs. This is logical in the sense that women smallholders with businesses are more likely to have higher demand for credit compared to those with no business ownership, as they need capital to run their business. The ownership of consumer durables is a significant negative determinant of access to credit from MFIs. Ownership of consumer durables indicates the wealthier status of women smallholders. As MFIs tend to target poor households, women smallholders with more consumer durables are, therefore, less likely to have access to credit from MFIs. Finally, women smallholders who belong to districts with higher labor participation rates are more likely to have higher access to credit from MFIs. The reason could be that districts with higher labor participation rates are more economically prosperous and have higher economic activity. Women smallholders are therefore also more likely to be economically active in these districts. This is probably why women smallholders in districts with higher labor participation rates demand more loans from MFIs, which in turn increases their likelihood of access to credit from MFIs. The results in Table 6 indicate that women smallholders' access to credit does not have any significant determinants. None of the varibale has come out as a significant determinant of access to credit from infomral financial sources (cloumn (4)).

4.2 Determinants of amount of credit

Table 6 presents the determinants of the total amount of credit of women smallholders. As the dependent variables in equation 2 are ratio level variables, the OLS technique has been applied. Four different OLS models have been estimated for four different dimensions of the total amount of credit presented in four different columns in Table 6 – (1) represents results for log of total amount of credit from all financial sources; (2) shows results for log of total amount of credit from commercial and development banks; (3) presents results for log of total amount of credit from MFIs; and (4) illustrates determinants of log of total amount of credit from informal financial sources.

The total amount of credit has the following significant determinants (column (1)). These are total household members aged 0–5 and 6–25, minority status, business ownership, and ownership of other assets. The total number of household members of women smallholders aged 0–5 and 6–25 significantly (and positively) determine the total amount of credit from all sources, which means that the total amount of credit increases with the increase in the number of household members in these two categories. Women heads of these smallholder households are younger and likely to be more economically active. Moreover, it also reflects that these households have more school- and college-going students and need more income for financing education. Therefore, they are more likely to demand credit from MFIs.

The minority status (i.e. belonging to Hindu, Buddhist, and Christian religions) of women smallholders significantly (and positively) determines the total amount of credit from all sources. It means that minority women smallholders are more likely to receive a higher amount of credit from different financial sources. The likely reason is that they do not have any religious restrictions in receiving interest-based credit from financial sources, as interest is not prohibited by their religion. Some Muslims are unable to seek interest-based credit from any financial sources as interest is prohibited in Islam. A

Table 6 Determinants of the amount of credit

	(1) Total loans	(2) CB loans	(3) MFI loans	(4) Informal loans
Age	0.0403	0.0294	0.0463	−0.0317
	(0.76)	(1.43)	(1.15)	(−0.94)
Age^2	−0.000546	−0.000243	−0.000565	0.000231
	(−1.06)	(−1.21)	(−1.44)	(0.70)
Married	−0.332	−0.0316	−0.436	0.136
	(−1.09)	(−0.27)	(−1.89)	(0.70)
Education	0.0496	0.0253	0.0334	−0.00892
	(1.37)	(1.80)	(1.21)	(−0.39)
HH members 0 to 5	0.430*	0.128	0.175	0.147
	(2.31)	(1.77)	(1.23)	(1.24)
HH members 6 to 25	0.251**	0.0516	0.175*	0.0278
	(2.72)	(1.44)	(2.50)	(0.47)
HH members 26 to 60	0.220	0.0310	0.180	−0.0112
	(1.27)	(0.46)	(1.37)	(−0.10)
HH members 60a	−0.199	0.0159	−0.121	−0.0735
	(−0.63)	(0.13)	(−0.51)	(−0.37)
Minority	1.346***	−0.134	1.507***	−0.0393
	(3.39)	(−0.87)	(5.00)	(−0.16)
Land (log)	0.0411	0.0406	−0.00854	0.00561
	(0.75)	(1.92)	(−0.21)	(0.16)
Financial assets (log)	−0.0100	−0.0101	−0.000409	−0.00102
	(−0.39)	(−1.02)	(−0.02)	(−0.06)
Business ownership	1.462*	0.797**	1.156*	−0.491
	(2.32)	(3.27)	(2.42)	(−1.23)
Other assets index	−0.262***	0.00246	−0.194***	−0.0680
	(−3.69)	(0.09)	(−3.60)	(−1.51)
District CB loans (log)	0.000420	−0.0260	−0.0210	0.0502
	(0.01)	(−0.87)	(−0.36)	(1.03)
District agri value addition (log)	−0.0584	0.00552	−0.0709	0.00315
	(−0.28)	(0.07)	(−0.45)	(0.02)
District labor participation (log)	0.790	0.0447	0.670	0.144
	(1.44)	(0.21)	(1.61)	(0.41)
Constant	−2.148	−0.864	−1.936	0.342
	(−0.70)	(−0.73)	(−0.84)	(0.18)
N	1000	1000	1000	1000
R^2	0.061	0.029	0.077	0.015

t statistics in parentheses.
* $p < 0.05$, ** $p < 0.01$, *** $p < 0.001$.

woman smallholder with a business ownership is more likely to receive a higher amount of total credit from different sources as business ownership is a significant (positive) determinant of the total amount of credit from different financial sources. The likely reason is that women smallholders owning businesses require more capital to run their businesses and are therefore more likely to demand and receive a higher amount of credit from different sources. Ownership of consumer durables is a significant (and negative) determinant of the total amount of credit from different sources, as ownership reflects the level of wealth of women smallholders. Wealthy smallholders are less likely to require credit for household expenditures as they have sufficient wealth. For this reason, the ownership of other assets is a significant (negative) determinant of the total amount of credit of women smallholders from all sources.

There is just one significant determinant of the total amount of credit from commercial and development banks of women smallholders, which is business ownership (column (2)). Women smallholders need more capital to run their businesses and are therefore likely to require more credit from commercial and development banks compared to women smallholders with no business ownership. Moreover, commercial and development banks in the formal sector prefer these borrowers as they have a regular income to repay the loans. As a result, women smallholders with business ownership are more likely to receive more credit from formal sector commercial and development banks.

There are four significant determinants of the total amount of credit from MFIs (column (3)). These are total family members aged 6–25, minority status, business ownership, and the ownership of other assets. The total number of household members aged 6–25 significantly (and positively) determines the total amount of credit from MFIs. Having more household members in this age category signifies that these women smallholders are more likely to be young, energetic, and economically active and therefore more likely to need credit from commercial and development banks. Moreover, they can fulfill the MFIs' mandatory requirement of attending weekly meetings and therefore receive credit. For these reasons, women smallholders with more household members aged 6–25 are more likely to receive higher amounts of credit from MFIs.

Minority women smallholders are more likely to receive a higher amount of credit from MFIs as they are not constrained by religious restrictions from receiving interest-based credit from financial sources. This is likely the reason behind non-Muslim women smallholders receiving a higher amount of total credit from MFIs. The ownership of other assets (including consumer durables) is a significant and negative determinant of the total amount of credit from MFIs. It means that women smallholders who own a lesser number of consumer durables are more likely to receive a higher amount of credit from MFIs. The higher ownership of consumer durables reflects the higher wealth of women smallholders. As MFIs target poorer households with collateral-free small loans, wealthy women smallholders are excluded from receiving loans from MFIs. Like access to credit from informal financial sources, the total amount of credit from informal financial sources does not have any significant determinants (column).

5 Conclusion

This chapter has assessed the extent of financial support for women smallholders in Bangladesh. Smallholders are identified on the basis of the definition given by the Bangladesh Bureau of Statistics (BBS) of the Government of Bangladesh. In its annual Statistical Year Book, the BBS identifies those households as smallholders that own operated land of more than 4 decimals and less than 250 decimals. In a patriarchal society, the senior male adults are the household heads in Bangladesh. The most senior adult female household members become heads when the male heads die, are sick, or are disabled. Therefore, women smallholders are identified as those households that have land ownership between 5 and 249 decimals and have female heads. Since male adults are more economically active and the main earners, and women's economic participation is socially constrained, women smallholders are more likely to be economically destitute. This chapter uses a nationally representative data set generated by BBS in 2010, popularly known as the Household Income and Expenditure Survey 2010 (HIES 2010). It covers 12,240 households from all districts in the country.

The descriptive results illustrate that there were 7343 smallholder households in the country, of which 1000 were owned by women. Just 173 (17.3% of all women smallholders) had access to credit from any financial source. The average total amount of credit of these women smallholders was approximately Taka 58 000. Just 18 (1.8%) of all women smallholders had access to credit from formal sector commercial and development banks. The average size of a loan from these sources was around Taka 200 000. From microfinance institutions (MFIs), 106 women smallholders (10.6%) received credit of an average amount of Taka 13 000. Fifty women smallholders (5%) received credit from informal financial sources, the average size of which was Taka 100 000.

The assessment of determinants of financial support for women smallholders was carried out in two dimensions – access to credit and total amount of credit from different financial sources. In the process of identifying access to credit, four different aspects of access to credit have been identified – access to credit from any sources; access to credit from commercial and development banks; access to credit from MFIs; and access to credit from informal financial sources. Similarly, four aspects of total amount of credit have been considered – total amount of credit from all sources; total amount of credit from commercial and development banks; total amount of credit from MFIs; and total amount of credit from informal financial sources. Access to credit is a binary variable and therefore the probit regression technique has been applied. As the total amount of credit is a ratio level variable, the OLS regression technique has been applied.

The results indicate that the marital status of the household head, the number of household members aged 0–5 and 6–25, minority status, business ownership, and ownership of other assets of the household, as well as district labor participation rate, were the major determinants of access to credit from different sources. Similarly, the number of household members aged 0–5 and 6–25, minority status, business ownership, and ownership of other assets of the household were the major determinants of the total amount of credit from different sources.

Women smallholders are credit-constrained, as only around 17% have access to credit. The prevailing policies in the financial sector are not conducive for women smallholders to receive credit from any of the financial sources in the country. As the agriculture sector significantly contributes to economic growth, employment creation, and food security in the country, access to credit for women smallholders needs to increase. Women smallholders are likely to be economically destitute, and improved access to credit will help enhance income levels, improve livelihoods, and reduce poverty. It is essential in terms of reducing gender disparity in the financial sector in Bangladesh. The prevailing policies related to access to credit in Bangladesh's financial sector must be redesigned to make the environment friendlier to smallholders, especially women smallholders, in the country.

6 References

Abdulai, A. & Rieder, P. (1996). Agriculture's contribution to overall economic growth: an application of Kaldor's first growth law to Ghana. *African Development Review*, *8*(1), 97–114.

Awokuse, T. O. & Xie, R. (2014). Does agriculture really matter for economic growth in developing countries? *Canadian Journal of Agricultural Economics/Revue Canadienne d'agroeconomie*, *63*(1), 77–99.

Banerjee, A. V. & Duflo, E. (2014). Do firms want to borrow more? Testing credit constraints using a directed lending program. *The Review of Economic Studies*, *81*(2), 572–607.

Arenius, P. & Minniti, M. (2005). Perceptual variables and nascent entrepreneurship. *Small Business Economics*, *24*(3), 233–247.

Aterido, R., Beek, T., & Iacovone, L. . (2013). Access to finance in Sub-Saharan Africa: Is there a gender gap? *World Development*, *47*, 102–120.

Bardhan, P. K. (1980). Interlocking factor markets and agrarian development: a review of issues. *Oxford Economic Papers*, *32*(1), 82–98.

Bangladesh Bureau of Statistics (BBS). (1992). *Statistical yearbook of Bangladesh*. Dhaka, Bangladesh: Statistics Division, Ministry of Planning.

Bangladesh Bureau of Statistics (BBS). (2011). *Household income and expenditure survey 2010*. Dhaka, Bangladesh: Statistics Division, Ministry of Planning.

Bangladesh Bureau of Statistics (BBS). (2013a). *Labour force survey 2013*. Dhaka, Bangladesh: Statistics Division, Ministry of Planning.

Bangladesh Bureau of Statistics (BBS). (2013b). *Statistical yearbook of Bangladesh*. Dhaka, Bangladesh: Statistics Division, Ministry of Planning.

Bangladesh Planning Commission (BPC). (2020). *8th five year plan*. Dhaka, Bangladesh: Ministry of Planning.

Bangladesh Bureau of Statistics (BBS). (2022a). *Labour force survey 2022*. Dhaka, Bangladesh: Statistics Division, Ministry of Planning.

Bangladesh Bureau of Statistics (BBS). (2022b). *Statistical yearbook of Bangladesh*. Dhaka, Bangladesh: Statistics Division, Ministry of Planning.

Bell, C., Srinivasan, T. N., & Udry, C. (1997). Rationing, spillover, and interlinking in credit markets: the case of rural Punjab. *Oxford Economic Papers*, *49*(4), 557–585.

Bhaduri, A. (1977). On the formation of usurious interest rates in backward agriculture. *Cambridge Journal of Economics*, *1*(4), 341–352.

Bruhn, M. (2009). Female-owned firms in Latin America: Characteristics, performance, and obstacles to growth. *World Bank Policy Research Working Paper* (No. 5122).

Carter, M. R. (1988). Equilibrium credit rationing of small farm agriculture. *Journal of Development Studies*, *28*, 83–103.

Chandio, A. A. & Jiang, Y. (2018). Determinants of credit constraints: evidence from Sindh, Pakistan. *Emerging Markets Finance and Trade*, *54*(15), 3401–3410.

Chowdhury, M. J. A., Ghosh, D., Wright, R. E. (2005). The impact of micro-credit on poverty: evidence from Bangladesh. *Progress in Development Studies*, *5*(4), 1–12.

Chowdhury, F. D. (2009). Theorising patriarchy: the Bangladesh context. *Asian Journal of Social Science*, *37*(4), 599–622.

Chowdhury, M. J. A. (2011). The determinants of entrepreneurship in a conflict region: evidence from the Chittagong Hill Tracts in Bangladesh. *Journal of Small Business and Entrepreneurship*, *24*(2), 265–281.

Chowdhury, M. J. A. (2014). Microcredit, micro-enterprises, and self-employment of women: Experience from the Grameen Bank in Bangladesh. In N. Perdikis & K. Ruziev (Eds.), *Development and financial reform in emerging economies* (pp. 183–200). Pickering & Chatto.

Chowdhury, M. J. A., Amin, S., & Farah, T. (2016). Access to microcredit and women's entrepreneurship: evidence from Bangladesh. *PEP working paper series* (No. 2016-13).

de Mel, S., McKenzie, D., & Woodfuff, C. (2008). Returns to capital in microenterprises: evidence from a field experiment. *Quarterly Journal of Economics*, *123*, 1329–1372.

Dupas, P. & Robinson, J. (2013). Savings constraints and microenterprise development: evidence from a field experiment in Kenya. *American Economic Journal: Applied Economics*, *5*, 163–192.

Fafchamps, M. (2014). Credit constraints, collateral and lending to the poor. *Revue d'économie du développement*, *22*(HS01), 71–89.

Fletschner, D. (2008). Rural women's access to credit: market imperfections and intrahousehold dynamics. *World Development*, *37*(3), 618–631.

Foltz, J. D. (2004). Credit market access and profitability in Tunisian agriculture. *Agricultural Economics*, *30*(3), 229–240.

Ghate, P. (1992). *Informal finance: some findings from Asia*. Oxford University Press for the Asian Development Bank.

Ghosh, D. (1986). Monetary dualism in developing countries. *Economies ET Sociétés*, 19–30.

Government of Bangladesh. (2020). *8th five-year plan July 2020 – June 2025*. Dhaka, Bangladesh: Ministry of Planning.

Gollin, D., Parente, S. L., & Rogerson, R. (2002). The role of agriculture in development. *American Economic Review*, *92*(2), 160–164.

Griffin, K., Khan, A. R., & Ickowitz, A. (2002). Poverty and the distribution of land. *Journal of Agrarian Change*, *2*(3), 279–330.

Guirkinger, C. & Boucher, S. R. (2008). Credit constraints and productivity in Peruvian agriculture. *Agricultural Economics*, *39*, 295–308.

Hussain, A. & Thapa, G. B. (2012). Smallholders' access to agricultural credit in Pakistan. *Food Security*, *4*(1), 73–85.

Hoque, A. (1988). Farm size and economic-allocative efficiency in Bangladesh agriculture. *Applied Economics*, *20*(10), 1353–1368.

Inderst, R. & Mueller, H. (2007). A lender-based theory of collateral. *Journal of Financial Economics*, *84*, 826–859.

Johnston, B. F. & Mellor, J. W. (1961). The role of agriculture in economic development. *The American Economic Review*, *51*, 566–593.

Khandker, S., Samad, H. A., & Ali, R. (2013). *Does access to finance matter in microenterprise growth? Evidence from Bangladesh*. Paper presented at the Agricultural & Applied Economics Association's 2013 AAEA & CAES Joint Annual Meeting, Washington, DC, August 4–6, 2013.

Jaim, J. (2021). Bank loans access for women business-owners in Bangladesh: obstacles and dependence on husbands. *Journal of Small Business Management*, *59*(sup1), S16–S41.

Khandker, S. R. & Faruqee, R. R. (2003). The impact of farm credit in Pakistan. *Agricultural Economics*, *28*(3), 197–213.

Khan, M. H., Chaudhry, M. G., & Qureshi, S. K. (1986). Landlessness and rural poverty in underdeveloped countries [with comments]. *The Pakistan Development Review*, *25*(3), 371–402.

Klapper, L. & Parker, S. (2011). Gender and the business environment for new firm creation. *World Bank Research Observer*, *26*, 237–257.

Lele, U. & Mellor, J. W. (1988). Agriculture and governments in an interdependent world. In *Proceedings of the Twentieth International Conference of Agricultural Economists held at Buenos Aires, Argentina, 24–31 August 1988*. Connecticut: West Hartford.

Lycette, M. & White, K. (1989). Improving women's access to credit in Latin America and the Caribbean: Policy and project recommendations. In M. Berger & M. Buvinic (Eds.), *Women's ventures: Assistance to the informal sector in Latin America* (pp. 19–30). Kumarian Press.

Murshid, N. S., Akincigil, A., & Zippay, A. (2016). Microfinance participation and domestic violence in Bangladesh: results from a nationally representative survey. *Journal of Interpersonal Violence*, 31(9), 1579–1596.

Petrick, M. (2005). Empirical measurement of credit rationing in agriculture: a methodological survey. *Agricultural Economics*, 33(2), 191–203.

Piza, C. & de Moura, M. J. S. B. (2015). The effect of a land titling programme on households' access to credit. *Journal of Development Effectiveness*, 8(1), 129–155.

Rao, J. M. (1980). Interest rates in backward agriculture: notes and comments. *Cambridge Journal of Economics* , 4, 159–167.

Richardson, P., Howarth, R., & Finnegan, G. (2004). The challenges of growing small businesses: insights from women entrepreneurs in Africa. ILO Geneva, SEED WP No. 47.

Stiglitz, J. E. & Weiss, A. (1981). Credit rationing in markets with imperfect information. *American Economic Review*, *71*(3), 393–410.

Schultz, T. W. (1964). Changing relevance of agricultural economics. *Journal of Farm Economics*, *46*(5), 1004.

Yunus, M. (1999). *Banker to the poor: micro-lending and the battle against world poverty*. Public Affairs.

Chapter 17

Approaches to strengthen women's land rights and tenure security through agricultural programming

Krista Jacobs, Caitlin Kieran, Tatiana Gumucio, Nayna Jhaveri, Landesa, USA

This chapter is an introduction for gender and agriculture practitioners on why and how to integrate approaches to strengthen smallholder women's land rights and tenure security into agricultural development programming.

Smallholder women farmers are critical for the productivity and resilience of food systems as women comprise 38% of the workers in primary agricultural production (FAO, 2023). Compared with men, smallholder women farmers have disproportionately limited and insecure land rights and also have less access to agricultural inputs and services (FAO, 2023). Without confidence that in the long-term they can access a given parcel of land, make decisions about the land and agricultural production on it, and control corresponding earnings, farmers (of all genders) have less incentive and ability to invest in agricultural productivity and sustainable practices (e.g. tree planting, terracing, fallowing). Efforts to ease both land tenure constraints and agricultural input/service constraints are essential for advancing women farmers' productivity and empowerment, their families' welfare and food security, and the performance and inclusivity of the food system.

http://dx.doi.org/10.19103/AS.2024.0148.21

Although integrating land interventions into agricultural programming is not the norm, we make the case that it is feasible and impactful for women and for agricultural projects and businesses. We use examples from development projects, government programs, and the private sector to draw lessons and show practices that agricultural projects and private sector actors can apply to improve both smallholder women farmers' land tenure and the sustained application of good agricultural practices. We separate development programs and private sector examples not because the approaches are meaningfully different but to increase the visibility of relatively nascent and less documented private sector efforts.

1 Review of concepts and evidence on rural women's land tenure

1.1 The bundle of rights and their security

Land tenure refers to a 'bundle of rights' that a person has over land (Von Benda-Beckmann et al., 2006). Doss and Meinzen-Dick (2020) organize land rights into

- use rights, including rights of access and taking or using resources on the land;
- rights to manage, build, or make physical changes to the land;
- rights to earnings from what is produced on land;
- rights to prevent others from using land;
- rights to transfer land, whether temporarily or permanently; and
- inheritance or other rights that can be realized in the future.

A person may hold one or more of these rights by themselves, with other people, or as part of a community. Different people or groups can have different rights over the same piece of land, and those rights may be separate, joint, overlapping, or competing.

Understanding women's and men's rights to use and manage land and to make decisions about agricultural earnings is key for developing interventions that sustainably increase agricultural productivity and increase women's economic empowerment.

Much of the existing literature and evidence on land tenure focuses on 'ownership,' which typically covers all or most of the rights in the bundle. Ownership does not have to be legally recognized for someone to have the rights associated with ownership. For example, in customary tenure systems, although much of the agricultural land is not legally registered, people may still be able to use, bequeath, rent, etc. the land (Slavchevska et al., 2021).

Joint ownership is especially important for women because land owned or held jointly with husbands is a substantial portion of women's landholding, and for many women, the only land they hold. Governments are increasingly promoting or requiring joint land registration of spouses as a tactic to increase legal recognition of women's land rights (Jackson, 2003). Over one-third of the countries that have reported on Sustainable Development Goal (SDG) indicator 5.a.2 – the proportion of countries where the legal framework guarantees women's equal rights to land ownership and/or control – allow for or require joint registration of land between spouses (FAO, 2023). However, having their names on land documents does not guarantee that wives will have the same decision-making power over land as their husbands or that wives will receive their share of the land or its value if the marriage dissolves (Alvarado et al., 2022; Jacobs and Kes, 2015; Doss et al., 2014; Agarwal, 2003).

Land tenure security is how confident a person feels that they will not lose any part of their bundle against their will. Tenure security is important because farmers make decisions based on their perceptions. Institutional and social conditions that affect women's land tenure security are

- legal recognition – statutory law recognizes women's land rights;
- sociocultural legitimacy – families, communities, and traditional and religious leaders recognize women's land rights;
- exercisability – women know their rights and do not need approval to exercise them;
- durability – the bundle of rights does not change if women's marital status changes; if there are changes in family composition or local traditional or political leadership; or if women do not use the land for some time;
- enforceability – women know where and how to make claims and can afford to do so. When women do make claims, authorities hear the claims and take them seriously, adjudicate fairly, and maintain a setting that is accessible to and feels safe for women; and
- Duration – rights in the bundle are for a known time period and long enough to benefit from the land.

(Jhaveri, 2021; Doss and Meinzen-Dick, 2020; Scalise and Giovarelli, 2020; Giovarelli et al., 2016)

Social norms are an underlying factor influencing many of the conditions. Norms influence how duty bearers choose to implement or enforce laws as well as the extent to which women and men choose to comply and so determine the sociocultural legitimacy, durability, and exercisability of women's land rights. Many norms are embedded in customary land systems that make women's land rights contingent on their male relationships can be barriers to women's tenure security (Doss and Meinzen-Dick, 2020; Tibenkana, 2020; Ghebru 2019;

O'Sullivan, 2017; Banerjee et al., 2014). Such norms and rules make marriage and its dissolution potentially dangerous inflection points for women's tenure security. The durability of smallholder women farmers' land rights tends to depend on their marital status, the quality or duration of her marriage, the actions of male family members, and the actions of local authorities – all of which depend on what people know and see as socially acceptable regarding women's land rights. Examples from Rwanda and Nepal highlight the importance of norms to sociocultural legitimacy and durability of women's land rights.

- Local authorities in Rwanda related that even after the national land registration program which emphasized joint titling 'Many men still have the thinking that the land belongs to them and that their wife doesn't have the right to that land ... [they] still believe that they don't have equal right on land' (Kloos et al., 2016).
- Bayisenge's (2018) research in Rwanda found that women noted that although husbands now involved them in decisions about land transactions because of the legal requirement for wives' signatures, daily management of the land had not changed. Several women expressed that land was for the husband or the family generally and that women having their own property would create conflict.
- In Nepal, Pradhan et al. (Pradhan et al., 2019) found that for some women exercising their claims on personal assets (e.g. small livestock) can damage their standing as a 'good wife' and weaken their claims on jointly held land. They also found that land registration efforts had little impact on women's empowerment because of the social risks women faced to register land in their own names.

1.2 Evidence on the state of women's land tenure

Studies on gendered land rights in developing countries, primarily in sub-Saharan Africa and Asia, consistently conclude that while there are substantial gaps in the availability of data, the existing data reveal large gender gaps in land rights (Slavchevska et al., 2021; Gaddis et al., 2018; Doss et al., 2015; Kieran et al., 2015; Gaddis et al., n.d.).

Data from the 46 countries reporting on SDG Indicator 5.a.1(a) – proportion of total agricultural population with ownership or secure rights over agricultural land, disaggregated by sex – convey that we are far from achieving global gender equality in ownership. *Figure* 1 presents the publicly available data on 5.a.1(a), with women's land rights on the *x*-axis, men's land rights on the y-axis, and a dotted line depicting equality between women and men. A higher proportion of men than women own or have secure rights to agricultural land in all countries to the left of the dotted line. In 87% of countries reporting, a larger

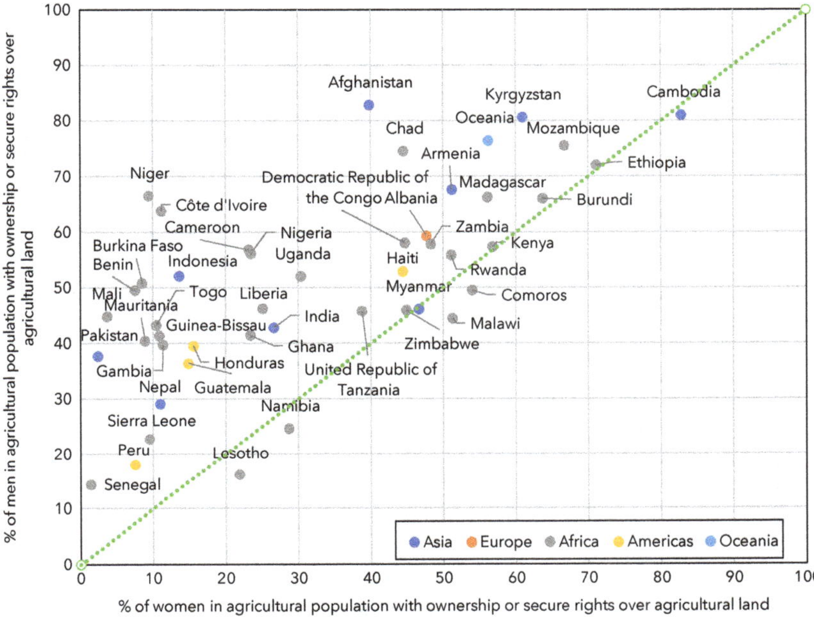

Figure 1 Proportion of total agricultural populations with ownership or secure rights over agricultural land. Note: Data available at https://unstats.un.org/sdgs/dataportal/database. *Figure* 1 presents the most recent data on indicator 5.a.1(a) for each country. *Figure* 1 re-creates a figure presented in 'The Status of Women in Agrifood Systems' (FAO, 2023). For this indicator, land ownership is defined as a legally recognized right to acquire, to use, and to transfer land, while secure rights include land ownership as well as the right to sell and the right to bequeath land.

share of men than women have ownership and/or secure rights to agricultural land (FAO, 2023).

Table 1 illustrates the variation in *which rights* women have over agricultural land within a country and also that the size and direction of gender gaps in any given right varies across countries. All data are nationally representative and focus on agricultural land. Within countries, the proportion of men and women with a right varies across the different rights. For example, in Malawi, both men's and women's rates of documented ownerships (e.g. land certificates, titles) are nearly zero, while over half of men and women have management rights. More men than women have land rights in Ethiopia, Nigeria, the United Republic of Tanzania, Uganda, and Niger. In both Cambodia and Malawi, a slightly higher proportion of women than men have land rights.

Globally, approximately 20% of rural women and 20% of rural men are tenure insecure in that they believe they could lose land against their will within

Published by Burleigh Dodds Science Publishing Limited, 2025.

Table 1 Percentage of women and men with rights over agricultural land

Country		Cambodia	Ethiopia	Malawi	Nigeria	United Republic of Tanzania	Uganda	Niger
Year		2019/20	2013/14	2013	2012/13	2012/13	2010/11	2011
Documented ownership	Male	67%	38%	1%	3%	10%	–	–
	Female	69%	30%	2%	0.4%	8%	–	–
Reported ownership	Male	–	48%	33%	23%	42%	43%	63%
	Female	–	45%	42%	4	32%	38%	35%
Reported sole ownership	Male	–	14%	24%	17%	23%	14%	40%
	Female	–	11%	33%	2%	13%	9.%	14%
Exclusion rights	Male	71%	63%	28%	50%	50%	–	–
	Female	74%	58%	30%	14%	38%	–	–
Right to sell	Male	70%	63%	28%	50%	50%	–	–
	Female	73%	58%	30%	14%	38%	–	–
Land Managers	Male	–	57%	56%	32%	44%	44%	72%
	Female	–	48%	58%	11%	41%	54%	50%
Sole land Managers	Male	–	16%	18%	25%	15%	5%	39%
	Female	–	8%	18%	6%	12%	10%	19%
Control over output	Male	–	52%	50%	30%	40%	45%	–
	Female	–	55%	51%	13%	37%	50%	–
Sole Control over output	Male	–	9%	19%	20%	12%	9%	–
	Female	–	14%	20%	4%	11%	15%	–

Note: Cambodia data are from the Living Standards Measurement Study – Plus. Data from Ethiopia, Malawi, Nigeria, United Republic of Tanzania, Uganda, and Niger are from the Living Standards Measurement Study – Integrated Surveys on Agriculture. Statistics on the land rights depicted in green were presented in Box 3.3 Table A of 'The Status of Women in Food Systems' (FAO, 2023), while statistics in yellow were presented in *Figure* 1 of 'Beyond Ownership: Women's and Men's Land Rights in Sub-Saharan Africa' (Slavchevska et al., 2021). Dashes denote estimates not reported in either source.

5 years.[1] Although the most common sources of insecurity were the same for women and men globally – being asked by the owner to leave, insufficient economic resources, and family disagreements – women are more likely than men to cite family disagreements, divorce, or death of a spouse as a source of insecurity and men are more likely than women to cite potential seizure by governments or companies (Prindex, 2020). Globally, 48% of married women worry that they would lose rights to land if their marriage ended compared with 34% of rural married men (Prindex, 2020).

1.3 Roles of women's land rights and tenure security in agriculture

Here we highlight the empirical evidence for women's land tenure affecting agricultural outcomes.[2] This evidence sits within a broader literature on the relationship between land tenure and agricultural outcomes that does not consistently differentiate between women's and men's tenure or between women's and men's agricultural decisions, practices, or production. Despite gendered data gaps, existing evidence and experience show that secure land tenure is foundational to women farmers' agricultural production and that there are approaches projects and companies can incorporate to expand women farmers' bundle of land rights and increase their security.

There is strong evidence that strengthening smallholder women's land tenure increases their investments in soil and water conservation, enables them to more securely rent out land, and increases their decision-making power within the household. Several studies find positive impacts of female-headed households or female farm managers having land titles on making investments in soil and water conservation (e.g. terracing, bunding, channels), fallowing, planting trees, and other agroforestry practices (Ali et al., 2014; Goldstein et al., 2018; 2015; Alvarado et al., 2022; Meinzen-Dick et al., 2019). Moreover, these impacts tend to be larger than for male-headed households or male farm managers who may already have more secure tenure. Quisumbing and Kumar (2014) find that increasing women's knowledge about their land rights also promotes women's uptake of soil and water conservation and agroforestry. Alvarado et al. (2022) found in the 12 years following receiving a land certificate in their name, female-headed that female-headed households in Ethiopia were 2–14 percentage points more likely to rent land out, potentially because the certificate gives legal recourse if tenants try to appropriate the land. This aligns with other findings from Ethiopia that more female-headed households began

1 Authors' calculations using Prindex's nationally representative survey data for 140 countries, available at https://www.prindex.net/.

2 For a more detailed theory of change and summary of evidence, see Lisher (2019) and Stanley and Lisher (2023).

renting out land after receiving certificates with their name (Deininger et al., 2011).

Evidence across several contexts indicates that having more secure land rights increases women's decision-making in her household (Alvarado et al., 2022; Meinzen-Dick et al., 2019). Researchers hypothesize that women having more secure land tenure, often in the form of documentation, shifts perceived gender roles and changes in women's relative power within the household. Meinzen-Dick et al.'s (2019) review of impacts of women's land rights notes positive relationships between women's land ownership and women having more control over food consumption decisions, greater household food availability, and lower risk of being underweight for children.

Evidence is mixed on the extent to which stronger land tenure and its effects on agricultural investments and practices translate into increased agricultural productivity, incomes, and food security (Stevens et al., 2020). The ambiguity at least partially results from the historical design of empirical studies. Much of the empirical evidence is from evaluations looking for impacts of land titling in contexts of missing credit markets, sparse agricultural extension services, and improved inputs being difficult to access. When agricultural services and inputs are difficult to access, land tenure interventions would not likely have widespread, large, or immediate impacts. Women smallholder farmers have less access to these services and inputs, making it more difficult to observe land interventions' impacts on women's agricultural productivity (FAO, 2023). Evaluations also have tended to look for effects within 1–6 years of the intervention, which may not be long enough to see impacts on productivity because it can take several seasons for farmers to change behaviors and to access and effectively apply inputs and services. Finally, historically, empirical data on land tenure and agriculture have been available mainly at the household level, meaning the majority of existing evidence focuses on female-headed households instead of women as individuals whose land tenure and agricultural options are influenced by their marital status, household dynamics, age, parental status, socioeconomic status, and other factors (Feyertag et al., 2021; Meinzen-Dick et al., 2019).

2 Approaches to promote and protect women's land rights for agricultural projects

Agricultural projects and private sector actors need to understand which land rights women smallholder farmers do and do not have, how secure they believe their rights to be, and the factors behind gendered differences in tenure and security. These rights, beliefs, and factors affect women farmers' incentives and abilities to engage with the project/company, apply desired practices, and otherwise benefit themselves and the project/company. Devoting attention to expanding women's land rights and tenure security is a means to shift local land and agricultural systems to perform better in general and for women in particular by increasing women's decision-making power in agricultural production and land use, women farmers' ability to adopt climate-resilient practices, and women farmers' benefit from agricultural production.

In this section, we describe approaches that agricultural projects could integrate or have integrated to expand the bundle of land rights women smallholder farmers have, increase their tenure security, or enable women to better leverage their land rights for agricultural production. For each approach, we draw from projects or government programs in the land or agriculture sectors to identify practices that were important in advancing women's land rights or tenure security. We also describe concrete examples of implementation and outline tenure, agricultural, and empowerment outcomes the approach achieved.

We reviewed peer review and gray literature (e.g. project reports, manuals, and evaluations) for development projects in the land and agriculture sectors and major land certification programs. After selecting efforts with well-described land tenure interventions or that documented impacts on women's tenure, we arrived at 20 projects spanning five general approaches:

- Awareness raising;
- Expanding women's land access and governance roles working with customary leaders;
- Land certification;
- Combining land and agricultural interventions; and
- Addressing social norms.

Appendix 1 provides an index of approaches programs in Sections 2 and 3 applied and general tenure and agricultural outcomes.[3] Appendix 2 summarizes the projects.

Although projects often applied multiple approaches, we separate them here to be able to delve into good practices and examples from each approach. Some approaches, such as land certification, may not be suitable for modestly sized agricultural projects/companies to undertake at a large scale. However, agricultural projects and agribusinesses can coordinate with national or large-scale certification efforts and for their participants, outgrowers, communities, etc., as several examples highlight. Other approaches, such as raising awareness of women's land rights or women's land literacy, are logistically simpler and can

3 Scalise and Giovarelli (2020) is also a valuable summary of evidence on programmatic approaches that improve women's land tenure. However, its target audience is land practitioners not agricultural practitioners.

still lay groundwork for the exercisability and social legitimacy of women's land rights.

2.1 Awareness-raising and legal literacy

Women and men typically have different knowledge and understanding about their land rights and about how to access land administration and governance services (e.g. reporting or resolving a land dispute or registering land that is inherited). It is also common for women, men, and leaders to have a limited understanding of women's land rights. These knowledge gaps are foundational to gender bias in land tenure in rural areas: if women do not understand their rights, they cannot fully demand them; if leaders do not know women's rights, they cannot fully uphold them. Understanding their land rights can also enhance women's decision-making authority in households and also improve local land governance bodies' inclusion of women in decision-making.

Awareness-raising and legal literacy activities are relatively simple to integrate into projects, and there are variety of modes, intensities, and scale that can be tailored to agricultural projects. In-person training, short plays, traditional media, and new digital technologies all can educate about women's land rights (United Nations, 2013). An example of a light-intensity activity is adding a women's land rights education session to farmer group meetings; a high-intensity example is creating mass media campaign or an edutainment radio or television program.

Whatever method is used, two lessons from several projects are the importance of having multiple, iterative interactions and for the awareness-raising activities to reach and be interesting to women, men, and leaders in order to create deeper understanding of women's land rights and expand women's options for exercising them (Sutz, 2021). The examples discussed in the following sections also illustrate these lessons.

2.1.1 Radio programming

The Rwanda LAND project collaborated with Radio Ishingiro on a 10-month campaign to enhance peoples' understanding of women's land rights and promote gender equality as part of good land governance (Chemonics, 2016). The project developed a series of radio dramas, talk shows, and quiz programs. Informed by results from a survey of women's and men's knowledge and attitudes about women's and girls' rights to land, the content emphasized equitable inheritance and rights of marital property, including for informally married women. A poetry, song, and dance competition was an opportunity for people to express what they had learned about women's and girls' land rights in their own words and create excitement and greater acceptance. The campaign produced substantial changes in knowledge and attitudes. For example, at the

Published by Burleigh Dodds Science Publishing Limited, 2025.

campaign's end, 91% of men surveyed supported daughter's rights to inherit equally with sons, compared to 71% before the campaign.

2.1.2 Talking books and town halls in Liberia

In six counties in Liberia over 2021–2023, the NGO Landesa and partners used a combination of pre-recorded digital messages and town hall meetings to educate communities on Liberia's recently passed Land Rights Law (2018). Messages covered fundamentals of land rights laws, community land formalization, land and property rights for women in customary marriage, services of the Liberia Land Authority, and alternative dispute resolution.

Example messages include the following:

- A woman who is married the traditional way can have her own property, whether she has had it before the man has paid her dowry, or after he has paid her dowry.
- When a community owns customary land, all the members of that community – women and men – together are the owners of that land, not one person or a few people.
- When the land-owning community decide[s] to go through the step that will make them get to their land deed, it is important that everybody in the community – including the men, women, young people, disable[d] people and even strangers, women who moved in the community to marry, single women, women who living with men but not married to the men and other people who have been living in the community for long – be a part of this decision.

The messages were recorded on small, rugged, portable digital devices called Talking Books[4] that households had for 3–4 weeks. Women-headed households, youth-headed households, women's groups, and youth groups were prioritized to receive the books. People could play messages as many times as they liked, listen on their own schedule, take the Talking Book with them wherever they worked or spent time, and listen by themselves or in groups. This flexibility enabled more people, and people who were not typically active in community land decisions, to learn about women's land rights and community land formalization than the typical practice of community meetings.

The percentage of women and men who correctly answered questions about alternative dispute resolution, who qualifies as a community member, married and single women's rights to hold and manage community land, and legal requirements for women in community land governance increased from

4 https://www.amplio.org/.

0–20% before the Books to 90–100% after the Books.[5] Anecdotally, multiple chiefs shared with project that they began inviting women to meetings about land after listening to the Talking Books.

Critical to the success of the Talking Books were (1) that the content was in local languages, (2) hiring community members to help with technical difficulties and encourage people to use the Talking Books, and (3) that after each set of households completed their time with the Talking Books, the project held town halls and also a separate meeting for women for participants to discuss the topics together and pose questions to legal experts.

2.2 Expanding women's land access and governance roles in customary tenure systems

Especially in areas with customary land tenure systems, it is important to tailor activities to the gender norms and traditional governance structures that affect women's access to land for farming, decision-making over production and earnings, and tenure security.

Several projects working in customary tenure systems applied one or more of the following strategies – all of which hinge on customary leaders' roles as gatekeepers, social authorities, and rule-makers – to increase women's land access or tenure security:

- Sensitization and dialogue with traditional leaders to make the case for women's land rights;
- Strengthen local governance by including women and promoting gender integration; and
- Facilitate group access to customary lands for women's use and management according to and motivated by existing culturally appropriate roles prescribed to women in the community.

As illustrated in several examples, projects often also incorporated awareness-raising among women community members on their land rights and leadership training for women. These trainings enable women to be active members of local governance bodies and to seek out promoted customary leaders for assistance in resolving disputes.

Findings across the projects indicate that the three strategies can

- directly improve women's access to land;
- shift communities' perceptions toward seeing women as competent land managers and farmers;

5 These figures are across Landesa's internal assessments across multiple rounds of Talking Books deployed in communities.

- increase women's decision-making in land use, agricultural production, and use of agricultural produce or income;
- increase women's active participation in land meetings and confidence in customary land governance;
- expand women farmers' application of improved technologies or cultivation of higher value crops; and
- increase women's self-confidence and sense of collective empowerment.

(Bessa et al., 2023; Dungumaro and Amos, 2019; USAID, 2019; Winrock International, 2019; Van Den Bold et al., 2015; Rolfes and Seitz, 2013; USAID, 2013)

2.2.1 Make the case to customary leaders

Key actions to influence change through customary leaders were dialogues that push for change but engage respectfully with cultural norms, collaborating with local organizations and actors familiar to the community, and using messages that resonate with community goals. For example, the Enhancing Customary Justice Systems in the Mau Forest, Kenya Project engaged traditional elders in a series of trainings, conversations, and mentoring sessions on women's constitutional rights. Having a female lawyer from the area's ethnic group on the project team also conveyed to participants that the project appreciated their community's customs and traditions. These activities shifted the elders' from feeling threatened to becoming supporters of women's land rights (USAID, 2013). In particular, sessions highlighting local gender inequalities in the division of labor helped illuminate to elders the critical role that women play in working the land and in ensuring the family's survival. The project team worked with elders to create a pie chart of the of the number of hours that women and men worked during a typical day, visualizing women's greater labor burden. Through these sessions, elders came to support that women ought to have a say in land use decision-making, incorporated gender considerations into the elders' constitution and code of conduct, and established a new requirement for written spousal consent for land transactions (USAID, 2013).

The Feed the Future Agricultural Development and Value Chain Enhancement II (USAID, 2019) and AgNRM projects in northern Ghana made the benefits of women's land access more visible to traditional leaders and helped achieve traditional leaders signing MOUs to set aside parcels of land totaling 1600 acres for 1000 women over a 10-year lease. ADVANCE II leveraged networks of local farm service providers[6] to raise traditional leaders' awareness on why women should have rights to manage land (USAID, 2019). Drawing

6 Local, farmer-run businesses that provide services to or buy produce from smallholder farmers.

from their experiences, the service providers conveyed women smallholder farmers' capabilities and contribution to the local agricultural economy, and the need for secure land rights to enable them to do so. ADVANCE II also collaborated with the local advocacy organization NORTHCODE, which held a workshop with traditional leaders to share research findings that women having land far from their homesteads negatively impacts their productivity and that having access to closer land would improve agricultural productivity. AgNRM identified male champions to advocate on behalf of women and disseminate messages highlighting how women's increased tenure security can enhance household food security, nutrition, and income (Winrock International, 2019).

2.2.2 Promote women's roles in land governance

Women-focused Women Rights and Leadership Fora (WRLF) were created in Tanzanian pastoralist communities with the aim of strengthening women's voice in land use decisions. Including customary leaders in the Fora was critical to raising their status and social legitimacy as was the local government's recognition of the Fora. All WLRF members receive training on women's land rights; customary leaders in the WRLFs receive specific training on women's social welfare challenges, and women members alone receive training on leadership and entrepreneurship skills (Dungumaro and Amos, 2019).

The USAID Integrated Land Resource and Governance (ILRG) Project supported women's participation in community land governance through training both women and men leaders. In Malawi, the project trained women and men members of Customary Land Committees together to promote male committee members' acceptance of women in the Committees. The project also provided stipends to committee members which promoted female members' husbands' acceptance of their service on the committee (Bessa et al., 2023). In Zambia, the project engaged traditional male leaders in community sensitization sessions before Community Resource Board elections to promote acceptance of women's participation and leadership on the Boards (Malasha, 2020). In both Malawi and Zambia, the project provided technical training and socioemotional skills to women in the Customary Land Committees and Community Resource Boards to more effectively exercise their new roles.

2.2.3 Allocate customary lands to women's groups

Keys to convincing communities and traditional leaders to allocate customary lands to groups of women were aligning the lands' use with women's locally defined sociocultural responsibilities and projects providing additional support to demonstrate that women can manage the land well.

The Millennium Challenge Corporation's (MCC) Alatona Irrigation Project in Mali facilitated allocation of land for women's market gardens, noting that market gardening was a socioculturally acceptable role prescribed to women in local communities (Rolfes and Seitz, 2013). The project also helped establish women's associations as legal entities to own and manage the land by developing model articles of incorporation and by-laws for the associations and a model agreement between the associations and members that granted members legal access to individual garden parcels. Once established, land titles were issued to the associations.

The Enhanced Homestead Food Production (E-HFP) project in Burkina Faso worked with landowners in communities to identify land that could be used for a Village Model Farm (VMF) run by the project's women lead farmers for women in the village. E-HFP provided initial agricultural and animal inputs to the VMFs where women received training on crop production and animal rearing. Women often applied what they learned in the VMFs in their own home production. Women's successful production in the VMFs positively shifted community members' opinions on women's capabilities to manage land (Van Den Bold et al., 2015).

2.3 Support gender-responsive, participatory certification of land rights

Certification of land rights can either solidify existing gender inequalities in land rights by legally registering the status quo or, if done in a gender-responsive manner, legally protect women's land rights and expand them in practice. Correspondingly, government certification programs can either limit or expand women's incentives to invest in agriculture and their access to financial services, government programs, and commercial opportunities that depend on being recognized as landholders.

Agricultural projects and agribusinesses have supported participatory land certification as a pathway to

- promote farmers' uptake of agricultural inputs, climate-smart agricultural practices, and other investments in productivity and sustainability through improving farmers tenure security; and
- lower business risk in the agriculture sector by clarifying land rights, reducing uncertainty and risk of disputes, and resolving existing disputes.

Participatory land certification typically includes the following steps that require several months and close coordination with local leadership:

- Raise public awareness about land rights and the certification process;
- Identify and resolving existing land conflicts;
- Visit land parcels to document their boundaries and occupancy;
- Create maps of parcels within a community;
- Public viewing of maps and opportunities for corrections and identifying additional disagreements;
- Verification of parcel data;
- Registration with the government/other authority; and
- Issuing certificates to landholders.

Whether and how an agricultural program incorporates aspects of land certification can be tailored to the project's scale, participants, and scope. There may be opportunities to coordinate with ongoing national certification efforts to assist project farmers (women and men) to certify their lands or enable women farmers to participate in one or more of the certification steps; to provide education on the steps of certification and women's rights to be certified landholders; to register newly allocated land to formerly landless women and men; or to encourage government to apply gender-responsive practices to land certification, including practices outlined in this section. Among the examples we examined, most projects that integrated gender aimed to increase women's access to information about the land certification process, increase women's participation in land certification, and promote supportive attitudes for women's land rights among community members and leaders.

Evidence across the programs suggests that the participatory certification efforts

- increased women's and men's knowledge of their land rights;
- increased women's and men's tenure security;
- increased women's and men's uptake of agricultural inputs, uptake of improved and sustainable practices, and ability to participate in commercial agriculture;
- had mixed effects on women's decision-making over land, agricultural production and earnings, and other measures of empowerment; and
- may have reduced violence against women.

2.3.1 Joint rights and joint participation

Certifying spouses' joint rights has become more common as governments adopt gender-equitable laws. Joint registration is an important step forward in legal recognition of women's land rights and moving toward their enforceability. Having both wives and husbands listed as owners/rightsholders is also often more socioculturally acceptable than married women having rights by themselves to land.

All of the examples in this section registered joint spousal rights, and most certified rights to land that women, men, and their households were already farming. Two examples are notable because they promoted joint rights for women and men who were not yet using the land. In allocating state lands to people who were landless, the Homestead Land Grants Program in Odisha State in India required joint titles in dual-/male-headed households, conveying that the state was giving land to both women and men. During the certification process, Rwanda's Land Tenure Regularization registered both female children and male children as having interests in their parents' land, strengthening legal recognition of daughters' inheritance rights in a context where inheritance favors sons (Hoza, 2012).

Projects also took action to make sure that husbands and wives were together at key points in the certification process and that both spouses received a certificate. Rwanda's Land Tenure Regularization requested both women and men to be present when people registered claims to a parcel. The Land Investment for Transformation (LIFT) Project in Ethiopia required that wives be present during meetings to publicly display maps and resolve land disputes. When wives were not present, husbands had to provide evidence for the reason (e.g. notes from health workers if a wife had recently given birth). The LIFT team did not release any certificates or other documents to the household until wives were present.

2.3.2 Build support for women's land rights in communities and among project staff

During trainings, meetings, and religious or social gatherings, projects used the following strategies to promote the social acceptability and perceived benefits of couples jointly registering land: male staff held conversations with male community members and leaders; emphasizing the security and well-being of family and children; and using real examples of consequences of women losing land rights or of being able to prosper or protect family interests because they had registered rights. Local staff led these discussions and trainings. Their understanding of social dynamics and how communities would receive messages was critical to crafting and delivering credible content.

LIFT and PROMAC II developed messages to encourage seeing women's land rights and joint registration as practical protection for children and as a moral good. LIFT tailored messages to all community stakeholders, including husbands, fathers, brothers, fathers-in-law, brothers-in-law, wives, guardians and caregivers, mediators, and government administrators. A selection of LIFT's target messages includes the following (DAI Guide, n.d.):

- Wives
  - Women work outside the home. They are farmers. Therefore, they should know about the land they are farming and have their rights to the land protected.
- Husbands
  - Families have greater land tenure security when wives are aware of household assets and can contribute to decision-making.
  - Allowing women to participate in land registration is an expression of manhood and modernity. It shows respect and acknowledges human dignity.
- Brothers-in-law
  - Do you want to see your sister-in-law and her children lose their land rights and live in destitution?

In Malawi, USAID ILRG provided gender equality and social inclusion training to all stakeholders at every step of the land registration process, including government officers, project data collectors and processors, members of customary land tribunals and district land tribunals, and members of the Customary Land Committees (Bessa et al., 2023). The training promoted buy-in and understanding the importance of gender integration throughout the registration process. Particularly critical was training on gender-based violence (GBV), recognizing that the land registration process can inadvertently create or exacerbate intrahousehold tensions.

2.3.3 Have a designated staff person to engage women throughout certification's steps

The LIFT Project's Social Development Officer (SDO[7]) spent time in communities to identify women (and other vulnerable persons) who were experiencing land disputes or who were vulnerable to either to harm during the certification process or to being left out of the process (e.g. widows, pregnant women, polygynous wives, women with disabilities, orphaned girls). SDOs and LIFT field teams could then better ensure these women and girls had the attention needed from the project to participate and to manage conflicts.

SDOs targeted awareness-raising to women by

- announcing meetings at community events and places where women were present (e.g. churches, mosques, and ceremonies) and communicating to men that women should attend certification meetings;

7 Based on their experience with SDOs in LIFT in Ethiopia, DAI created a SDO manual to promote the use of SDOs in other certification programs. https://www.dai.com/uploads/DAI%20SDO%20Guide-3ddf0d.pdf.

- making house visits to women identified as vulnerable to personally invite them to events and provide basic information about the certification process and women's land rights; and
- holding women-only meetings.

During awareness-raising activities with communities, local leaders, government, and other LIFT staff, SDOs explained the law behind women's legal rights. They also facilitated conversations about land rights violations that women face, drawing from local cases and encouraging participants to share their own experiences.

Having SDOs increased women's and men's understanding of and confidence in the certification process and benefits of certificates. They enabled women to have their disputes settled and raise questions and reduced violence toward vulnerable landholders. By increasing women's and men's participation from the beginning, identifying conflicts and problems early, and ensuring all landholders were able to present their interests on a parcel, were present for mapping boundaries, and were present for final decisions on parcels and rights to them, SDOs improved the efficiency of the overall certification process (DAI Guide, n.d.; LIFT Final Report, n.d.).

2.4 Combining land tenure and agricultural interventions

Although secure land tenure and access to resources are important for all farmers to be productive, women smallholder farmers, more than their male counterparts, tend to have a slimmer bundle of rights, less secure land rights, and less access to financial and extension services, accessible and timely inputs (FAO, 2023). Projects that both expand women's access to agricultural inputs and services and strengthen their land rights and tenure security may have better outcomes both for women and for productivity than projects that address only one.

We highlight different ways, five projects pursued a combined approach to address land tenure and access to agricultural inputs, training, or financial services, often with a focus on women. Across the projects outcomes included

- smallholder women farmers either feeling more tenure secure or having access to more land;
- women having greater access to financial services;
- women having greater roles in decisions about production, land use, or using income; and
- greater use of agricultural inputs (e.g. improved seeds or fertilizer) and application of sustainable agricultural practices in general.

(Pieters, 2023; Landesa, 2016, 2020; Nathan London Associates and SIAPAC, 2020; LIFT, 2019a, 2019b; Van Den Bold et al., 2015; Fletschner and Kenney, 2014; LIFT, n.d.).

A lesson from projects that included a land certification or registration component is that efforts to get women's names as rightsholders on land certificates were essential to women being able to access agricultural inputs and services, especially financial services.

2.4.1 Land Investment for Transformation Project, Ethiopia

LIFT in Ethiopia was the first large-scale land certification project to deliberately include a way to leverage land certificates to access loans. LIFT worked with local microfinance institutions (MFIs) and the Government of Ethiopia to pilot a loan linked to the land certificates. People listed on the certificate as landholders could apply for loans as individuals (instead of as part of a group) on the basis of their land certificate, business plan, and personal creditworthiness. This enabled flexibility in loan size and repayment options. The land use rights, validated by the certificates, or the produce of their land served as collateral. If borrowers defaulted, lenders would have temporary use rights to the land until the outstanding value was covered or up to 2 years.

The combination of more women having land certificates in their name with the certificate-linked loan expanded agricultural finance to smallholder women farmers, especially female household heads. Female household heads were often previously unable to access credit, including through group loans because group members saw them as a default risk. More than 22 000 people received loans using their land certificate, with only a 2% default rate. Eleven percent were women in female-headed households; 23% were women in male-headed households. All borrowers were first-time borrowers for formal credit. Farmers on average received loans two-to-three times the size of loans typically offered through groups (LIFT, n.d.). MFIs benefited through an expanded customer base and improved liquidity and had more demand for the certificate-linked loans than they could initially meet (LIFT, 2019a, 2019b, n.d.).

2.4.2 Promotion of Climate Smart Agriculture Project II (PROMAC II), Mozambique

PROMAC II directly provided land certificates and agricultural inputs to the same set of project farmers, rather than linking services or making them available in the same geography. The project viewed the certificates and inputs as a necessary combination to build farmers' sustainable productivity and restore depleted soils. Preliminary evaluation findings[8] suggest that the land certification process increased demand for agricultural inputs and willingness

8 Internal project materials shared with authors. The World Bank Gender Innovation Lab is conducting the evaluation, with results forthcoming.

to make agricultural investments only among women farm managers, who may have initially been more tenure insecure. There were no similar effects among male farm managers or women working on farms their husbands managed.

2.4.3 My Home My Land (Nijo Griha Nijo Buhumi) Microplots, West Bengal, India

The NGNB Microplots Program shows that a very small portion of land combined with agricultural inputs and training can have substantial impacts. West Bengal's state and local government institutions allocated and registered small parcels (mostly between 0.10 and 0.16 hectares) to landless households, prioritizing female-headed households and households with daughters and no sons. Since 2007, 73% of titles included women's names. Households that received microplots were also targeted to receive a variety of inputs and government services, including seeds and saplings, micro-irrigation, chicks and small livestock, extension and animal husbandry services, and microcredit. Agricultural training programs intentionally targeted women to enable them to produce for home production and their own earnings (Landesa, 2016). Results of this integrated program included

- women felt secure that they would be able to reside and farm on their microplots in the future;
- women in households that received a microplot were more likely to be involved in decisions about the use of the microplot, decisions about other land the family used, household food consumption, whether to take out loans, and whether to buy productive assets; and
- households that received microplots were more likely to use fertilizer and pesticides, rent agricultural equipment, and plant seedlings.

(Landesa, 2016; Fletschner and Kenney, 2014)

2.4.4 Support to the Government of Myanmar's Land to the Landless Program

An example of less intensive, informal coordination of land, financial, and agricultural extension projects generating impacts for farmers comes from Myanmar. From 2017 to 2021, Landesa partnered with the Government of Myanmar's Land to the Landless program to allocate and register farmland to rural and landless households.[9] Connected by a common funder, Landesa, the Helen Keller Institute (HKI), and the microfinance institution Proximity Designs identified common sites and coordinated scheduling of activities to encourage

9 Content for the Myanmar example comes from Landesa reports to the private funder, internal project records, and personal communication from Landesa staff.

farmers working with each organization to take advantage of the services offered by the others.

In Ayeyarwady Region, Landesa introduced Proximity Designs to villages where Landesa supported the government to allocate farmland. Proximity Designs then held meetings with farmers in these villages explaining Proximity Designs' services and criteria. Landesa discussed the merits of women taking out loans with men and women villagers. Messages concentrated on the household functioning well as a unit if wives and daughters manage finances because men are busy in the fields and encouraging women to take out loans and attend capacity-strengthening events to support managing the loans. Landesa's and Proximity Design's common funder partially guaranteed the loans. Approximately 30% of borrowers were women.

HKI and Landesa directly provided seeds, tree seedlings, fertilizer, farm equipment, and poultry to farmers who had received farmland. Landesa also encouraged women in its project villages to attend HKI meetings that provided training for cultivating home gardens to support child nutrition. Over 1014 women farmers and 1440 male farmers in households that were recently allocated farmland received agricultural training or inputs.

2.4.5 Expanding the PepsiCo potato supply chain to women farmers in the USAID ILRG Project

In West Bengal, India, a partnership funded by USAID and PepsiCo implemented under the USAID ILRG project supported of approximately 40 landless women in seven groups to collectively rent land and farm potatoes for PepsiCo's supply chain. The women were mainly from scheduled castes, scheduled tribes, and religious minorities. The women's land-leasing groups were formed from existing self-help groups with women who had prior farming experience. Women in the leasing groups received agronomic training through PepsiCo's extension services, which USAID supported to make more gender-responsive by training farmer-facing staff in basic concepts of gender, barriers and benefits to women producing PepsiCo potatoes, and reasons and impacts of women having less access to agricultural training.

The land leasing groups demonstrated that women could manage a farm, perform to commercial quantity and quality standards, and engage with aggregators and suppliers. Over the course of the project, women in the groups increasingly saw themselves as farmers in their own right rather than helpers or laborers, as did their families and other community members (Brooks and Sommerville, 2023).

2.5 Social norms and behavior change interventions

A woman's ability to exercise her land rights is often influenced by social norms in her community. Land and agricultural projects run the risk of not only failing to benefit women but potentially harming women if they do not address norms that present barriers to women's land rights.

We outline three good practices from projects that focused specifically on strengthening women's land tenure and tenure security: engage men, identify and support champions, and facilitate local solutions. In addition to these practices, most trainings involved repeated interactions with participants to build trust, create opportunities for participants to practice and reflect on new behaviors, and influence a core group of people before pursuing widespread reach.

2.5.1 Engage men

Securing Your Family's Future (SYFF) was one of the first projects to engage men in an intervention to shift social norms related to women's land rights. Launched in 2016, the SYFF course enrolled men in Uganda, Kenya, and Tanzania to build positive norms and attitudes, accurate knowledge, and self-efficacy among men to encourage specific behaviors: (1) fathers allocate land to daughters and sons equally; (2) husbands include wives as joint owners and/or managers of land; (3) husbands and fathers bequeath wives and daughters land in their wills; and (4) husbands seek wives' consent before selling family land or making other major decisions about land. Each of these behaviors had specific activities and key messages integrated throughout the course. The course, comprised of six 2-hour sessions conducted over several weeks, employed interactive storytelling, role playing, and small- and large-group discussions. For example, small groups prepared and presented skits about men calling into a radio station for advice on land questions, followed by group discussion. Participants also received a take-home activity every session to reflect on and practice skills from the session in their daily lives.[10]

The Rising Up! Project, implemented in the Democratic Republic of Congo by Women for Women International (WfWI) and Innovation and Training for Development and Peace (IFDP), aimed for women to safely receive customary land titles. The project adapted WfWI's women's empowerment and GBV prevention approaches to the land sector and adapted IFDP's existing Customary Decentralized Land Management Resource Model to incorporate GBV prevention and response. IFDP and WfWI used these tools to train male community leaders about women's land rights and how to identify and address

10 This description is based on a review of the curriculum materials, which were shared with Landesa but are not publicly available.

Published by Burleigh Dodds Science Publishing Limited, 2025.

GBV. The project encouraged male leaders to reflect upon and shift norms and customs that limited women's access to land, to use gender-sensitive practices in dispute resolution, and provide referrals for GBV survivors (Women for Women International, 2022).

In turn, trained male leaders facilitated a 16-week 'step-down' training for the husbands of women interested in obtaining customary land titles. The curriculum covered gender concepts, men as role models, positive masculinity, managing couple relationships, land rights and women's access to land, customary land management, land conflict management, and personal behavior change. All male leaders served as role models by granting land to their spouses to register in their names, and 95% of the men they trained did the same (Women for Women International, 2022).

2.5.2 Identify and support champions

There will be pushback to any project that attempts to change the status quo. It can be more effective to invest resources in champions of a cause instead of spending resources to change the attitudes of people very resistant to the change. Champions should be influential people with enough social capital that they are unlikely to experience social harm when they question social norms or encourage others to do so.

As part of customary land registration, the USAID ILRG project partnered with the Government of Malawi to facilitate household dialogues on harmful gender norms. Over five participatory sessions, household members discussed gendered roles and responsibilities, power dynamics, GBV, and decision-making. After the dialogues, the project supported the most engaged participants to serve as community gender champions. After an orientation on facilitating discussions about the benefits of gender-inclusive land ownership and decision-making, community champions received a discussion guide, FAQ document, comic booklet, posters t-shirts, and a small stipend. The champions worked in pairs to spread messages at community gatherings and by going door-to-door, reaching over 4500 people (Bessa et al., 2023)

2.5.3 Facilitate local approaches

USAID ILRG engaged *indunas*, customary chiefs and their advisors, to design actions to strengthen women's land rights in Zambia's Eastern Province. *Indunas* not only allocate land but also serve as cultural custodians. The project facilitated three dialogues over 1 year in which *indunas* analyzed gender inequalities in land ownership, access, and control; envisioned the change they wanted to see and how they could contribute to this change; and discussed achievements and challenges. Although some *indunas* were resistant to changes in norms, others embraced them. In five of the seven chiefdoms, *indunas* facilitated community discussions about harmful traditions such as forcing divorcees and widows to leave the villages where they lived. In four chiefdoms, *indunas* drafted by-laws to protect women's land rights and ban property grabbing. *Indunas* also encouraged men to include women's names on land documents, and some led by example. This locally led process promoted ownership, accountability, and sustainability (Malasha, 2021).

3 Private sector examples

Companies, cooperatives, and other private sector actors can decrease risk, expand their supplier base, and encourage investments in productivity through actions to strengthen women's land tenure. Several of the approaches outlined in Section 2 are relevant for private sector actors. Because private sector action on women's land tenure is relatively new and undocumented in agriculture, it is useful to make these examples more visible. The following examples of certification, activities to address norms, and a company rule change show that companies and cooperatives, large and small, can take action to strengthen women's land tenure and security.

3.1 Promoting joint land rights in law and practice in a coffee cooperative

The Bukonzo Joint Cooperative Union (BJCU), a microfinance and coffee cooperative union in Uganda with mostly female membership, began in 1992 as a savings self-help group. At the time BJCU was forming, few women had land ownership (sole or joint) despite doing the bulk of agricultural work. Without registered ownership, women were ineligible for membership in coffee cooperatives and not able to directly access the credit or savings services cooperatives provided. Women often strategically harvested coffee cherries before their husbands returned from migratory work so they could directly control the income, resulting in low quality coffee harvests.

In 2007, BJCU began applying the Gender Action Learning System (GALS) across its entire coffee value chain (from farming workshops to board meetings) (Bukonzo Joint Coffee, 2015). GALS engages both women and men in participatory activities such as drawing, discussions, and role plays. In one activity, women and men listed the activities of women and men in coffee farming and placed them on a picture of a tree, with women's land rights at the foundation and income-generating activities on branches and leaves. This activity created a community vision of women's land rights as a foundation for the benefits coffee production generates, emphasized the benefits of cooperation between women and men, and opened opportunities for women's greater involvement in household and cooperative decisions. Greater cooperation and women's increased involvement in decisions eventually enabled BJCU to shift production into higher qualities of specialty coffee. GALS activities combined with awareness-raising on gendered inheritance rights also laid the groundwork for BJCU to support wives and husbands to legally register their land as jointly owned. One of BJCU's mottos is, 'Balanced trees bear richer fruit.'

3.2 Succession of outgrower contracts

Women's ability to participate in outgrower schemes is often limited because they are less likely to have recognized rights to land and to have the economic resources and social acceptance to lease land as individuals. The KASCOL (Kaleya Smallholder Company Limited) that produces fair trade sugar cane in Zambia developed an arrangement whereby family members can inherit an outgrower contract for a 14-year renewable sub-lease on company-leased land if the household head dies (Wonani et al., 2013). Once women join the outgrower scheme, they can also access fire and rain insurance and join the outgrower association that holds equity in KASCOL.

Contract holders have increasingly opted for female successors because they believe women will use the earnings to care for family (Manda, 2021). Over time, this has led to many women joining the scheme and increased the stability of KASCOL's outgrower base in an area with a high prevalence of HIV/AIDS. Benefits from the contracting change included marginally higher yields among women outgrowers than men and greater food security in households with women outgrowers because they had stronger decision-making over income. However, these changes were not accompanied by efforts to enhance women's roles in the sugar associations (Manda, 2021).

3.3 Cooperative-issued land certificates to reduce risk

A partnership between Illovo Sugar Company and USAID in Mozambique issued land certificates through agricultural cooperatives as precursors to fully legally registered certificates, which took additional time and effort to obtain. Illovo and other sugar companies accepted the cooperative-issued certificate as validation of landholdings required for farmers to register with them as outgrowers. Farmers reported that the certificates were easier to obtain than other documents companies accepted to register as an outgrower. The cooperative-issued certificates increased women (and men) farmers' perceived tenure security and their ability to pursue commercial production while strengthening the breadth and stability of the company's supply base (Persha and Taha, 2022).

4 Recommendations for integrating women's land tenure

Sections 2 and 3 show practical steps and lessons to strengthen women farmers' land tenure that agricultural programs and private sector actors can integrate into their own activities. Examples demonstrate how these approaches increase not only women's land rights and tenure security but also women's uptake of agricultural inputs and sustainable practices, access to information and services, commercial participation, and economic empowerment – goals for development actors and companies.

Programs and companies should match the intensity, scale, and nature of activities for smallholder women's land tenure to their engagement with farmers. The goal is to identify activities that will have meaningful impact on women smallholder farmers' land rights and tenure security that are also feasible given available resources. The following paragraphs are lessons emerging across the five approaches.

To determine entry points and avoid harm, research existing laws, customary rules, and norms to understand barriers and opportunities to women's land tenure. Also review project/company practices and policies related to land that may affect women's eligibility or ability to engage. Adding women's land tenure as a topic to the gender analyses required by many funders is one place to begin. Private sector actors similarly can include women's land tenure as part of their gender, sustainability, or responsible investment analyses. (Toolkits in Section 6 outline key questions to address in your analysis.) In examining their rules and practices, projects and companies should pay attention to requirements for land ownership, land area, minimum production values, being named on land documents, and cooperative membership that may disproportionately exclude women. Also, interventions do not always have to be about land ownership or access rights – advancing women smallholders' ability and power to manage land, production, and agricultural earnings are natural choices for agricultural programming.

In these analyses and during design and implementation, engage land experts to advise on or to lead activities focused on land rights literacy, mapping and certification, land governance, or gender norms around land. Agricultural practitioners and land practitioners who were collaborating found it helpful to have a designated liaison and to understand key terminology that the other used.

Combine multiple approaches as feasible. Applying every approach outlined in this chapter or addressing every barrier to women's land tenure is unrealistic. However, most examples applied at least two approaches – most often awareness-raising with another approach. Awareness-raising about women's land rights is often necessary before or alongside other approaches so that women know their rights and how to exercise them and so that leaders know and can uphold women's land rights. In certification programs, awareness-raising was often necessary for people to understand that women can be on certificates and what the benefits are. Training government and traditional leaders on women's land rights combined with registering land certificates with local authorities can increase the enforceability of women's land rights, making them matter of public record and public responsibility.

Awareness-raising and norm change activities need to be tailored, iterative, and locally led. Both land rights and gender are sensitive and deeply felt topics that affect agricultural systems. Establishing trust and accompanying people, communities, and institutions to understand new content and practice new behaviors takes time. Successful efforts were conducted over several months, had multiple in-person interactions, created opportunities for people to learn together and reinforce each other, and were guided by leaders or people from the local area. Tailoring messages to the local context and to specific stakeholders or social roles (e.g. recognizing women as farmers or highlighting community mediators' responsibility to be equitable in LIFT or encouraging fathers to supportive of daughters' rights and well-being in SYFF, LIFT, and Rwanda LAND) conveyed respect those roles and were relevant to the people in them. Similarly, building on socially acknowledged and valued roles for women – such as growing food for the family, gardening or keeping small livestock to supplement children's diets or household income as in the E-HFP project in Burkina Faso and NGNB Program in India – created opportunities to secure or expand women's land access that communities could see as legitimate or even beneficial.

It is important to include men, especially traditional leaders, regardless of the approach because smallholder women farmers' access to land, access to inputs and resources to cultivate it, and power to decide how to use earnings and other benefits from land are greatly dependent on their male relatives and (mostly male) leaders. Most examples in Sections 2 and 3 engaged men in activities related to women's land rights. Applying any of the approaches needs male engagement for women to be able to safely participate. Even community meetings for awareness-raising, a relatively simple activity, benefits from men in the community encouraging women to attend or at least not impeding women's participation. As active champions, men can model and support each other in positive behaviors – from seemingly small behaviors like encouraging their wife to attend or speak at a land certification meeting to allocating wives land for their own use or jointly registering land. Over time, these changes can increase the sociocultural legitimacy of women's land rights; reduce the risk of women exercising land rights; reduce the burden of asking husbands' permission to exercise rights; and increase the likelihood that leaders or local institutions provide equitable services and dispute resolution.

Land certification can benefit agricultural programs' and companies' farmers tenure security and access to services and has multiple entry points for agricultural programs and companies to choose their involvement. Involvement can vary from linking farmers pursuing or who recently received land certificates with agricultural services or vice versa, financially and logistically supporting a project's women farmers to certify their land or incorporating training on women's land rights or certification into the project.

For combined land and agricultural interventions, fundamental good gender practices meant that more women could participate fully, to their benefit and to the project's or company's benefit. Practices from programs in this chapter included, but are not limited to, holding meetings at convenient times for women, inviting women to public meetings, having separate meetings as needed for women, having female project staff or lead farmers, providing childcare during trainings, and having channels for women's input throughout the project. They engaged a larger number of women overall who then had opportunities to learn about their rights, gain certificates in their name, participate in community land governance, and be in a better position to access agricultural loans and apply agricultural knowledge and inputs gained through the project.

Finally, agricultural projects are well-placed to contribute evidence on women's land tenure and practical ways to strengthen it. Including gendered measures of land rights and tenure security in project evaluations, documenting lessons from efforts to address social norms affecting women's land tenure, and examining gendered rights in addition to ownership, including managing agricultural land and earnings, are valuable additions to the evidence base that programs of any scale can contribute. Programs can also address two important limitations of this chapter – few examples outside sub-Saharan Africa[11] and no exploration of gender's intersections with age, Indigenous identity, or other aspects of identify – both of which are partially due to a lack of documented or accessible information. Sharing examples of *how* communities, projects, and companies approached addressing women's land tenure and the corresponding challenges, successes, and lessons is incredibly valuable for creating more effective programs and policies and influencing not only practitioners but governments and funders.

11 That the majority of examples are from sub-Saharan Africa reflects that research on women's land tenure programing draws heavily from sub-Saharan Africa (Meinzen-Dick et al. 2019) and reflects recent donor attention to land tenure in sub-Saharan Africa. For example, over half of active or completed single-country projects listed on USAID's LandLinks and dating back to 2003 are in sub-Saharan Africa (accessed https://www.land-links.org/usaid-land-projects/ November 12, 2023). Although the approaches and lessons are likely to be relevant globally, specific activities and messages may be less relevant in Latin American or Asian contexts.

5 Conclusion

Integrating interventions for women's land tenure into agricultural programming is feasible and impactful in a variety of settings. The examples in this chapter – of varying scales and drawn from rural development projects, government programs, and agribusiness – demonstrate that enhancing women's land tenure in agricultural programs can not only increase women's access to land and perceived tenure security but also increase women farmers' use of agricultural inputs and good agricultural practices and their participation in commercial agriculture, which can create a larger and more stable supplier base for agribusinesses. Combined, women's enhanced land tenure, women's increased agricultural production, and women's greater role in decisions about land, production, and income are recognized building blocks of women's economic empowerment, overall food security, and healthy rural economies.

While there remains much to learn about the practicalities of how to implement integrated agricultural and land tenure programming, the examples in this chapter illustrate that it is possible and valuable to address underlying gendered land tenure issues using approaches and degrees of intensity that align with programs' and companies' contexts and available resources. Key lessons from the examples are that good practices for gender-responsive agricultural programming (e.g. conducting a gender analysis and engaging men) are critical to identifying and creating opportunities to increase women's participation and everyone's benefit. Also critical are collaboration between land tenure and agricultural development practitioners and ensuring that communities and other stakeholders understand key aspects of gendered land rights.

Integrating women's land tenure and agricultural programming is not effortless and may not result in the full range of potential positive outcomes, especially in the short term. However, because women's land rights are part of the underlying foundation of agrifood systems, it is critical to address them in the iterative, long-term, multi-sector endeavor of making agrifood systems more effective and equitable.

6 Acknowledgments

The authors thank the following people for availing their expertise, programmatic experience, and time to inform the chapter: Rashid Alhassan, Christine Anderson, Ohnmar Myo Aung, Thais Bessa, David Bledsoe, Jennifer Brown, Naw Dah Htee, John Leckie, Workwoha Mekonen, Carolina Reynoso Pieters, Leonard Rolfes, Jolyne Sanjak, Matthew Sommerville, Darryl Vhugen, Mar Mar Wai, and Aye Chan Wynn. Lucas Belury, Megha Panigrahi, and Samantha Pickett provided excellent research assistance.

7 Where to look for further information

Concepts and definitions for land tenure

- USAID. Property Rights and Resource Governance Issues and Best Practices Module 1: Introduction to Land Tenure and Property Rights (LTPR) Concepts Selected Land Tenure Definitions. https://www.land-links.org/wp-content/uploads/2017/02/USAID_Land_Tenure_2012_Washington_Course_Module_1_LTPR_Concepts_and_Definitions.pdf.

Summaries of evidence on women's land tenure

- Stanley, Victoria, and Jennifer Lisher. 2023. *Why Land and Property Rights Matter for Gender Equality*. Washington, DC: World Bank. https://doi.org/10.1596/39990.
- Program & policy approaches to strengthen women's land tenure: Scalise, E. and Giovarelli, R. (2020). Scalise, E, and R Giovarelli. 2020. "What Works for Women's Land and Property Rights? What We Know and What We Need to Know." https://resourceequity.org/record/3170-what-works-for-womens-land-and-property-rights-what-we-know-and-what-we-need-to-know/.

Tools

- World Bank. 2022. "Integrating Gender in Land Projects: A Toolkit." Washington, DC: World Bank. https://thedocs.worldbank.org/en/doc/7212d6d4bd7ea638c9fa4a2a16323809-0090072022/original/Integrating-Gender-in-Land-Projects-A-Toolkit-October2022-rev.pdf.
- Hannay, L. and E. Scalise. 2014. "Improving Land Tenure Security for Women: A Women First Approach." Landesa Center for Women's Land Rights: Seattle. https://www.landesa.org/wp-content/uploads/landesa-toolkit-a-women-first-approach.pdf.
- Integrated Land and Resource Governance (ILRG). (2022). Training Manual for Household Dialogues on Gender Norms in the Context of Land Rights in Malawi. USAID Integrated Land and Resource Governance Task Order under the Strengthening Tenure and Resource Rights II (STARR II) IDIQ. https://www.land-links.org/wp-content/uploads/2023/08/ILRG-Malawi-Household-Gender-Norms-Dialogues-Manual-USAID-Review_508.pdf.
- Responsible Investments in Property and Land – Guidebooks for businesses, governments, and communities. https://ripl.landesa.org/.

- LandAssess - a risk management and assessment framework and checklists to help agricultural companies to assess and manage how they respect land rights. https://www.landesa.org/what-we-do/landassess-tool/.

National-level data on women's land rights

- Prindex. Gender-disaggregated, nationally representative data on perceived tenure security and documentation of land rights and analytical reports based on the data. https://www.prindex.net/.
- LandEx Global Land Governance Index. https://www.landexglobal.org/en/datasets/view-and-use-data/.

Projects and places to watch

- The Responsible Land-Based Investment (RLBI) Navigator 2.0, developed by the International Institute for Environment and Development (IIED) and the Land Portal Foundation, in another useful source created to help government, civil society, and private sector actors to identify and access guidance and examples on responsible land investments. https://landinvestments.org/.
- Securing Women's Resource Rights through Gender Transformative Approaches is a joint project by the International Foundation for Agricultural Development (IFAD), the Center for International Forestry Research and World Agroforestry Centre, the International Food Policy Research Institute, and the Alliance of Bioversity International, and the International Center for Tropical Agriculture (CIAT) to pilot gender-transformative approaches to strengthening women's land rights in a portfolio of IFAD's agricultural projects. https://www.cifor.org/wlr/.
- The International Institute for Environment and Development's land acquisitions and rights collection. https://www.iied.org/theme/land-acquisitions-rights.
- Landesa's Center for Women's Land Rights. https://www.landesa.org/what-we-do/womens-land-rights/.

8 References

Agarwal, B. 2003. Gender and land rights revisited: exploring new prospects via the state, family and market. *Journal of Agrarian Change* 3(1–2): 184–224. https://doi.org/10.1111/1471-0366.00054.

Ali, D. A., Deininger, K. and Goldstein, M. 2014. Environmental and gender impacts of land tenure regularization in Africa: pilot evidence from Rwanda. *Journal of Development Economics* 110: 262–275. https://doi.org/10.1016/j.jdeveco.2013.12.009.

Published by Burleigh Dodds Science Publishing Limited, 2025.

Alvarado, G., Kieran, C., Jacobs, K., Beduhn, J., Heidenrich, T. and Linkow, B. 2022. *Ethiopia Strengthening Land Tenure and Administration Program Follow-On Report: an Impact Evaluation of Long-Term Effects of Second-Level Land Certification*. https://www.land-links.org/wp-content/uploads/2022/10/ELTAP-ELAP-2021-Evaluation-Final_9.23.22.pdf.

Banerjee, R., Gulati, K., O'Sullivan, M. B., Rao, A. S., and Vinez, M. L. 2014. *Levelling the Field: Improving Opportunities for Women Farmers in Africa*. Washington, DC: World Bank Group. http://documents.worldbank.org/curated/en/579161468007198488/Levelling-the-field-improving-opportunities-for-women-farmers-in-Africa

Bayisenge, J. 2018. From male to joint land ownership: Women's experiences of the Land Tenure Reform Programme in Rwanda. *Journal of Agrarian Change* 18(3): 588–605. https://doi.org/10.1111/joac.12257.

Bessa, T., Dooley, M., Makhuva, S. and Sommerville, M. 2023. Lessons learned: integrating gender equality and social inclusion into customary land documentation in Malawi. Washington, DC: *USAID Integrated Land and Resource Governance Task Order under the Strengthening Tenure and Resource Rights II*. https://www.land-links.org/wp-content/uploads/2023/08/ILRG-Malawi-GESI-lessons-learned_June_2023-508.pdf

Brooks, S. and Sommerville, M. 2023. *The Business Case for Women's Economic Empowerment in PepsiCo's Potato Supply Chain in West Bengal: Key Results and Recommendations*. USAID.

Bukonzo Joint Coffee 2015. *Bukonzo Joint Cooperative Union Wins Specialty Coffee Association of America Sustainability Award 2015 for Its Pioneering Approach to Gender Justice*. https://bukonzojointcoffee.wordpress.com/2015/03/14/bukonzo-joint-cooperative-union-wins-specialty-coffee-association-of-america-sustainability-award-2015-for-its-pioneering-approach-to-gender-justice/.

Chemonics 2016. *From Policy to Practice: Exercising Gender-Equal Land Rights in Rwanda*. https://chemonics.com/impact-story/policy-practice-exercising-gender-equal-land-rights-rwanda/.

DAI Guide n.d. Inclusive land registration and the critical role of social development officers: a handbook for program design and implementation. Bethesda, Maryland. https://www.dai.com/uploads/DAI%20SDO%20Guide-3ddf0d.pdf

Deininger, K., Ali, D. A. and Alemu, T. 2011. Impacts of land certification on tenure security, investment, and land market participation: evidence from Ethiopia. *Land Economics* 87(2): 312–334.

Doss, C., Kim, S. M., Njuki, J., Hillenbrand, E. and Miruka, M. 2014. *Women's Individual and Joint Property Ownership*. http://dx.doi.org/10.2139/ssrn.2483968.

Doss, C., Kovarik, C., Peterman, A., Quisumbing, A. and van den Bold, M. 2015. Gender inequalities in ownership and control of land in Africa: myth and reality. *Agricultural Economics* 46(3): 403–434. https://doi.org/10.1111/agec.12171.

Doss, C. and Meinzen-Dick, R. 2020. Land tenure security for women: a conceptual framework. *Land Use Policy* 99: 105080. https://doi.org/10.1016/j.landusepol.2020.105080.

Dungumaro, J. and Amos, M. 2019. *Pastoral Women's Rights and Leadership Forums, Tanzania: Experience, Impact and Lessons Learned*. Rangelands Research Report. Nairobi, Kenya: ILRI.

FAO 2023. *The Status of Women in Agrifood Systems*. Rome, Italy: Food and Agriculture Organization. https://doi.org/10.4060/cc5343en.

Feyertag, J., Childress, M., Langdown, I., Locke, A. and Nizalov, D. 2021. How does gender affect the perceived security of land and property rights? Evidence from 33 countries. *Land Use Policy* 104: 105299. https://doi.org/10.1016/j.landusepol.2021.105299.

Fletschner, D. and Kenney, L. 2014. Rural women's access to financial services: credit, savings, and insurance. In *Gender in Agriculture*. Netherlands: Springer. http://dx.doi.org/10.1007/978-94-017-8616-4_8.

Gaddis, I., Lahoti, R. and Li, W. 2018. *Gender Gaps in Property Ownership in Sub-Saharan Africa*. Washington, DC: World Bank. https://doi.org/10.1596/1813-9450-8573.

Ghebru, H. 2019. *Women's Land Rights in Africa* (0th edn.). Washington, DC: International Food Policy Research Institute. https://doi.org/10.2499/9780896293649_04.

Giovarelli, R., Richardson, A. and Scalise, E. 2016. *Gender and Collectively Held Land: Good Practices and Lessons Learned from Six Global Case Studies*. http://www.landesa.org/wp-content/uploads/2016-Best-Practices-Synthesis-Report.pdf.

Goldstein, M., Houngbedji, K., Kondylis, F., O'Sullivan, M. and Selod, H. 2015. *Formalizing Rural Land Rights in West Africa: Evidence from a Randomized Impact 38 Evaluation in Benin*. World Bank Policy Research Working Paper: World Bank Group.

Goldstein, M., Houngbedji, K., Kondylis, F., O'Sullivan, M. and Selod, H. 2018. Formalization without certification? Experimental evidence on property rights and investment. *Journal of Development Economics* 132: 57–74.

Hoza, T. N. 2012. *Empowering Women through Land Tenure Reform: The Rwandan Experience*. New York: UN Women.

Jackson, C. 2003. Gender analysis of land: beyond land rights for women? *Journal of Agrarian Change* 3(4): 453–480. https://doi.org/10.1111/1471-0366.00062.

Jacobs, K. and Kes, A. 2015. The ambiguity of joint asset ownership: cautionary tales from Uganda and South Africa. *Feminist Economics* 21(3): 23–55. https://doi.org/10.1080/13545701.2014.926559.

Jhaveri, N. 2021. *Gender, Tenure Security, and Landscape Governance* (0th edn.). Washington, DC: International Food Policy Research Institute. https://doi.org/10.2499/p15738coll2.134790.

Kloos, K., McClung, M. and Huang, S.-T. 2016. Insights from Rwanda: analysis of a land conflict resolution project in eastern province. Paper prepared for presentation at the 2016 World Bank Conference on Land and Poverty.

Landesa 2016. *"Micro-plot Allocation to the Poor: Experience in West Bengal, India*. landesa.org.

Landesa 2020. *"Land Rights Matter for People and Planet Policy Recommendations for Climate Action Through Land Tenure Security*. www.landesa.org.

LIFT 2019a. *Building Healthy Portfolios: Technical Assistance on Default Management System for Partner Microfinance Institutions*. Land Investment for Transformation. London: Foreign, Commonwealth and Development Office (FCDO).

LIFT 2019b. *Effecting Change through an Innovative Financial Product*. Land Investment for Transformation. London: Foreign, Commonwealth and Development Office (FCDO).

LIFT n.d. *Life Project Completion Report*. Land Investment for Transformation. London: Foreign, Commonwealth and Development Office (FCDO).

Malasha, P. 2020. *Increasing Women's Participation in Community Resources Boards in Zambia: Outcomes and Lessons Learned from the Election Process*. USAID Integrated

Land and Resource Governance Task Order under the Strengthening Tenure and Resource Rights II. Washington, DC: USAID.

Malasha, P. 2021. Traditional leaders in Zambia shift gender norms and strengthen women's land rights. https://www.land-links.org/2021/07/traditional-leaders-in-zambia-shift-gender-norms-and-strengthen-womens-land-rights/.

Manda, S. 2021. *Sugarcane Commercialisation and Gender Experiences in the Zambian 'Sweetest Town:' Feminist Economics*. Washington, DC: New Academia Publishing.

Meinzen-Dick, R., Quisumbing, A., Doss, C. and Theis, S. 2019. Women's land rights as a pathway to poverty reduction: framework and review of available evidence, agricultural research for rural prosperity: Rethinking the pathways. *Agricultural Systems* 172: 72–82. https://doi.org/10.1016/j.agsy.2017.10.009.

Nathan London Associates and SIAPAC 2020. *The Impact of the EEU Access to Finance and Land Rental Interventions on Rural Ethiopian Households (Third Round Data Collection in 2020)*. LIFT. London: Foreign, Commonwealth and Development Office (FCDO).

O'Sullivan, M. 2017. *Gender and Property Rights in Sub-Saharan Africa: A Review of Constraints and Effective Interventions*. Washington, DC: World Bank. https://doi.org/10.1596/1813-9450-8250.

Persha, L. and Taha, N. 2022. *Follow-On Performance Evaluation of the Responsible Land-Based Investment Pilot in Mozambique: Final Report*. Washington, DC: United States Agency for International Development Communications, Evidence and Learning (CEL) Project. https://pdf.usaid.gov/pdf_docs/PA0211HT.pdf.

Pieters, C. 2023. *Promotion of Conservation Agriculture (PROMAC II) Final Program Report*. Mozambique: NCBA CLUSA.

Pradhan, R., Meinzen-Dick, R. and Theis, S. 2019. Property rights, intersectionality, and women's empowerment in Nepal. *Journal of Rural Studies* 70: 26–35. https://doi.org/10.1016/j.jrurstud.2019.05.003.

Prindex 2020. *A Global Assessment of Perceived Tenure Security from 140 Countries*. https://www.prindex.net/.

Rolfes, L. and Seitz, V. 2013. *Providing Farmland Ownership Rights to Women in Rural Mali*. Washington, DC: Millennium Challenge Corporation.

Scalise, E. and Giovarelli, R. 2020. *What Works for Women's Land and Property Rights? What We Know and What We Need to Know*. Washington: Resource Equity.

Slavchevska, V., Doss, C. R., Paula, A. P. and Brunelli, C. 2021. Beyond ownership: women's and Men's land rights in sub-Saharan Africa. *Oxford Development Studies* 49(1): 2–22. https://doi.org/10.1080/13600818.2020.1818714.

Stanley, V. and Lisher, J. 2023. *Why Land and Property Rights Matter for Gender Equality*. Washington, DC: World Bank. https://doi.org/10.1596/39990.

Stevens, C., Panfil, Y., Linkow, B., Hagopian, A., Mellon, C., Kulkarni, N., Bouvier, I., Brooks, S., Lowery, S. and Green, J. 2020. *Land and Development: A Research Agenda for Land and Resource Governance at USAID*. Washington, DC: USAID. https://www.land-links.org/research-publication/land-and-development-a-research-agenda-for-land-and-resource-governance-at-usaid/

Sutz, P. 2021. *Why Simple Solutions Won't Secure African Women's Land Rights*. London: International Institute for Environment and Development.

Tibenkana 2020. Inclusive land rights: tackling legal pluralism to reduce its effect on Ugandan women. *Global Policy Review*, no. Realizing women's rights to land and other productive resources.

UN 2013. *Realizing Women's Rights to Land and Other Productive Resources*. Geneva and New York: United Nations Human Rights Office of the High Commission, and UN Women.

USAID 2013. *USAID Land Tenure Profile: Sudan*. Washington, DC: United States Agency for International Development.

USAID 2019. *USAID Advance Initiatives of Women's Access to Farm Land*. United States Agency for International Development (USAID).

Van Den Bold, M., Dillon, A., Olney, D., Ouedraogo, M., Pedehombga, A. and Quisumbing, A. 2015. Can integrated agriculture-nutrition programmes change gender norms on land and asset ownership? Evidence from Burkina Faso. *The Journal of Development Studies* 51(9): 1155–1174. https://doi.org/10.1080/00220388.2015.1036036.

von Benda-Beckmann, F., von Benda-Beckmann, K. and Wiber, M. 2006. *The Properties of Property*. In K., von Benda-Beckmann, F., von Benda-Beckmann, M. G., Wiber. (Eds.). *The Changing Properties of Property,* New York: Berghahn Books.1–39.

Winrock International 2019. *Ghana Agriculture and Natural Resource Management Project*. Winrock International. Virginia: Arlington.

Women for Women International 2022. *"Rising Up!* Promoting Congolese Women's Land Access and Preventing GBV in Eastern DRC: Women for Women International Final Report." USAID. https://portals.iucn.org/union/sites/union/files/doc/women_for_women_final_report.pdf.

Wonani, C. M., Mbuta, W. S. and Mkandawire, A. M. 2013. *The Gender and Equity Implications of Land-Related Investments on Land Access, Labour, and Income-Generating Opportunities. A Case Study of Selected Agricultural Investments in Zambia*. Rome: Food and Agriculture Organization.

Appendix 1 Programmatic examples and outcomes by approach

Approach	Awareness raising of women's land rights	Land access and governance in customary systems	Gender-responsive, participatory land certification	Combining land tenure and agricultural interventions	Addressing social norms
Outcomes for agriculture and women's empowerment	(Not consistently assessed)	Women farmers' increased application of agricultural knowledge, improved technologies, or higher-value crops Women's increased decision-making in land use and agricultural production Higher self-confidence and empowerment among women	Women farmers' increased application of agricultural inputs and improved/ sustainable practices More women farmers enter commercial supply chains Expanded and more stable supply base for agribusiness Mixed outcomes for: women's decision-making over agricultural production and agricultural earnings; women's empowerment	Women farmers' increased application of agricultural inputs and improved/ sustainable practices Women's increased decision-making in land use, agricultural production, and agricultural earnings Increased access to credit / financial / insurance services Expanded customer base for microfinance institutions More women farmers enter commercial supply chains Expanded and more stable supply base for agribusiness	(Not consistently assessed)
Improved outcomes for women's land rights and tenure security	Rights: – Conditions for security: Exercisability, Enforceability <u>Notable outcomes</u> Women, men, and leaders have improved knowledge of women's land rights	Rights: Use, Manage, Earn from land Conditions for security: Sociocultural legitimacy, Enforceability <u>Notable outcomes</u> Women's increased access to land Women's increased participation in land meetings and decision-making bodies	Rights: Use, Transfer, Use as collateral, Ownership Conditions for security: Legal recognition, Sociocultural legitimacy, Durability <u>Notable outcomes</u> Women perceive their tenure to be more secure Women, men, and leaders have improved knowledge of women's land rights	Rights: Use, Manage, Use as collateral, Earn from land Conditions for security: Legal recognition <u>Notable outcomes</u> Women's increased access to land Women perceive their tenure to be more secure	Rights: – Conditions for security: Sociocultural legitimacy, Enforceability <u>Notable outcomes</u> Male leaders and community members become champions for women's land rights

(Continued)

Appendix 1 (*Continued*)

Approach	Awareness raising of women's land rights	Land access and governance in customary systems	Gender-responsive, participatory land certification	Combining land tenure and agricultural interventions	Addressing social norms
Program / private sector examples highlighted in chapter	Rwanda LAND Talking Books / LSRD II, Liberia	Enhanced Homestead Food Production, Burkina Faso Enhancing Customary Justice Systems in Mau Forest, Kenya Project Feed the Future ADVANCE II and AgNRM Projects, Ghana MCC Alatona Irrigation Project, Mali USAID ILRG (Malawi and Zambia) Women's Rights and Leadership Fora, Tanzania	Homestead Land Grants Program, India LIFT, Ethiopia PROMAC II, Mozambique Rwanda's Land Tenure Regularization Program USAID ILRG (Malawi) USAID and Illovo Sugar Company	KASKOL, Zambia LIFT, Ethiopia NGNB Microplots, India PROMAC II, Mozambique Support to the Government of Myanmar's Land to the Landless Program USAID ILRG and PepsiCo, India	Bukonzo Joint Cooperative Union Rising up!, Democratic Republic of Congo SYFF, Kenya, Tanzania, Uganda USAID ILRG (Zambia)
Programs / private sector applying the approach not highlighted in chapter	Bukonzo Joint Cooperative Union Enhancing Customary Justice Systems in Mau Forest, Kenya Project LIFT, Ethiopia PROMAC II Rising Up!, DRC Rwanda's Rural Land Transformation Program SYFF, Kenya, Tanzania, Uganda USAID ILRG (Malawi, Zambia, PepsiCo) Women's Rights and Leadership Fora, Tanzania	Talking Books / LSRD II, Liberia	Bukonzo Joint Cooperative Union Support to the Government of Myanmar's Land to the Landless Program Talking Books / LSRD II, Liberia	Enhanced Homestead Food Production, Burkina Faso MCC Alatona Irrigation Project, Mali	Enhancing Customary Justice Systems in Mau Forest, Kenya Project Rwanda LAND USAID ILRG (Malawi, PepsiCo)

Appendix 2 Summaries of project and private sector examples

Program Name: Alatona Irrigation Project (MCC Mali)
Countries: Mali
Years: 2007–2012
Implementers: MCA-Mali; MCC; AECOM; ACDI/VOCA; G-Force; Nyeta Conseils, *Institute Géographique du Mali*; and the *Direction Regionale des Domaines et du Cadastre*.
Funders: Millennium Challenge Corporation
Funding amount: $253 000 000
Description: The Alatona Irrigation Project sought to convert nearly 5000 hectares of Sahel scrubland in the Segou region of Central Mali into high-value irrigated farmland. The project consisted of various components including irrigation infrastructure development, resettlement, land allocation, agricultural services, and financial services. To promote gender equal land allocation, the project followed a three-part strategy: titling of women's market gardens, encouraging allocation of 5-hectare farmlands to women, and gender-responsive awareness-raising to encourage joint titling. Of 4940 hectares of irrigated land allocated, 1648 were titled jointly and 174 were titled to women alone.

Company Name: Bukonzo Joint Coffee
Countries: Uganda
Description: Bukonzo Joint Coffee is a majority-women, member-owned coffee cooperative union in Uganda. The cooperative provides technical training, financial services, marketing assistance, and organic certification training to its members. They have undertaken and train other cooperatives in Women's Empowerment Mainstreaming and Networks and Gender Action Learning Systems approaches.
Links: https://bukonzojointcoffee.wordpress.com/author/bukonzojointcoffee/

Program Name: Enhanced-Homestead Food Production Programme (E-HFP)
Countries: Burkina Faso
Years: 2010–2012
Implementers: Helen Keller International
Funders: USAID
Description: E-HFP aimed to improve maternal and child health and nutrition through a nutrition-sensitive agriculture program. The project helped communities identify land that women could communally farm, provided women participants with agricultural inputs and small animals, and trained women participants in optimal agriculture, animal-raising, health, and nutrition

practices delivered via a behavior change communication strategy. E-HFP reached over 1500 women in the Eastern region of Burkina Faso.

Program Name: Enhancing Customary Justice Systems in the Mau Forest, Kenya (Kenya Justice Project)
Countries: Kenya
Years: 2010–2014
Implementers: Landesa
Funders: USAID
Funding amount: $490 000
Description: The project piloted an approach to improving women's access to customary justice, particularly related to land rights, to help implement new guarantees in the 2010 Constitution on gender-equal land rights. The project developed and delivered training to traditional leaders and strengthened their capacities in gender-responsive land governance and dispute resolution. It also supported women's and communities' learning understanding of the 2010 Constitution and National Land Policy, and women's land rights within them.
Links: Project site – https://www.land-links.org/project/enhancing-customary-justice-systems-in-the-mau-forest-kenya/#project-content; Implementation guide – https://www.land-links.org/tool-resource/kenya-justice-project-implementation-guide/.

Program Name: Feed the Future Ghana Agriculture and Natural Resource Management (AgNRM)
Countries: Ghana
Years: 2016-2021
Implementers: Winrock International, Nature Conservation Research Centre, Center for Conflict Transformation and Peace Studies
Funders: USAID
Funding amount: $24 900 000
Description: Using an integrated landscape approach that aligned development with conservation, AgNRM sought to integrate food security and water systems, highlighting their importance to nutrition and resilience. The project was implemented under four outcomes: (1) Increased Incomes from Natural Resource Products; (2) Improved Food and Nutritional Security; (3) Increased Farmer and Community Security/Access to Land and Natural Resources; and (4) Strengthened Environmental Stewardship. The project sought to secure land tenure rights for women and vulnerable groups through negotiation and respect for local norms, focusing on use and management rights. The project sought to enhance tenure security through land agreements, but also through approval and support from husbands and other male household members. The project trained 9620 individuals on improved agricultural practices, increased

incomes of 3,507 rural people, and supported the proposal, adoption or implementation of 30 land use plans.
Links: https://winrock.org/project/protecting-food-security-and-natural-resources-in-ghana/

Program Name: Feed the Future Agricultural Development and Value Chain Enhancement II (Advance II)
Countries: Ghana
Years: 2014–2018
Implementers: ACDI/VOCA, Technoserve, ACDEP, PAB Consult
Funders: USAID
Description: As part of USAID/Ghana's Feed the Future program, ADVANCE II sought to enhance the livelihoods of 113 000 smallholder farmers by increasing the productivity of rice, maize, and soy value chains. The project sought to improve women's rights to use and manage farmland, recognizing that limited land rights inhibited women's value chain participation. The project used existing outgrower business networks to influence, collaborated with other Feed the Future projects, and worked with local advocacy groups to facilitate farmland allocations to women in customary communities.
Links: https://www.land-links.org/2021/08/increasing-womens-access-to-land-through-public-private-partnerships-in-ghana/

Program Name: Homestead Land Grants
Countries: India
Implementers: Government of Odisha, Landesa, Action Aid, Vasundhara
Funders: The World Bank
Description: The homestead land grants program sought to settle landless persons on government lands, increase women's participation in land policy-making processes and institutions, and increase women's access to land dispute resolution services. The program worked with civil society, local government, and funders to reach ensure the program benefited vulnerable women, including women who were wives, widows, divorced, and single.
Links: The Voluntary Guidelines and The World Bank: Gender Equitable Land Governance in Odisha. https://pubdocs.worldbank.org/en/836001470747254790/WEB-Good-Practices-Brief-3-FINAL.pdf
Company Name: Kaleya Smallholders Company Ltd (KASCOL)
Countries: Zambia
Description: KASCOL is a private company with approximately 160 smallholder sugar farmers.
Links: https://kascol.co.zm/about; Wonani, CM, WS Mbuta, and AM Mkandawire. 2013. "The Gender and Equity Implications of Land-Related Investments on Land Access, Labour, and Income-Generating Opportunities. A Case Study

of Selected Agricultural Investments in Zambia." FAO. https://www.fao.org/3/aq536e/aq536e.pdf

Program Name: Land Investment for Transformation (LIFT)
Countries: Ethiopia
Years: 2013–2021
Implementers: Development Alternatives Incorporated (DAI), Nathan Associates Incorporated, NIRAS A/S
Funders: Foreign and Commonwealth Development Office (Government of the United Kingdom)
Funding amount: £72.7 million
Description: LIFT operated in the Amhara, Oromia, Tigray, and Southern Nations, Nationalities and Peoples Regions, where it legally registered and certified over 15.7 million land parcels in support of the Government of Ethiopia's second-level land certification process. LIFT combined land certification with several interventions aimed to help farmers to leverage their new certificates or invest in improved agricultural inputs and practices. Interventions included working with local microfinance institutions to pilot loans for smallholders linked to the certificates; creating a standard land rental contract and training licensed facilitators to verify information, facilitate connections and negotiations between landlords and tenants, and to register the transaction with local government; partnering with government extension and local suppliers to make agricultural inputs more accessible to smallholders; and provided policy support and review of regulations related to land tenure and allocation to the Government of Ethiopia.
Links: https://www.dai.com/our-work/projects/ethiopia-land-investment-transformation-lift

Program Name: Nijo Griha Nijo Bhumi (NGNB) / My Home My Land
Countries: India (West Bengal State)
Years: 2007–2015
Implementers: State Government of West Bengal, Landesa
Funders: State Government of West Bengal
Description: Through NGNB (and its predecessor Cultivation and Dwelling Plot Allotment (CDPA)) the Government of West Bengal and local governments to allocate and register small parcels (mostly between 0.10 and 0.16 hectares) to landless households, giving priority to female-headed households, to households with daughters and no sons, households that were landless for two or more generations, households dependent on agricultural labor or artisanal production. The program targeted households that received microplots to also receive a variety of inputs and government services for home agricultural production, including seeds and saplings, micro-irrigation, poultry and small

Published by Burleigh Dodds Science Publishing Limited, 2025.

livestock, extension and animal husbandry services, and microcredit. The program reached over 275 000 families that had been landless.
Links: https://wb.gov.in/government-schemes-details-nijogriha.aspx

Program Name: Promotion of Climate Smart Agriculture Project II (PROMAC II)
Countries: Mozambique
Years: 2017–2022
Implementers: NCBA/CLUSA
Funders: The Royal Norwegian Embassy in Maputo
Funding amount: US$14 500 000
Description: PROMAC II operated in Zambezia and Manica Provinces to reach over 31 500 smallholder farmers with the aim of promoting climate-smart agriculture in areas where soil had become degraded. The project combined a lead-farmer model of extension for training on climate-smart practices; community-based service providers to increase access to inputs and markets through aggregation points; promoted small-scale mechanization; functional adult literacy and numeracy education for smallholder farmers; land rights education and land certification for project farmers.
Links: https://ncbaclusa.coop/project/mozambique-conservation-agriculture-promotion-promac-ii/

Program Name: Rising Up! Promoting Congolese Women's Land Access and Preventing GBV in Eastern DRC
Countries: DRC
Years: 2020–2022
Implementers: Women for Women International (WfWI) and Innovation and Training for Development and Peace (IFDP)
Funders: USAID
Description: This project aimed to increase women's land rights while preventing GBV. WfWI and IFDP trained male community leaders, community Land Reflection Groups, and women Change Agents on land rights, gender-sensitive approaches to resolve land disputes, and how to identify and address GBV, including providing referrals to support services when necessary. The project integrated GBV prevention into IFDP's existing Customary Decentralized Land Management Resource Model. This project reached 2,859 project beneficiaries and trained 173 people on GBV prevention.
Links: https://genderandenvironment.org/rising-up-promoting-congolese-womens-land-access-and-preventing-gbv-in-eastern-drc/

Program Name: Rwanda LAND
Countries: Rwanda
Years: 2012–2017

Implementers: Chemonics, Radio Ishingiro
Funders: USAID
Funding amount: US$12 million
Description: Overall Rwanda LAND aimed to increase understanding of the legal framework around land among government, civil society, and citizens and improve Rwandan institutions' capacity to generate evidence for public sector and civil society use. The project developed and facilitated local implementation of a national land research agenda (which included a theme on gender and land), trained community mediators, increased cooperation among government actors to improved land dispute resolution services, and supported communications and public awareness campaigns to enhance citizens understanding their land rights and the land research findings.
Links: https://www.land-links.org/project/rwanda-land-project/

Program Name: Rwanda's Rural Land Transformation Program
Countries: Rwanda
Years: Approximately 2008–2013
Description: The Government of Rwanda's Land Tenure Regularization Program was a national effort to map and register all parcels of land in Rwanda, recorded in a national registry. It is considered to be one of the most cost-effective land certification and registration programs that also build administrative capacity to reduce the time from parcel identification, mapping, and adjudication, to issuing of certificates.
Links: https://www.afdb.org/fileadmin/uploads/afdb/Documents/Publications/anrc/AfDB_RwandaLandReform_FA.pdf; Ali, D.A., Deininger, K. and Goldstein, M. (2014) 'Environmental and gender impacts of land tenure regularization in Africa: Pilot evidence from Rwanda', *Journal of Development Economics*, 110, pp. 262–275. Available at: https://doi.org/10.1016/j.jdeveco.2013.12.009.

Program Name: USAID and Illovo Sugar Partnership Pilot
Countries: Mozambique
Years: 2016-2019
Implementers: Conducted under the USAID Evaluation, Research, and Communications Task Order and then under USAID ILRG.
Funders: USAID, Illovo Sugar Company
Description: The pilot included land rights sensitization, participatory mapping and parcel registration, and a local cooperative issuing eligible participants certificates to prove ac USAID, Illovo Sugar Company acquisition of Land Use and Benefit Rights (DUATs). The project took place in an sugarcane growing area.
Links: Persha, L. and N. Taha. (2022). Follow-on Performance Evaluation of the Responsible Land-Based Investment Pilot in Mozambique: Final Report.

Washington, DC: USAID Communications, Evidence and Learning (CEL) Project. https://pdf.usaid.gov/pdf_docs/PA0211HT.pdf

Program Name: Securing Your Family's Future (SYFF)
Countries: Kenya, Tanzania, Uganda
Years: 2016-2021
Implementers: Kenya Legal and Ethical Issues Network on HIV & AIDS (KELIN), the Pastoral Women's Council in Tanzania, the Uganda Community Based Association for Women and Children's Welfare (UCOBAC), and Lori Rolleri Consulting
Funders: Wellspring Philanthropic Foundation
Description: The goal of the course was to increase women's access to, use, ownership, and control of, as well as decision-making about land. The course aimed to achieve this by promoting behaviors that strengthen women's land rights, which required shifting key determinants of the behaviors, including social norms, knowledge, attitudes, self-efficacy, and future intentions. The specific behaviors targeted by the course for men include: (1) fathers allocate land to daughters (regardless of marital status) as they do for sons; (2) husbands include wives as joint owners and/or managers of land; (3) husbands and fathers bequeath wives and daughters land in their wills (inheritance); and (4) husbands seek women's consent before selling family land or making other major decisions about land. Across the three countries, 412 men were trained (192 in Kenya, 100 in Tanzania, and 120 in Uganda). The course included a series of 6 2-hour sessions over 3–6 weeks using a variety of interactive teaching methods.

Program Name: Support to the Government of Myanmar's Land to the Landless Program
Countries: Myanmar
Years: 2017–2021
Implementers: Landesa
Funders: King Philanthropies
Description: In close collaboration with the government of the Republic of the Union of Myanmar, in addition to other King Philanthropies grantees, Landesa worked to reduce landlessness and bolster agricultural productivity through increasing land access and agricultural extension and other services for landless families and other smallholders; strengthen land tenure security; and support legislation and policy development. Efforts provided Ministries, parliamentary committees, and regional governments with legal and policy analysis and training and learning opportunities on issues of land access, customary tenure, gender integration and smallholder support services; provided technical assistance to support government allocation of public land

to the landless, return of land to smallholders, and protection of customary use rights in accordance with Myanmar laws and policies; and gathered accurate and gendered information regarding the prevalence of landlessness and experiences of landless people, available state land, and other critical matters to support design and modification of government programs.
Links: https://www.landesa.org/what-we-do/asia/myanmar/

Program Name: Talking Books, part of the larger Land Rights for Sustainable Development II Project (LRSD II)
Countries: Liberia
Years: 2021–2023
Implementers: Landesa, Amplio, Development Education Network – Liberia
Funders: King Philanthropies
Funding amount: US$3 million (LRSD II Project as a whole)
Description: The LRSD II Project focuses on implementing the 2018 Land Rights Law by supporting communities, policymakers, government officials, and civil society organizations. The Talking Books activity is part of the LRSD II workstream of legal awareness-raising and capacity-strengthening of community-based organizations and civil society. This workstream also included radio programming and technical and financial support to local organizations and working groups with emphases on women's and youths' land rights. The Talking Books also support LRSD II's workstream to accompany communities throughout the process of community land formalization. The process included awareness-raising, supporting community self-identification, drafting by-laws, mapping and harmonizing boundaries, and training community land governance committee members. The Talking Books initiative in the LRSD II project seeks to raise awareness at scale among customary communities on their land rights within the 2018 Land Rights Law. Talking Books are battery-powered audio devices that can play messages on demand. Each month, 400-600 participants use the Talking Books to learn about various topics along with their household members, after which Talking Books are rotated to new households. Community animators visit households in use of the Talking Books to provide additional technological support, and town hall meetings are used to reinforce and review recipients' engagement with the Talking Book messages. The activity seeks that 60% of Talking Book recipients are female, and it has a target number of beneficiaries of 50 400.
Links: https://www.landesa.org/what-we-do/sub-saharan-africa/liberia/

Program Name: USAID ILRG Malawi
Countries: Malawi
Years: 2021–2023
Implementers: Tetra Tech; Government of Malawi

Funders: USAID
Description: 'The Government of Malawi enacted a series of land laws in 2016, including the Customary Land Act that allows customary land holders to formalize ownership by registering their parcels. Building on this new legal framework, the United States Agency for International Development (USAID) Integrated Land and Resource Governance (ILRG) program partnered with the Government of Malawi to register all customary land within a traditional land management area (TLMA). The program applied a strong gender equality and social inclusion (GESI) lens to ensure that women, youth, the elderly, people with disabilities, and other marginalized groups were included throughout the process.' The ILRG project successfully documented 9272 parcels. 42 620 people had their customary land rights documented under the project, including 17 536 landowners (47% women) and 25 084 beneficiaries (50% women/girls).
Links: Lessons Learned: Integrating Gender Equality and Social Inclusion into Customary Land Documentation in Malawi – https://www.land-links.org/wp-content/uploads/2023/08/ILRG-Malawi-GESI-lessons-learned_June_2023-508.pdf;
ILRG MALAWI FINAL REPORT – https://www.land-links.org/wp-content/uploads/2023/10/ILRG-Malawi-Final-Report-508.pdf

Program Name: USAID ILRG – Women's Economic Empowerment in PepsiCo's Potato Supply Chain in West Bengal
Countries: India
Years: 2019–2023
Implementers: TetraTech, Landesa, PepsiCo
Funders: USAID, PepsiCo
Description: This partnership under the USAID ILRG Project aimed to test the business case for engaging women farmers in PepsiCo's potato supply chain. In eleven communities, the project provided agronomy training and to women farmers and established women-led demonstration farms; provided land literacy training to and held dialogues on gender norms with women and men community members; supported women's groups to collectively rent and farmland; and provided introductory training to PepsiCo staff on gender equality and women's empowerment in general and in their supply chain in particular.
Links: https://www.land-links.org/wp-content/uploads/2023/10/ILRG-India-Final-Business-Case-Brief.pdf

Program Name: USAID ILRG Zambia
Countries: Zambia
Years: 2018–2023

Implementers: Tetra Tech ARD
Funders: USAID
Funding amount: US$3.8 million
Description: 'The ILRG project builds the capacity of local organizations, district and national government, traditional leaders, and communities to document land and resource rights and to use this data to better inform land use planning and development decisions. The project supports a decentralized natural resources management model and encourages active engagement between government authorities and customary leaders on wildlife, forestry, and land. ILRG integrates best practices into land and natural resource policy and legislation through consultative processes. ILRG also works with local partners to pilot implementation of progressive legislation and regulations that promote community-based management of natural resources.' The project supported improved management of over 783 000 hectares of biologically significant land and documentation of land rights for more than 30 000 parcels of land, including 155 000 rights holders, of which 50% are women.
Links: https://www.usaid.gov/zambia/fact-sheet/integrated-land-and-resource-governance

Program Name: Women's Rights and Leadership Forums (WRLFs)
Countries: Tanzania
Years: The first WRLFs were established in 2009 and the initiative has been growing and ongoing among pastoralist communities in Northern Tanzania since.
Implementers: Pastoral Women's Council (PWC), Ujamaa Community Resource Team (UCRT), local communities themselves
Description: WRLFs are established with the aim of encouraging women to take a more public and leading role in securing resources, especially land, for their livelihoods. UCRT and PWC provide legal training and technical support to WRLFs. WRLFs work with both women and men to promote changes in existing norms around the treatment of women's rights. They also typically raise the funds needed to take action and participate in public decision-making. Over 350 women have received individual land titles through WRLFs mobilization and PWC support to date.
Links: https://pastoralwomenscouncil.org/rights-and-leadership/

Published by Burleigh Dodds Science Publishing Limited, 2025.

Chapter 18

Supporting women smallholder farmers: the role of the public and private sector

Margaret Alston, University of Newcastle and Monash University, Australia; and Zhaoen Pan, Monash University, Australia

1 Introduction

This chapter focuses on the role of the public and private sectors in supporting women on smallholder farms. For the purposes of this chapter, I am defining smallholder women farmers to include those working on small-acre farms as well as those who have established niche industries on larger farms. Context is significant and this chapter is written from an Australian perspective and is very much shaped by the vast continent supporting a range of farms in often hostile environments. In Australia, bigger farms are the norm, and agricultural developments are shaped by a 'get big or get out' mentality (Newsome 2020) that fosters farm production units expanding in order to enhance production and remain solvent (Cassidy et al. 2019; Becot and Inwood 2022). Agricultural production in Australia is dominated by a neoliberal mantra focused on economic growth that precludes significant attention to the people engaged in production. This market focus leads to an overemphasis on the economic drivers of production and a consequent lack of attention to the social structures that shape the industry. Not surprisingly, there is a lack of focus on smallholder farms and an ignorance of the needs of producers by governments and other

http://dx.doi.org/10.19103/AS.2024.0148.22

institutions and of the types of policies and structures that would assist those who are doing the work of farming.

While agricultural production remains a strength of the Australian economy contributing significantly to the Gross Domestic Product (GDP), the people who produce the goods are not only rendered invisible but are also largely unsupported. Elsewhere (Alston et al. under review), we have argued that this ignorance leads to a curious lack of attention to the social structures inherent in family farming and to the types of policies and supports offered by governments and institutions that might assist these families in their work roles. When we add gender to our analysis of productivist agriculture, the invisibility of women is more extreme, and the lack of support more stark. It is therefore arguable that the support for women on smallholder farms is even more tenuous (Zakaria 2017).

In Chapter 3, we introduced data on the significance of farming families in Australia noting particularly that the mitigation efforts adopted on Australian farms have had two significant outcomes. These include the introduction of technological advances that have reduced the need for strength as a dominant factor and made agriculture more accessible to women (Hay and Pearce 2014); and the adoption of mitigation efforts by farming families that address the increasingly worrying extreme climate events now evident in Australia. These catastrophic disasters are causing a major reshaping and expansion of the agricultural industries and of the agricultural workforce in Australia. Against this background, it is little wonder that the efforts of women on smallholder farms are lost in the background noise of expansionist agriculture. Yet there is increasing evidence of women moving into small acre and/or niche industry agriculture in Australia (Ball 2020, SBS News 2019), and the lack of institutional support for their efforts has a significant impact on their success.

Drawing on our 2021 study of Australian women engaged in agriculture who have been recognized as leaders in their industry, in this chapter we discuss the efforts of women on smallholder farms and the role of the public and private sector institutions that might assist their development. We note that while gender inequalities dominate agriculture in Australia (see, e.g. Alston et al. 2017), this further cements women's disempowered position and highlights the lack of attention given to smallholder farms in general and women smallholder farmers in particular. We begin by discussing women's precarious attachment to agriculture as a result of their lack of ownership of farms before discussing women who either work on smallholder farms or are producing a niche product on larger farms. Finally, we discuss the way public and private sector organizations can assist small these women.

2 Understanding the role of women in agriculture: the 'invisible farmer' project

As part of a much larger project on women in agriculture, the 'invisible farmers' project, research was conducted in 2021 with women who had been recognized as leaders in their agricultural pursuits. The research participants were drawn from a pool of Australian women who had either been selected for leadership development or had received national awards in the previous few years. Sixteen women were interviewed at length about their lives in relation to agriculture, the factors that shaped their success, and the perceived barriers to their productive capacity and to women in general in the context of agriculture. Of the 16, five had established niche enterprises on their farms.

Our sample of 16 includes 12 women who are actively engaged in farming, four are co-owners of farms, three bought their farms together with their husbands, eight are partially engaged in agriculture while also working off the farm, two have bought their own farms, two continue to work on farms owned by their parents. Six of the participants are on high-level boards, two hold senior positions in agricultural institutions, five have won the Rural Woman of the Year award and one has won the Farmer of the Year award.

Our participants discussed the barriers to, and the drivers of, their success and the factors that would assist them to work more effectively. Several had formal leadership positions including at the state, industry and commodity levels. In broad-ranging interviews, the participants were asked about their work, their engagement in agriculture and their leadership. Drawing from these interviews provides insights into the barriers for women farmers, the constraints on their achievements in general, and the factors they identify that restrict women's engagement. While there is limited information on the numbers of women on smallholder farms in Australia, this sample allows insights into the issues facing women in agriculture and, in particular, on small holder farms.

3 Women on smallholder farms in Australia

In Chapter 3, we noted the factors that shape the gendering of agriculture. Globally, despite women undertaking between 40% and 60% of agricultural labour, ownership of land by women remains precariously low. In Australia, approximately 10% of farms are owned by women reducing their status and power in an industry dominated by men. For women to engage in agriculture requires the stamina needed to overcome this inherent bias. As noted in Chapter 3, despite being denied inheritance rights to farms, women now occupy 55% of undergraduate positions in Australian universities (Sheridan et al., 2023) including in university agricultural programs and agricultural majors (Pratley 2017) indicating quite clearly that women are eager to work in agriculture and have the skills and knowledge to do this effectively. Yet the barriers to their entry remain and are most heavily policed by their own families, their communities, and their industry bodies. Why is this important in the context of this chapter? It is critical because the barriers to women's engagement on smallholder farms exist long before they begin farming.

The dominance of family farming in Australian agriculture implies a traditional family structure including an expectation that it is women who do the caring and household tasks. Yet, as noted in Chapter 3 and elsewhere (see, e.g. Alston et al. 2017 and Alston et al. 2018), these expectations on women have escalated to include significant on-farm work and off-farm income generation. On traditional family-based farms, there is now a widespread expectation that women will work on farms doing physical labour facilitated by technological advances, and off the farm for much needed income. These expectations without the power that goes with public acknowledgement create a sense amongst farming women that they are outsiders not only on their farms but for some, also in their communities (see, e.g. Alston et al. 2018). Further, the idealized family farm image of the male farmer remains and has several consequences including the structural exclusion of women from farming (Cassidy et al. 2019); the dominance of male farmers in the language and imagery of farming; the lack of inheritance by daughters in succession planning; the lack of ownership of farms by women; the typecasting of women as 'farmers' wives'; and greater expectations on women to adopt gendered roles including caring and household work.

3.1 Historical neglect

Historically, women were only involved in agriculture largely through marriage and, despite the active engagement in farm and off-farm work women remain secondary players in the public imagination. Participant 8 for example noted that, historically at least, roles were very much typecast by gender.

> *whilst Mum did all the housekeeping and Dad did the majority of physical farming there was always a sense that that was their roles, and there wasn't – I don't know, it wasn't seen as being neither positive or negative, that's just how it was. And having said that, Mum still was very active in the business in keeping the books, and always had an opinion about what was going on.*

Participant 10, in discussing the role, her grandmother played on her farm noted:

> *Now my grandma would never have described herself as a farmer she would have called herself a farmer's wife, that's what she was. But of course, when she was on the farm, the labour in preparing meals, washing with the old copper, doing the chooks,*

the veggies, the fruit trees, yeah the physical labour associated with running a house was so much greater than it is now.

Our participants agreed that the historical typecasting of women as absent from agriculture despite their work, denied the very complex interaction of family, farm business, enterprises and community engagement undertaken by women and men on farms. The awakening consciousness of women's active exclusion was noted by participants and is significant in explaining the absence of any acknowledgement of women on smallholder farms or of women developing niche products on larger farms. Yet, on smallholder blocks and on larger farms where they are part of family enterprises, our sample indicates women are establishing niche industries in a diverse range of products. Amongst our 16 recognized rural leaders, these enterprises included growing flowers, beekeeping, cheese making, the production of niche dairy products, growing citrus and vegetables, winemaking and egg production.

3.2 Inheritance practices and gender disadvantage

Critical to understanding the historical neglect and disregard of women are the inheritance practices adopted in farm family succession planning. Of our sample of 16, despite more than half (9) being from a farm family background and having a love of agriculture, none had inherited their family farms. This was a source of much dismay amongst participants. One of the responses often given to justify women's exclusion from agricultural inheritance is that girls are not interested, and they should therefore seek a career. Participant 2 noted the bias against girls inheriting the family farm.

As a child there was always the unspoken expectation that my brother would take over the farm even though us three kids were very active on the farm. … and it was not until I lived [overseas] that I thought 'Why can't I be involved in agriculture?' and that was a seminal moment for me and that's when I came back to Australia and got heavily involved in agriculture.'

Others noted they had entered their farm partnership through marriage leading to what Participant 7 describes as the precariousness of women's engagement in family farming:

A lot of women are the daughters-in-law … so they have a really tricky place.

3.3 Link between land ownership and niche market production

When we examine our sample of women, it is illuminating to note that, of those who are actively engaged in hands-on farming, all have a stake in property ownership either through a partnership agreement with their spouse or parent or having bought their own farm. While this may take various forms, there is an evident link between ownership and active engagement in production. For example, Participant 5 noted she is in partnership with her husband and that they work as a team not only in their dairy but also in their niche industry – cheese making.

> *I am the co-owner, co-founder, and I do milking and all of the – we split the jobs down the middle, literally; I do cheese making, milking, work on the tractor, fixing fences, everything that the farm entails. So, Michael and I probably have our areas of expertise: he's got a little bit more technical knowledge than I do in the cheesery, and I've got a bit more technical knowledge, than he does, with the sheep. So, we work well together, we have areas of expertise and we'll be able to make the final call if there's a difference of opinion.*

Participant 5 goes on to note that they are developing a cheese and wine bar concept on their farm.

> *I anticipate that I will be setting it (cheese and wine bar) up and spending a fair bit of time there, and [my husband] will be taking on more of the cheesemaking.*

Meanwhile, participants 15 and 16, who both own their properties, discuss their deep engagement with their farm and their production. Of particular interest is that both participants 15 and 16 note the 'support' of their husbands making it clear that while they are the dominant owner-operator, their partners are the support crew. Participant 16 noted:

> *Although I am, perhaps, the driven one, I can't do it without him as my cheer leader and doing the day to day, and mopping up behind me... he's putting everything together behind me.*

Participant 15 also noted that while she is the owner-operator of a vineyard, her partner is her support crew.

> *He's obviously there for support in different ways during vintage. He'll drive the forklift and help me with vintage, so we have an extra person. He can be on the tractor if we're running short of staff, He helps me with working out finances and just a second person to tilt ideals back and forth with to understand whether it's profitable, whether it should be done. but he doesn't do very much in terms of the business itself.*

These examples demonstrate how profoundly ownership influences who is the driver in a farm enterprise and how significant land ownership is for women in gaining the confidence to establish a farm enterprise as a principal owner.

Participants 15 and 16 both own and operate their farm business and both demonstrate extraordinary capacity and success in their business undertaking.

4 Opportunities for female empowerment on smallholder farms

As noted, for women on smallholder farms, there is a strong link between ownership and a sense of empowerment very evident in the responses of the women in our research. By having their own enterprise, they can move away from gendered arrangements and embrace the empowerment inherent in owning and running their own business (see also Sheridan and Newsome 2011). Sheridan and Newsome (2011) note that smallholder farming enables a more amenable balance between work and care responsibilities and gives women more power over their time and management. This is supported by our research which indicates that women running their own small-scale enterprises are building their own businesses at the same time as they are balancing their family and farming commitments.

One of the evident factors shaping the success of the women in our study is their ability to be creative and responsive in the marketing of their product. For example, Participant 16 noted how she responded to a flood disaster that devastated infrastructure and road access. Commenting on the agility of smallholder farms, and the need to be creative and responsive to changed conditions, she noted:

> *... at the time we were very focused on selling to the public, and when the bridge got knocked out the road got closed for six months, and so we could not anymore sell to the public at that time. So, we pivoted towards wholesale and trade sales, and that was fine, that was a learning. But obviously, when you don't learn properly, you'll either make the same mistake or something else will happen. So, in between the flooding and COVID we had bushfires, so in 2020 we lost all of our crop to smoke taint, so we dropped everything to the ground. It was a good learning in a way. Obviously, farming is that. I think, in a 10-year period, you'll have one ideal year, and then three or four good ones, and then a couple of disasters, and it all evens out. We're very fortunate, Margaret, because we did this off the back of our own savings, so we don't have the debt that others would have.*

Her innovative response indicates how nimble smallholder farmers must be to be successful.

Participant 5 endorsed this position noting how smallholder farms can react quickly.

> *I think, first and foremost, farmers need to have really strong profitable businesses in order to be able to weather those storms of the economy and the markets, and climate. So, there will always be a need for the bigger farms, yes, but I think the middle size and smaller farms, the dynamic nature of those businesses and their ability to shift and*

move, and be flexible, and react quickly to what they need to react to, is what we really need to focus on, I believe, in agriculture. I suppose that's where I see my skill base.

Participant 16 also commented on the way she adopted innovative marketing techniques during COVID.

I put together a little pack called my lockdown pack, and I did a video, and a friend of ours who does Facebook marketing said, "Can I help you?" The video did well on our social media and so he took that video and made an advertisement out of it and put it onto Facebook. We've never had so many sales in one month, it was just amazing. But then, off the back of that, Dan Murphy's approached as and said, "What can we do to help?" so they bought around about 50 or 60% of what we would have otherwise sold to wholesale, just that particular year, and that was amazing for us

Participant 15 also noted that being a female winemaker has its advantages.

People want wine because you're a female wine maker, there are so many reasons why people buy wine...

5 Addressing the barriers for women on smallholder farms and in niche industries in agriculture

Despite their ability to react quickly, there are a number of constraints against women in agriculture generally that affect women on smallholder farms. These include

- inheritance practices;
- the lack of ownership of land and the resources associated with agricultural production;
- general societal attitudes associated with men as farmers and women as 'helpers';
- attitudes of banks and other lending authorities;
- government policies associated not only with farm support but general supports such as child care;
- access to agricultural supports, loans, input subsidies, and crisis grants;
- inadequate Information and Communications Technology (ICT) services in rural areas;
- the failure of representative bodies to address the poor representation of women in general and smallholder farmers in particular; and
- a lack of attention by industry bodies to the complex gendered interactions amongst farming families.

There is little doubt that the women engaged in our research demonstrate extraordinary success. Those undertaking niche market development demonstrate initiative and are dynamic and creative in responding to the significant challenges they have faced. Given the rough ride women have in entering agriculture from a natal family, a married-in family, or as an individual buying their own small block, how can we address the factors that shape the inherent bias against women achieving in agriculture? What can we learn from these leading women about the barriers that restrict women from wanting to engage in smallholder farming or niche product development on larger farms? What are the factors that cripple women's expectations and/or weigh them down under the expectations of their assigned roles and what are the identified needs for women? The following section will attempt to draw out these factors noting that there is a complexity of responses. What is clear is that there are a number of ways that the public and private sectors can do better to address these needs.

Critical to understanding and addressing these factors is attention to the social systems that shape agriculture. Australian farm families operate in a political, economic and social system that is firmly focused on the economic and technological needs of farm businesses with little if any attention to the social needs. Writing from a US perspective, Becot and Inwood (2022) note the lack of attention to the social needs of farm families. This holds true for Australian families and has led to a decline in rural infrastructure and a lack of essential services such as child and aged care, internet and infrastructure services, and, in remote areas, a lack of access to birthing services (Dietsch et al. 2009). An appraisal of the social needs of farm families, and of women farmers in particular, is long overdue.

There are numerous factors that shape and unnecessarily restrict women working on farms across Australia. One of the most significant barriers for women is the lack of recognition they have in the agricultural industry. If we are to support women on smallholder farms, gaining entry into agriculture remains the most critical barrier for women. In the case of inheritance practices, easing access for women is essential. Convincing families to canvas the interests of all their potential inheritors remains a work in progress. While this requires attention to family dynamics, it is also necessary to examine the way agriculture is represented by industry bodies, the way governments support those engaged in agriculture, the way the media presents 'farmers' and the dominant voices that represent the industry. Focusing attention not only on the attitudes of families in relation to inheritance practice, but also addressing the imagery of farming that reinforces the dominant male farmer, and focusing on the practices of institutions such as banks, farming organizations and industry bodies and governments would provide a good start. Critically, government policies that fail to attend to the requirements of all members of farm families in general and women on smallholder farms in particular require urgent attention. For example, addressing the social structures that shape women's lives in rural areas requires governments to ensure that women on smallholder farms and in niche industry development have access to input subsidies (see, e.g. Smale and Theriault 2022) and other government support schemes available to farmers. Governments should also examine very basic factors such as the provision of adequate services including child and aged care services, adequate internet, infrastructure and transport services.

Participant 8 noted that basic services such as a lack of childcare severely restrict women's productive capacity.

> *but friends that do have kids it still falls mostly to the female to do that and that's really restrictive, obviously, on their ability to be productive outside when they've got the shorter hours.*

Our research participants raised a number of additional barriers that restrict women's engagement in agriculture. These include the failure of industry bodies to focus on issues of critical importance to women. For example, Participant 6 noted the high levels of sexual harassment that she and other women have experienced in the course of their day-to-day lives as a major barrier for women in the industry.

> *The social context still allows the level of sexual harassment that we have of women … the agricultural sector … that still allows it to continue unchallenged … to me is a huge indicator of how any intrinsic shift between the genders hasn't happened, that really fundamental shift where it would never enter your mind to actually do or say the things that you're doing and saying, or would never enter your mind to actually to let someone get away with doing and saying that. … to me its almost like the canary in the mineshaft – it's the fundamental thing that keeps indicating that the notion of gender and how it plays out – we haven't gotten to where we need to be … deep under the surface is that fundamental gender stuff that seems – it's just there. It's so big that people don't want to even look.*

Participant 2 noted the need for women to have adequate skills training.

> *But how is it that we can develop the skills to be more not just innovative, and adaption is part of innovation, but more adaptive, and I feel as though we have to have a shift around how it is that we relate to change. I think the pandemic has really thrust us forward to be uncomfortable with uncertainty, and I think if we don't develop those skills enough I have huge concern about mental health broadly, but especially in the regions because mental health provision is so terrible, it's terrible.*

Participant 3 noted the attitudes of banks towards women as a significant constraint. Her bank would not lend to her without a male guarantor!

> *So, I rented a site, I owned for the first time nothing in my life, I spent all my savings on plants, I rented a computer, I just had to go an asset purchase a ute; the bank wouldn't give me any funding at all because I was female and single, in a business, unless I had a male guarantor.*

6 Conclusion

There are a number of barriers facing women on smallholder farms. These include the overall context including country location and in-country infrastructure. It includes cultural factors such as inheritance practices that restrict women's entry to farming by denying the claims of daughters and it includes community perceptions relating to who are legitimate farmers. There are socioeconomic barriers relating to the profitability of small-scale enterprises and there are barriers that are shaped by gender inequalities and gendered perceptions of work roles on farms, particularly when there are care roles. There are further restrictions shaped by the lack of government and private sector service infrastructure including childcare facilities and health care – largely the responsibility of governments. There are barriers relating to lending facilities that view women as inadequate borrowers and there are barriers in the way industry bodies shape their agenda and often fail to address the issues of significant concern to women. Overcoming these barriers requires attention from government, private sector, media, community and families. This is not an easy task. However, on hearing the stories of the women who contributed to the research presented here, what is most surprising is not only the success of these women but also the enthusiasm they bring to smallholder farming. These women not only provide exemplars but are indicating through their efforts that women can be exceptional farmers when given the chance to excel.

7 References

Alston, Margaret, et al. (2017). Gender relations, livelihood strategies, water policies and structural adjustment in the Australian dairy industry. *Sociologia Ruralis*. https://doi.org/10.1111/soru.12164

Alston, Margaret, et al. (2018). Limits to adaptation: reducing irrigation water in the Murray-Darling Basin dairy communities. *Journal of Rural Studies*. 58, 93–102.

Alston, Margaret, et al. (under review). 'Out here no one can hear you' – Addressing the Social Needs of the Farm Household. Paper submitted to Agriculture and Human Values, January 2024.

Ball, Jennifer A. (2020). Women farmers in developed countries. *Agriculture and Human Values*. 37(1), 147–160.

Becot, Florence A. and Inwood, Shoshanah M. (2022). Examining access to health insurance and health care along the life course to shed light on interactions between farm households' social needs, social policy and the farm business. *Sociologia Ruralis*. 62, 485–508.

Cassidy, Anne, et al. (2019). Generational transmission of smallholder farms in late capitalism. *Canadian Journal of Development Studies / Revue canadienne d'études du développement*, 40(2), 220–237. https://doi.org/10.1080/02255189.2019.1592744

Dietsch, E., et al. (2009). *'Luckily we had a torch': contemporary birthing experiences of women living in rural and remote NSW*. School of Nursing and Midwifery, Faculty of Science, Charles Sturt University, Wagga Wagga.

Hay, R. and Pearce, P. (2014). Technology adoption by rural women in Queensland, Australia: women driving technology from the homestead for the paddock. *Journal of Rural Studies*, 36, 318–327. https://doi.org/10.1016/j.jrurstud.2014.10.002

Newsome, Lucie (2020). Beyond 'get big or get out': female farmers' responses to the cost-price squeeze of Australian agriculture. *Journal of Rural Studies*, 79, 57–64.

Pratley, Jim (2017). *Agriculture: from macho to gender balance*. Proceedings of the 18th Australian Society of agronomy conference. 24th–28th September, 2017. Ballarat. https://www.agronomyaustraliaproceedings.org/

SBS News (2019). Australian farmers turn to niche markets 6 March. Data accessed on 10 December from https://www.sbs.com.au/news/article/australian-farmers-turn-to-niche-markets/wfxhrcw97

Sheridan, A., et al. (2011). Making visible the 'space of betweenness': understanding women's limited access to leadership in regional Australia. Gender, Place & Culture, 18(6), 732–748. https://doi.org/10.1080/0966369X.2011.617909

Sheridan, Alison, et al. (2023). Changing scripts: gender, family farm succession and increasing farm values in Australia. *Journal of Rural Studies*, 100, 103024.

Smale, M. and Thériault, V. (2022). Input subsidy effects on crops grown by smallholder farm women: the example of cowpea in Mali. *Oxford Development Studies*, 50(3), 244–258. https://doi.org/10.1080/13600818.2021.2008892

Zakaria, Hudu (2017). The drivers of women farmers' participation in cash crop production: the case of women smallholder farmers in Northern Ghana. *The Journal of Agricultural Education and Extension*, 23(2), 141–158. https://doi.org/10.1080/1389224X.2016.1259115

Part 5

Case studies

Chapter 19

Women smallholder farmers and water resource management

Sandra Nereida Barrera Galvis, University of Bogotá Jorge Tadeo Lozano, Colombia; Stephanie Buechler, Pennsylvania State University, USA; and Magnolia Longo, University of Bogotá Jorge Tadeo Lozano, Colombia

1 Introduction

Women smallholder farmers globally are confronting social, economic and environmental challenges related to water resource management with impacts on access to energy, drinking water and food. Furthermore, it is common for women in rural regions to be subjected to physical, economic and slow violence. This study was carried out in a region extending north of the Andean foothills to the Caribbean Sea in South America. The study sites are located along high, middle and low elevation gradients within the South, Center and North of Colombia's Magdalena River watershed. We utilize a water-energy-food (WEF) nexus framework with a gender lens to analyze women smallholders in the fields and the home in three rural communities. The WEF nexus framework helps reveal integral connections between these elements to show how women gain access to these resources to produce food and reproduce lives. They must use

http://dx.doi.org/10.19103/AS.2024.0148.25

a variety of different sources of water and food, energy, each of which requires different management techniques.

Because water for domestic and agricultural use is so intertwined, we do not make false distinctions between the two, but rather address their intersections. We situate gender dimensions of WEF nexus issues within the context of slow violence that includes land grabs, water infrastructure that should be publicly provided but instead is community-organized and financed and the contamination and overuse of water and land resources by large private enterprises. Women also experience slow violence in their agricultural labor with lower pay than men. They experience more physical violence due to this vulnerable position of lower remuneration in their work, in addition to a lack of control over household income that is in the hands of men. We also examine gendered impacts of climate change on rural women's lives and livelihoods.

The study reveals issues common globally, i.e. uneven access between men and women and between large and smallholders to land and water. It also presents a case where most women and men do not have land of their own for agricultural use. This is set in a context of a country with one of the highest socioeconomic inequality rates globally (World Bank, 2023).

This inequality in landholding has deep repercussions for farming. Women in these communities in Colombia especially face unequal access to water, land and other means of production partly because of patriarchal structures that mostly precludes women from gaining access even to community managed land and partly due to land tenancy (and consequently, water resources on that land) in areas where land is in the hands of large landowners or the community. Like many women globally, they farm without legal access to the land. Similarly, women engage in multiple livelihood activities that include growing crops for home consumption and sometimes for small-scale sale or barter, working on plantations and working in activities like fishing. Women receive no pay for their work on land that their male household members have access to and receive lower daily pay than men for plantation work. This is set in a national context in which 42.5% of the population lives below the nationally determined poverty line, among the highest globally (World Bank, 2023: 6). In addition to food, land and water challenges, energy supply is insufficient and inefficient in the study sites. Agricultural production requires energy for irrigation. Women's work inside the home is also affected by this lack of access to a continuous, affordable energy supply.

2 The water-energy-food (WEF) gender nexus in a context of violence

The WEF nexus is a framework utilized to illuminate linkages between resources once considered separately, that is, water, energy and food

(Brouwer, 2022). WEF analyses can be useful in policy and program design. A nexus signifies a dynamic process in which a change in one of these resources will exert a change in the other two resources. For example, demand or supply pressures on one resource, such as water, will impact energy and food. The WEF nexus was originally focused on physical aspects of these resources; it was later expanded to include social, economic, environmental, policy and programmatic issues. In parallel, WEF nexus analysis began to be used as a tool to achieve equity and sustainability. WEF nexus became a useful conceptual approach in understanding water, energy and food security, which in turn revealed that security is not possible to achieve without improving equity and environmental, social and economic sustainability within the nexus. When examined from the perspective of equity, the WEF nexus can also reveal vulnerabilities within the nexus and areas of policy and/or program intervention.

Recent studies have reviewed methodologies that can be used to help integrate a WEF perspective into research and policy (Albrecht et al., 2018; Kurian, 2017; Yung et al., 2019). There is also specific literature on integrating WEF into research and policy for Latin America and the Caribbean (Mahlknecht et al., 2020; Santos Da Silva et al, 2019); on South and East Asia (Buechler et al., 2016; Nepal et al., 2021; Mukherji, 2022; Katyaini et al., 2021); the Middle East (Zarei, 2020); Africa (Botai et al., 2021; Jacobson & Pekarcik, 2022) and Europe (Voelker et al., 2022). Recent WEF literature has also examined actors' perceptions related to how WEF nexus resources shape livelihoods (Wolde et al., 2020), including in Colombia (Barrera et al., 2024).

The WEF nexus framework is a tool for enhancing the efforts of governments in conflict-affected countries that are interested in protecting and managing WEF resources. It is essential to situate WEF nexus analysis within the geopolitical and social context of the study. Our study locations in Colombia are characterized by significant acts of economic and physical violence (Boelens et al., 2021) and by what Nixon (2011) has termed "slow violence." Women in the region are at high risk of economic and physical risk for several reasons, including a patriarchal social structure with men controlling critical livelihood resources. Women are also concentrated in the poorly remunerated informal labor market with little to no protections and in jobs that pay women less, such as in agricultural day labor. The implementation of the WEF nexus framework in countries like Colombia, which has experienced decades of armed conflict, presents several challenges, including significant political, social and economic instability, a lack of infrastructure and governance systems to support integrated resource management, as well as high levels of corruption. Additionally, relevant data and information to support planning and decision-making processes in the face of climate change are scarce for many regions of the country (Abbot et al., 2017).

Therefore, analyzing and implementing various programs under the WEF nexus framework in conflict-affected countries is crucial to addressing the underlying causes of conflict, promoting peace, stability and building resilience in affected communities. The framework offers an opportunity to integrate sustainable management of biodiversity and its ecosystem services with reconciliation, peacebuilding and development measures, all aimed at resolving conflicts and building a sustainable and prosperous future for present and future generations.

It is well known that armed conflict, corruption and the actions of unlawful groups in any country can have marked negative repercussions on the unsustainable use of biodiversity and ecosystems, putting water, energy, and food security at risk (Karatayev et al., 2017). This is due to the destruction of infrastructure, illegal land possession, forced displacement, and the disruption of trade and supply chains, directly affecting the social aspects and well-being of local communities. Zarei (2020) points out that although all WEF resources face threats and high vulnerability, water resources face the most significant threats with a high risk to both quality and quantity for human use due to global socio-economic, political and climatological challenges. According to Levine (2024), WEF resources have been targeted by terrorism at least 9398 times in 50 years, including incidents in Colombia (Global Terrorism Database [GTD] from the University of Maryland).

The concept of slow violence, as developed by Nixon is violence like pollution, deforestation and climate change occurring over a longer period and often less visibly than a single event that is cataclysmic. This type of violence is not as often recognized as inducing great harm over time and as fueling conflict. Slow violence causes both the environment and poor people to become more vulnerable and disempowered. Nixon stated that environmental writers should attempt to make this slow violence more visible:

> A major challenge is representational: how to devise arresting stories, images, and symbols adequate to the pervasive but elusive violence of delayed effects. Crucially, slow violence is often not just attritional but also exponential, operating as a major threat multiplier; it can fuel long-term, proliferating conflicts in situations where the conditions for sustaining life become increasingly but gradually degraded (2011: 3).

Nixon also called attention to the displacement of local populations as a form of slow violence:

> Attritional catastrophes that overspill clear boundaries in time and space are marked above all by displacements–temporal, geographical, rhetorical, and technological displacements that simplify violence and underestimate, in advance and in retrospect, the human and environmental costs. Such displacements smooth the way for amnesia, as places are rendered irretrievable to those who once inhabited them, places that ordinarily pass unmourned in the corporate media (2011: 7).

Rarely is slow violence examined in the context of Colombia. For the Magdalena River basin, O'Bryen's work is among the few. O'Bryen notes the extreme contamination of the basin's rivers and wetlands by agribusiness and mining, but the focus is on literary works, including poetry about the area (2020).

Economic and physical violence interact with slow violence in various ways. Slow violence increases the exposure of household members, especially women, to physical and economic violence. The contamination of ecosystems and the variable availability of natural resources make it more challenging for women to access water of sufficient quality and to grow subsistence crops for their families. Thus, slow violence can create conditions of livelihood insecurity for women, increasing the likelihood of physical violence against them.

When studying WEF nexus issues, it is important to examine a range of social groups' experiences. Gender and other intersecting characteristics, such as social class, age, ethnicity, race, occupation and urban versus rural location and other geographical issues shape individuals' lived experiences with the WEF nexus. A growing body of literature examines gender issues within WEF (e.g. Bruns et al., 2022; Buechler et al., 2020, 2021; Jalonen et al, 2022; Villamor et al., 2020). This chapter examines differences between women and men's access to and management of water and land resources for agricultural production and household use and the energy implications of this water use in three Colombian communities. It reveals the close linkages between patriarchy, capitalism and colonialism in Latin America (Quiroga and Castelblanco, 2024) and in the study sites. Nixon's work on slow violence is pertinent for the gender analysis of WEF nexus issues made here; he examined, e.g. the feminist environmental activism of Wangari Maathai, which helped illuminate the importance of linking environmental rights, women's rights and property rights. In our analysis of WEF resources in the study sites, we consider these rights, including their intersections.

3 Case study: the Magdalena Basin, North Colombia

This study was carried out in the Department of Magdalena, located in the north of Colombia, in the Caribbean Plains (Figure 1). It covers an area of 2314438 ha (2.03% of the national territory and 11.35% of the Caribbean Plain). The estimated population in 2024 was 1513782; 47% are small farm (campesino) families (DANE, 2023b). It comprises 30 municipalities; 38% (878721 ha) are suitable for agriculture and 29% (674244) for livestock (UPRA, 2023). The weather in the region is tropical with dry to monsoon seasonal conditions. The plain region has annual precipitation of 1000–2000 mm. This region experiences an annual bimodal cycle with two wet seasons: September–November (major rainy season) and April–May, and two dry periods, December–March (major dry season), and June–August (Arregocés et al., 2024). Additionally, interannual

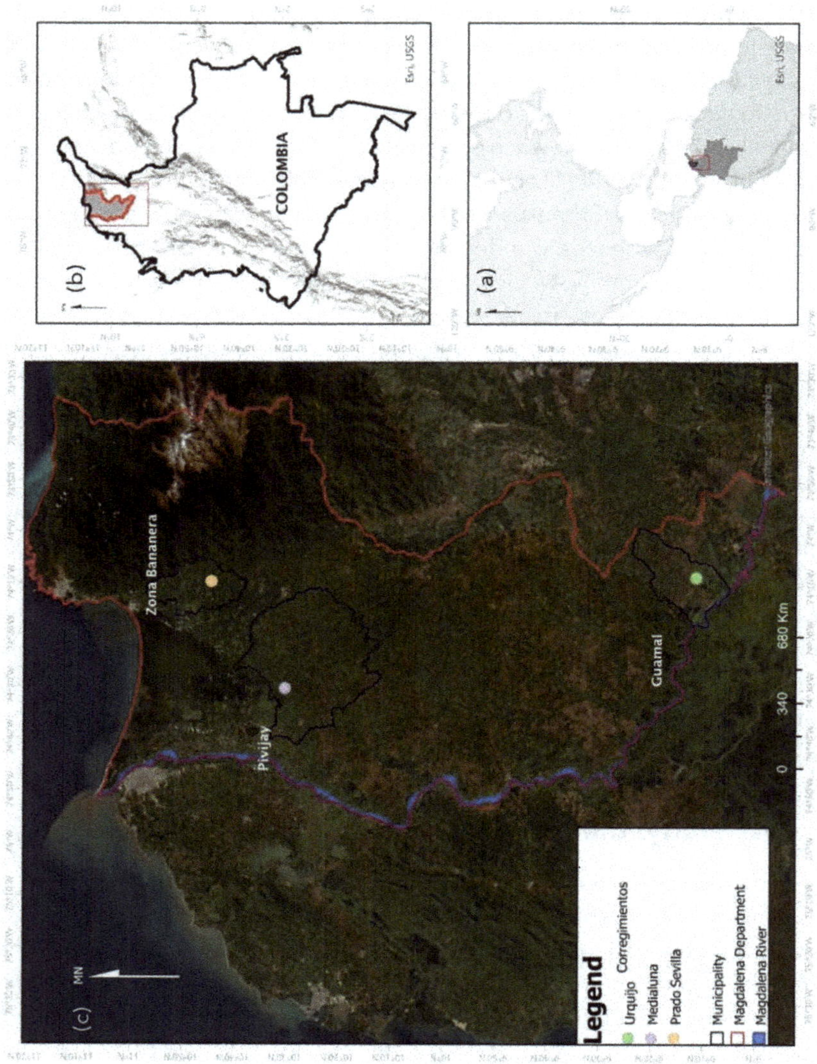

Figure 1 Location of the study area with the villages/corregimientos. The brown outlined areas represent the municipalities. Credit: César Rojas.

rainfall anomalies are strongly related to El Niño Southern Oscillation (ENSO) (Morales-Acuña et al., 2021). The department is rich in wetlands with swamps and "caños" (streams) associated with the Magdalena-Cauca basin and coastal swamps (Duque Escobar, 2021).

Three villages were selected to obtain a good representation of the most traditional way of life in the agricultural area, within a transect along the Magdalena River and wetland Ciénaga Grande de Santa Marta, where most people live, to the rural, more isolated inland areas (Figure 1). Urquijo (Guamal Municipality) and Medialuna (Pivijay Municipality) are two smaller villages, and Prado Sevilla is larger. The mainly low-income Urquijo and Prado Sevilla are surrounded by big wetlands ("ciénagas"), secondary forest patches, savannas for cattle that frequently flood and where wild rice grows and is harvested mainly for livestock feed (DANE, 2023a).

Urquijo is located at an elevation of 30 m above sea level, covering an area of 565 km^2, with temperatures of 24°C–38°C. The population is approximately 2139 (44.4 inhabitants/km^2) (DANE, 2023a). This community is in the southern part of the Magdalena department, about 50 km from Guamal city, in Guamal municipality. Its landscape is characterized by a tropical dry forest ecosystem, composed mainly of grasslands (40%), trees (29%), farmland (13%) and shrubs (11%) (Hernández, 2021). Economic activity depends on medium-scale crops and extensive cattle ranching; lower-income inhabitants mainly subsist on small-scale agriculture and fishing.

Medialuna is located at an elevation of 18 m above sea level. The average temperature is 32°C, and the rainfall pattern is bimodal (May–June and August–October), 850–1500 mm per year. The population is approximately 3612 inhabitants (DANE, 2023a). This area is part of a large wetland connected to the Magdalena River delta. The economy revolves around cattle ranching and fishing.

Prado Sevilla, considered the municipal center, is located at an elevation of 30 m above sea level, covering an area of 471 km^2 with an average temperature of 29°C. The population in 2023 was approximately 2291 inhabitants (Zona Bananera Mayor's Office, 2024). This municipality is the most diverse of the three studied, with higher percentages of indigenous and Afro-Colombian populations (44.4%) (DANE, 2022a). Bananas and oil palm occupy large tracts of land.

In all study sites, small farmer livelihoods consist of fishing (artisanal fishing) and diversified agricultural systems, principally for subsistence. Prevailing crops are beans, plantain, maize, mango, watermelon, sesame, yucca, chili, corn, corozo, topito pepper, sour guayaba, zapote, pumpkin and others. Only in Prado Sevilla are there banana and palm plantations owned by national and international companies. Other subsistence livelihoods there include bee farming, hunting, domestic employment, and jobs for workers on banana and palm plantations.

Landowners in Magdalena represent only 1.5% of Colombia's landowners. In this department, the median size of properties owned or possessed is 13 ha, while the average is 36.8 ha. Additionally, based on the Gini index, income inequality is high at the national level (0.86) and at the departmental level in Magdalena (0.71) (Contreras et al., 2021), with a poverty index of 33.4% (DANE, 2022a). Although no published data exist for the villages studied, it is known that most men and women are not owners of agricultural land.

The communities were informed about the purpose and scope of this study, and their voluntary participation, with the freedom to continue as part of the study or to withdraw at any time. Informed consent was shared, along with formats for data management and their voluntary participation. Transect walks were first conducted in areas that are sources or sinks of water, like the wetland area, streams, wells (for drinking water), and septic systems, to help in the design of the interview questions. Transect walks to kitchen gardens were also conducted. Two semi-structured surveys were conducted: the first had 48 questions, and the second had 24 questions (the first focused on water sources for domestic and agricultural use, types of crops, and the other on women's experiences with water management). Additionally, interviews and photo documentation were conducted across the three study areas. The survey was divided into three sections: the first part focused on socio-economic information, the second on water use and agriculture as part of livelihoods and the third on women's relationship with water. Half of the questions were open-ended, allowing respondents to express their perceptions. The surveys were designed and reviewed by experts and local professionals to adjust terminology according to the study area. The surveys were conducted with village residents. The sample size consisted of 28 women and 21 men for a total of 57 surveys: 20 in Medialuna, 18 in Urquijo and 19 in Prado Sevilla; respondents were 20–76 years old. Additionally, 30 in-depth interviews were conducted across the three villages, each lasting about 45 min with questions about their roles in agriculture and water use.

4 Case study results: agricultural activity in the Magdalena Basin

Each area within the Magdalena Basin has different environmental contexts that impact agricultural and fishing activities; each village grows different crops, and livestock and fishing activities are carried out differently (Figure 2). One of the main products in all study sites is mango, with two harvests per year. These harvests benefit the entire community, including local schools. People harvest the fruits collaboratively from trees located in yards, streets, schools and farms.

The fruits are taken to a collection center, from where they are sent to more populated areas of the country and food companies. In the Magdalena

Figure 2 Agricultural production systems in the study area. Illustration by Sergio Pereira.

department, approximately 25 varieties of mango are produced, with azúcar mango being the most prevalently exported (Lastre M. et al., 2022). In 2020, 3000 ha of mango crops were reported for export (Yacomelo-Hernández, 2021). In addition, unestimated land areas supply fruit to domestic markets and national juice manufacturers.

In the three villages, women participate in agricultural activities involving cassava, maize and bananas, grown on their own (<1 ha), rented, or borrowed land. Women are responsible for tasks such as land preparation, weeding, planting and harvesting (Figure 3). Some interviewees mentioned that income from crop sales is managed by men.

In Prado Sevilla, women primarily work for banana companies. In the three study communities, some women have small areas in their home compounds dedicated to raising pigs and chickens, used both to generate income and for self-consumption. Livestock activities, including the sale of meat and milk, are carried out and managed mainly by men; however, women participate in some of the animal care tasks without any remuneration. Most of the interviewees stated that they do not own land for farming due to socioeconomic difficulties that limit land ownership. Men are responsible for planting, while women and children handle irrigation and weeding. Men and children also hunt wild game for home consumption.

Figure 3 Woman harvesting yuca in Urquijo. Photo Credit: Sandra Barrera.

WEF nexus literature on Colombia focuses much more on farmers who own land for cultivation (Barrera et al., 2024; Caucci et al., 2023; Rodríguez-de-Francisco et al., 2019). Our case study addresses the situation of three populations located in the upper basin of the Magdalena River in the Magdalena department, where the majority do not own land. What we will discuss further is that this lack of access to land strongly impacts WEF nexus dynamics, including control over their access, quality, quantity and protection. Most of this literature also does not examine the WEF nexus in Colombia from a gender perspective, obscuring that although the entire population is resource vulnerable, there are gender inequalities.

In Prado Sevilla, income is mainly obtained from day labor in extensive banana and oil palm production systems. In 2020, the area planted with bananas in Magdalena was 13 291 ha; through 2019, the Zona Bananera municipality contributed 70% of this production (Minagricultura, 2020). Regarding palm cultivation, in 2020 it covered 46 739 ha, distributed across 17 municipalities in Magdalena (Fedepalma, 2024).

Although all three study areas face socioeconomic poverty, Zona Bananera has slightly better economic conditions due to these agribusinesses. Women are employed in these plantations, performing jobs stratified by gender. These jobs include harvesting, washing, bagging and packing fruits. Men are hired to perform tasks such as planting and cutting, transporting baskets of fruit, loading trucks and fieldwork involving tools. As of 2024, the daily wage in these companies for women is USD$10.00–13.00 (COP$40 000–50 000), while for men it ranges between USD$13.00–15.00 (COP$50 000–60 000) even though women and men both work 8 hours per day.

Lack of pay for family-based agricultural work and unequal pay for agricultural day labor are examples of the kinds of slow violence endured by women in the study sites.

Fishing is also an important activity in all three study sites: in Urquijo and Medialuna, there are nearby wetlands, and in Prado Sevilla there is a proximate river. Fishing expeditions (1–2 days) take place in canoes, both with and without motors, and involve men and women, generally from the same family group (Figure 4). In the canoes, the women clean the fish, removing the entrails and scales. Both men and women handle sales of larger raw fish to urban restaurants and smaller fish to local villages. Bartering also occurs, exchanging fish for

Figure 4 A woman buying fish from a river boat. Medialuna. Photo Credit: Sandra Barrera.

other types of food. Some women prepare and sell lunches from their homes, consisting of beans, cassava, rice, wild game meat and fish.

In addition to agricultural and fishing-related activities, some women do domestic service. In Medialuna, a day's (8 hours) work is paid at US$2-5 (COP $10 000–20 000), while in Urquijo and Prado Sevilla, the daily pay is US $5–8 (COP $20 000–30 000). Girls receive lower pay than women. Women's wages are low in general and lower than men's across the three villages, whether employed by large companies or small-scale employers.

Men obtain work on large-scale cattle ranches. Men earn many times what women earn in homes as maids. Men are hired for the legally stipulated USD 325/month (COP $1 300 000 in October 2024) (Mintrabajo, 2024). From this salary, the employer deducts a portion for rent of the house and land around it used by the worker's family, leaving them with approximately USD 225/month (around COP $900 000); mostly without health or pension benefits. The family uses this land for small gardens or subsistence crops, and for raising animals such as pigs and chickens. One woman interviewed stated that neither she nor any other family member receives compensation for fence and property maintenance. In exchange, they are allowed to live in a house owned by the employer and reside with their family, as well as grow crops like cassava and fruit trees, and raise a few chickens, all for self-sustenance. Women are responsible for housework and preparing meals for their male partners who work in cattle farming; however, they do not receive any financial compensation for these tasks.

The labor market in rural areas is unfavorable for women; they face lower employment and higher unemployment rates than men. On average, women work more hours per day than men, with most of that work being unpaid. In Colombia in 2019, the Global Participation Rate (GPR) of rural women was 39.1%; for rural men it was 75% and for urban women 56.7%; rural women´s GPR was the lowest. Most of the rural population is male, with women in the minority (Magdalena: 47.19% women, 52.87% men) (DANE, 2020). In Colombia, there is higher female outmigration from the region to areas internal to Colombia (in 2022, 51.75% female compared to 48.57% male internal migration) (DANE, 2022c). Although this information is unavailable for the Magdalena department, unequal population figures for men and women indicate this holds for this department as well. In the Magdalena department, in 2022, there was outmigration of 6.4–8.4% of the population, with 63.8% of women responding that the driver of their migration was to improve their labor market opportunities or improve their quality of life (DANE, 2022c).

5 Case study results: land access for agricultural use

Women in the region face inequalities in access to land ownership and compensation for agricultural work. In Colombia, land tenure statistics reveal significant gender gaps, with only 26% of women responsible for agricultural units with their own tenure compared to 62% of men. In the department of Magdalena, the figures are close to the national average (IGAC, 2024). For most of the agricultural work women do, they receive no compensation. This reflects statistics reported by the National Time Use Survey (DANE, 2021a), which shows that 92.9% of rural Colombian women engage in unpaid labor, dedicating an average of 9 h and 57 min per day to these activities. These gender gaps in paid labor limit women's access to land ownership, other assets, services and decision-making power.

One of our interviewees explained her role in the plot her husband was granted access to as part of a communal land system or "ejido":

Q: Do you participate in farming activities?

A: Yes, I've participated. I've gone to remove small plants. Yes, that's right, to clean.

Q: Oh, you mean pulling weeds around the crops?

A: Yes, that's right, the things that prevent them from growing. At least the weeds.

Q: Do you receive any compensation for this activity?

A: No, it's normal [for women to do this type of work without pay].

In Colombia, land ownership laws grant equal rights to men and women for buying and inheriting land. However, when women hold titles, they often lack decision-making autonomy over the land, as men typically make the decisions (Contreras et al., 2021). According to DANE in the statistical note on rural property (2022b) for the department of Magdalena in 2019, it was recorded that only 23% of women owners made decisions about productive units, which shows that control is not the same as legal ownership.

The above is an example of restrictive gender norms that are more evident in rural areas (DANE, 2020: 33); these relegate women to a secondary role in decision-making and land ownership. This can be conceptualized as structural violence over time against women, or slow violence.

5.1 Case of Medialuna

In Medialuna, there is a model of communal usufruct of state-owned lands. These land plots are known as "*ejido* La Colorada" and belong to the municipality. This land area is approximately 3365 ha, and benefits around 900 farming families. According to local testimony, this land has been serving the community for about 100 years, although it originally covered approximately

6000 ha. It is possible that a minor part of this land loss has occurred due to boundary fences of the property being gradually shifted over time, reducing the size of La Colorada. However, La Colorada has also survived attacks from armed conflict, as this territory has been targeted by illegal groups. These actions have led to civilian resistance as part of the broader struggle for land and the legalization of untitled lands. There has been violence associated with the armed conflict, causing suffering and fear among peasant organizations and farming associations (Hernandez, 2019). In Colombia, these types of power struggles–sometimes involving physical violence–aim to take control of land and political systems (Grajales, 2021) and have led to the displacement of many inhabitants from their territories. It is part of the violence, including the slow violence of land grabs over time, that has occurred in Medialuna.

Another example of slow violence is that women in Medialuna are involved in production for self-consumption, the national market and for export without receiving pay for their work or the crops sold. Male farmers are assigned a plot that ranges from 0.5 to 1 ha, with the boundaries of these plots marked by mango trees and other trees. Women are not assigned any land. Fencing and building on the land are prohibited to avoid legal disputes over land ownership. To organize agricultural work, these male farmers have formed an association. The crops grown on this land include beans, sesame, yucca, chili peppers, corn, squash, and *guineo topocho* (a type of plantain), among other species, for domestic consumption and small-scale trade. They also grow fodder grass.

Women participate in agricultural activities mainly during the sesame harvest season. La Colorada is home to many mango trees that have grown there naturally, and the farmers harvest them for sale. The male landholders are organized to sell the mango harvest, primarily to Barranquilla city, where it is processed at the Tutifruti bottled juice plant (Personal communication). The men receive the money from this trade, although women are involved in harvesting.

5.2 Case of Prado Sevilla

In Prado Sevilla, lands adjacent to wetlands, considered untitled state lands, are utilized by 28 families. These families formed an association to organize their agricultural and livestock activities. The land is divided into plots of approximately 4.5 ha; only 40% can be used for planting. As wetland areas, they don't fully dry out and can only be used during the dry season, which lasts from December to March. Thus, only short-term crops such as watermelon, melon, chili peppers and squash can be planted, as they are harvested in about three months, before the rains start in early to mid-April, and the lands flood again. These plots are marked with sticks, and those with livestock build fences around their designated areas during this season. This practice has been ongoing for about 30 years, with land possession rights exclusively held by men.

Women harvest the crop and carry the baskets to the collection points to transport them to points of sale. These lands provide economic resources and enhance the food security of these families. Most must also work on the plantations.

Women from all three communities studied, although they work in family harvests, report that they do not directly receive any of the profits from crop sales. They mentioned that the man of the household manages the family's money, following gender norms reported by Zorio (2016). Thus, women, despite being involved in agriculture, do not have access to agricultural income.

The livelihoods and food security of the rural communities under study primarily depend on agriculture and fishing. This section highlights the close relationship between livelihoods, land access, and food security in these rural communities. We analyze the vulnerability of these households to climate change in relation to the linkages between livelihoods, land access and food security.

6 Case study results: access to water sources

6.1 Domestic use: water management for cooking and drinking

Governance of potable water systems is a priority for public agencies as water is an essential resource for human life and well-being. However, none of the communities in the study have constant access to sufficient potable water for domestic and agricultural needs. Water quality is also often a concern. According to Greenwood et al. (2024), 4.4 billion people, or half of the world's population, lack access to clean drinking water. In the study sites, many lack access to safe drinking water. Also, water sources for consumption and domestic tasks in the study communities are the same as those for small-scale irrigation. Zwarteveen (2008, 2017) addresses the issue of water and gender, and Jepson et al. (2023) discuss water management for low-income populations; both studies stress that worldwide, water infrastructure design must consider this dual use.

In all the study sites, sources of water for domestic use are surface water sources such as irrigation canals, rivers and streams (Table 1) and rainwater that is collected from roofs. Some also use groundwater from dug or drilled wells. Within households, the task of collecting water for domestic use is typically carried out by women of all ages and children. The women possess in-depth knowledge of the best uses for each water source. As also documented by Corredor-Rodríguez (2024) for northern Colombia, women use their skills to classify water based on its taste, feel, and smell.

Table 1 Surface water and groundwater sources by village and distance from populated area

Medialuna	Urquijo	Prado Sevilla
Harvested rainwater	Harvested rainwater	Harvested rainwater
Ciénaga (swamp) Don Alonso or Caño Ciego (creek) (41m from village)	Ciénaga de Chilloa (South) (5 m) (swamp)	Rio Sevilla (0 m) (river), Río Frío (20 km)
Caño Shiller (North) (creek) (166.49 m)		Quebrada Limoncito (stream) (106.63 m)
Caño Concepción (400 m) (creek)		Irrigation canals of the banana and palm plantations
Caño Agua Viva (121 m) (creek)		Storage reservoirs in the rivers for plantation irrigation
Caño El Playazo (1 km) (creek)		
Concrete-lined dug wells (in some home compounds)	Drilled wells (home compounds, schools, community)	Drilled wells (home compounds, schools, community)

In Medialuna, all wells are concrete-lined dug wells excavated using rudimentary methods by local people, while in Prado Sevilla and Urquijo, they are deeper, drilled wells. If the groundwater is considered good quality, it is used for all domestic purposes. In other cases, the groundwater does not meet minimum standards for consumption or food preparation. Community members describe this water as cloudy, salty and soapy. They only use this water for domestic chores, and women and children must continue to carry water for drinking and food preparation from neighbors, schools or community wells. People with fewer economic resources must use this water for all purposes, even though they suffer from skin rashes (as seen in Prado Sevilla due to plantation pollution). Over time, their bodies become accustomed to the water, and they report fewer immediate effects such as gastrointestinal illness, but long-term effects are less well known. Some women also mentioned that certain grains, like beans, become tough and never fully cook when using well water, and their taste changes. Therefore, they must buy water for cooking these foods. Other women with more financial resources purchase water in bags or buy large containers of water collected from streams far from the villages.

In Prado Sevilla and Urquijo, there are rudimentary aqueducts that pump well water for domestic use. For these community aqueduct services, users must cover the cost of the piping to their homes and pay a minimum monthly fee of USD 2.00, as there are no meters to measure the volume of water consumed. When a family wants to connect to the system, they must purchase construction materials and install it themselves. Residents living far from the village centers, however, must rely on water from the river, swamps or streams, as the rural aqueduct does not reach them.

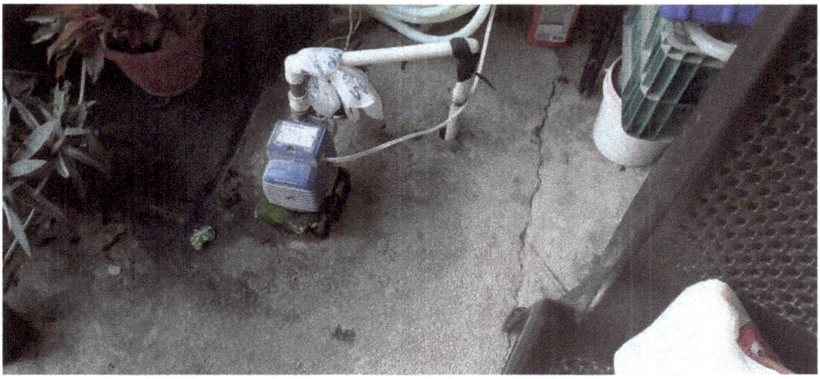

Figure 5 Electric pump connected to a pipe leading to the aqueduct that brings water to the house. Photo by Sandra Barrera.

One interviewed woman explained that they must use an electric pump to get water from the aqueduct pipe into the house because of low pressure (see Figure 5). Water is supplied in three-hour shifts, in the morning to one sector of the community and afternoon to another sector of the community, sometimes only twice a week. Residents, according to one man, have to "fetch water from the river. People who live on the hill must walk 45 minutes down to the river to carry buckets of water. The government water utility association says it's over budget, so the water supply cannot be improved".

In Medialuna, there isn't a rural aqueduct, and most homes and the school in the central area of the village have wells (Figure 6). Households with higher incomes have wells 6–10 meters deep in their yards. Poorer families who do not have wells at home get water from the main school's well, filling cans or buckets to take home, or from community wells dug by groups of neighbors and located on the streets. The rudimentary wells provide water with varying characteristics; some yield fresh water, while others produce salty water unsuitable for consumption or cooking. The interviewed residents are unsure why this occurs. In this type of context, Zwarteveen et al. (2024) emphasize the need to implement local groundwater governance policies to help communities understand the hydrogeological dynamics and conditions of the aquifers in their region.

In the studied communities, some households treat their water before drinking it. This adds work for women, who are responsible for this task. Most people in Medialuna do not treat the water before consumption. In Urquijo, some households practice rudimentary water treatment. One mother explained: "The water cannot be consumed directly, you have to add an alum tablet [containing aluminum potassium sulfate] the night before, and then it's

Figure 6 Woman in Medialuna using a bucket to manually lift water out of a well. Photo by Sandra Barrera.

drinkable the next day." This type of tablet is actually not suitable for such use, as it only removes visible particles, not all chemical or biological contaminants.

In Prado Sevilla, some women treat water for direct consumption. A mother head of household explained: "I decontaminate the drinking water with chlorine; I add 24 drops of chlorine to the tank in the morning and by the afternoon, it's safe to drink." This practice is followed due to the high incidence of gastrointestinal diseases and skin rashes. Health problems increase women's burden, since they are responsible for caring for children and adults, including the elderly.

Gastrointestinal illnesses are largely caused by the contamination of the well due to the proximity of private wells to septic tanks. The compounds are approximately 4 meters x 5 meters (Figure 7). Only toilet waste is channeled into septic tanks. The remaining household wastewater (graywater) is thrown into the yard or street.

Most women in all three villages, even those with wells, also use rainwater, especially during drought, because the water level drops so low that the water becomes muddy. There is no advanced system for collecting rainwater; it is collected from rooftops with clay tiles arranged in vertical channels or zinc sheets. These channels are connected to a single pipe directing the water into a container on the ground. The water is then stored in plastic or cement containers. Rainwater is used for all domestic activities, including cooking

Figure 7 Septic tank in the yard of a house, Urquijo. Photo credit: Sandra Barrera.

(Figure 8). Interviewees consider it to have better quality and characteristics than groundwater. Rainwater (or purchased bottled water) is used to cook beans and lentils because they claimed that this water cooks them better.

Some families have the financial means to buy water in large containers (collected locally using artisanal methods) and transported by carts. Twenty liters of this bottled mountain water costs an average of USD 1.00. Some families buy potable water in plastic bags from local factories, which creates significant environmental pollution.

In the Zona Bananera, there are irrigation systems used exclusively for banana and palm oil plantations. However, some women in Prado Sevilla with canals that run through their properties use the water to wash clothes, bathe and cook (Figure 9).

According to some testimonies, this water is not suitable for human consumption due to the high levels of nutrients, salts, and toxic elements from aerial spraying. People report health issues related to the use of this water, such as skin rashes, lung diseases, stomach illnesses, and hair loss. Women and children are more exposed to this water, which results in more severe health effects from the fungicides, fertilizers and other chemicals present in it.

Figure 8 A woman washing dishes on top of plastic rainwater storage containers in Medialuna, Pivijay. Photo credit: Sandra Barrera.

In Prado Sevilla, despite two nearby streams, people who can afford it or who have wells drink water from wells (aquifers) because of the belief that the streams and rivers are contaminated. The banana and palm companies provide information warning residents to avoid consuming this water to prevent illness.

The case of Prado Sevilla reveals that exposure to polluted water varies by geographical location, social class, gender and age. The contamination of local surface and groundwater sources by agribusiness represents a form

Figure 9 Women washing clothes in a polluted irrigation canal and caring for children. Prado Sevilla, Zona Bananera. Photo credit: Sandra Barrera.

of slow violence. Feminist earth scientist Jill Schneiderman warns that such environmental degradation often "masquerades as inevitable" (2012: 90).

6.2 Water management for clothes washing

In the three communities, laundry is washed manually or with washing machines. Women in the household are responsible for washing clothes. Most respondents said they wash clothes manually due to a lack of resources to buy a washing machine. Manual washing is done on a washboard in the house or, as mentioned above, in streams and canals, where they lay a flat, wide plank across and place the clothes on it. They put soap on the clothes and beat them with a flat stick called a "manduco," then rinse them in the river water and finally carry the clothes home in containers. Another method involves washing clothes on a washboard, using well water stored in cement or plastic containers.

Washing machines used regionally are not fully automatic due to an inconsistent water supply. This adds to women's labor. Washing is done partly by hand and partly by machine. The first compartment of the manual washer is filled with water, soap and clothes, and the soaking time is set. Afterward, clothes are rinsed by hand on the washboard or in buckets, then placed into the second machine compartment for spinning. Water from all stages of the laundry process in the house is reused for flushing toilets, housecleaning and watering trees, plants, and home garden crops. This is an indication of relative water scarcity in and near homes and the value women place on water.

6.3 Water: access for irrigation and livestock

In the study locations, small-scale agriculture is primarily rainfed; only manual irrigation using buckets and water from wells is accomplished by the physical labor of men, women, and children. In schools, girls are learning to use drip irrigation and fertigation systems (Figure 10). In the future, as drought becomes more frequent, small-scale agriculture will need to be modernized.

Among the four farmer associations in Urquijo, one is composed of 18 women. They produce citrus fruits and vegetables for sale, and soaps, oils and other citrus-based products; chickens are also raised. These women (almost all household heads) joined together and bought one hectare of land after two years of hard work selling empanadas, lunches and other products. They do not have an irrigation system and depend on rainwater and the community aqueduct. They must carry water twice a day during drought and power outages for crop irrigation and water for the chickens. As an association, they can access government training for association management and production and marketing.

Figure 10 Local school girls are learning about drip irrigation on their school's demonstration plots. Medialuna, Pivijay. Photo credit: Sandra Barrera.

For medium-scale agriculture, such as mango and citrus production for agro-export, electric pumps are used for drip irrigation. In Prado Sevilla, large banana and palm companies are implementing drip and sprinkler irrigation systems, with some models using solar panels to supply energy. Additionally, large-scale banana and palm oil farms collect river water during the rainy season and store it in reservoirs dug into the ground and lined with plastic, mainly for use during the dry season. These reservoirs are locally known as "jahuelles" or "aljibes." Large companies also have irrigation systems drawing water from the Sevilla and Frio rivers. However, as mentioned, residents also use this water for irrigation and domestic use and large agribusinesses use and contaminate large volumes of local water supplies. During the dry season, the Sevilla and Rio Frio rivers almost completely dry up. These dynamics add to women's work, carrying, purifying and purchasing water.

The extensive livestock model in the area is managed with free-roaming cattle on lands equipped with wells, "jahuelles" or drinking troughs. These water sources are supplied by streams, creeks or deep wells. Men dig this water infrastructure for the cattle. In Medialuna, buffalo farming (water buffalo) has been successfully introduced and is competing with traditional cattle farming. Buffalo farming produces more milk with higher fat content. However, this livestock activity is negatively affecting water quality because buffaloes need to bathe to regulate their body temperature, and they contaminate the water.

6.4 Water: challenges due to climate change

World Bank SSP3-7.0 projections indicate that Colombia's annual mean temperatures nationwide are expected to increase further. From an approximate mean of 24.50°C during 1995–2014, temperatures are projected to rise to 25.20°C during 2020–2039, and to 25.99°C during 2040–2059. Colombia's high atmospheric moisture content will lead to an increase in the days with a Heat Index greater than 35°C by mid-century. The Caribbean region, where all study sites are located, has experienced higher temperature increases in summer and fall than any other Colombian region (World Bank, 2023: 2).

In the Magdalena department, as part of climate change, rainfall has decreased by 90.79 mm (World Bank, 2023: 18). This department has become significantly drier in the spring and fall (World Bank, 2023: 3), and it is more prone to frequent intense floods and droughts related also to the El Niño-Southern Oscillation (ENSO) (World Bank, 2023: 4). These changes in temperature and rainfall negatively impact crops. Local populations must adapt to increasingly extreme environmental changes. Shifting rainfall and drought patterns disrupt harvest schedules, increasing crop loss risks. For fruit trees, these changes alter flowering patterns, delaying fruit harvests by up to two months.

During drought, access to water from wells and rivers decreases as water levels drop and water quality worsens. Floods and strong winds make accessing fields difficult and cause damage to access roads. Given the intensity of these extreme events, all family members are involved in maintaining crops. A woman in Medialuna exclaimed: "Here, the wind took the zinc sheets from the kitchen roof, and it toppled our small mesh-covered crops." She also mentioned that prolonged droughts "affect us; the well dries up, leaving only mud. We must go to a neighbor's farm 2 kilometers away on a donkey once in the morning and once in the afternoon, bringing about 160 liters each trip. It's also for the animals and plants that are drying up. The cassava crop dies because it doesn't rain." Another woman from Medialuna, who does not have a well and lives further away, explained that she must collect rainwater in a rudimentary manner (Figure 11a). Together with her neighbors, in a similar situation, they built a "jagüey" (a small reservoir), and although the water is not suitable for consumption, both humans and livestock must drink from it during drought (Figure 11b).

Internationally, it has been documented that women's labor in family farming increases more than men's during extreme weather events such as floods and droughts. During extreme heat, women spend less time farming and more on care and domestic work, but both responses lengthen their already long workdays. Women farmers also invest more in agriculture but lose more when adverse climate events occur. Globally, women receive only 5% of

Figure 11 Top: A rural home with a rainwater collection system from a zinc roof. Bottom: A jagüey constructed to collect water for human and animal consumption. Photo credit: Woman Interviewee, Resident of this Home.

agricultural extension services (USAID, 2023), highlighting the need for more women-oriented extension services that address climate change (FAO, 2024).

7 Case study results: access to energy

7.1 Challenges in pumping water to fields and homes

In the study area, residents need energy to pump water for domestic and agricultural use. Women rely more heavily on electricity for domestic purposes, for irrigating crops in their home compounds and, in some cases, agricultural plots. As documented in other Latin American countries, women are disproportionately affected when they lack access to reliable electricity sources (Amigo-Jorquera et al., 2019). Most families use electric pumps powered by the municipal grid provided by the AIR-E company (with energy produced from coal and hydroelectric projects) (Olivares et al., forthcoming). In some cases, they use gasoline or diesel pumps as an emergency alternative. Most common is manual extraction of water from wells; the most common method in Medialuna, where dug wells are concrete-lined. Manually drawing water for domestic use is time-consuming and labor-intensive. Women and children are usually responsible for this and for irrigating, normally unpaid tasks. Irrigation systems requiring energy are not available for most small and some medium-scale agriculture, such as for cassava and sesame crops. However, drip irrigation systems are used for certain crops like mango and citrus for export (Figure 12). In Prado Sevilla and Urquijo, water pumping requires electric pumps, especially for tube wells.

Electricity access in Colombia, particularly in the study area, is complicated. According to the DANE Quality of Life survey 2020 (DANE, 2021b), 98.2% of the Colombian population has access to electricity; in rural areas, this figure is 92.9%. However, in the study area, the electrical grid suffers from coverage, connectivity and cost issues. For homes connected to the grid, there is an average of one hour of power outage per day. Connection is also disrupted during rainstorms due to problematic wiring and transformers. Climate change is making storms more intense, leading to more frequent cuts. Technical assistance is slow. Business owners, like storekeepers, often have gasoline and diesel-powered motor pumps, which pollute both air and soil, to run freezers and refrigerators.

Electricity rates are extremely high compared to local incomes. The department of Magdalena has the most expensive energy tariffs in the country, with steep annual increases. For 2024, household energy rates rose by 30–50%, and for businesses and companies 100% (CCSM, 2024). According to the August 2024 energy tariff bulletin, AIR-E had the highest KWH rate for stratum 1, of 488.30 COP (0.12 USD) per KWH (Superservicios, 2024). These

Figure 12 An electric pump supplies a drip irrigation system for medium-scale citrus farms.

high rates deepen the existing energy inequalities, pushing a large percentage of the population to connect to the service illegally and intermittently. Some women stated they had no other choice, as they could not afford to pay the rates. Others go without electricity. One woman mentioned that she owns a small fridge, a fan, a radio and four light bulbs, and her bill still comes to 120 000 COP (28.80 USD).

Women, who spend more time at home and carry the burden of domestic work that requires water, are more affected by the lack of access to a reliable and affordable energy source. Women must store water due to the inconsistent water service and the lack of power to pump water to their homes. They endure high temperatures inside their homes when they can't afford cooling appliances or pay electric bills. Given the region's high temperatures and humidity, residents constantly use refrigerators, freezers, fans and air conditioners, appliances that consume large amounts of energy and drive up costs. During the day, only fans are used while women and children are at home.

At night, when everyone is home, air conditioning is used–if available. One interviewee's perception of why there were continuous power outages, yet high tariffs, was that the company needed to recover losses from residents' illegal grid connections. Challenges related to access to energy to supply adequate water for domestic use include improving the electrical grid in both coverage and quality. More investment is required to improve socioeconomic conditions and allow the acquisition of water pumps and fuel.

The climate change challenges that the study area's residents, particularly women, face due to issues with energy, are linked to food preservation, enduring increasingly intense heat, non-functioning water pumps that led to lack of water, frequent appliance breakdowns caused by power cuts, and the overall inability to enjoy technologies that keep them informed. An Urquijo woman agricultural producer association member stated that if the power outage occurs during her three-hour water access turn, she loses the chance to store water for her crops. She must then carry buckets of water for irrigation, increasing her workload. She irrigates twice a day due to high temperatures. Similarly, a woman from Medialuna said, "We suffer from a lot of heat, and my small fridge has been damaged. During the power cuts, my food goes bad. I can't use the washing machine either, which makes more work for me."

8 The WEF nexus, climate change and violence: impacts on women's livelihoods

Living in rural areas of tropical dry forest, where the annual temperature exceeds 24°C, with seasonal rainfall and periods of flooding caused by high flows from large rivers, is not easy for women. The study area is also characterized by a patriarchal social structure, where men decide whether women can work and men are in charge of financial, land, water and energy resources. This positions women in a state of significant psychological, emotional and economic dependency (Minsalud-DANE-UNFPA, 2020). Such dependency fosters conditions that facilitate economic violence and physical violence, mostly in the form of domestic violence.

The labor market is unfavorable, especially for women, who not only face fewer job opportunities but also receive lower wages. Most women who work outside the home do so outside of the formal labor market. As informal workers, these women are more vulnerable to experiencing physical violence than any other category of workers in the labor market (INMLCF, 2022). Increased exposure to domestic violence is also directly associated with level of education therefore women with low levels of education are more at risk (INMLCF, 2022). These are the main factors influencing the livelihoods and well-being of women in the Magdalena region.

In this context, establishing a link between access to the resources of the WEF nexus, types of gender-based violence, and means of livelihood under the effects of climate change enables progress in analyzing and improving women's livelihoods (Figure 13). Climate change impacts every domestic and agricultural

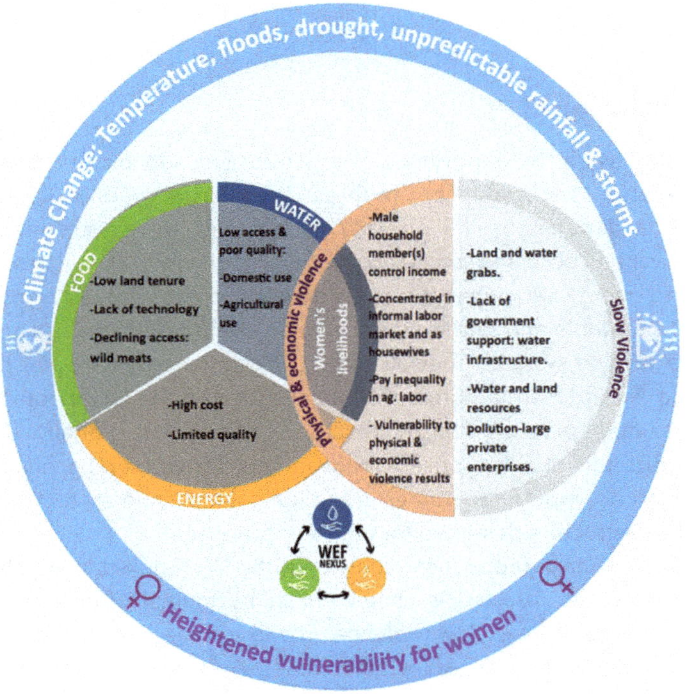

Figure 13 Climate change-violence-gendered livelihoods through the Lens of the WEF nexus.

activity carried out by women and increases the incidence of domestic violence against them due to heightened physical, psychological and economic stress among community and household members. This trend has been documented by Castañeda-Camey et al. (2020) and is observed worldwide, putting women's physical and financial security at greater risk.

With respect to water, the WEF nexus framework allows for the evaluation of water governance and the proposal of community-based solutions that can improve management, benefiting the economy and overall well-being, particularly for women who face daily challenges of securing water. This focus on water reveals that agroindustry controls and overuses water, at the same time that it pollutes water, which creates scarcity conditions and contamination risks for women who gather water and who use water for household chores like laundry. The very high cost of electricity creates a barrier for women related to home energy resources because of their greater time spent at home and their responsibility for maintaining perishable food items. There is also only intermittent energy available for women's tasks in pumping water for domestic and small-scale agricultural use. Women's food production is integrally linked to their ability to control water and energy resources. In this context of climate change, agriculture becomes more difficult and women are further disadvantaged due to their lack of control over land, water and energy resources.

9 Conclusion

This study reveals the importance of a gender focus in WEF nexus analyses. The case study was conducted in three different geographical areas of the Magdalena department. It highlights gender and social class inequities in control over and access to sufficient quantity and quality WEF resources of land, water and energy. Women in the study area lack access to their own land and reliable, affordable and safe sources of water for domestic use and agriculture and safe sewage systems. They also lack sufficient energy for water pumping, domestic use and thermal comfort. All water sources are used for both domestic and agricultural purposes; this is the case in many areas globally. The failure to recognize this dual use poses disadvantages for women because, despite significant labor inputs into domestic and agricultural spheres, they must make do with infrastructure designed for only one purpose.

The social and geopolitical context of the WEF nexus in which women in Magdalena are inserted is characterized by multiple forms of violence: economic, physical and slow violence. For example, men in the study sites control household income, and women receive unequal pay (for agribusiness work) and low pay (in informal sector work) and lack of pay (in family agriculture), making them more vulnerable to economic and physical violence.

Slow violence, set in a patriarchal context, exacerbates economic and physical violence risks for women because it makes the resources they depend on less reliable and less safe, heightening multidimensional stress within households. These interconnections must be recognized in Colombia and used to shape social programs that positively impact women. This includes creating jobs with equal access and pay for men and women and it entails working with male farmers to convey the importance of women's pay within family agriculture. Such programming could translate into more equitable incomes, in turn facilitating land and food security. Within governmental land allocation and formalization programs, control mechanisms and specialized technical teams with gender expertise are required to determine land productivity and water availability.

Slow violence also includes displacement, land struggles and consequences of the country's armed conflict. More powerful groups are able, through time, to engage in overt and more covert land grabs, affecting the viability of small-scale agricultural production. The national land agency, the national land registry (Catastro) and geographical land database within the governmental Agustin Codazzi Institute (IGAC) need to work together to strengthen boundary clarification and update land registries. Greater security is also needed from the Ministry of Defense for the protection of land properties.

Slow violence is also apparent in toxic pollution from plantation agriculture. Women and children are more exposed due to routine involvement in domestic activities with contaminated water. Women must attempt to use alternative sources for drinking and cooking, often expensive to access. Slow violence also includes contamination of water by cattle ranchers. Clear policies are needed to ensure higher quality water is available not only for agriculture but also for communities. In Randullabad, India, groundwater-reliant farmers have had to make decisions and organize with government support during drought. Water-intensive crops like bananas and sugarcane are no longer cultivated (Alsekar et al., 2023); as seen in Prado Sevilla, these crops also cause contamination of scarce water resources. It is crucial for leaders in the study areas, particularly Prado Sevilla, to explore new agricultural economies using WEF nexus and slow violence lenses.

Slow violence also includes human-caused climate change that is impacting all types of agricultural and livestock husbandry activities, largely through changes in flood and drought incidence. Drought-resistant crops cultivated locally, like sesame and yuca, help local populations adapt to climate stressors. Efficient, subsidized irrigation systems and additional high-value, drought-resistant, low water-consuming crop varieties may assist farmers; women farmers would need to be included in training and outreach. Associations can help women farmers access government assistance.

In the energy context, the intermittent yet expensive electricity service in homes for both domestic and agricultural use primarily affects women and children. Solar parks established regionally are expected to improve energy access and impact tariffs.

However, the energy sector's regulatory bodies must intervene to ensure that quality energy service is provided. Energy must be recognized as being deeply linked to water for domestic and agricultural use. Women's linked water-energy needs must be prioritized.

Ensuring livelihoods and food security for rural communities is a global challenge that requires the development of strategies to measure vulnerability and support communities to be more resilient to the effects of climate change. Therefore, it is necessary to create programs that help women diversify their livelihood activities and adopt climate-resilient strategies for agriculture and fishing that promote food security for these communities into the future.

Research undertaken from a WEF nexus perspective is very scarce in Colombia and the Latin America and Caribbean region. The WEF nexus, a powerful tool that provides information and analysis, could support decision-making regarding WEF resource management and governance, including socioeconomic aspects like gender, applied at different scales. This information would stimulate more efficient and integrated management of WEF resources and the implementation of national and regional strategies in the face of climate change challenges.

10 Where to look for further information

- Buechler, S. & Hansen, A. (2015). *A Political Ecology of Women, Water and Global Environmental Change*. Routledge.
- Acevedo-Guerrero, T., et al. (2024). *Routledge Handbook of Gender and Water Governance* (p. 456). Taylor & Francis.
- Empowering Women in the Water Energy Food Nexus
- https://www.youtube.com/watch?v=2y6EQOTmhlE
- Living Waters Museum. https://www.livingwatersmuseum.org/about

11 References

Abbott, M., et al. (2017). Examining the food–energy– water and conflict nexus. *Current Opinion in Chemical Engineering*, *18*, 55–60.

Albrecht, T. R., et al. (2018). The water-energy-food nexus: a systematic review of methods for nexus assessment. *Environmental Research Letters*, *13*(4), 043002.

Alcaldía Zona Bananera. (2024). Alcaldía Zona Bananera. Mi municipio. Consultado 10 agosto de 2024 en: https://www.zonabananera- magdalena.gov.co/MiMunicipio/Paginas/Informaci%C3%B3n-del-Municipio.aspx.

Alsekar, U., et al. (2023). The Participatory Groundwater Management initiative in Randullabad, India. https://www.t2sgroundwater.org/narrative-reports-randullabad-india.

Amigo-Jorquera, C., et al. (2019). Does energy poverty have a female face in Chile? *Tapuya: Latin American Science, Technology and Society*, *2*(1), 378–390. https://doi.org/10.1080/25729861.2019.1608038.

Arregocés, H. A., et al. (2024). Annual and monthly precipitation trends: an indicator of climate change in the Caribbean region of Colombia. *Case Studies in Chemical and Environmental Engineering*, *10*, 100834. https://doi.org/10.1016/j.cscee.2024.100834.

Barrera, L. T., et al. (2024). La Gestión del Nexo Agua-Energía-Alimentos-Género: entre la Subsistencia y la Sostenibilidad en Isla Fuerte, Colombia//Managing the Water-Energy- Food-Gender Nexus: between Subsistence and Sustainability on Isla Fuerte, Colombia. Research Report. https://acrobat.adobe.com/link/review?uri=urn%3Aaaid%3Ascds%3AUS%3Aa6438f77-3b69-366d-82bb-2d37b4589dd7&viewer%21megaVerb=group-discover.

Boelens, R., et al. (2021). River lives, river movements: fisher communities mobilizing local and official rules in defense of the Magdalena River. *The Journal of Legal Pluralism and Unofficial Law*, *53*(3), 458–476. https://doi.org/10.1080/07329113.2021.2012376.

Botai, J. O., et al. (2021). A review of the water–energy–food nexus research in Africa. *Sustainability*, *13*(4), 1762.

Brouwer, F. (2022). Introduction to the water-energy-food nexus. In F. Brouwer, et al. (eds.). *Handbook on the Water-Energy-Food Nexus* (pp. 1–14). Edward Elgar Publishing.

Bruns, A., et al. (2022). Nexus disrupted: Lived realities and the water-energy-food nexus from an infrastructure perspective. *Geoforum*, *133*, 79–88.

Buechler, S., et al. (2016). Re-linking governance of energy with livelihoods and irrigation in Uttarakhand, India. *Water*, *8*(10), 437.

Buechler, S., et al. (2020). Patriarchy and (electric) power? A feminist political ecology of solar energy use in Mexico and the United States. *Energy Research & Social Science*, *70*, 101743.

Buechler, S. & Martínez-Molina, K. G. (2021). Energy justice, renewable energy, and the rural-urban divide: Insights from the Southwest US. *Energy and Climate Change*, *2*, 100048.

Calvo R., et al. (2021). Desarrollo de indicadores de pobreza energética en América Latina y el Caribe", serie Recursos Naturales y Desarrollo, N° 207 (LC/TS.2021/104), Santiago, Comisión Económica para América Latina y el Caribe (CEPAL).

Camara de comercio de Santa Marta (CCSM). (2024). Alerta del Sector Productivo del Magdalena por aumento en la tarifa de energía tras la reciente Resolución de la CREG. Consultado 20 de Agosto del 2024 en: https://www.ccsm.org.co/es/alerta-del-sector-productivo-del-magdalena-por-aumento-en-la-tarifa-de-energia-tras-la-reciente-resolucion-de-la-creg-EV245.

Castañeda-Camey, I., et al. (2020). *Gender-Based Violence and Environment Linkages: The Violence of Inequality*. IUCN, Gland.

Caucci, S., et al. (2023). Vulnerability reduction in post-conflict areas through a Nexus approach to sustainable food production systems: a case study in Colombia. Pre-print. https://assets-eu.researchsquare.com/files/rs-2183999/v2/21bdfc19-aa3b-469e-9df9-dbb48ca3145c.pdf?c=1673905402.

Contreras J., et al. (2021). Análisis de distribución de la propiedad rural en Colombia. Resultados 2018. Bogotá: UPRA (accessed 18 September 2024).

Corredor-Rodríguez, S. (2024). Women's Bodily Experiences: Accessing and treating water in the Colombian Caribbean. In T. Acevedo-Guerrero, et al. (eds.). *Routledge Handbook of Gender and Water Governance* (pp. 79–88). Routledge.

DANE - Departamento Administrativo Nacional de Estadística. (2020). Mujeres rurales en Colombia. Consultado en: https://www.dane.gov.co/files/investigaciones/notas-estadisticas/sep-2020-%20mujeres-rurales.pdf (accessed 17 July 2024).

DANE. - Departamento Administrativo Nacional de Estadística. (2021a). Encuesta nacional de uso del tiempo (ENUT) 2021. https://www.dane.gov.co/index.php/estadisticas-por-tema/pobreza-y-condiciones-de-vida/encuesta-nacional-del-uso-del-tiempo-enut.

DANE - Departamento Administrativo Nacional de Estadística. (2021b). Encuesta Nacional de Calidad de Vida (ECV) 2020: servicios de los hogares. Acceso a servicios públicos, privados o comunales. https://www.dane.gov.co/files/investigaciones/condiciones_vida/calidad_vida/2020/Bolet in_Tecnico_ECV_2020.pdf (accessed 08 August 2024).

DANE - Departamento Administrativo Nacional de Estadística. (2022a). Colombia en cifras: Perspectiva general del departamento del Magdalena. https://www.dane.gov.co/files/investigaciones/planes-departamentos-ciudades/220317-Universidad_del_Magdalena.pdf (accessed 17 September 2024).

DANE - Departamento Administrativo Nacional de Estadística. (2022b). Nota estadística de propiedad rural en Colombia. https://www.dane.gov.co/files/investigaciones/notas-estadisticas/jul_2022_nota_estadistica_propiedad_rural.pdf (accessed 10 September 2024).

DANE - Departamento Administrativo Nacional de Estadística & UNFPA - United Nations Population Fund. (2022c). Una aproximación sociodemográfica de la población migrante interna a partir de la estructura poblacional de migración interna (ECV). https://www.dane.gov.co/files/investigaciones/poblacion/informes-estadisticas-sociodemograficas/2022-02-15-aproximacion-sociodemografica-poblacion-migrante-ECV.pdf (accessed 08 July 2024).

DANE - Departamento Administrativo Nacional de Estadística. (2023a). Necesidades Básicas Insatisfechas (NBI). Censo Nacional de Población y Vivienda (CNPV) 2018. Geoportal Dane. micrositio. https://microdatos.dane.gov.co/index.php/catalog/643 (accessed 22 July 2024).

DANE - Departamento Administrativo Nacional de Estadística. (2023b). Proyecciones y retroproyecciones de población municipal para el periodo 1985–2019 y 2020–2035 con base en el CNPV 2018. https://www.dane.gov.co/index.php/estadisticas-por-tema/demografia-y-poblacion/proyecciones-de-poblacion (accessed 11 September 2024).

Duque Escobar, G. (2021). *La gran cuenca Magdalena-Cauca*. Documento para el Contexto de CTS. de la U.N. de Colombia. Manizales. https://repositorio.unal.edu.co/bitstream/handle/unal/84643/la_gran_cuenca_magdalena-cauca..pdf?sequence=2.

FAO. (2024). *The unjust climate – Measuring the impacts of climate change on the rural poor, women and youth: summary*. FAO. https://openknowledge.fao.org/server/api/core/bitstreams/5f714e89-2bdd-4313-8b72-45504d00953a/content.

Fondo de cultivadores de palma-FEDEPALMA-(2024). Informe de Gestión 2024. https://fedepalma.org/wp-content/uploads/2025/06/informe-de_gestion_2024-digital_.pdf

Fondo Nacional del Arroz (FEDEARROZ). (2024). Servicio climático. El clima al día. https://clima.fedearroz.com.co/magdalena/.

Grajales, J. (2021). *Agrarian Capitalism, War and Peace in Colombia: Beyond Dispossession*. Routledge.

Greenwood, E. E., et al. (2024). Mapping safe drinking water use in low-and middle-income countries. *Science*, 385, 784–790.

Hernández, I. (2021). Análisis de los impactos climáticos sobre la producción agropecuaria del municipio de Guamal del departamento del Magdalena. Trabajo de grado. Facultad de Ciencias Empresariales y Economía. Universidad del Magdalena. Santa Marta. 30 pp.

Hernandez Polo, N. (2019). Autodefensas Unidas de Colombia en Pivijay Magdalena 1996–2006. Universidad de Cartagena Facultad de Ciencias Humanas Programa De Historia. (2019). Tesis De Grado Para Obtener el Título De Historiador. https://repositorio.unicartagena.edu.co/server/api/core/bitstreams/065b77f6-9e00-4933-b6b2-a9506e4c33b8/content.

Instituto de medicina legal y ciencias forenses (INMLCF), 2022. Observatorio de violencia contra la mujer. Informe técnico 001-2022. https://www.medicinalegal.gov.co/documents/20143/656206/InformeTecnico_01_2022_Magdalena.pdf.

Instituto geográfico Agustin Codazzi (IGAC). (2024). IGAC y ONU Mujeres, lanzan su cartilla Diálogos Geográficos N° 2 Enfoque de Género en el Catastro Multipropósito. Recuperado Agosto 22, 2024 de: IGAC y ONU Mujeres, lanzan su cartilla Diálogos Geográficos N° 2 Enfoque de Género en el Catastro Multipropósito | Instituto Geográfico Agustín Codazzi.

Jacobson, M. & Pekarcik, G. (2022). Water-energy-food nexus approaches and initiatives in Africa. In F. Brouwer (ed.). *Handbook on the Water-Energy-Food Nexus* (pp. 211–230). Elgar Online.

Jalonen, R., et al. (2022). Gender equity and social inclusion in the water-energy-food-ecosystems (WEFE) nexus: frameworks and tools for moving from resource-centric to people-centric WEFE nexus approaches.

Jepson, W., et al. (2023). Revaluing multiple-use water services for food and water security. FAO Land and Water Discussion Paper No. 19. Rome, FAO. https://doi.org/10.4060/cc7317en.

Karatayev, M., et al. (2017). The water-energy-food nexus in Kazakhstan: challenges and opportunities. *Energy Procedia*, *125*, 63–70.

Katyaini, S., et al. (2021). Water–food nexus through the lens of virtual water flows: the case of India. *Water*, *13*(6), 768.

Kurian, M. (2017). The water-energy-food nexus: trade-offs, thresholds and transdisciplinary approaches to sustainable development. *Environmental Science & Policy*, *68*, 97–106.

Lastre, M. J., et al. (2022). Análisis de la exportación de mango desde el departamento del Magdalena. Universidad Cooperativa de Colombia. Tesis de grado. https://repository.ucc.edu.co/entities/publication/6b114aa0-c175-438d-9af8-b10638d0910a.

Levine B. (2024). *Water, Energy, Food, and Terrorism: Towards a Nexus Approach*. https://saisreview.sais.jhu.edu/water-energy-food-and-terrorism-towards-a-nexus-approach/.

Mahlknecht, J., et al. (2020). Water-energy-food security: a nexus perspective of the current situation in Latin America and the Caribbean. *Energy*, *194*, 116824. https://doi. org/10.1016/j.energy.2019.116824.

Minagricultura (Ministerio de agricultura y desarrollo rural). (2020). Cadena de Banano. https://sioc.minagricultura.gov.co/Banano/Documentos/2020-12-31%20Cifras%20Sectoriales.pdf.

Ministerio del trabajo (Mintrabajo). (2024). https://www.mintrabajo.gov.co/comunicados/2023/enero/en-el-2024-el-salario-minimo-es-de-un-millon-300-mil-pesos-y-auxilio-de-transporte-de-162-mil-pesos (accessed 16 September 2024).

Morales-Acuña, E., et al. (2021). Assessment of precipitation variability and trends based on satellite estimations for a heterogeneous Colombian region. *Hydrology*, *8*(3), 128.

Mukherji, A. (2022). Sustainable groundwater management in India needs a water-energy-food nexus approach. *Applied Economic Perspectives and Policy*, *44*(1), 394–410.

Nepal, S., et al. (2021). Achieving water security in Nepal through unravelling the water-energy-agriculture nexus. *International Journal of Water Resources Development*, *37*(1), 67–93.

Nixon, R. (2011). *Slow Violence and the Environmentalism of the Poor*. Harvard University Press.

O'Bryen, R. (2020). Untangling the Mangrove: Slow Violence and the Environmentalism of the Poor in the Colombian Caribbean. In L. Blackmore & L. Gómez (eds.). *Liquid Ecologies in Latin American and Caribbean Art* (pp. 73–88). Routledge.

Personal communication. Community members in Medialuna. June 2024.

Quiroga, C. & A. Castelblanco. (2024). Beyond water justice and water security- debates on water, women, and climate change in Latin America. In T. Acevedo-Guerrero, et al. (eds.). *Routledge Handbook of Gender and Water Governance*. Routledge. https://doi.org/10.4324/9781003100379.

Rodríguez-de-Francisco, J. C., et al. (2019). Payment for ecosystem services and the water-energy-food nexus: Securing resource flows for the affluent? *Water*, *11*(6), 1143.

Ruiz-Agudelo, C. A. (2024). de Paula Gutiérrez-Bonilla, F. The remnant natural capital of the Magdalena-Cauca basin: immense losses for the 80% of Colombian inhabitants. *Journal of Environmental Studies and Sciences*, *14*, 135–153. https://doi.org/10.1007/s13412-023-00873-2.

Santos Da Silva, S. R., et al. (2019). The Paris pledges and the energy-water- land nexus in Latin America: exploring implications of greenhouse gas emission reductions. *PLoS One*, *14*(4), e0215013.

Schneiderman, J. S. (2012). Awake in the anthropocene. *Contemporary Buddhism*, *13*(1), 83–97.

Superservicios. (2023). Instructivo para el reporte del formato de estratificación y coberturas- REC. Versión 3 - Junio 2023. https://sui.superservicios.gov.co/sites/default/files/inline-files/Instructivo-Reporte-de- Estratificaci%C3%B3n-Coberturas-REC%20%281%29.pdf.

Superservicios. (2024). Información tarifaria. Servicio público de energía eléctrica. Principales comercializadores integrados al operador de Red Agosto 2024. https://www.superservicios.gov.co/Empresas-vigiladas/Energia-y-gas- combustible/Energia/Tarifas.

UPRA. (2023). Unidad de Planeación Rural Agropecuaria. Planificación del ordenamiento productivo y social de la propiedad rural en el Departamento del Magdalena. https://upra.gov.co/Kit_Territorial/2-%20Informaci%C3%B3n%20por%20Departamentos/MAGDALENA/ (accessed 16 September 2024).

USAID. (2023). Food security and gender. Factsheet. https://pdf.usaid.gov/pdf_docs/pnadr706.pdf.

Villamor, G. B., et al. (2020). Gender specific differences of smallholder farm households perspective of food-energy-land nexus frameworks in Ethiopia. *Frontiers in Sustainable Food Systems*, *4*, 491725.

Voelker, T., et al. (2022). The role of metrics in the governance of the water-energy-food nexus within the European Commission. *Journal of Rural Studies*, *92*, 473–481.

Wolde, Z, et al. (2020). Local community perceptions toward livelihood and water–energy–food nexus: a perspective on food security. *Food and Energy Security*, 9e, 207. https://doi.org/10.1002/fes3.207.

World Bank Group. (2023). Climate Risk Country Profile: Colombia. https://climateknowledgeportal.worldbank.org/sites/default/files/country-profiles/16698- WB _Colombia%20Country%20Profile-WEB.pdf.

Yacomelo-Hernández, M. J., et al. (2021). Análisis nutricional y niveles de extracción del mango cv. azúcar en el Departamento del Magdalena, Colombia. *Bioagro*, 33(2), 79–90.

Yung, L., et al. (2019). How methods for navigating uncertainty connect science and policy at the water-energy-food nexus. *Frontiers in Environmental Science*, *7*, 37.

Zabala, E. (10 de Junio de 2014). Medialuna, Magdalena. Wordpress. Obtenido de https://medialunamagdalena.wordpress.com/2014/06/10/media-luna-magdalena/.

Zarei, M. (2020). The water-energy-food nexus: a holistic approach for resource security in Iran, Iraq, and Turkey. *Water-Energy Nexus*, *3*, 81–94.

Zona Bananera Mayor's Office, Magdalena (2024). Datos de interés. Accessed 15 September 2024. https://www.zonabananera-magdalena.gov.co/Paginas/default.aspx.

Zorio, S. (2016). Anexo Temático II: Sobre derechos de las mujeres a la tierra. Bogotá: Disponible en. https://antiguo.igac.gov.co/sites/igac.gov.co/files/normograma/anexo_tematico_ii_- sobre_derechos_de_las_mujeres_a_la_tierra.pdf.

Zwarteveen, M. (2008). Men, masculinities and water powers in irrigation. *Water Alternatives*, *1*(1), 111–130.

Zwarteveen, M. (2017). Hydrocracies, engineers and power: questioning masculinities in water. *Engineering Studies*, *9*, 78–94.

Zwarteveen et al. (2024). Caring for Groundwater. *International Journal of the Commons*, *18*(1), 384–396.

Chapter 20

Women smallholder farmers growing horticultural crops

Francis Denisse McLean-Rodríguez and Arie Sanders, Zamorano University, Honduras

1 Introduction

Women's involvement in the agri-food sector in Honduras is steadily growing in significance. Various factors, such as global and local labor migration, women's increased mobility, agricultural commercialization, and climate adaptation, are driving the feminization of agriculture. These factors contribute to changes in gendered labor patterns in rural areas, resulting in more women working in agriculture. Men leaving rural areas may also contribute to the feminization of agriculture as women become more involved in farming (Kawarazuka et al., 2022). Men's migration, whether temporary or permanent, results in women bearing a heavier workload in both farm and household spheres. Conversely, it is contended that this allows for a greater emphasis on women's participation in agriculture and provides them with more opportunities for decision-making and generating income. However, women frequently encounter limited access to land, labor, capital, credit, and agricultural extension services compared to men. Additionally, they often find themselves in precarious wage employment positions within the agricultural sector. Insufficient access to vital resources hinders women from progressing beyond small-scale agriculture and assuming influential roles in the value chains.

We investigate the impact of this agricultural feminization process by examining the case of an indigenous women's organization in Intibucá, a

http://dx.doi.org/10.19103/AS.2024.0148.26

department in Western Honduras. The Association of Renewed Intibucan Women (AMIR, by its Spanish acronym), has actively empowered indigenous women to take on leadership roles in their communities. Through training programs and access to resources, these women have been able to increase their agricultural productivity and improve their livelihoods. Additionally, the organization has fought to recognize women's rights in land ownership, challenging traditional gender norms and creating a more equitable agricultural landscape (Velasco Palacios, 2022). Through their efforts, AMIR is an interesting example of how a grassroots organization can forge its path for its development despite an adverse context.

We analyze the experience of AMIR through the lens of a "new materialist sociology" framework as systematized by Fox and Alldred (2017) that rejects the traditional sociological dualities such as agency/structure, nature/culture and micro/macro, in favor of a flat or "monist" social ontology. This monist ontology has considerable significance concerning gender issues in agriculture. It challenges traditional and fixed ideas that have traditionally influenced gender roles. Moving away from essentialist perspectives and rejecting fixed dichotomies enables a more nuanced understanding of gender dynamics. In the context of AMIR, this means acknowledging that gender identities and power relations are not predetermined but rather emerge from complex interactions within assemblages of agricultural practices. The monist ontology prompts reconsidering how gender is constructed and experienced in AMIR members' daily activities and interactions. The question we address is: How do women farmers affect or are affected by the production of horticultural crops? This chapter centers on organized women farmers residing in remote rural areas characterized by an agricultural context and indigenous Lenca culture. The interviewed participants actively produce horticultural crops to improve their livelihoods, while at the same time building self-esteem and strengthening their identity as farmers.

2 Theoretical framework

Gender-sensitive analysis is critical for understanding women's roles in agrifood systems and ensuring equitable and sustainable development. Women are farmers, entrepreneurs, processors, and caregivers in food systems. Their contributions and needs, however, are frequently overlooked or undervalued. Gender-sensitive analysis reveals the unique challenges that women face, such as limited access to resources, discriminatory practices, and social norms that limit their empowerment. It also emphasizes the transformative potential of empowering women, who can be powerful change agents in food security, poverty reduction, and resilience.

According to the FAO (2023), to achieve a more significant and influential method of empowering women in agriculture, it is crucial to recognize the interconnection between reproductive and productive aspects in economic activities linked to the market. This recognition will help create food systems more attuned to gender-related issues. According to feminist economic theory, non-monetized reproductive activities, such as childcare, cooking, and other domestic tasks, are essential for functioning productive activities and maintaining and reproducing the labor force (Njuki et al.,2016). Reproductive responsibilities profoundly affect women's engagement in productive activities, influencing their circumstances and choices within the food system. From a constructivist viewpoint, it becomes evident that women are intricately woven into and shaped by the dynamics of a gendered society. The gender roles assigned to men and women in agrifood systems affect their access to resources. They also influence women's role and degree of participation, control over decisions affecting the management of the agrifood system, and use of benefits generated by the activity (Sachs, 1996).

The utilization of agrifood systems and gender theory to examine the participation of women in agricultural value chains poses specific ontological challenges. There is a tendency to prioritize the distinction of tasks carried out by women and men within the system and ascertain whether women or men were the primary beneficiaries and gained access to the newly generated income. These studies commonly investigate the distribution of decision-making power among various levels and the individuals who assume leadership positions. While many studies reach nuanced conclusions, using a "gendered" approach carries the danger of portraying agrifood systems as mechanistic entities controlled exclusively by macro-level structuring forces. It is thus crucial to approach relationships involving dominance and submission from a critical feminist standpoint. The exclusion of women (as well as other marginalized groups) from agrifood systems cannot be uniquely attributed to factors such as market capitalism or patriarchy. This analysis examines the core arguments and outcomes of new materialism in relation to gender-sensitive analysis within agrifood systems. Through this approach, we explore the complex interaction between human and non-human matter, challenging traditional concepts of power dynamics and the capacity to act within these systems.

Our theoretical and methodological framework is based on the work of Fox and Alldred (2017) on the subject of new social materialism and their ethological toolkit (Fox & Alldred, 2022b). Additionally, we integrate their concepts of material dis/advantage, as articulated in their more recent work (Fox & Alldred, 2022a). The concept of dis/advantage refers to the differential opportunities and constraints that individuals or groups experience regarding social, economic, and physical aspects of their lives. It recognizes that advantages and disadvantages are not fixed or essentialist properties of individuals but rather relational divergences across a population. Dis/advantage can be influenced by various socio-material factors such as income, health, and well-being. The emphasis lies on the daily interactions between humans and non-human matter in the "field" within the framework of the production of advantages or disadvantages that extend beyond human involvement. These interactions can lead to what Fox and Gavrilyuk (2022) call "tiny dis/advantages," which are the various opportunities and limitations that arise from these encounters. Although certain dis/advantages may be temporary, others can endure and accrue over time, resulting in enduring dis/advantage, social stratification, and disparities. The concept of dis/advantage disrupts conventional understandings of social hierarchy and class analysis by highlighting the detailed scrutiny of everyday interactions and the intricate dynamics of emotions that contribute to the varying levels of advantages or disadvantages individuals encounter.

To translate the ontology into more methodological terms, Fox and Alldred (2022b) use four ethological concepts – affects, capacities, assemblages, and micropolitics – which are key components of the new materialist approach used in our study. Affects refer to the capacities for affecting and being affected that exist within matter, both human and non-human. Affects are not fixed attributes but emerge in specific contexts when bodies or other matter interact with each other. For example, in the context of horticulture production, land can have different affects depending on the context, such as being a factor of production, guarantee for a loan, source of livelihood, or cultural heritage. Affects can be physical, psychological, social, cultural, economic, or political.

Capacities are the abilities or potentials that emerge from an assemblage's affects. They are the things that bodies or non-human matter can do in a specific event or encounter. Capacities are context-dependent and can either enhance or limit opportunities for action or interaction. In a horticulture production assemblage, for example, capacities can influence the growth cycle or the quality of the product.

Assemblages are complex and unpredictably connected networks of human and non-human matter. They are the result of specific actions and events that bring matter together. Tools, technologies, infrastructure, and social relations are all examples of assemblages. Affects within an assemblage determine what bodies and non-human matter are capable of doing in a given situation. For instance, consider the assemblage of a drip irrigation system where technological infrastructure (pump, tubing, and drippers) and social relations (irrigation board) contribute to the dis/advantages experienced by different groups in terms of accessibility and effective participation.

Power relations and dynamics that emerge within assemblages and affect the capacities of bodies are referred to as micropolitics. They are concerned with affect flows and the capacities they generate within a specific context, such as a horticulture value chain or community. Micropolitics shapes what bodies can do and helps to deploy powers and resistance. Dis/advantages can be created through micropolitics, as certain capacities can enable or constrain individuals' actions and opportunities. Micropolitics, for example, can determine the access to knowledge and technologies in a horticulture value chain, which can have implications for women´s crop yield, income, and well-being.

Our conceptual framework for the case study about AMIR centers around these four ethological concepts. This framework serves as our lens for understanding how the interplay of more-than-human materialities can shape social dis/advantage within the context of horticulture production by indigenous women in Honduras. It underscores the inherently relational and context-specific nature of these influences, emphasizing the significance of exploring the interactions between human and non-human matter in unpacking the dynamics of the advantages and disadvantages that women farmers producing horticultural crops experience.

3 Materials and methods

Our analysis focuses on the case of AMIR, on members who are women smallholder farmers growing horticultural crops. AMIR originated in 1980 within the Catholic Church in La Esperanza, Intibucá, as a grassroots women's movement focused on defending the interests of Lenca women and their families. The group eventually parted ways because the interests of AMIR seemed incompatible with church assembly authorities. The separation process led to AMIR losing around 2200 of its 2500 members. Over time, AMIR has undergone a process of resistance and reconstruction, evolving from a self-organized group hosting training on leadership, empowerment, and women's rights in members' own houses, to a formalized NGO implementing development projects on organic horticulture, offering microcredit, and owning small grocery stores and plots for collective administration and cultivation. Training now encompasses gender issues, food security, value chain integration, and the inclusion of women in public spaces for decision-making and public policy advocacy processes. The organization has a long history of cooperation with municipal authorities and international NGOs. Currently, 650 Lenca women are part of the group, organized into 25 subgroups from various communities from Intibucá. We chose the case of AMIR because it is a feminist organization, and it is now one of the few organizations in Honduras working with horticulture integrated exclusively by women. Furthermore, although the organization has chosen not to be certified as organic, while working with horticulture it has chosen to prioritize organic practices. As such, we were interested in understanding how the production of horticultural crops fitted in the process of AMIR's evolution and feminist struggle.

Our study set out to understand from a new materialist perspective how the assemblage of relations surrounding AMIR members and horticultural crops is constructed and plays out. To elicit narratives about horticultural crop production, we asked interviewees about how they started growing horticultural crops, how the crops they cultivate have changed over time, and how their plots, production and consumption practices, homes, families, communities, buyers, networks, and sense of self have changed over time. To get a better sense of positive and negative affects, and the enabling and constraining capacities produced, we asked interviewees which were for them the positive and negative aspects of growing horticultural crops, what plans they have for the future and what aspects could they change in relation to horticultural crop production if they could. Such questions do not seek to privilege human agents and the social construction of their world over other agents in the assemblage, but rather to approach the assemblage from the point of view of key informants (Fox et al., 2018). Consistent with the de-privileging of human subjects, no sociodemographic data about interviewees other than the community of origin is reported. The interview guide used for this chapter can be found in Annex 1.

We report data from 27 adult women smallholder farmers growing horticultural crops. To reach AMIR members, interviewers joined AMIR technicians for a week during their training and follow-up visits to various subgroups. AMIR technicians helped identify the group members who could provide the most articulate accounts of their experience growing horticultural crops, including former and current group leaders and board members. While this did not produce a random sample, but rather biased the study toward the experiences of extroverted members, the broadest accounts were needed as, in this case, these accounts are the entry point to understanding the assemblage. After the training, interviewers explained to the selected participants the study's objective and their right to withhold responses or withdraw from the interview at any given moment. Members who agreed to participate and gave their verbal consent for the record were interviewed. Interviewees came from seven communities, with at least two participants per community.

Recorded interviews were transcribed and coded on Nvivo software version 14, following the coding framework provided by Fox and Alldred (2022b). According to this framework, we identified within the narratives:

- What relations between human and non-human matter can be identified within the assemblage?
- What kind of affects draw these relations into the assemblage (physical, psychological, sociocultural, economic, or political)?
- What are the capacities produced or constrained by these affects?

- What micropolitics emerge from the interplay of these changing capacities within relations?
- What outcomes of these dynamics can accumulate into long-lasting dis/advantages?

4 Findings

4.1 Material relations in the AMIR-horticulture assemblage

Table 1 provides an overview of the complex network of human and non-human matter relations that comprise the dynamic assemblage surrounding AMIR members and their involvement with horticultural crops. This assemblage includes human actors, institutional entities, non-human matter, natural and economic resources, inputs, pests and diseases, infrastructure, food items, locations, and practices.

Interviewees alluded to two key features that characterize the relationship between AMIR and horticultural crops. The first is that this relation was not part of the assemblage that gave origin to AMIR. AMIR's origin as a group of women interested in addressing women's concerns differs from other organized groups of farmers working in horticulture:

Table 1 Relations in the AMIR - horticultural crops assemblage

Type of relation	Example
Humans	
People	Women, men, children, neighbors, brokers ["coyotes"], direct buyers, foreign buyers, AMIR technicians
Institutions	AMIR, church, local government, NGOs
Non-human matter	
Horticultural crops	Cilantro, carrots, radishes, cabbage, beetroot, squashes, green beans, and broccoli, among others
Staple crops	Corn, beans
Cash crops	Potatoes, coffee
Natural resources	Land, water, climate
Economic resources	Money, labor
Inputs	Seeds, ridges, soil amendments, organic or chemical fertilizers and pesticides
Pest and diseases	Flies, other unspecified pests, frostbite
Infrastructure	Water reservoir, net houses, solar driers, roads
Foodstuff	Rice, eggs, improved recipes (with added vegetables)
Places	Plot, house, community, La Esperanza, Green Market ("Mercadito Verde"), schools
Practices	Agronomic, organic, culinary, nutritional

First it was a religious group, from there the organization [AMIR] was born, and it was not born because of a [rural development] project but because that space [for women participation] was not there. For example, they [the assembly authorities] made the agenda in the meetings and said, "we are going to discuss these points", and then there was never the participation of women in the agenda, nothing appeared. We asked for the floor, "we want to speak" [we said], "no, it does not appear in the agenda, you cannot speak, just raise your hand if you agree or disagree, just raise your hand" [they said], so we could not speak. E26

The second is that horticultural crops were brought into the assemblage as a strategy for AMIR members to justify their participation in the group while complying with their caregiving role. When AMIR started, members' husbands found that the benefits from women's participation in the trainings were not evident for the families and that women were neglecting their roles in the household – such as cooking and caring for children – to attend the meetings. Thus, women applied for a project to produce horticultural crops to create a tangible benefit for their families:

We made a proposal. Being on the board of directors I was seeing a lot. I told other members: "if we are in training sessions, if we are often in meetings, we need to eat, we need food. Why don't they make a proposal to obtain support for inputs, seeds, and we work. When the production is ready, we sell. We come to the meeting, we bring back home the earnings from the sales, and from there we take what to eat. That way our husbands will believe us. If not, they don't believe us. Because they are already tired of so many meetings. Of course, because [they say]: "my wife is the one who knows how to cook corn, watch over the animals, the children". And I was over there [in AMIR meetings] and I didn't bring back anything. So that idea opened our mind and we thought, this is going to benefit us, to have something to eat so that at least this sacrifice we make can be seen. E26

4.2 Affects and capacities produced in the AMIR-horticulture assemblage

Non-human matter involved in horticultural crop production had physical, economic, and socio-cultural affects on AMIR members, creating and constraining for them a series of capacities. Physically, horticultural crops allowed AMIR members to produce food for their families with resources within their reach. For example, small-scale organic horticulture requires fewer commercial inputs than conventional or large-scale horticulture or other cash crops. AMIR hired technicians who taught members how to produce organic soil amendments, fertilizers, and pesticides using their own labor and ingredients available at their own households or plots:

These [organic crops] take more time. But you invest less, because you get the organic material right here. If we make organic fertilizer, we only have to gather cattle

manure, dry grass or green grass, whatever it is, we only have to buy a few things, because we can get the organic material right here. E10
We use, for example, the ashes, the cow manure, in short... all the waste from the house, everything is compost. E16

As a result, horticultural crops allow women to produce food with less initial capital compared to other cash crops, such as potatoes. While other crops require farmers to buy seeds, fertilizers, and pesticides, seeds are the only major cash expense for organic horticultural crop production:

You have to choose cheaper crops, what you can maintain [...] Carrots for one "tarea" (436.81 m2) you spend only L 5,000. We are going to make L 15,000 [...] Carrots are profitable. For potato, you invest L 10,000 and you get half, L 5,000. It is not enough to plant potatoes and sell cheaply. The pesticides and fertilizer are very expensive. E21

Additionally, a relatively small plot of horticultural crops can provide sufficient produce to supply the family and a surplus for sale. This limited land requirement increased the capacity for women to negotiate with their husbands the access to the land, specially to the plots that are adjacent to the household:

I told my husband, no, the corn and bean plot over there and my crops here because I am here in the house and I am in the kitchen for a little while and then I come back [to the plot], I clean it, I add organic fertilizer, I water it, that makes it easier for me. E26

Accessing these plots is essential to make the work in the field time compatible with the reproductive activities that women are expected to perform in the household:

I had to get up in the morning, help my mother to leave it [the corn] grinded, as that is the first thing you do, and then you go to do the rest. What you do is that you work from 9 o'clock until 3 o'clock, and after 3 o'clock you go back to the house to do the kitchen. From 9 to 3 in the plot, you go there to clean the land to continue planting, and if there is anything to clean, you go to clean it. E18

Horticultural crop production also increased the availability of fresh food at members' houses and within the communities, thus the capacity of families to access these foods:

In the past, there was a sad situation because if you needed a bell pepper you had to go to La Esperanza, if you needed cilantro, if you needed a bunch of carrots. Imagine that now, if one member doesn't have it but the other one does, you don't go to town because here you have it. And that is the big difference, that we save time, we save money too because there [in town] things are more expensive, there you will get it at a price, and here you will get it at a more affordable price because it is within the community. E01

Economically, horticultural crops increased women's capacity to contribute to household finances either through savings in food expenses or through income from sales. Interviewees perceived this capacity as supplementary to the capacity to supply food for their family, and certainly as valuable.

> *That has improved the economic part, because imagine if I have to spend L 80 (US$ 1 roughly equivalent to L24.5 in 2023) for transportation to bring something from La Esperanza, I already have it here and I don't spend it, those L 80 are already useful to buy something else that is needed for the house, or I can save them to have them at home. E01*

Socio-culturally, because women cultivate horticultural crops mostly organically, they perceive these crops as safer and healthier foodstuff, more suitable for their families than conventional crops:

> *The engineer told us that crops had to be organic, because those chemicals are affecting our health. Because if you eat a product with chemicals, you are going to have many diseases. But if you eat organic, it will be different, it will not affect you at all. That helps your health a lot. E16*

Of course, horticultural crops and the conditions and practices required to cultivate them organically also constrain the capacities of AMIR members in other aspects, mostly mediated through physical affects. For example, climate's affect allows AMIR members to grow horticultural crops in their plots, but it constrains them to identify a supplementary water source for irrigation or to grow horticultural crops only during the rainy seasons:

> *We plant where there is water to irrigate with a motor. We must buy gasoline and irrigate. We irrigate cabbage, carrots, and potatoes. If we didn't have irrigation, we wouldn't plant anything, only when it rains, like now. Here and now, we can plant anywhere because it rains. E21*

The climate also constrains women to face losses when changes in weather patterns or extreme weather events occur:

> *When [hurricanes] ETA and IOTA happened, we were sad because the landslides took everything we had: irrigation, plots. The plots we had sank, they broke. E02*

Likewise, microclimatic conditions constrain the production of specific crops in some plots:

> *What I would like to plant here on this side, but I think that it was tested in this zone for a while and it did not work, is broccoli. Broccoli works, but in cold soils and here it is a bit too hot, so here broccoli tends to remain hollow and wither. E01*

Land's affect enables AMIR members to grow horticultural crops, but at the same time limits how much women can plant. Families must allocate enough land during the rainy season to grow staple crops to secure their basic diet throughout the year, which conditions the land available for horticultural crops:

> *Right now, we don't have enough land, so we leave a limited space to plant vegetables in this period, from May to November, because the corn harvest is underway. But from November to May, since there is more space for vegetables, we can expand our plantings and we can sell at a good price. But it is not possible at this time of the year because corn is planted. E01*

Organic horticultural crops require women to implement new agronomic and organic practices. Women did not apply these practices before participating in AMIR, either because it was their husbands who practiced agriculture, or because they applied only conventional practices. While AMIR members acknowledge the benefit of these practices, they also recognize that some are labor intensive:

> *Before we did not make ridges [...] We use the ridges for all crops because the soil is softer. [...] For rooting, it is better to have lots of loose soil... But it is a more difficult job [...]*
> *Before we burned what we called "garbage". But now with the training we have been told that it is better to make dead barriers. Because if we leave it [the plant residues] there and let it decompose, it becomes an organic fertilizer that helps the soil. The benefits we have seen are that the soil is more nutritious because we add a little bit of chicken manure or the same organic fertilizer that we have been taught to prepare here [...]*
> *The barriers help sloping terrains to not wash off when it rains, because the rain washes away everything that is there. But now, with live barriers, the fertile soil remains. There are fewer losses. E10*

Organic horticulture also constrains women to produce inputs such as natural pesticides by themselves to deal with pests and diseases:

> *They [AMIR technicians] have trained us in [how to make] the medicines as well. The one that goes with chili, onion, and everything they taught us. We apply this and it is good for the flies, because the flies do not get close to the chili. Before we did not do that. Before, we used to apply poison that we bought, but there was poison that did not work. Now the flies don't come close. Not long ago they gave us training on how to control the flies with nylon with oil or lard, and with that we can see that the flies stick to the nylon and the flies do not go to the plant. E18*

Finally, as perishables, horticultural crops constrain women's capacity to store any product for more than a few days. Instead, women must harvest produce when it reaches maturity and take the price they find at that moment in the market. In addition, the distance between the communities and the fresh market in La Esperanza constrains women to pay for transportation or accept the lower prices brokers offer to pick up the product. This physical affect is intensified by the precarious state of the unpaved roads:

Here the buyers buy from you cheaply. And because of the distance to go out to the road, what can you do? This is another disadvantage for us who are very involved. What do we do? We sell it at the price we are paid. Because if not the brokers say "well, if you want, I'll take it, and if not, we leave it". We just saw a carrot that we had. Here they were paying at L 8 a bunch and [...] in La Esperanza they were buying at a good price, which was double what they were paying here. E11

4.3 Micropolitics in the AMIR-horticulture assemblage

AMIR members did not just passively accept the newly found constraints in their capacities emerging from their relation with horticultural crops. Instead, with the support from AMIR technicians, they identified strategies to broaden to some extent their capacity to grow horticultural crops in the spaces they occupy, contributing to the micropolitics of the assemblage. For example, women learn and consciously apply agronomic practices such as crop rotations to help them break the cycle of pests and diseases.

Sometimes I sow one kind of vegetable, sometimes I sow another. For example, now that I have coriander, cabbage, and carrots, then I have radish and squash, to sow differently, to change them. That stops pests from spreading because if we sow only one crop then the pest spreads and sometimes it is difficult to destroy. E01

When the pressure of pests and diseases intensifies, women may apply small doses of synthetic pesticides to complement the effect of the organic pesticides:

Yes, we apply a bit of chemicals [pesticides], because sometimes with pure organic it is hard for the plant to get strong... so we always need a little bit, less of the chemical than usual, and then we also apply the organic part. E10

To choose which horticultural crops to plant, as well as when and where to plant them, women take the microclimatic conditions and changes in weather patterns they are aware of into account. For example:

I am now waiting for October and I am hoping perhaps to sow earlier because I usually sow the green beans from the 15th onwards to be able to sell them at Christmas, but today remember that the weather is not lasting so the harvests are lengthening the production [...] For that reason I am waiting for October 8th to start sowing about two "tareas" of green beans, to be able to achieve the price in December. I have been noticing that for about 2 or 3 years that it [the weather] is no longer the same. E01

Also, women use staggered planting to soften the effect of price fluctuation. Staggering planting also enables women to use horticultural crops for household consumption through the season, soften the effect of income fluctuation, and make more efficient use of the available land:

We sell radish and cilantro, but we plant very little, because if we plant too much it [the market] is supplied and we don't sell. That is why right now I sowed a ridge, and I am going to sow another one, so that it will come out in stages. I am also finishing the radishes and I am going to sow another ridge to sell. Now this is common, because one has to stagger, because one has to be sowing and when it [the harvest] is coming out the other little ridge is being prepared, to continue harvesting. And with green beans too, that's how I did. I sowed to sell and to eat. E19

Through AMIR and other development institutions, women have built various physical infrastructures that affect horticultural crops positively. For example: water reservoirs and irrigation systems enable women to store rainwater and irrigate horticultural crops:

Before, my mom, as she also received support to install a reservoir, she used to collect water there to irrigate, because then she filled it up when it rained, and that reservoir was used to supply the two plots. Before, we had to go and draw water to irrigate the plots so that they wouldn't dry out, but not now. E18

Meanwhile, net houses allow women to cultivate horticultural crops in plots that are exposed to frost, while solar driers allow them to harvest staple crops earlier in the season, freeing the land for horticulture:

I have built a net house. Right now, the harvest has just passed, I have had a great benefit because I work selling tomatoes and cucumbers, quite a lot. And the tomato is a vegetable that removes humidity, because of the climate, I had tried it outside the net house, and it burned when I was already at the peak of the harvest. At dawn, it was black. Once inside the net house, I was able to obtain production. E08

We have the dryer that will allow us to remove the corn and to put it there, and then to cut the stalks in the field and leave the residues in the soil and then to plant [horticultural crops] maybe at the end of November. E01

For commercialization, women apply negotiation strategies and rely on foreign buyers to obtain better prices for their crops. Less perishable crops such as grains allow women to store the product while they wait for the best buyer, increasing their bargaining capacity.

Right now, we have two "tareas" of carrots ready to sell. As I go to La Esperanza, we find the "coyotes" ready to grab the carrots. I tell them "L 20", because you cannot let yourself be outsmarted. "No", they say, "not L 20, L 15 yes". And since I already know the fair price is L 15… [I sell]. If I tell them L 15, they lower it to L 12. E21

Today with the knowledge we have, when we have enough beans, we sell to buyers who come from far away because they are more considerate, they pay a little better. E19

Finally, AMIR has implemented a biweekly fresh market known as Green Market ("Mercadito Verde") in La Esperanza. While the number of members and the size of the market constrains the volume of products women can sell, the market has made it possible for members to sell their harvest directly to consumers and obtain better prices:

> *Also, what we have improved now is that AMIR is helping us to sell on Saturdays at the "Mercadito Verde". There we have been able to sell better. We also go to the regular market. Since the "Mercadito Verde" is every two weeks, if a product comes out before the two weeks, it overripens. What I do is that, for example, my mother sells it in the market in La Esperanza. In the "Mercadito Verde", it sells better, not in large quantities, but it always sells. E18*

4.4 Reproduction and interruption of social dis/advantages

Interviews also showed how, when repeated over time, some of the capacities horticultural crops enabled in women in AMIR reproduce long-lasting social dis/advantages that translated into inequalities between women and men, while others have the opposite effect. In the first case, for example, horticultural crops have a socio-cultural affect. Through culinary and nutritional practices that AMIR members and government and school authorities promote, horticultural crops increase women's capacity to provide diverse, nutritious, and flavorful food to their families, supplementing in volume and nutritional content the staple ingredients usually available in the households.

> *When there is harvest, we eat it. For example, the radish is mixed with the egg and the carrot is mixed with the rice [...] the radish leaf is fried and with that we eat the tortillas. [...] We have changed, because we already know how to prepare the vegetables, or how to prepare a main dish. Because if we get a couple of eggs, it is not enough for the whole family, so we have to mix them with the available vegetables and then it is enough. And the family likes it. If I forget, they will bring the vegetables from the plot and tell me: "Make it the way you made it the other day, it was delicious". E21*

However, interviewees explained how women continue to assume most reproductive tasks, of

which meal preparation is an important part:

> *My daughter is the one who cooks, she is the one who spends her time in the kitchen [...] Right now she is not studying until next year, so she is the one who helps me. It has always been like that. Sometimes she has a meeting, then it's my turn, whether I want to or not, I must get into the kitchen. E16*

Women also continue to assume the task of caring for children, including the preparation of school meals. Local schools are organized so that every two weeks each family must bring the school meal for all the children in the grade. Women use horticultural crops as ingredients in these meals, reproducing in the process their role as the primary caretakers in the family:

> *At the school meal [...] when it's my turn, I incorporate a vegetable in the meal. From school they send you rice, so I make rice with green beans, peppers, carrots... the kids think it tastes good. And you can also see... personally, if I come to school to leave a meal, the kids don't leave leftovers, they even have seconds. Sometimes I see that when others prepare the meal there are leftovers. You have to look for ways to make the children feel good too. E10*

However, interviews also showed how horticultural crops enabled capacities in women that over time reduced dis/advantages and social inequalities between women and men. For example, women's participation in a productive activity traditionally performed by men, and their control over the income from sales has strengthened their individual position in their household. As a result, women can better distribute rights, roles, and responsibilities between them, their husbands, and children:

> *Before we didn't work, we didn't pick up the hoe, it was only the men. I didn't work [in the field] but now I do. I have my own plot, I work. Before I was only in the kitchen, making tortillas and taking care of the children. I wasn't in charge of production. [...] Now I sow, I harvest. I work together with my husband [...] Before I used to ask my husband. I used to tell him I was going to such and such training. But now I don't. I just tell him I'm leaving, but I don't ask him for permission. Before I was embarrassed to ask him because I didn't have money. But now it is not the case. E04*
>
> *That has changed a lot, because it involves everyone, from husband to children [...] Everyone has been involved, from the children, we work on the plot. Before it was just the husband working and me at home. Not anymore, now we are equal, you go out to work and, in the afternoon, everyone is also involved in the house. E08*

Horticultural crop production combined with training in leadership and women's rights had a psychological affect, as it enabled AMIR members to empower themselves to make decisions about their own lives:

> *One of the things I didn't do before was to go out to meetings, to training... you devoted 100% of your time to work at home. Now I feel more independent. You go out, you do anything, and we have also had that thanks to the support of AMIR, which has made us know that women are not only meant to be at home, but to perform in various aspects [...] Most AMIR members only have a sixth-grade education. So even to speak, to express ourselves, to claim our rights was difficult for us. Thanks to AMIR we have been able to empower ourselves and defend our rights. E08*

As outcomes became evident for other people in the community, horticultural crops had a socio-cultural affect, as they increased the capacity of AMIR to attract new members, including members who had left the group. Request for membership expanded so much over time, that the board decided to only accept daughters of current members as new members. As this transition unfolds, interviewees explain how, as an outcome of these accumulated capacities and micropolitics in the assemblage, their daughters had a different experience to them joining the organization:

> *You had to get permission [from your father or husband to join the organization]. If he gave you permission you would go, if he did not, you would not, and most of us women went through the same thing. Now when I said to my daughter, "Are you going to join?" "Yes, I already have the money to join" she said. Then she asked for the floor and said, "Will you accept me as a member?" Now there is a different mentality, clearer. She did not ask for our permission to join but she just said: "I am going to join". E26*

5 Discussion and conclusion

In this chapter, we applied a new materialist perspective to characterize and understand the assemblage around AMIR members and the production of horticultural crops. Applying an ethological conceptual framework comprising relations, affects, capacities, and micropolitics, to interpret interview data provided a more thorough understanding than a conventional sociological perspective would have provided of the forces at play in a social phenomenon such as the struggle of a group of rural indigenous women for autonomy, empowerment, and development. This new materialist perspective provided insight into three key points on which we will focus our discussion.

First, an understanding of everyday events as the elements that assemble and produce the social world, rather than to perceive it as the product of structural forces at a macro level, allowed us to understand women not as passive subjects of these forces but as agents of their own development through their everyday choices and actions (Fox et al., 2018). First, AMIR members had a political affect over church assembly authorities and created their own capacity to discuss and take action on the issues of their concern. Later, AMIR members had a political affect on their families and found an alternative to resolve their husbands' resistance to their participation in the group. Finally, AMIR members had and continue to have a physical affect on the non-human matter around them and enhance their own capacity to produce, consume, and sell horticultural crops.

As a tiny dis/advantage, the fact that AMIR members continue to assume the primary responsibility for meals preparation for their families and for school children, and for making these meals diverse, flavorful, and nutritious, contributes to the long-lasting dis/advantage between women and men in which women assume most reproductive roles in the households. However, AMIR members chose to maintain these long-lasting dis/advantages to enable their capacity to discontinue others. Accumulated over time, the capacities of AMIR members to produce horticultural crops are building a long-lasting equality, so that women as well as men can participate in and economically benefit from horticulture or any other productive activities. Eventually, AMIR members and other groups seeking economic opportunities for women need to consider the cumulative effect of these double reproductive and productive roles on women's empowerment, as jointly they constrain women's time and energy. Relations and affects that broaden the capacity of men and other family members to participate in reproductive activities are needed to move towards full equality.

Second, by acknowledging and analyzing the agency of non-human matter, we were able to understand how non-human matter affects humans, and constrain and broaden their capacities (Fox & Alldred, 2016). In this assemblage, horticultural crops, staple crops, pests and diseases, and natural resources such as land and water had a physical affect on AMIR members and constrained when and where women could practice horticulture. This analysis goes beyond the understanding of buyers and markets as the main forces shaping what farmers cultivate. But horticultural crops also had other physical, economic, and sociocultural affects on AMIR members, broadening their capacity to secure nutritious food and a source of income for their families and communities. As the benefits of the capacities horticultural crops enable for AMIR members continue to outweigh the drawbacks from the constrained capacities, AMIR members will continue to devise strategies to maintain their relation with horticultural crops.

Third, a relational understanding of the social world in which the affective capacities of relations emerge as a property of the assemblage, rather than being an essential quality of humans and non-human matter, allows us to consider alternative scenarios (Fox et al., 2018). Our findings illustrate the capacities that horticultural crops enabled and constrained for women in AMIR, which were an important part of the reason why AMIR members brought horticultural crops into the assemblage in the first place. However, the described affects of horticultural crops should be understood as specific for the studied assemblage. Our theoretical perspective implies that horticultural crops may not have produced the same capacities in a different assemblage comprising a different set of relations. This is evident in the fact that horticultural crops do not necessarily produce the capacity to provide nutritious food for families and communities for farmers in other assemblages in the same geographical area; see, for example, the work of Fischer and Benson (2006) about the production of broccoli in Guatemala. In other assemblages of horticultural crop production, women are not or only partially involved, as in the case of "Vegetables Lenca," a local producers association that produces vegetables for supermarkets (Ullaguari et al., 2023). Thus, these assemblages do not produce the capacity for women to become involved in productive activities and control the income obtained from sales.

Importantly, this relational understanding also implies that choices of economic activities other than producing horticultural crops may not have produced the same capacities in the assemblage of women in AMIR. For example, local cash crops such as coffee and potato require more external inputs, including seed or planting material, fertilizers, and pesticides, than organic horticultural crops. They require access to larger plots to be produced commercially, which women would have more difficulty negotiating with men in their families. Coffee and potatoes are high-value crops with less demand than horticultural crops within the community. Other economic activities that women tend to perform in the region, such as weaving, craft making, or food processing, require significant initial investment and have a smaller local market than horticultural crops. They contribute to the household's strategy of income diversification but their impact on household food security is indirect (Michel & Walker, 2020).

As relationships within the assemblage undergo shifts, with some connections fading away and new ones being incorporated, the affective capacities discussed in this chapter are poised for transformation. A clear illustration of this evolution can be observed in the changing interests of AMIR members over time. With the founding members aging and the subsequent entry of their daughters into the organization, a growing demand for productive activities that are less physically demanding than horticulture emerges. Simultaneously, the aspirations of the youth to pursue higher education and engage with modern technologies introduce a need for activities aligned with these goals. The once central role of horticultural crops may undergo alterations, and their capacity to meet the evolving needs of AMIR members could shift. Consequently, it is likely that AMIR members will adapt their choices of productive activities or even redefine the trajectory of their organization to better align with the emerging needs within a new assemblage. The crucial takeaway from this examination of affective capacities lies in emphasizing the significance of contextual knowledge. This knowledge is instrumental in guiding the design of public or private development interventions to support women smallholder farmers and mitigating inequality, recognizing the dynamic nature of the socio-material landscape.

6 Where to look for further information

For an introduction to the new materialist sociology including examples of varied applications see:

- Fox, N. J., & Alldred, P. (2017). *Sociology and the new materialism: Theory, research, action* (1st edition). Thousand Oaks CA: Sage. For additional scholarly work on the case of AMIR as women smallholder farmers growing horticultural crops in Honduras see:

- Reyes, A., Velasco, H., García, M., & Pérez, O. (2022). Facing COVID-19 in rural Honduras: experiences of an indigenous women's association. In C. E. Sachs, P. Castellanos, & A. R. Tickamyer (Eds.), *Gender, Food and COVID-19* (pp. 63–72). London: Routledge. https://doi.org/10.4324/9781003198277-10
- Sanders, A. (2021). Ensamblando la cadena de cultivos de alto valor en el occidente de Honduras. *Eutopía. Revista De Desarrollo Económico Territorial*. (20), 97–112. https://doi.org/10.17141/eutopia.20.2021.5160
- Velasco Palacios, H., Castellanos, P., Jensen, L., Larson, J., Reyes Rocha, F. A., Sachs, C., Sexsmith, K. (2024). Perceptions and Experiences of Gender Transformative Approaches in Rural Honduras. *Rural Sociology*. Advance online publication. https://doi.org/10.1111/ruso.12567

Annex: Interview guide (English version)

Production

1. When did you start growing horticultural crops? How did you start? After that, did you continue growing/planting until now? If you stopped growing for some time, how did you get back to growing/planting?
2. How have the crops you have grown changed over the course of your life? What do you currently grow? How did you change from one crop to another?
3. How has your life changed with horticultural crop production so far?
   a. How has your plot changed: What was there before you planted horticultural crops on your plot? What is there now? What did you do before you planted horticultural crops on your plot that is different from what you do now? What do you do now?
   b. How have your production practices changed: What did you do before you planted horticultural crops in your production? What do you do now?
   c. How has your house changed: What was there before you planted horticultural crops in your house? What is there now? What did you do before you planted horticultural crops in your house? What do you do now?
   d. How has your family changed: What was there before you planted horticultural crops in your family? What is there now? What did you do before you planted horticultural crops in your family? What do you do now?
   e. How has your community changed: What was there before you grew horticultural crops in your community? What is there now? What did

you do before you grew horticultural crops in your community? What do you do now?

- f How have your buyers changed: What buyers did you have before you planted horticultural crops? What buyers do you have now? What did you do with buyers before you planted horticultural crops? What do you do now?
- g How have other people you interact with changed: What other people did you interact with before you planted horticultural crops? What people do you interact with now? What did you do with these other people before you planted horticultural crops? What do you do now?
- h How have the groups you relate to changed: What groups did you relate to before planting horticultural crops? What groups do you relate to now? What did you do with these groups before planting horticultural crops? What do you do now?
- i How have the organizations (NGOs, churches, government, etc.) you interact with changed: Which organizations did you interact with before planting horticultural crops? Which organizations do you interact with now? What did you do with these organizations before planting horticultural crops? What do you do now?
- j How have you changed: What did you do before planting horticultural crops? What do you do now?
- k How did you feel before you planted horticultural crops? How do you feel now?
- l What did you think about yourself before you planted horticultural crops? What do you think now?
- m What did other people think or say about you before you planted horticultural crops? What do they think or say about you now?

4 What have been the positive/good aspects of growing/planting horticultural crops in your life?

5 What have been the negative/bad aspects of growing/planting horticultural crops in your life?

6 If you could change anything about the way you produce/plant horticultural crops, what would you change? How would this help your life?

7 If you could change anything about the way you sell/market horticultural crops, what would you change? How would this help your life?

8 When did you join AMIR? What activities have you been involved in working with AMIR? What is your opinion about your work with AMIR?

9 What plans do you have for the future for your horticultural crop production: What do you plan to do? What would you like to do? How will this help your life?

10 What plans do you have for the future other than horticultural crop production: What do you plan to do? What would you like to do? How will this help your life?

Consumption

1 Which of the horticultural crops you grow do you *consume at home*? How do you prepare them? What does your family think about these foods?
2 Which of the horticultural crops you grow do you *not eat at home*? What does your family think about these foods? How do you know these foods can be prepared?
3 Which of the horticultural crops you grow *are consumed in your community*? For example, at school meals or among your neighbors. How do you know these foods can be prepared? What does your community think about these foods?
4 Which of the horticultural crops you grow *are not consumed in your community*? What does your community think about these foods? How do you know these foods can be prepared?
5 Which of the horticultural crops you grow *do you sell outside your community*? Who are your buyers? Where do they come from? What do your buyers think about these foods? How do you know these foods can be prepared?
6 On a typical day, what do you usually eat at different mealtimes in your household? Can you give us an example of what you would eat at home on a typical day? (Probe for all mealtimes and snacks, but do not assume that all families eat 3 meals a day and/or snacks) How do you prepare these foods? What does your family think about these foods?
7 On a typical day, what ready-made grocery shop products (sodas, chips, cookies, candy, etc.) do you eat in your household? (This question may have already been at least partially answered in the previous question, but probe specifically for these products.) Who in your household consumes these products? How often do you consume these products (how many times a week)? What does your family think about these foods?

7 References

FAO. (2023). *The status of women in agrifood systems*. FAO. https://www.fao.org/documents/card/en/c/cc5343en https://doi.org/10.4060/cc5343en

Fischer, E. F. & Benson, P. (2006). *Broccoli and desire: Global connections and Maya struggles in postwar Guatemala*. Stanford, CA: Stanford University Press.

Fox, N. J. & Alldred, P. (2016). Sociology, environment and health: A materialist approach. *Public Health*, *141*, 287–293. https://doi.org/10.1016/j.puhe.2016.09.015

Fox, N. J. & Alldred, P. (2017). *Sociology and the new materialism: Theory, research, action* (1st edition). Thousand Oaks CA: Sage.

Fox, N. J. & Alldred, P. (2022a). Bodies, non-human matter and the micropolitical production of sociomaterial dis/advantage. *Journal of Sociology*, *58*(4), 499–516. https://doi.org/10.1177/14407833211002641

Fox, N. J. & Alldred, P. (2022b). Doing new materialist data analysis: a Spinozo-Deleuzian ethological toolkit. *International Journal of Social Research Methodology*, *25*(5), 625–638. https://doi.org/10.1080/13645579.2021.1933070

Fox, N. J. & Gavrilyuk, T. (2022). The more-than-human production of material dis/advantage: a Russian case study. *International Social Science Journal*, *72*(246), 1033–1051. https://doi.org/10.1111/issj.12373

Fox, N. J., et al. (2018). The micropolitics of obesity: materialism, markets and food sovereignty. *Sociology*, *52*(1), 111–127. https://doi.org/10.1177/0038038516647668

Kawarazuka, N., et al. (2022). Myths about the feminization of agriculture: implications for global food security. *Global Food Security*, *33*, 100611. https://doi.org/10.1016/j.gfs.2022.100611

Michel, V. & Walker, I. (2020). *Honduras jobs diagnostic* (Job Series No. 17). Washington, DC. http://hdl.handle.net/10986/33304

Njuki, J., et al. (2016). Introduction – gender, agriculture and food security: where are we. In J. Njuki et al. (Eds.), *Routledge studies in food, society and environment. Transforming Gender and Food security in the Global South* (pp. 1–18). Oxon: Routledge.

Sachs, C. E. (1996). *Gendered fields: Rural women, agriculture, and environment. Rural studies series of the Rural Sociological Society*. Boulder, CO and Oxford: Westview Press.

Ullaguari, J. X., et al. (2023). Farmer-producer organizations during the COVID-19 pandemic: examining the role of organization-producer relationships on coping, resiliency, and effectiveness. *Advancements in Agricultural Development*, *4*(3), 131–145. https://doi.org/10.37433/aad.v4i3.331

Velasco Palacios, H. (2022). *Assessing the effectiveness of participatory gender-responsive development projects in rural Honduras* (Thesis). State College.

Chapter 21

Women smallholder farmers growing beverage crops

Alissa Bilfield, Tulane University, USA

1 Introduction

Coffee and tea are two of the most ubiquitous beverage crops that are grown globally, and together they represent a significant share of the beverage industry overall. Even with the rise of packaged sodas, tea remains the second most popular beverage in the world, just below water. Coffee, though less sipped than tea by volume, is second only to crude oil in terms of its value (International Coffee Organization, 2015). At the center of these two industries are two crops that are cultivated by farmers around the world. Today 80% of all coffee is produced by 25 million smallholders, from the two domesticated coffee plants that have been cultivated for consumption, which include, *Coffea arabica* and *Coffea canephora*. Overall, around 125 million people worldwide depend on coffee for their livelihoods (Fairtrade UK, 2022). Smallholder farmers produce 60% of the world's tea, which is derived from the singular plant *Camellia sinensis* (FAO, 2022). Due to the value of these beverage crops, the production of coffee and tea represents an important point of focus for addressing global inequalities in agriculture for women in smallholder farming.

The geographic origins of coffee and tea are separated by thousands of miles. Unlike coffee, tea has a longer domesticated history by a few thousand

http://dx.doi.org/10.19103/AS.2024.0148.28

years. Tea historians have traced the human domestication of the tea plant to China according to the botanical record (Meegahakumbura et al., 2018). The original wild tea trees that predate domestication likely spanned a larger geographic area that includes not only China but the modern-day Chinese border in Yunnan through Myanmar, Laos, and Vietnam (Drew, 2019). Historically, tea in China and its other countries of origin had been predominantly managed by small family farms. More labor-intensive aspects of production were then coordinated at the village level (Vega, 2008).

The botanical history of coffee has been traced to the highlands of Ethiopia, Southeastern Sudan, and Northern Kenya, at some point making its way to Yemen (Wild, 2005). In Northern Africa, during the height of the Ottoman Empire, it was coffee that fueled the creativity, genius, and culture of the time. During this epoch, the popularity of coffee spread throughout the region where it became established as a key beverage from Cairo to Damascus and then to Istanbul. This resulted in the introduction of the coffee house as a meeting place where news, ideas, and politics were exchanged.

Beyond their botanical geographies and origins of domestication, these two beverages share a common modern history linked to colonization. While tea was being planted by colonizers in the 1800s, coffee was being planted in tandem. One of the first colonial powers to plant coffee as a cash crop during this period were the Dutch. They brought coffee seeds from the Mocha region in Yemen to Indonesia, where the Dutch East India Company first started growing coffee as a colonial crop in Java during the 1690s (Fasseur, 1992). From this point onward, coffee began to spread to other colonial outposts beyond Asia to the Caribbean and Central and South America. The modern coffee belt now spans the equator, with the top ten largest global producers split between Latin America, Africa, and South Asia (Szenthe, 2019).

The colonization of tea developed in tandem with the colonization of coffee. In some instances where coffee had failed to flourish more widely, it was supplanted with tea. The British East India Company had become intertwined in the Chinese tea trade in the 1800s. At this point, the British began to establish the plantation-style model for tea production throughout the British empire in Asia, and then about a century later introduced tea to Kenya (Rappaport, 2017).

In the context of colonialization, which initiated the widespread cultivation of both coffee and tea around the equator, land was usurped and local populations were enslaved and indentured (Law et al., 2013). During this period, the structural inequities of colonization directed the form of the early era of the global tea and coffee trade. Both of these crops were reliant on many human hands for the harvest. As a result, the profitability of these early industries was heavily dependent on the exploitation of local people and imported laborers. These workers lived on site in extremely impoverished conditions, as many still do today, their lives governed by the quantity of tea leaves or coffee cherries plucked. The plantation model was immensely profitable in its exploitative architecture, and it has laid the foundation of a system of deeply entrenched inequity that is only now beginning to be acknowledged, addressed, and altered (Curtin, 1998).

Many of the structural inequities of colonial attitudes, trading practices, and infrastructure remain in both the coffee and tea industries. Gender inequity is one element of the various facades of inequity. As a result of the gendered dimensions of land dispossession during this period, and historically gendered division of labor in agriculture, gender inequity has been embedded within the post-colonial agricultural system of coffee and tea production (Levien, 2019). However, a variety of forces are restructuring how these beverages are produced, traded, and consumed. Two primary forces that have most impacted women's experiences operating within these industries in the post-colonial era have been the concurrent rise of the fair-trade movement and the feminization of agriculture.

2 Fair trade

The incubation of the fair-trade movement is one of the early markers of this incremental shift in business practices within the coffee and tea industries away from the colonial model in the production of beverage crops. While originally focused on the coffee industry, the modern fair-trade system has spread to tea and other cash crop commodities. The goal of the fair-trade system is to provide a safety net for smallholder farmers by offering a fair price, as its name suggests. In addition, the system originally sought to promote environmentally sustainable growing practices, human rights (from gender equity to ending child labor), and it incentivized the creation and support of farmer-led agricultural cooperative organizations. In fact, it still does today. Fair trade has grown from a social movement of patchwork grassroots organizations and non-profits in the early twentieth century to a market-based industry that serves more than 1.6 million farmers and workers with global sales that have surpassed 8.95 billion dollars (Fairtrade International, 2018).

In its modern form, the fair-trade movement can be traced to the late 1940s, where parallel models were developing in the United States and Europe. First motivated by individuals and then by organized groups, individuals involved in the early movement were altruistically motivated and saw an opportunity to provide better markets to artisans and producers working in the developing world. Throughout the second half of the twentieth century, non-profit organizations across the globe perceived the need for fair-trade marketing organizations to support the burgeoning *trade not aid* dialogue within the international development community, which resulted in the creation of a variety of Southern Fair Trade Organizations that were linked to newly established fair-trade organizations in the Global North, with a goal of creating equity in international trade through partnership, dialogue, transparency, and respect (World Fairtrade Organization, 2018).

While the labeling standards have evolved, the main focus has been on providing a niche specialty market to small farmers and producers in an effort to assuage the inequities in the global agricultural market while building democratically enhanced community resilience. Up to this point, the main focus of fair trade had been on supporting smallholder producer cooperative organizations. These producer organizations would receive the fair-trade minimum price for their products, and in addition, they would also receive an annual social premium reimbursement that the cooperative could reinvest into their community. This could take the form of additional technical assistance, establishing small-scale agricultural facilities, developing an education scholarship fund for cooperative school children, or other projects focused on health, nutrition, or gender equity. The most important aspect of the premium was that the democratically organized cooperative associations would decide how to use the funds.

The fair-trade movement's focus on labor issues related to equity and justice has included fair wages, livelihood diversification initiatives, and the support of smallholder farmers and their cooperatives. As an integral part of the development of fair trade was also the support and development of the cooperative movement amongst smallholders. While the history of the cooperative movement predates the modern rise of the fair-trade movement, the rise of modern cooperative agricultural organizations across the Global South during the 1960s and 1970s has provided an institutional infrastructure to support smallholder farmers at each stage of the production process, while also enabling their eligibility for the formalized systems of fair-trade certification.

Over the last several decades, the fair-trade model has changed and evolved. Originally the program was limited to certifying smallholder farmer cooperative organizations to create a niche market. In this context, the system was broadly envisioned as a development mechanism to address entrenched rural poverty amongst agricultural communities by supporting economic, social, and environmental justice for cooperative organizations, smallholder farmers, and their families. In the case of industries where cooperative organizations have been more of the exception, as has been the case in the tea industry, the fair-trade system provides for exceptions and alternative requirements. More recently, to scale the system more widely, different organizations within the fair-trade movement have opted to also certify larger producers with more industrialized models, to create wider impact for not just smallholders and their cooperatives, but for agricultural workers who also stand to benefit from the system.

Whereas coffee cooperatives began proliferating in the 1960s and 1970s, stimulated by the interested in fair trade within the coffee industry, this shift to cooperative organization has only just recently begun to happen in the tea industry (Fairtrade International, 2021). The majority of tea today is still produced using a plantation style model, where wage workers pluck the tea and live onsite. In tea, the beginning of a new economic reorganization can be traced to various political and economic factors that include the rise and subsequent demand of fair trade a few decades after cooperative organizations began to proliferate in the coffee industry. Coffee was the initial focus of the fair-trade movement, paving the way for a plethora of other cash crops. In both coffee and tea, a byproduct of the fair-trade movement has been an increased focus on supply chain transparency, especially as it relates to social, economic, and environmental issues 'at origin'. Here, smallholder producers are managing the most labor-intensive and critical parts of growing and processing of these crops.

A critical catalyst of the fair-trade movement has been the rise of ethical consumerism. Although mutually reinforcing, this is another one of the driving forces behind the evolution of not just the coffee and tea industries but the rise in the consumer preferences that value sustainability (World Fair Trade Organization, 2017; Kituyi, 2014). As the momentum of fair trade and other associated movements has increased, collective consumer awareness of the impacts of industries like coffee and tea on farmers, workers, and the environment has also grown exponentially. This has included concern for gender equity across the supply chain in these industries. As a result, more consumers want not only higher quality coffee and tea but to also drive transformation through their purchasing patterns.

3 The feminization of agriculture

Alongside the fair-trade movement and institutionalization of various fair-trade certification programs has been the concurrent phenomena of more women participating in the formal economic sphere of agriculture. While men have historically been involved in cash crop agriculture a combination of factors has led to a 'feminization of agriculture' within coffee, tea, and other cash crop industries (FAO, 2010; Lastarria-Cornheil, 2008; Lyon, Bezaury, and Mutersbaugh, 2010). This trend has been defined by an increase in women's participation rates in the formal agricultural sector, either as self-employed producers or as agricultural wage workers, increasing the percentage of women who are economically active in rural areas. This also represents an increase in the percentage of women in the agricultural labor force relative to men, due to either more women working and/or fewer men working in agriculture (World Bank, 2006). The Food and Agriculture Organization estimates that while the proportion of the labor force working in agriculture began to decline during the 1990s, the proportion of women working in agriculture increased (FAO, 2010).

The feminization of agriculture has been fueled by a variety of national and transnational phenomena. Women producers in coffee and tea have long been marginalized as invisible labor in the production process. However, in some instances, internal conflicts due to civil wars, drug trafficking, and genocide have left women widows in charge of their land. In other instances, unemployment and the search for better opportunity elsewhere has led to the migration of family members to cities or outside the country, leaving women to tend to the family's land. As agriculture has become less profitable and more uncertain, research indicates that men are seeking alternative livelihoods, which often involves outmigration from rural areas into urban areas, or even transnationally (Lyon, 2007). While women have arguably been involved in agricultural production since it's advent, over the last few decades, women's contribution in the formal agricultural sector has become more visible, partially due to research and data collection aiming to measure this topic, and partially due to their actual increased involvement in the sector (FAO, 2010; Leder, 2022). As they have begun taking on more formal roles in these supply chains as cooperative members and leaders, partly as a result of the influence of the fair-trade movement.

The feminization of agriculture has also been tied to the intentional inclusion of women in agricultural development strategies that address rural poverty. Since the 1970s there has been an evolution in focus within the international development community toward more integrated approaches that address the interrelated conditions of poverty, health, and nutrition at the household level. This focus was expanded during the 1980s and 1990s to include building food sovereignty as a strategy within the context of sustainable livelihoods. Described in the literature as 'a means of gaining a living,' a focus on livelihoods involves the development of a complex web of activities that may involve not only cash crop agriculture but other small-scale enterprises, activities, and interactions (Chambers and Conway, 1992; Scoones, 2009). A key feature of the concept of sustainable livelihoods involves the diversification of assets and income sources combined with an improvement in abilities and well-being, which when woven together, augment individual and household resilience to external natural and market-based shocks (Scoones, 1998).

Within the broader field of international development, women's key role in agriculture and the importance of gender inclusion for achieving agricultural productivity in developing countries has been on the agenda for decades. As a result, gender equality and empowerment was first prioritized as the third millennium development goal and has now been listed as the fifth sustainable development goal (Grave, 2012; Kabeer, 2005; United Nations, 2013; United

Nations Development Program, 2015). A vast body of literature on gender, agriculture, and development has amassed over the last half century, pointing toward the critical importance of ending gender-based discrimination and empowering women as a way to achieve a 'multiplier effect' in development that addresses the intertwined determinants of poverty (Blumberg, 1988; Cole, Kantor, and Heddad, 2014; Heddad, Pena, Quisumbing, and Slack, 1995; Kabeer, 2013; Sarapura and Rajaratnam, 2014; Smith, Ramakrishnan, Ndiaye, Heddad, and Martorell, 2003; World Bank, 2006). This research has focused on the connection between women's empowerment in agriculture and improvements in health, nutrition, and livelihoods at the household level. A large portion of this research has and continues to focus on intra-household resource distribution and control over household income, as a mirror to the broader development literature (Blumberg, 1988; Dwyer and Bruce, 1988; Gummerson and Schenider, 2013; Thomas and Chen, 1994; Udry, 1994).

Women have always played a central role in agricultural production in developing countries as unpaid household labor and wage workers. As they step into formal roles as producers and cooperative members, they continue to have less access to productive resources such as land, inputs, information, credit and technical assistance, which has resulted in the 'gender gap' in agriculture (FAO, 2010; Rocheleau, 1988; Puskur et al., 2023). A variety of studies and reports have demonstrated that closing this gap could not only benefit agricultural production and lead to economic growth and increased food security but that it could also contribute to community development through gender-sensitive interventions that have been shown to improve women's overall well-being, in addition to enhancing child welfare and education (Fair trade, 2015; Malapit et al., 2015). Research has also shown that addressing the gender gap in agriculture requires more than just a focus on improving women's access to enhanced economic opportunities and material goods and services. It is also essential to concurrently address the underlying socio-cultural and political phenomena that perpetuate gender inequality and constrain women's capabilities (Apothekar, Pyburn, and Laven, 2012; Meinzen-Dick and Quisumbing, 2013; Kirk, 2012).

Addressing the underlying socio-cultural and political phenomena that perpetuate gender inequality requires a multilevel approach that includes women and men. Historically, women's inclusion in agriculture was operationalized through the Women-in-Development agenda of the 1970s, and then the Gender-in-Development agenda of the 1980s and early 1990s. Since the 1990s, there has been a shift in focus toward gender transformative approaches to women's involvement in development, and more specifically in agricultural growth in developing countries (Cole et al., 2014; Razavi and Miller, 1995). This has extended into the coffee and tea industries, where women have been participating more formally in higher rates as the feminization of agriculture has extended into the coffee and tea producer communities.

Gender transformative approaches have been articulated conceptually since the 1990s and embrace a more complex understanding of gender as a social construct, embedded in how societies define gender roles (Martin, 2004; Morgan, 2014; Young, 1993). These approaches have evolved out of Gender Theory, which is concerned with the intersectionality of gender with race, ethnicity, and other social constructs that contribute to oppression (Risman, 2004; Wharton, 2011). Gender Theory also focuses on analyzing and understanding gender inequality, predominantly through qualitative and participatory research. Gender transformative approaches address not only gender roles and power dynamics but also institutional and legal exclusion as a key barrier to equality, justice, and the achievement of global development outcomes. A main goal of gender-transformative approaches involves the creation of an enabling social environment and more equitable inclusion in formal and informal institutions that support expanded choices for women and men that move beyond superficial gender integration.

The next part of this chapter will discuss efforts to create more equitable inclusion in the formal institutions that support women smallholder producers in the coffee and tea industries. These changes can be observed in nested levels, with increased participation of women smallholders in cooperatives being one of the most catalytic changes. Further along the supply chain, greater representation of women as exporters, importers, roasters, retailers, and baristas also represents an important systemic shift that reinforces the position of women farmers in coffee and tea. Finally, national and global institutions that focus on providing technical assistance and production differentiation through certification highlights ways in which this emerging ecosystem of women-centered businesses in the coffee and tea industry synergistically propels gender transformative change in these industries.

4 Women in coffee and tea cooperatives

One of the major areas of change that has shifted the nature of women's participation in the feminization of agriculture within the coffee and tea industries has been the rise of cooperatives and the integration of women into these organizations as formal members. Whereas women have been involved in coffee and tea production since their domesticated advent, only recently have they participated equally in cooperative institutions in the same way as men. This change has been noted even beyond these industries, where at the global level, the International Labor Organization has demonstrated that there has been a 20% increase in women's participation as cooperative members globally (International Labor Organization, 2015). As institutional arrangements, agricultural cooperatives offer numerous benefits to members and can play an important role in supporting small-scale agricultural producers and marginalized groups including indigenous minorities and women. They seek to empower their members economically and socially and to create sustainable rural employment through business models that may be more resilient to economic and environmental shocks. In many cases, they also offer small agricultural producers a wide range of services in addition to improved access to markets, which may include improved access to natural resources, facilities, information, communication, technologies, credit, and training programs. Cooperatives also provide smallholder producers with the opportunity to participate in collaborative decision-making, support them in securing land-use rights, and negotiate better terms for their engagement in contract farming (FAO, 2010).

In the most recent survey by the International Labor Organization (ILO) and the International Cooperative Association (ICA) on gender equity and cooperatives, results indicate that cooperatives advance the status of women in a variety of ways, ranging from the opportunity structures of the organization to the guiding principles that create and reinforce a supportive social environment. Overall, findings emphasized that the noticeable advantages of co-operatives for women fall into three overlapping and mutually reinforcing categories: access to employment, improved conditions of work and social benefits. By offering financial, legal, and marketing services specifically tailored for women, co-operatives enable women to start and grow their own businesses. In addition, when women establish a co-operative, they gain access to (self-) employment, as well as provide jobs to other women (and men) as employees. Furthermore, some co-operatives specifically aim to employ women who are disadvantaged by factors such as poverty, gender-based violence, and limited access to education and training. This extends inclusion to the most marginalized women.

Extensive research has also shown the ways in which co-operatives directly affect women's access to employment, as well as improve their conditions of work. Co-operatives inherently provide a structure that promotes resource pools, which women use to produce and market their goods and services. Women also gain collective bargaining power in co-operatives and gain access to resources necessary for work, such as contracts with public authorities and other private sector institutions. Collective organizing within co-operatives also helps to create transparency in financial transactions and coordination of activities, improving enterprise efficiency and profitability – benefits which are in turn extended to all members by virtue of collective ownership. As noted previously, participation in co-operatives improves women's income and livelihoods. In fields such as food security, finance, housing, healthcare, childcare, and eldercare, co-operatives provide women with affordable and accessible services, which enable women to work while meeting their basic needs. With enhanced empowerment, for instance, women may be better positioned to address personal and communal needs such as freedom from violence and environmental protection.

In many cases, women's associations that operate within co-operatives may assist members in better identifying their interests, organizing and advocating for improved conditions of work. In a similar vein, member control affords flexibility and self-determination within the organization and with regards to the scheduling of work. This enables women to balance participation in co-operatives with other commitments and responsibilities, such as childrearing. Co-operatives lend women respect and recognition while developing their confidence to take on leadership roles within the workplace, family, and community at-large. Wider research has also demonstrated that members also gain training, skills, and leadership experience to which they would not otherwise have access (Bilfield, 2022; Hansen and Asmild, 2023). In some cases, co-operatives are at the forefront of innovation in the application of social, technological, and institutional innovation. In joining co-operatives where such innovation exists, members gain access to new skills, which help to develop their competitive advantage in the labor market.

While membership has the potential to provide access to resources and create fundamental changes for women, this hinges on the values and approaches of cooperative organizations themselves. Where power dynamics and social norms and stigmas pervade around gender equity, barriers still exist as they do outside of the institutional context. However, where cooperative organizations have the potential to harness the values of inclusivity and equity core to the fair-trade movement and the United Nations Sustainable Development Goals, observable progress is being made in support of women coffee and tea producers. There are many examples of coffee and tea cooperatives that have focused on gender equity as a core part of their organizational strategy. These cooperatives may be motivated by a variety of factors ranging from certification requirements that prioritize equity and inclusion to the practicalities of managing organizations with growing numbers of women in societies where this may be less common. At the same time, there has also been the rise of women specific cooperatives amongst coffee and tea producers. Two notable examples include social entreprises like the Kyambura Women's Coffee Cooperative in Uganda, and COMUCAP, an organization of women growers in Honduras. In both cases, these producer organizations were developed and uniquely tailored to empower women farmers. Kyambura was developed as a community-based initative to provide training to women and an alternative and sustainable source of income. Many of the women involved are HIV positive and widowed. COMUCAP was formed by women growers who had suffered domestic violence and economic insecurity due to gender discrimination. Through the cooperative, women have been able to purchase their own land, which has allowed them to generate greater economic stability while breaking a cycle of abuse.

5 Global supply chain support

Support for addressing global inequities in agriculture for women smallholders in coffee and tea extends beyond the context of the cooperative as an institution and can be observed along the global supply chain and within the coffee and tea industries overall. Over the last few decades, as more women have advanced their positions as farmers and cooperative members, so too have women become powerful business owners and leaders in the coffee and tea industries at other levels of the supply chain. This participation has resulted in the creation of several notable international organizations comprised of women business members in coffee and tea who have advocated for and created programs and support structures for women coffee and tea farmers. One of the major organizations that has been at the center of this work in the coffee industry has been the International Women's Coffee Alliance (IWCA) which was created in 2003. Since then, the organization has grown to represent over 33 different coffee-growing countries with a mission to empower women across the global coffee industry. Their work is focused on addressing the multiple challenges that women face in the coffee value chain, including underrepresentation in leadership roles.

The organization also provides expertise to address institutional and industry challenges related to gender equity, facilitates educational programming, and advocates for a more equitable industry. Through an international network of independent, self-organized, self-governing chapters, women in the coffee industry collaborate on empowerment initiatives through leadership development, strategic partnership, and amplified market visibility. The organization also serves as a conduit for partnerships with other multi-stakeholder organizations, including the African Fine Coffees Association, International Coffee Organization, National Coffee Association, Specialty Coffee Association, World Coffee Alliance, International Labor Organization, and the International Trade Center. In this role, IWCA serves as a critical connection point, centering the needs of women in the coffee industry. Through these partnerships, various organizations have been brought together to support improvements at the farm level, for farmer communities, and within the production process. Grant programs, events, and trainings are also essential ways that the IWCA network has helped coffee-producing communities and the coffee industry at large integrate gender equity and create real improvements (IWCA, 2023).

To date, no parallel organization currently exists in the tea industry at this moment, although IWCA presents an excellent transferrable model. Unlike the coffee industry, where there is a greater diversity of farm size and business structure, the tea industry remains primarily structured by larger scale plantation models. This highlights the fact that the tea industry, though it has made strides in the last few decades, still has much room for growth in terms of structural organization, and the institutional support available to women producers, business owners, and pluckers in tea-growing regions and across the globe where tea is sold and sipped.

6 Women-grown certifications

Around the same time that the IWCA was created in 2003, a model program called 'Café Feminino' was also taking shape. At the time, 464 women farmers in northern Peru decided to separate their coffee production from the men's to create their own product and income. Their coffee cooperative, CECANOR, joined their commercial partner Organic Products Trading Company to create Café Feminino, a gender-focused program to support social justice and empowerment for women coffee producers worldwide. Today, in addition to the participating producer cooperatives, there are 125 participating roasters across the United States, Canada, United Kingdom, and Australia.

To participate, cooperatives must provide women farmers with access to a range of political and economic benefits, which include the legal right to the land on which they farm, leadership positions within their cooperatives, financial and business decision-making power, and direct payment for their coffee, including a Café Feminino premium. Through their participation, women coffee farmers receive recognition from the work they do on the home and the farm, increase in family incomes, increase in gender parity in the home, and improvements in school attendance among girls in the family. Other country-based initiatives exist as well, in addition to branded coffees that demonstrate less formal arrangements yet still give credit to women coffee producers and raise awareness about women in coffee (Café Feminino, 2023).

To date, no parallel certification program currently exists in the tea industry at this moment, although Café Feminino presents an excellent transferrable model. This highlights the fact that the tea industry, though it has made strides in the last few decades, still has much room for growth not just in terms of institutional support available to women in tea, but through the creation of a formal certification program like Café Feminino that is focused on addressing gender inequities.

7 Conclusion

Over the last few decades, an increasing number of women smallholder beverage producers in the coffee and tea industry have participated as cooperative members and leaders alongside their peers. Their participation has not only icreased the ratio of women formally participating in these sectors, but more importantly they have worked collaboratively to address gender inequities in their particular agricultural sphere. This shift has occurred as a result of two main factors, the rise of the fair-trade movement, and the feminization of agriculture. In combination with other associated trends, these two global phenomena have accelerated the focus on gender equity in agriculture for important cash crops such as coffee and tea. In coffee, advanced efforts at the level of the global supply chain to create supportive organizations like the International Women's Coffee Alliance and certification programs like Café Feminino. While efforts in the coffee industry may currently be more advanced in terms of infrastructure and development, this progress points to what is possible not only for the other prominent beverage crop, tea, but for other. While the coffee and tea industries share a common colonial history that has shaped issues related to labor and equity.

8 Future trends

While there has been impressive progress to address global inequities in agriculture for women smallholder farmers growing beverage crops such as coffee and tea, there is still huge potential for future improvement. In coffee, where fair trade has been mainstreamed, efforts at the industry level have moved further beyond to include a plethora of other certifications, some like Café Feminino that specifically focus on gender equity. In that regard, there is future potential to enhance and mainstream this type of effort in coffee and to scale similar programs to the tea industry and beyond. Additional information on this topic can be found by referring to international and country-based organizations that are currently working on issues related to women in agriculture and industry groups focused on beverage crops, including many of the entities referenced in this chapter.

9 References

Apotheker, R., et al. (2012). Why focus on gender equity in agricultural value chains? In: *Challenging chains to change* (1st ed.). Accessed online: https://www.cordaid.org/en/publications/challenging-chains-change-gender-equity-agricultural-value-chain development/.

Bilfield, A. (2022). *Brewing sustainability in the coffee and tea industries: from producer to consumer*. London: Routledge.

Blumberg, R.L. (1988). Income under female versus male control: hypotheses from a theory of gender stratification and data from the third world. *Journal of Family Issues*, 9(1): 51–84.

Café Feminino. (2023). Our Story. Accessed online: https://www.cafefemenino.com/story

Chambers, R. and Conway, G. (1992). Sustainable rural livelihoods: practical concepts for the 21st century. In *IDS Discussion Paper*, Vol. 296, Brighton: IDS.

Cole, S., et al. (2014). *Gender-transformative approaches to address inequalities in food, nutrition and economic outcomes in aquatic agricultural systems*. Penang, Malaysia: CGIAR Research Program on Aquatic Agricultural Systems. Working Paper: AAS-201-42.

Curtin, P. 1998. *The rise and fall of the plantation complex: essays in atlantic history* (2nd ed., Studies in Comparative World History). Cambridge: Cambridge University Press. https://doi.org/10.1017/CBO978051181914

Drew, B. (2019). The growth of tea. *Nature* 566: S2–S4.

Dwyer, D. and Bruce, J. (1988). *A home divided: women and income in the third world*. Stanford, CA: Stanford University Press.

Fairtrade UK. (2022). Accessed online: https://www.fairtrade.org.uk/farmers-and workers/coffee/#:~:text=Coffee%20is%20one%20of%20the,on%20coffee%20for%20th r%20 livelihoods

Fairtrade International. (2018). "Coffee." Accessed online: https://www.fairtrade.net/product/coffee

Fair trade International. (2018). Coffee Farmers. Accessed online: http://www.fairtrade.org.uk/Farmers-and-Workers/Coffee

Fairtrade International. (2021). Time for a Stronger Fairtrade Tea Standard. Accessed online: https://www.fairtrade.net/news/time-for-a-stronger-fairtrade-standard-for-tea

Fair Trade International. (2015). *Equal harvest: removing the barriers to women's participation in smallholder agriculture*. Fairtrade Foundation Report.

Fasseur, C. (1992). *The politics of colonial exploitation: Java, the Dutch, and the Cultivation System* (No. 8). SEAP Publications.

Food and Agriculture Organization. (2010). *Gender dimensions of agricultural and rural employment: differentiated pathways out of poverty – status, trends, and gaps*. Rome, Italy: FAO.

Food and Agriculture Organization. (2022). *International tea market: market situations, prospects, and emerging issues*. Rome, Italy: FAO.

Grave, S. (2012). An empirical examination of women's empowerment and transformative change in the context of international development. *American Journal of Community Psychology*, 49: 233–245.

Gummerson, E. and Schneider, D. (2013). Eat, drink, man, woman: gender, income share and household expenditure in South Africa. *Social Forces*, 91(3): 813–836.

Hansen, H. O. and Asmild, M. (2023). Women's participation on the boards of farmer-owned cooperatives. *Frontiers in Veterinary Science*, 10: 1060817.

Heddad, L., et al. (1995). *Poverty and nutrition within households: review and new evidence*. Washington, DC: World Health Organization.

International Coffee Organization. (2015). Country Data on International Coffee Trade. Accessed online: http://www.ico.org/profiles_e.asp

International Labor Organization. (2015). Accessed online: https://www.ilo.org/global/aboutthe-ilo/newsroom/news/WCMS_349679/lang--en/index.htm

International Women's Coffee Alliance. (2023). What We Do. Accessed online: https://www.womenincoffee.org/what-we-do

Kabeer, N. (2005). Gender equality and women's empowerment: a critical analysis of the third millennium development goal. *Gender and Development*, 13(1): 13–24.

Kabeer, N. and Natali, L. (2013). Gender equality and economic growth: is there a win-win? IDS working paper No. 417.

Kirk, L. (2012). Gender Equality: It's Your Business. Briefings for Business. Oxfam International. https://www.oxfam.org/en/research/gender-equality-its-your-business

Kituyi, M. (2014). Fifty years of promoting trade and development. *International Trade Forum*, 2014(2): 28–31. https://doi.org/10.18356/db1c4549-en

Lastarria-cornheil, S. (2008). *The feminization of agriculture: trends and driving forces*. Washington, DC: The World Bank.

Law, R., et al. (Eds.). (2013). *Commercial agriculture, the slave trade and slavery in Atlantic Africa* (Vol. 4). Woodbridge: Boydell & Brewer Ltd.

Leder, S. (2022). Beyond the feminization of agriculture: rural out-migration, shifting gender relations and emerging spaces in natural resource management. *Journal of Rural Studies*, 91: 157–169.

Levien. (2019). Gender and land dispossession: a comparative analysis. In *Gender and generation in Southeast Asian agrarian transformations* (pp. 9–32). London: Routledge.

Lyon, S., et al. (2010). Gender equity in fair trade-organic coffee producer organizations: cases from Mesoamerica. *Geoforum*, 41: 93–103.

Lyon, S. (2007). Maya coffee farmers and fair trade: assessing the benefits and limitations of alternative markets. *Culture and Agriculture*, 20(2): 100–112.

Malapit, H., et al. (2015). *Instructional guide on the abbreviated women's empowerment in agriculture index*. Washington, DC: International Food Policy Research Institute.

Martin, P. (2004). Gender as social institution. *Social Forces*, 82(4). 1249–73.

Meegahakumbura, M. K., et al. (2018). Domestication origin and breeding history of the tea plant (Camellia sinensis) in China and India based on nuclear microsatellites and cpDNA sequence data. *Frontiers in Plant Science*, 8: 2270. https://doi.org/10.3389/fpls.2017.02270

Meinzen-Dick, R.S. and Quisumbing, A.R. (2013). Women in agriculture: Closing the gender gap. In: *2012 Global Food Policy Report Chapter 4* (pp. 37–47). Washington, DC: International Food Policy Research Institute. https://hdl.handle.net/10568/153443

Morgan, M. (2014). Measuring gender transformative change.

Puskur, R., et al. (2023). Closing gender gaps in productivity to advance gender equality and women's empowerment. CGIAR GENDER Impact Platform Working Paper.

Rappaport, E. (2017). *A thirst for empire: how tea shaped the modern world*. Princeton, NJ: Princeton University Press. https://goi.org/10.1515/9781400884858

Razavi, S. and Miller, C. (1995). *From WID to GAD. Conceptual shifts in the women in development discourse. Vol. 1*. Geneva: United Nations Research Institute for Social Development.

Risman, B. (2004). Gender as a social structure: theory wrestling with activism. *Gender and Society*, 18(4): 429–50.

Rocheleau, D. (1988). *Gender, resource management and the rural landscape: implications for agroforestry and farming systems research. Gender issues in farming systems research and extension*. Boulder, CO: Westview Press.

Scoones, I. (1998). *Sustainable rural livelihoods: a framework for analysis. IDS working paper*, 72, Brighton: IDS.

Scoones, I. (2009). Livelihoods perspectives and rural development. *The Journal of Peasant Studies*, 36(1): 171–196.

Smith, L., et al. (2003). *The importance of women's status for child nutrition in developing countries*. Washington, DC: International Food Policy Research Institute.

Szenthe, A. (2019). *Top coffee producing countires. WorldAtlas*.

Thomas, D. and Chen, C. (1994). *Income shares and shares of income: empirical tests of models of household resource allocations*. Santa Monica, CA: Rand Corporation.

Udry, C. (1994). *Gender, agricultural production and the theory of the household*. Evanston, IL: Department of Economics, Northwestern University.

United Nations. (2013). *The millennium development goals report*. New York, NY: United Nations.

United Nations Entity for Gender Equality and the Empowerment of Women. (2015). *A transformative, stand-alone goal on achieving women's rights and empowerment*. UN Women.

Vega, F.E. (2008). The rise of coffee. *American Scientist, 96*(2): 138.

Wharton, A. (2011). *The sociology of gender*. Oxford: Wiley-Blackwell.

Wild, A. (2005). *Coffee: A dark history*. WW Norton & Company.

World Bank. (2006). *Gender issues in agricultural labor*. Washington, DC: World Bank. Accessed online: http://siteresources.worldbank.org/INTGENAGRLIVSOUBOOK/Resources/Module8.pdf

World Fair Trade Organization. History of Fair Trade. (2018). Accessed online: https://wfto.com/about-us/history-wfto/history-fair-trade

World Fair Trade Organization. Annual Report 2017. (2017). Accessed online: https://wfto.com/sites/default/files/WFTO%20Annual%20Report%202017.pdf

Young, K. (1993). *Planning and development with women: making a world of difference*. London: Macmillan.

Chapter 22

Enhancing gender equality in agroforestry systems

Marlène Elias[1], Bioversity International, Italy; Gloria Adeyiga, Bangor University, United Kingdom, and World Agroforestry (ICRAF), Kenya and Forestry Research Institute of Ghana (FORIG), Ghana; Elisabeth Simelton, Swedish International Development Cooperation Agency, Sweden; Yovita Ivanova, Bioversity International, Peru; Ana Maria Paez Valencia, Bioversity International, Costa Rica; Barbara Vinceti, Bioversity International, Italy; and Tim Pagella, Bangor University, United Kingdom

1 Introduction

Agroforestry is practised by almost a third of rural smallholders (Miller et al., 2017) on 43% of the world's agricultural lands (Zomer et al., 2014). Agroforestry systems can take many forms, ranging from agrisilvicultural (trees combined with crops) to silvopastoral (trees combined with pastures or animals) and agrosilvopastoral systems (trees combined with animals and crops) (Nair et al., 2021). Agroforestry systems contribute to provisioning food and nutrition, medicine, and income to rural households and play critical cultural roles and regulatory functions (e.g. water regulation, soil fertility, biodiversity conservation, pollination) around the world (ibid.). As such, agroforestry is central to the resilience of rural people and landscapes globally.

Yet these socio-ecological systems embed and (re)produce significant gender differences and inequalities. These range from gender-specific

1 Corresponding author: marlene.elias@cgiar.org.

http://dx.doi.org/10.19103/AS.2024.0148.31

labour, knowledge, and skills to inequalities in decision-making, access to and control over resources, and costs and benefits related to agroforestry. Gender relations interact with other social relations, such as those based on ethnicity, age and generation, socio-economic status, and more, to produce differential experiences, outcomes, and inequalities (Colfer et al., 2018) in relation to agroforestry.

The Sustainable Development Goals (SDGs) recognize gender equality as a goal in its own right, and as a key lever for achieving other SDGs. Similarly, gender equality may be its own goal in agroforestry interventions, and a lever for achieving other desirable agroforestry outcomes. Yet gender must be intentionally addressed in agroforestry initiatives if they are to hold promise for social justice and sustainable development (Elias et al., 2023).

This chapter explores the nexus of gender and agroforestry. Specifically, we recognize that gender shapes agroforestry systems and that agroforestry systems can either (re)produce, or alternatively, challenge gender inequalities; and we explore projects and initiatives that have sought to do the latter. To support this analysis, we draw on two frameworks – the Gender at Work framework, and the Reach–Benefit–Empower–Transform framework – to understand, respectively: (1) gender transformative change in and through agroforestry; and (2) a range of programmatic approaches for working with gender to support this change. To bring the discussion to life, we draw on examples of agroforestry projects that have intentionally addressed gender relations and examine how they have fared. In so doing, we highlight entry points for agroforestry interventions to promote greater equality.

2 Conceptualizing gender in agroforestry

2.1 Gender (in)equality in agroforestry systems

To structure our analysis of the gender–agroforestry nexus, we draw on Rao and Kelleher's (2005) 'Gender at Work' framework of 'what needs to change' to achieve gender equality. Adapted from Wilber (1999), the framework highlights the formal and informal barriers to gender equality that must be lifted together, from the individual- to the system level, to support equality (Fig. 1). As shown by the arrows, changes across levels and quadrants are mutually reinforcing.

Broadly explained, in the top left quadrant (quadrant 1), we include aspects related to gendered knowledge, self-awareness, decision-making, and agency (the ability to make strategic life decisions and act upon them in situations where this ability was previously denied (Kabeer, 1999)), which are linked with women's (and men's) capacities, leadership, voice, and influence. These contribute to achieving a rich and reflective understanding of the world (critical consciousness) that underpins action to redress the inequalities

Published by Burleigh Dodds Science Publishing Limited, 2025.

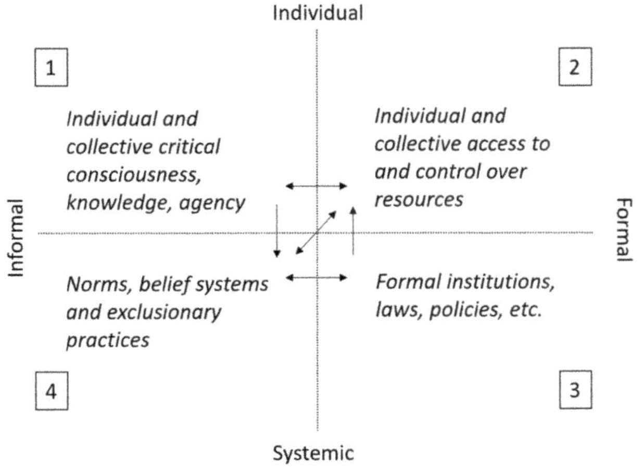

Figure 1 Gender at Work framework: 'What are we trying to change?'. Source: Adapted from: Rao and Kelleher (2005, p. 60).

described in quadrants 2–4. Women's access to and control over resources (quadrant 2) includes control over assets, income, and information. Following Morgan et al. (2023), we highlight that between the individual and the systemic levels, changes in women's and men's critical consciousness and rights to resources also occur at the collective level. The bottom-right quadrant 3 refers to the changes needed in formal policies and regulations, including those that condition governance processes. Finally, the bottom-left quadrant 4 includes the informal 'rules of the game' (North, 1991), such as gender norms and discriminatory beliefs, traditions, and practices that shape the conditions and processes (decision-making, control over resources, policy-making, and implementation) described in the other quadrants. We apply the framework below to demonstrate some of the ways in which gender and agroforestry are mutually shaped and summarize these in Table 1.

2.1.1 Critical consciousness, knowledge, and agency

Due to socially prescribed roles and responsibilities, rural women and men are often tasked with collecting, using, and managing different tree products (Catacutan and Naz, 2015; Gelinas et al., 2015; Kiptot, 2015; Villamor et al., 2015). This gendered division of labour influences the perceptions, knowledge, and skills women and men acquire about tree resources (e.g. Müller et al., 2015), and the value they place on diverse tree species, traits and products, and environmental services (Kiptot et al., 2014; Degrande and Arinloye, 2015; Gutiérrez Velásquez, 2016; Karambiri et al., 2017).

Table 1 Understanding gender in agroforestry through the 'Gender at Work' framework

Quadrant	Examples of manifestations in agroforestry systems
1. Critical consciousness, knowledge, and agency	Agroforestry knowledge, decision-making overuse, management, and sale of agroforestry products and over priority tree species, perceptions of one's own contributions to agroforestry (i.e. recognition of who is a farmer or resource manager)
2. Access to and control over resources	Land tenure, tree tenure, and access to tree products, inputs, labour, information, and income
3. Formal institutions, laws, and policies	National agricultural and agroforestry policies, agricultural extension systems, financial services, rules for membership in farmers' associations, or for participation in agroforestry initiatives
4. Norms, belief systems, and practices	Norms around decision-making, access to land and other resources, distribution of labour, and more, related to agroforestry

Despite these gendered knowledge systems and preferences, women are often excluded from decision-making processes related to agroforestry (Kiptot and Franzel, 2011; Gonçalves et al., 2021). The capacity to make decisions about land use, which tree species to safeguard or plant, how to use different parts of trees, whether or not to adopt agroforestry innovations, and more, are shaped by gender and other social identities, such as marital status, lineage, and stage in the life cycle (Assé and Lassoie, 2011; Colfer et al., 2015; Gumucio et al., 2017). Women's influence over decision-making processes is often limited in restoration, climate action (e.g. Reducing Emissions from Deforestation and Forest Degradation (REDD+)), and other agroforestry programmes and initiatives (Larson et al., 2018; Sarmiento Barletti et al., 2019; Kristjanson, 2020; Elias et al., 2022). This partly owes to a lack of recognition of women as agroforesters and resource managers, despite their multiple contributions to managing trees and collecting, processing, trading in, and otherwise using tree products (Islam, 2015). Challenging this misperception, which rural women themselves have frequently internalized (Elias, 2015), is an important step for increasing women's individual and collective self-awareness, critical consciousness, and agency in and through agroforestry.

2.1.2 Access to and control over agroforestry resources

Gender inequalities are also manifest in relation to access to and control over land, trees, and other agroforestry resources. Customary and formal laws, which embed and reproduce gender inequalities, mediate rights to land and incentives and capacity to manage tree resources. In patrilineal systems, women's rights hinge upon their relationship with male relatives – usually a husband, father, or uncle – who are considered primary rightsholders (Colfer et al., 2016; Meinzen-Dick et al., 2019). Rights to plant, harvest, or fell trees are also formed at the intersection of gender and other factors, such as marital and residence status (Kiptot and Franzel, 2011; Elias et al., 2017). In the many contexts where tree planting signals a claim to land, there are commonly prohibitions against women planting trees because of their insecure rights to land (Kiptot and Franzel, 2012; Colfer et al., 2017). Gendered land and tree tenure regimes affect tree management strategies, as tenure insecurity limits women's incentives to plant and manage trees to which they may not have long-term rights (Howard and Nabanoga, 2007).

Although they are dynamic and interrelated, there can also be distinctions between rights to land and to the trees it harbours. For example, in Burkina Faso, rights to shea trees (*Vitellaria paradoxa*) are not automatically conferred to the borrower when land is loaned (Kuusaana, 2022), and access to néré (*Parkia biglobosa*) pods follows a hierarchy based on women's lineage, order of marriage (in polygamous households), and migrant or resident status (Pehou et al. 2020). Rural women's more limited access (as compared to men) to other types of agroforestry assets, such as agricultural inputs, new technologies, means of transport, as well as to labour and to complementary services (information, credit, etc., described below) also hinder women's access to benefits from agroforestry (Degrande and Arinloye, 2015).

2.1.3 Formal institutions, laws, policies, etc.

As mentioned above, women's exclusions from decision-making and from accessing and controlling agroforestry resources are frequently enshrined in formal institutions, laws, and policies. Extension services and input delivery systems, financial services, laws regulating land tenure, and other formal institutions may fail to recognize women as agroforesters, farmers, and resource managers (Kiptot and Franzel, 2011; Degrande and Arinloye, 2015; Quisumbing et al., 2019). Other times, these institutions embed the problematic assumption that rights and services delivered to men will automatically reach and benefit other household members (Quisumbing et al., 2019). Extension systems are themselves largely staffed with men, which complicates contact with rural women when norms discourage interactions across gender groups with non-family members (Ragasa, 2014).

Gender biases in private and public services, such as credit and extension services (Manfre et al., 2013; Kosec et al., 2023), limit women's capacity to participate in agroforestry markets and to build remunerative tree-based enterprises (Catacutan and Naz, 2015; Ingram et al., 2016; Davis et al., 2019). Where benefits from agroforestry are formally linked to land rights, such as in many Payments for Ecosystem Services schemes, women who are not registered on titles receive limited direct benefits from these initiatives (Sunderlin et al.,

2018; Kariuki and Birner, 2021; Tseng et al., 2021). When rules of the entry limit the number of members in farmers' associations or cooperatives to one per household, women are usually excluded in favour of their husband, which limits women's direct access to services and benefits from these groups (Kaaria et al., 2016).

2.1.4 Norms, belief systems, and discriminatory practices

Norms and belief systems underpin these exclusions and inequalities. For example, as noted above, gender norms shape the division of labour in agroforestry systems. As Sunderland et al. (2014) show in their global comparative study, although there are regional specificities, rural women typically dominate the collection of non-timber tree products for food, fuel, fodder, medicine, and small-scale trade, whereas men are primary harvesters of structural fibre, such as timber for construction or sale. There are also normative specifications about where women and men should collect tree products, which species they can plant or harvest, which parts of the tree they can draw from, and more (Elias et al., 2017). Claims to land, the most fundamental resource in land-based production systems, are also normatively mediated. For example, even when women's and men's rights to land are equal before the law, social sanctions and stigmas may strongly deter women from claiming these rights (Doss and Meinzen-Dick, 2020).

As noted above, gender norms shape decision-making patterns, including in agroforestry initiatives (Colfer et al., 2015; Gumucio et al., 2017). In many contexts, such norms discourage women from attending public meetings or speaking up in front of their male counterparts (Agarwal, 2001; Elias et al., 2020). At the household level, too, women's participation in certain decision-making processes, including those related to land and tree management, may be frowned upon and limited – though the issue is complex and context-specific (Mwangi et al., 2011; Kiptot, 2015; Gumucio et al., 2017). Taboos, social restrictions on mobility, and time deficits resulting from the gender division of labour further limit women's capacity to engage in agroforestry initiatives and markets (Agarwal, 2011; Kiptot and Franzel, 2011; Degrande and Arinloye, 2015).

2.2 Programmatic approaches for supporting gender equality through agroforestry

Many agroforestry initiatives – programmes, projects, and research – seek to recognize, respond to, and challenge these gender-based constraints. The 'Reach, Benefit, Empower' (RBE) framework (Johnson et al., 2018) offers clear practical guidance for understanding and developing effective strategies to achieve and monitor progress towards these goals. The original RBE framework was adapted to include the 'Transform' (T) dimension (CGIAR Research Program on Fish Agri-Food Systems, 2017), which captures programmatic aims and strategies to lift the barriers to gender equality, moving beyond a focus on individual women and their groups towards reforming the informal and formal discriminatory structures (norms, policies – quadrants 3 and 4 of Fig. 1) that maintain inequality.

Table 2 presents the key elements of the framework, aligning progressively more ambitious objectives for women's empowerment and gender equality with strategies for achieving these, and with indicators that can be used to monitor their achievement. The four dimensions of RBET are not mutually exclusive, but rather reinforcing. As shown in Table 2, goals and strategies to reach, benefit, or empower women, or to transform unequal gender relations and norms are often distinct, however. In this chapter, we draw upon the RBET framework to demonstrate some of the approaches that agroforestry initiatives have taken to reach and benefit women, support their empowerment, and transform discriminatory structures to advance gender equality.

3 Case studies

3.1 Recognizing women as legitimate stakeholders, and addressing gendered knowledge and priorities in agroforestry systems (quadrant 1)

One of the most common ways agroforestry projects attempt to reach and benefit women is by involving them in the selection of tree species. In so doing, projects recognize that women and men are both legitimate stakeholders with overlapping but also distinct knowledge and priorities for tree species. Two illustrative examples of such efforts are presented below.

The first is the study **'Gender and climate mitigation through coffee agroforestry'**, conducted as a complementary research component within four partly overlapping projects implemented in Northwest Vietnam (Dien Bien and Son La provinces) by World Agroforestry, CARE, and *Centre de coopération internationale en recherche agronomique pour le développement* (CIRAD). The northern upland region of Vietnam is populated by ethnic minority groups, now sedentary and characterized by diverse cultural practices and norms. The study focused on the Thai ethnic group, which typically occupies the lower segments of valleys, with access to both paddy fields and some sloping land with forestry and home gardens. Decades of monoculture on deforested hills have resulted in severe land degradation, and in response there has been a recent shift in government policies (e.g. Nationally Determined Contributions) towards more

Table 2 Programmatic objectives, strategies, and indicators related to gender, as understood through the RBET framework

	Reach	Benefit	Empower	Transform
Objectives	Include women and ensure their participation in programme activities	Increase women's well-being (e.g. food security, income, health)	Strengthen women's ability to make strategic life choices and put those choices into action	Create deep and enduring change towards gender equality, including more equitable gender norms
Potential strategies (examples)	- Invite women as participants - Reduce barriers to participation - Implement a quota system for women's participation in events	- Address gendered needs, preferences, and constraints to ensure that women benefit equitably from project activities	- Enhance women's decision-making power in households and communities - Address areas of disempowerment	- Encourage critical awareness among men and women about social inequality (e.g. in control over resources, gender-based allocation of responsibilities, and women's and men's access to and influence in decision-making) - Engage women and men as change agents
Potential indicators	- Number or proportion of women participating in a project or activity - Rate of women versus men who speak up during activities and influence project agendas	- Sex-disaggregated data for positive and negative outcome indicators (e.g. productivity, income, assets, nutrition, time use, etc.)	- Women's decision-making power (e.g. over agricultural production, income, and/or household food consumption) - (Reduction of) outcomes associated with disempowerment (e.g., gender-based violence, labour burdens)	- Women's and men's attitudes towards discriminatory gender norms (measured by normative gender attitude assessments)

Source: Dick Frederiksen et al. (2021), who adapted it from Johnson et al. (2018) and CGIAR Research Program on Fish Agri-Food Systems (2017).

tree planting, both in forests and agroforestry systems, to address livelihood, degradation, and climate change concerns (Mulia et al., 2020).

Project staff had been implementing agroforestry in the region since 2011 and observed that forestry was typically considered men's purview. Men worked in the forests and were therefore assumed to be more knowledgeable about trees in general. Forestry extension officers were men, who tended to talk to male farmers, thereby reinforcing the exclusion of women. Within some ethnic groups, however, fruit trees in home gardens were under women's care; and despite limited dedicated attention to gender issues, project staff observed that men and women had different priorities for trees.

A complementary applied research component (2018–2020) was thus developed to bring women's tree preferences to the fore. The research aimed to: (1) fine-tune the methodology for collecting data on local socio-ecological knowledge about tree species and related preferences for coffee agroforestry systems, and (2) provide evidence on the importance of a gender-sensitive understanding of agroecological knowledge and priorities. Through structured and semi-structured interviews and participatory group exercises with women and men separately, the project team explored local women's and men's knowledge of priority tree species, and their valuation of ecosystem services associated with each species in *Arabica* coffee agroforestry systems compared to coffee monocultures. Simelton et al. (2021a) provide more detail on how the study followed the prompts from the 'ShadeTree Advice' methodology (Rigal et al., 2022) to identify gender-specific tree species preferences.

Findings, presented in Simelton et al. (2021a) and Nguyen et al. (2020), showed that men prioritized tree species for inclusion in coffee agroforestry systems according to three main criteria, namely, soil quality (species contributing to soil fertility, soil moisture, low root competition, litter provision, erosion control), coffee production (species that do not negatively impact coffee yields), and income generation (Fig. 2). In comparison, women valued more types of benefits related to: labour requirements, shade provision, micro-climate, pest and disease resistance, coffee production, income generation, and soil quality (Fig. 3). Despite its importance to women, income generation was not a primary criterion for them to prioritize tree species. Instead, they considered fruit tree species and others that reduce household expenditures because their products are consumed within the home as priorities.

The overall preferred species for men was honey locust (*Fructus gleditschiae (Gleditsia)*), whereas for women it was plum (*Prunus salicina*), a multipurpose species (Table 3). The three species that men prioritized were also included among women's priorities so this rendered no conflicting result (Table 3). Instead, because women's priorities were based on a wider range of criteria, also consulting them resulted in a larger number (nine) of potential tree species for inclusion in agroforestry systems.

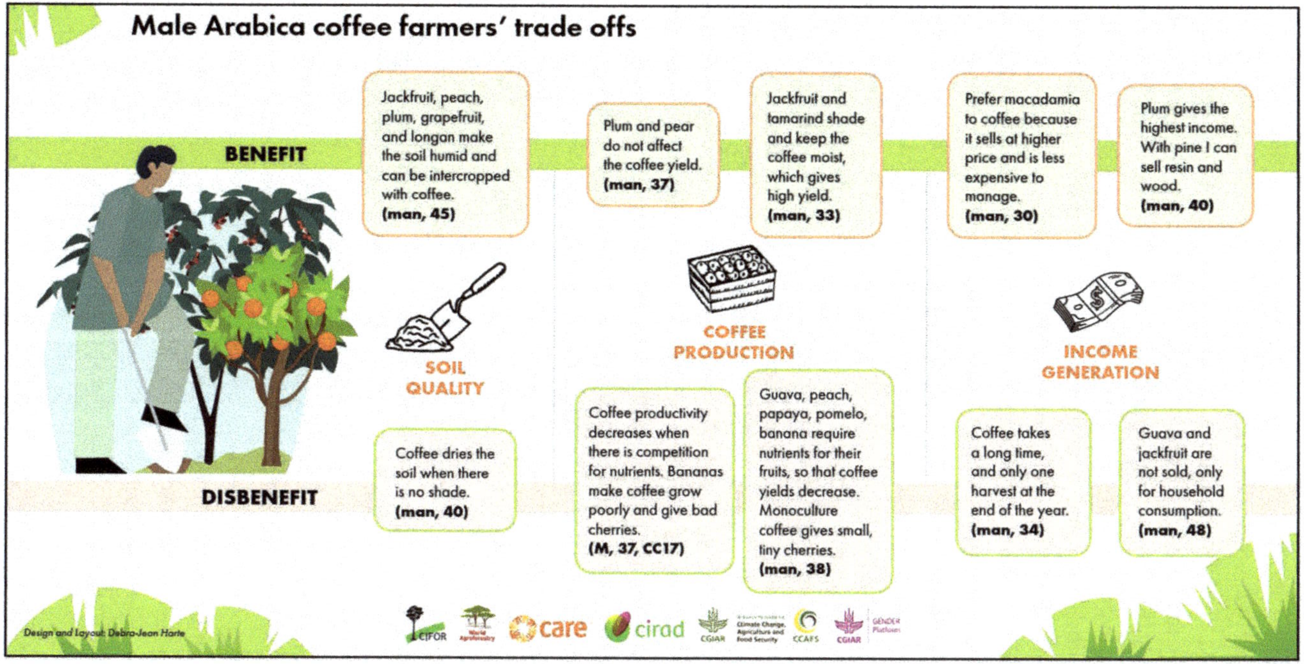

Figure 2 Criteria shaping men's preferences for tree species for inclusion in coffee agroforestry systems. Source: Simelton et al. (2021a) (Design and layout: Debra-Jean Harte).

Figure 3 Criteria shaping women's preferences for tree species for inclusion in coffee agroforestry systems.Source: Adapted from: Simelton et al. (2021a) (Design and layout: Debra-Jean Harte).

Table 3 Gender-specific preferences for tree species (excluding *Arabica* coffee) to incorporate in coffee agroforests and reasons for these preferences

	Preferred benefit cited	
Species	Men (*n* = 53)	Women (*n* = 53)
Honey locust *(Fructus gleditschiae (Gleditsia))*	Income generation ***Overall preference***	Soil quality Coffee production Pest and disease resistance
Candle bush (*Senna alata*)	Soil quality	Shade Labour requirements
Tamarind (*Leucaena leucocephala*)	Soil quality Coffee production	Soil quality Coffee production
Tenasserim pine (*Pinus latteri)*	–	Shade
Longan (*Dimocarpus longan*)	–	Shade
Eucalyptus (*Eucalyptus* spp.)	–	Labour requirement
Chinese black olive (*Canarium nigrum*)	–	Pest and disease resistance
Jackfruit (*Artocarpus heterophyllus*)	–	Soil quality
Plum (*Prunus salicina*)	–	Multiple benefits ***Overall preference***

Source: Adapted from: Table 1 and Figure 1 in Simelton et al. (2021a).

Several reasons account for differences in men's and women's tree species preferences. For one, rural men from ethnic minorities in Vietnam generally have more years of formal schooling than their women counterparts and are more likely to be bilingual and speak Vietnamese, the national language. They are therefore better able to access public information and communicate with extension and project officers, which are predominantly done in Vietnamese. Preferences are also shaped by gender roles. Because women weed and pick coffee cherries under the sun, they value trees with canopies that provide suitable shade for coffee (such as locust and candle bush) and for themselves. Moreover, their heavy farming and domestic responsibilities, including their caretaking (of children, elders) roles, leave them little time to follow courses or attend meetings, which often restricts their participation in value-addition activities of the family enterprise. This contributes to women's preferences for fruit trees, from which they can sell fruits directly.

By exploring gender-specific tree species preferences, the project helped to reveal, document, and discuss these normative gender roles with study participants (quadrant 4 in Fig. 1). Findings have also informed subsequent projects targeting ethnic minority women in developing marketing and economic skills for coffee and agroforestry business (Simelton et al., 2021b). Results have additionally been shared with investors, government, and extension officers, who usually decide what tree seedlings to select for agroforestry projects (see Nguyen et al 2020).

Reinforcing results from the first case study in a different regional context, the second example – the '**Agroecological Regenerative Cocoa**' (ARC) project – focuses on another tree crop: cocoa. Cocoa is the second most important agricultural product in Peru after coffee, and the Peruvian cocoa sector employs more than 100000 families, generating 13 million wages annually (Soto et al., 2022). Cocoa forms part of the ancestral Amazon culture, knowledge, and heritage of Indigenous peoples. Yet the expansion of cocoa plots is one of the main threats to forest ecosystems in the region. Cocoa agroforestry (versus monoculture) is considered a way to strike a balance between profitability and forest conservation.

The ARC project (2020–2025) aims to contribute to the establishment of productive and sustainable cocoa agroforestry landscapes to improve the quality of production, ecological resilience, socio-economic sustainability, and sustainable access to markets. Led by the Alliance of Bioversity International and CIAT in collaboration with KAOKA, Colpa de Loros cooperative, CECAO cooperative, Conservation International, and World Agroforestry, the project's gender equality objective focuses on the equal participation and leadership of women and men. Emphasis is on 'reaching' and 'benefiting' (Table 2) women by including them and ensuring their participation in the co-design and implementation of activities, and designing activities 'according to the context and the necessities of women and men' for the benefit of all.

The project recognizes that traditionally women in the Peruvian Amazon have played a crucial role in the preservation and sustainable use of tree species, non-timber forest products, and medicinal plants, drawing on their knowledge of ancestral practices and its intergenerational transmission. In Peru's Amazonian region of Ucayali, the project is working with the Colpa de Loros cooperative that includes 30% of women members. The project has taken an intersectional approach by including women of different age groups (a group of 40- to 60-year-olds and another of 60- to 80-year-olds) and ethnicities (Quechua from the Andean regions and mestizos from the coastal region).

In the project site, a situational analysis showed important gender gaps in household headship (Fig. 4), which is linked to decision-making authority. Although this is a common traditional household structure in the Peruvian Amazon, the project emphasizes more egalitarian decision-making in relation to tree species selection in agroforestry plots to benefit not a specific gender group, but the entire community and the environment.

The situational analysis surfaced differences in women's and men's tree species preferences, which were based on different criteria. Men wanted to plant a greater number and proportion (85%) of timber trees, such as mahogany (*Swietenia macrophylla*) and caipirona (*Calycophyllum spruceanum*), because of their high market value. In contrast, although financial benefits were important for women, they wanted to plant a large proportion (50%) of fruit tree species,

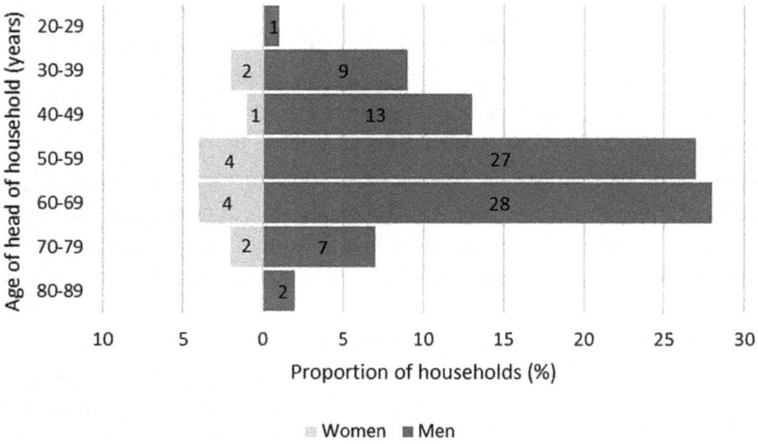

Figure 4 Household headship by gender and age in the ARC project site. Source: Adapted from: ARC (2021).

and of species providing medicinals, firewood, and other products related to household maintenance and well-being.

By applying a gender-sensitive methodology for tree species selection, the project was able to respond to the preferences of women and men, to provide benefits of importance to both. Planting multiple tree species supported diverse income streams to alleviate financial constraints between cocoa harvesting seasons, and greater dietary diversity, including the production of nutritious seeds, nuts, and fruits. As in the Vietnam case, assuming that all species are ecologically compatible, cocoa agroforestry systems informed by women's and men's priorities would thus promote greater agrobiodiversity while promoting greater equity in agroforestry. In both cases, prior information, group consultations, and debriefing sessions were important for supporting women's full participation in research and other project activities.

3.2 Strengthening women's access to and control over agroforestry resources (quadrant 2)

The project **'Nutrition-sensitive forest restoration to enhance the capacity of rural communities in Burkina Faso to adapt to change'** (2016–2019) aimed to increase the utilization and restoration or planting by smallholder farmers of tree diversity with optimal characteristics for local use, adaptation to climate change, and nutrition. Coordinated by Bioversity International, the project was implemented in the Sahelian and Sudanian regions of Burkina Faso, in collaboration with the Burkinabè NGO *Association tiipaalga*, the Burkina Faso National Tree Seed Centre (CNSF), the Institute of Research in Applied Sciences and Technologies (IRSAT) in Ouagadougou, and the University of Natural

Published by Burleigh Dodds Science Publishing Limited, 2025.

Resources and Life Sciences, Vienna (BOKU). In the provinces of Oubritenga and Kourwéogo, on the Central Plateau of Burkina Faso, one component of the project described in Tiendrébéogo et al. (2020) focused specifically on the constraints women face in implementing land restoration activities, and the potential that gender-responsive restoration holds for improving women's lives and livelihoods. In that region, most of the farm and forest land has been degraded by overexploitation, and prevalent shallow and relatively infertile soils are vulnerable to erosion and runoff. The Mossé, who are the dominant ethnic group in these areas, are patrilineal and practice agropastoralism.

Mixed methods, including interviews with women and men farmers and key informants, focus group discussions with women and men and members of women's self-help groups, and direct observation, were used to gather information on experiences with restoration, types of restorative practices, and factors supporting or hindering women's and men's adoption of such practices. Participants were purposively sampled to represent a diversity of backgrounds and experiences. Data were additionally gathered on social networks through which innovative practices are shared and adapted, taking into account the economic and social heterogeneity of different actors involved.

Research showed that farmers in the study sites are involved in different restoration activities on their agroforestry plots, such as creating *zaï* pits, half-moon catchments, stone bunds, composting, mulching, charcoal making, farmer-managed natural regeneration (FMNR), grass strips, fencing, and firebreaks. The characteristics of these techniques influence requirements for their adoption, such as access to and control over specific types of assets (quadrant 2 in Fig. 1). For instance, agroforestry practices including FMNR and fencing as practised in the study sites (exclosures of 3 hectares) require large areas of land and secure land tenure, which women lack in the area. Composting requires access to organic waste, and the creation of stone bunds demands access to stones and means of transporting them to desired field areas, all of which are also difficult for women. Some techniques, such as fencing and charcoal making, require financial capital to purchase materials. Those requirements pose different constraints to adoption for men and women of different socio-economic and generational groups.

As noted above and shown in this research, women face multiple constraints to implementing restoration practices, as socio-cultural norms (quadrant 4) and gendered power relations shape their access to assets (quadrant 2), labour, and benefits from restoration, as well as their recognition as legitimate stakeholders who should have influence and decision-making power in restoration processes (quadrant 1). Gender, however, is not the only factor that determines who will implement and potentially benefit from restoration practices. Whether a woman is married, where her husband resides, the size of the plots she can access, and even whether the woman has adult male children all influence the probability of

Table 4 Adoption of selected restoration practices among different groups of women and men

	Male heads of household (*kãsmdambã*)	Male youth (*yapã*)	Female farm heads (*pagb sẽn koodb ye*)	Widows (*pug kõapã*)	Wives of migrant husbands (*yap-pagbã*)
FMNR	X		X	X	
Zaï pits	X		X		X
Half-moons	X				
Stone bunds	X		X		
Grass strips	X		X		

Source: Tiendrébéogo et al. (2020).

a woman implementing restoration practices and gaining some of the benefits. Notable differences exist among groups of women, such as married women living with their husbands, widows, and wives of migrant husbands (Table 4).

Widows who live with the families of their deceased spouse do not systematically benefit from their late husband's fields. They cultivate land that his family has given them and often find themselves with only small, degraded plots. This group has the most difficulty adopting restoration techniques. They are excluded from practising fencing techniques because they do not have enough land (quadrant 2). Those who are not part of farmer organizations targeted by NGOs also lack the knowledge to practise FMNR. Other widows have left the household of their deceased spouse and manage their own lands. They have more autonomy over decisions related to the adoption of restoration techniques, but are typically asset-limited and farm relatively small plots of land.

Wives of emigrant husbands were mostly 30–40 years old in the study, and cultivated fields that their husband had inherited. These fields are larger and more fertile than those of widows, and their situation allowed them to acquire a certain autonomy in managing agricultural work in their husband's absence. This group of women practices *zaï* pits and has some rights to use compost if living in the compound of their husband's parents. Yet, because they lack of membership in farmer organizations, do not have access to their husband's labour, and have limited land, they do not practice FMNR, fencing, or implement stone bunds or half-moons.

To address these constraints, women's self-help groups, with some support from NGOs such as *Association tiipaalga*, are working to secure communal access to land for widows and young women whose husbands have migrated. *Association tiipaalga* has acted as an intermediary between landowners and women's groups to help the latter acquire rights to land to be able to 'benefit' from agroforestry. Once acquired, women fence off this land to promote natural regeneration and plant income-generating tree and crop species. Working as a group, women can thereby improve their collective land rights as well as their access to other resources and farming implements while strengthening their social capital and collective action.

NGOs such as *Association tiipaalga* also train women on different land restoration techniques, which participants have applied to their household fields and personal plots to increase their yields. In the study villages, restoration activities have resulted in women's employment and income through the sale of non-timber forest products. Micro-credit provided to women's groups by *Association tiipaalga* and other NGOs have further enabled members to carry out other income-generating activities and to use their income to pursue micro-entrepreneurial opportunities.

3.3 Challenging gender norms that exclude women in agroforestry (quadrant 4)

The '**Regreening Africa**' project (2021–2023) aimed to reverse land degradation to improve livelihoods, food and nutrition security, and climate resilience by scaling evergreen agricultural practices among smallholder farming households. The project was coordinated by World Agroforestry and implemented in eight sub-Saharan African countries, including Ghana. Ghana's Bawku West district is a degraded area, with a largely agrarian population that is heavily dependent on tree resources to fill annual food gaps. Tree product-based livelihoods are especially vital for women, who have discretionary access to land in a patrilineal landscape where women's rights are embedded within men's rights. Cultural norms exclude women from land use and management decisions, including those related to restoration and to the benefits of restorative practices.

In this district, World Vision Ghana and Catholic Relief Services took an intentionally transformative approach focused on redressing restrictive gender norms as a precursor to achieving effective restoration with more balanced outcomes for men and women. Over 22 months, the project adapted and implemented several methodologies described as 'gender transformative approaches' (GTAs) (Mayoux, 2014; Mohanraj and Hillenbrand, 2015; CARE, 2019; Wong et al., 2019). These methodologies seek to unpack the underlying causes of gender inequality, such as discriminatory gender norms (quadrant 4 in Fig. 1) that underpin unequal access to productive resources (quadrant 2), participation in decision-making (quadrant 1), labour burdens (quadrant 4), and a lack of recognition of women's roles and contributions (quadrant 1). Taking a GTA, the project encouraged critical self-reflection and awareness of the influence of gender relations, gender roles, and stereotypes on women's access to resources and information (quadrants 1, 2, and 4).

A gender analysis preceded the design of GTA modules. The analysis highlighted imbalances around land access and control, work burdens, as well as women's exclusion from land decisions (Adeyiga et al., in prep.; Adeyiga,

2021; Bourne, 2022). This baseline directly fed into the design of GTA modules, outlined below; effectively setting the priorities for the first phase of the initiative.

The methodology combined work at a landscape scale with work at the community and household scales. The Regreening Africa project involved 30 villages in Bawku West, all of which received technical capacity strengthening in evergreen practices, including composting, FMNR, and tree nurturing. Of these communities, 15 were randomly selected, after which 10 households per community were randomly selected to participate in the GTA initiative. Two members of each household took part in activities: usually the man household head and their partner (one spouse), or in the few women-headed households, the woman head with an adult man from her household. Gender training was facilitated by local experienced man–woman pairs, and community champions were selected and received additional training to support and encourage other participating households.

GTA modules focussed on equipping men and women with the capacities and opportunity to critically reflect on and navigate their sensitive and tightly held beliefs around gender in relation to assets, labour, finances, and decision-making. The training consisted of adapted reversed role play and visioning exercises to bring the gender norms that provide constraints or opportunities for women and men as individuals, and for their households and communities, to the fore (quadrant 1). Participants discussed what social changes they desired, prioritized behaviours to change, and committed to sharing decisions (quadrant 1), asset control, workloads, and expenditures (quadrant 2) to achieve more equitable households.

Following a participatory outcome mapping methodology (Mohanraj and Hillenbrand, 2015), men and women participants developed indicators to capture the changes that they were committed to making over the research period (Table 5). These were categorized as immediate changes, which were expected to happen quickly, and long-term changes requiring sustained engagement and action.

These indicators were used for monitoring every fifth month on three occasions. At each monitoring meeting, men and women discussed what was changing in their lives and what was not changing, such as women's access to land. Strategies were proposed to encourage men to share homesteads (home gardens or cultivated land near the dwelling) with women. Community champions played a key role in supporting households along this journey, visiting each household at least once a month to discuss the household's particular situation and changes.

In the initial two years of project implementation, the GTA resulted in important changes in gender norms. The most significant outcome was women's increased agency (quadrant 1, Fig. 1) to make changes in homesteads. Prior to

Table 5 Indicators developed by participant women and men to monitor changes in equality and household resilience

	Women	Men
Expect to see	Men and sons help with washing clothes	Men provide money for buying condiments
	Men and sons help with fetching water	Men give women money for the grinding mill
	Men bathe children	Men pay for health insurance and medications for women and children
	Men take young children to school	Men pay school fees for children
	Women receive money from men for trading	Men weed women's farms after men's farm (family farm)
Like to see	Women gain complete control over their own livestock income	Women take care of livestock
	Women's farm is given priority and ploughed after the family farm, before ploughing for other men	Men support and encourage women to attain leadership positions
	Men help with taking sick children and older people to the hospital	Men actively participate in bringing up the girl child
	Men reduce alcohol intake	Men receive support from women to buy farm inputs for men's farm (family farms)
	Women discuss family planning with husband without judgement	Men are concerned about women's general health: e.g. accompany women to the hospital and support them on antenatal visits
Love to see	Women participate in key household decisions (e.g. income expenditure)	Men include women in key decisions: e.g. income expenditure
	Women access fertile family land for farming	Wives are accepting of additional wife
	Spouses jointly control family farm produce: e.g. what to sell, when to sell, how much to sell, and how to use cash from crop sales	Men allocate fertile land to women for farming
	Women freely express grievances to their husband without ridicule, shame, or fear	Men apologize to their spouse when wrong
	Women have control over their own body in decisions of sex	The family is more peaceful and united

the GTA initiative, women had been unable to make decisions in this area and spent long hours daily gathering resources far away from this central area. At the end of 2 years, 80% (n = 120) of the women who took part in the GTA initiative were able to access up to 0.5 acres near their homes to cultivate crops of their choice (quadrant 2). There were also significant shifts in household labour, with men challenging existing norms (quadrant 4) by contributing more often to collecting water and fuelwood, bathing children, and taking them to school. Participating households reported that the gender training had improved household relations, and that even though women could not participate in land decisions openly, they felt consulted in most decisions by their spouses.

Due to the positive feedback from participating households, word quickly spread within the communities about the GTA and several people asked the project team to include them in future training. With internal funding, World Vision Ghana scaled out gender training to an additional 25 households in each research community. The project also designed an interactive radio programme in the local language (Kusaal) that reached an estimated 16 000 listeners in five districts across the Upper East region. In this programme, gender specialists discussed ways to address unequal gender norms to improve household relations and wellbeing. By popular request, the programme extended its run and involved significant listener participation through phone-ins. Based on listener responses, there was an appetite for more GTAs.

GTA modules went beyond 'reaching' men and women. The reflection activities enhanced men's and women's critical consciousness and women's capacities to negotiate and navigate the complex interactions and normative prescriptions that exclude them from decisions (i.e. 'empower'). By explicitly engaging men in challenging discriminatory norms, these activities also sought to 'transform' the structures that perpetuate inequality.

4 Conclusion

The case studies demonstrate progressively more ambitious ways in which agroforestry projects have integrated gender considerations to reach and benefit differentiated groups of women and men, and support their empowerment as well as changes towards greater gender equality. In the first (Vietnam and Peru) cases, projects 'reached' women by engaging them in meetings, research, and listening to their priorities; and used this information to develop tree species portfolios that would 'benefit' women as well as men. The cases surfaced inequalities in women's and men's access to resources (quadrant 2 in Fig. 1), and sharing results back with farmers served as a launchpad for discussions that raised their awareness of gender relations (quadrant 1 in Fig. 1). In both cases, considering gender at the planning stage, when tree species are selected, was important for ensuring that the chosen species fulfil the breadth of social, economic, and environmental functions that both women and men farmers seek. Local perspectives, and rural women's influence in the selection of trees, showed promise for enhancing synergies among livelihoods, nutrition, biodiversity, and other socio-economic and environmental goals.

In the Burkinabè case, social inequalities based on gender, family structure, and other intersecting factors of social differentiation influenced the ability of farmers to adopt land restoration techniques. Although there were differences across groups of women, rural women on the whole faced several constraints in carrying out land restoration activities. These constraints include a lack of perceived legitimacy as stakeholders in restoration initiatives (quadrant 1) and discriminatory norms (quadrant 4) that limit their access to and control over land, labour, and other productive resources (quadrant 2). Through a group-based approach, the project intentionally addressed some of these constraints to allow women to benefit from agroforestry, strengthen their social capital, and enhance their land rights and livelihood opportunities – all of which further sustain women's interest and investments in restoration.

Finally, the Ghana case showed that it is possible to challenge discriminatory norms (quadrant 4) in the context of an agroforestry project. Supporting women's and men's equitable inputs into land use and management decisions requires 'transforming' deep-rooted gender inequalities in households and communities. Applying a GTA supported several changes in gender dynamics, such as improving intra-household relations and women's confidence to negotiate for their strategic interests, over a relatively short period of time.

Research is only beginning to examine how gender equality can support other desirable agroforestry outcomes, and how to achieve synergies among these multiple goals (Elias et al., 2021). Further long-term research should explore how changes in gender equality, such as those described above, can be sustained over time and scaled beyond the household and community levels. Finally, research is needed on how simultaneously pursuing changes across the four quadrants described above in the context of agroforestry projects can accelerate progress towards gender equality and other agroforestry outcomes, to benefit people and the planet.

5 Where to look for further information

The following publications provide a good overview of the subject:

- CGIAR FTA (2020). *Gender Equality and Social Inclusion: A Revised Agenda and Action Plan for the CGIAR Research Program on Forests, Trees and Agroforestry 2020–2021*. Bogor: CGIAR Research Program on Forests, Trees and Agroforestry (FTA).
- Colfer, C. J. P., Catacutan, D., Naz, F. and Pottinger, A. J. (Eds) (2015). Gender in agroforestry: special issue. *International Forestry Review* 17(Suppl 4). https://www.ingentaconnect.com/content/cfa/ifr/2015/00000017/a00404s4.
- Elias, M., Paez Valencia, A. M., Ihalainen, M. and Monterroso, I. (2021). *Advancing Gender Equality and Social Inclusion*. FTA Highlights of a

Decade 2011–2021 series. Highlight No.15. Bogor: The CGIAR Research Program on Forests, Trees and Agroforestry (FTA).

- Fortmann, L. and Rocheleau, D. (1985). Women and agroforestry: four myths and three case studies. *Agroforestry Systems* 2(4), 253–272. https://doi.org/10.1007/BF00147037.
- Ingram, V., Haverhals, M., Petersen, S., Elias, M., Sijapati Basnett, B. and Phosiso, S. (2016). Gender and forest, tree and agroforestry value chains: evidence from literature. In: Colfer C. J. P., Sijapati Basnett, B. and Elias, M. (Eds) *Gender and Forests: Climate Change, Tenure, Value Chains and Emerging Issues*. New York: Routledge, pp. 221–242.
- Kiptot, E., Franzel, S. and Degrande, A. (2014). Gender, agroforestry and food security in Africa. *Current Opinion in Environmental Sustainability* 6, 104–109.
- Kiptot, E. (2015). Gender roles, responsibilities, and spaces: implications for agroforestry research and development in Africa. *Forest Policy and Economics* 17(4), 11–21.
- Mathez-Stiefel, S-L, Ayquipa-Valenzuela, J., Corrales-Quispe, R., Rosales-Richard, L. and Valdivia-Díaz, M. (2016). Identifying gender-sensitive agroforestry options: methodological considerations from the field. *Mountain Research and Development* 36(4), 417–430.

Key research in this area can be found at the following organizations:

- AFR100 (https://afr100.org/).
- Alliance of Bioversity International and CIAT (https://alliancebioversityciat.org/).
- CIFOR-ICRAF (https://www.cifor-icraf.org/).
- IUCN (https://www.iucn.org/).
- World Resources Institute (https://www.wri.org/).

6 Acknowledgements

We thank Haley Zaremba for her editing assistance, as well as the CGIAR Initiative on Gender Equality, the CGIAR GENDER Impact Platform and all the funders who supported this research through their contributions to the CGIAR Trust Fund: https://www.cgiar.org/funders/trust-fund/. The 'Gender and climate mitigation through coffee agroforestry' research was implemented with funding from the CGIAR GENDER Platform, the CGIAR Research Program on Climate Change, Agriculture and Food Security (CCAFS), and bilateral grants by the Australian Department of Foreign Affairs and Trade (DFAT) (for the TEAL-project, implemented by CARE) and EU Horizon 2020 (for the Breedcafs project, implemented by CIRAD and ICRAF). The 'Agroecological Regenerative Cocoa' (ARC) project was financed by the French Facility for Global Environment (FFEM). The project 'Nutrition-sensitive forest restoration to enhance the capacity of rural communities in Burkina Faso to adapt to change' was funded by the Austrian Development Cooperation, with co-financing from the CGIAR Research Program on Forests, Trees and Agroforestry. 'Regreening Africa' is funded by the European Union.

7 References

Adeyiga, G. (2021). Gender-transformative pathways in the regreening landscapes of Ghana. *World Agroforestry*. Retrieved on 24 October 2023. Available at: https://www.worldagroforestry.org/blog/2021/08/23/feature-gender-transformative-pathways-regreening-landscapes-ghana.

Agarwal, B. (2001). Participatory exclusions, community forestry, and gender: an analysis for South Asia and a conceptual framework. *World Development* 29(10), 1623–1648. https://doi.org/10.1016/S0305-750X(01)00066-3.

ARC (2021). Socioeconomic assessment of the cocoa producers of the cooperative Colpa de Loros, in Ucayali, Peru. Unpublished report.

Assé, R. and Lassoie, J. P. (2011). Household decision-making in agroforestry parklands of Sudano-Sahelian Mali. *Agroforestry Systems* 82(3), 247–261. https://doi.org/10.1007/s10457-011-9395-2.

Bourne, M. (2022). Can traditional gender norms shift? *Regreening Africa*. Retrieved on 24 October 2023. Available at: https://regreeningafrica.org/in-the-news/can-traditional-gender-norms-shift/.

CARE (2019). *Gender-Transformative Adaptation*. Geneva: CARE.

Catacutan, D. and Naz, F. (2015). Gender roles, decision-making and challenges to agroforestry adoption in Northwest Vietnam. *International Forestry Review* 17(4), 22–32. https://doi.org/10.1505/146554815816002266.

CGIAR Research Program on Fish Agri-Food Systems (2017). *CGIAR Research Program on Fish Agri-Food Systems (FISH): Gender Strategy*. Penang: CGIAR Research Program on Fish Agri-Food Systems.

Colfer, C. J. P., Achdiawan, R., Roshetko, J. M., Mulyoutami, E., Yuliani, E. L., Mulyana, A., Moeliono, M., Adnan, H. and Erni (2015). The balance of power in household decision-making: encouraging news on gender in southern Sulawesi. *World Development* 76, 147–164. https://doi.org/10.1016/j.worlddev.2015.06.008.

Colfer, C. J. P., Sijapati Basnett, B. and Elias, M. (2016). Introduction. In: Colfer, C. J. P., Sijapati Basnett, B. and Elias, M. (Eds). *Gender and Forests: Climate Change, Tenure, Value Chains, and Emerging Issues*. Earthscan London: London, pp. 1–25.

Colfer, C. J. P., Elias, M., Sijapati Basnett, B. and Hummel, S. S.(2017). *The Earthscan Reader on Gender and Forests*. Earthscan: London

Colfer, C. J. P., Sijapati Basnett, B. and Ihalainen, M. (2018). Making sense of 'intersectionality': a manual for lovers of people and forests. Occasional Paper 184. Bogor: Center for International Forestry Research (CIFOR).

Davis, K., Franzel, S. and Spielman, D. J. (2019). Extension options for better livelihoods and poverty reduction: a selected review 2012–2015. *Gates Open Research* 3(386), 386. https://doi.org/10.21955/gatesopenres.1115095.1.

Degrande, A. and Arinloye, D.-D. A. (2015). Gender in agroforestry: implications for action-research. *Nature & Faune Journal* 29(1), 6–11. Available at: https://www.worldagroforestry.org/publication/gender-agroforestry-implications-action-research.

Dick Frederiksen, S., Elias, M., Zaremba, H. and Aynekulu, E. (2021). Developing gender-equitable ecological restoration initiatives: a synthesis of guidance to improve restoration practice. Rome: The Alliance of Bioversity International and CIAT, 58 p. ISBN: 978-92-9255-209-1. Available at: https://hdl.handle.net/10568/117868.

Doss, C. and Meinzen-Dick, R. (2020). Land tenure security for women: a conceptual framework. *Land Use Policy* 99, 105080. https://doi.org/10.1016/j.landusepol.2020.105080.

Elias, M. (2015). Gender, knowledge-sharing and management of shea (*Vitellaria paradoxa*) parklands in central-west Burkina Faso. *Journal of Rural Studies* 38, 27–38. https://doi.org/10.1016/j.jrurstud.2015.01.006.

Elias, M., Hummel, S. S., Basnett, B. S. and Colfer, C. J. P. (2017). Gender bias affects forests worldwide. *Ethnobiology Letters* 8(1), 31–34. https://doi.org/10.14237/ebl.8.1.2017.834.

Elias, M., Grosse, A. and Campbell, N. (2020). Unpacking 'gender' in joint forest management: lessons from two Indian states. *Geoforum* 111, 218–228. https://doi.org/10.1016/j.geoforum.2020.02.020.

Elias, M., Kandel, M., Mansourian, S., Meinzen-Dick, R., Crossland, M., Joshi, D., Kariuki, J., Lee, L. C., McElwee, P., Sen, A., Sigman, E., Singh, R., Adamczyk, E. M., Addoah, T., Agaba, G., Alare, R. S., Anderson, W., Arulingam, I., Bellis, S. ḵ. V., Birner, R., De Silva, S., Dubois, M., Duraisami, M., Featherstone, M., Gallant, B., Hakhu, A., Irvine, R., Kiura, E., Magaju, C., McDougall, C., McNeill, G. D., Nagendra, H., Nghi, T. H., Okamoto, D. K., Paez Valencia, A. M., Pagella, T., Pontier, O., Post, M., Saunders, G. W., Schreckenberg, K., Shelar, K., Sinclair, F., Gautam, R. S., Spindel, N. B., Unnikrishnan, H., Wilson, G. N. and Winowiecki, L. (2022). Ten people-centered rules for socially sustainable ecosystem restoration. *Restoration Ecology* 30(4), e13574. https://doi.org/10.1111/rec.13574.

Elias, M., Zaremba, H., Tavenner, K., Ragasa, C., Paez Valencia, A. M., Choudhury, A. and de Haan, N. (2023). Beyond crops: toward gender equality in forestry, fisheries, aquaculture and livestock development. CGIAR GENDER Impact Platform Working Paper #012. Nairobi, Kenya: CGIAR GENDER Impact Platform. Available at: https://hdl.handle.net/10568/129708.

Gelinas, N., Lavoie, A., Labrecque, M.-F. and Olivier, A. (2015). Linking women, trees and sheep in Mali. *International Forestry Review* 17(4), 76–84. https://doi.org/10.1505/146554815816086462.

Gonçalves, C. D. B. Q., Schlindwein, M. M. and Martinelli, G. D. C. (2021). Agroforestry systems: a systematic review focusing on traditional indigenous practices, food and nutrition security, economic viability, and the role of women. *Sustainability* 13(20), 11397. https://doi.org/10.3390/su132011397.

Gumucio, T., Twyman, J. and Clavijo, M. (2017). Gendered perspectives of trees on farms in Nicaragua: considerations for agroforestry, coffee cultivation, and climate change. Working Paper. CIAT Publication No. 432. Cali, Colombia: International Center for Tropical Agriculture (CIAT); CGIAR Research Program on Climate Change, Agriculture and Food Security (CCAFS); CGIAR Research Program on Forests, Trees and Agroforestry (FTA). Available at: https://cgspace.cgiar.org/handle/10568/78670.

Gutiérrez Velásquez, C. (2016). Determinación de medios y estrategias de vida, nutrición rural y uso de agro biodiversidad con enfoque de género para el desarrollo sostenible en cuatro comunidades del municipio de Waslala, región autónoma del Atlántico Norte, Nicaragua. Master's thesis. Turrialba, Costa Rica: Centro Agronómico Tropical de Investigación y Enseñanza (CATIE). Available at: http://repositorio.bibliotecaorton.catie.ac.cr/handle/11554/8595.

Howard, P. L. and Nabanoga, G. (2007). Are there customary rights to plants? An inquiry among the Baganda (Uganda), with special attention to gender. *World Development* 35(9), 1542–1563. https://doi.org/10.1016/j.worlddev.2006.05.021.

Ingram, V., Haverhals, M., Petersen, S., Elias, M., Sijapati Basnett, B. and Phosiso, S. (2016). Gender and forest, tree and agroforestry value chains: evidence from the literature. In: Colfer, C. J. P., Sijapati Basnett, B. and Elias, M. (Eds). *Gender and Forests: Climate Change, Tenure, Value Chains, and Emerging Issues*. London: Earthscan, pp. 221–242.

Islam, M. S. (2015). Women in the homestead agroforestry-contribution to sustainable development. *Journal of Natural Sciences Research* 5(1), 129–139.

Johnson, N., Balagamwala, M., Pinkstaff, C., Theis, S., Meinsen-Dick, R. and Agnes, Q. (2018). How do agricultural development projects empower women? Linking strategies with expected outcomes. *Journal of Gender, Agriculture and Food Security* 3(2), 1–19. https://doi.org/10.19268/JGAFS.322018.1.

Kaaria, S., Osorio, M., Wagner, S. and Gallina, A. (2016). Rural women's participation in producer organizations: an analysis of the barriers that women face and strategies to foster equitable and effective participation. *Journal of Gender, Agriculture and Food Security* 1(2), 148–167. https://doi.org/10.22004/ag.econ.246035.

Kabeer, N. (1999). Resources, agency, achievements: reflections on the measurement of women's empowerment. *Development and Change* 30(3), 435–464.

Karambiri, M., Elias, M., Vinceti, B. and Grosse, A. (2017). Exploring local knowledge and preferences for shea (*Vitellaria paradoxa*) ethnovarieties in Southwest Burkina Faso through a gender and ethnic lens. *Forests, Trees and Livelihoods* 26(1), 13–28. https://doi.org/10.1080/14728028.2016.1236708.

Kariuki, J. and Birner, R. (2021). Exploring gender equity in ecological restoration: the case of a market-based program in Kenya. *Ecological Restoration* 39(1–2), 77–89. http://doi.org/10.3368/er.39.1-2.77.

Kiptot, E. (2015). Gender roles, responsibilities, and spaces: implications for agroforestry research and development in Africa. *International Forestry Review* 17(4), 11–21. https://doi.org/10.1505/146554815816086426.

Kiptot, E. and Franzel, S. C. (2011). Gender and agroforestry in Africa: are women participating? *ICRAF Occasional Paper No. 13*. Nairobi: World Agroforestry Centre. Available at: https://www.worldagroforestry.org/publication/gender-and-agroforestry-africa-are-women-participating.

Kiptot, E. and Franzel, S. (2012). Gender and agroforestry in Africa: a review of women's participation. *Agroforestry Systems* 84(1), 35–58. https://doi.org/10.1007/s10457-011-9419-y.

Kiptot, E., Franzel, S. and Degrande, A. (2014). Gender, agroforestry and food security in Africa. *Current Opinion in Environmental Sustainability* 6, 104–109. https://doi.org/10.1016/j.cosust.2013.10.019.

Kosec, K., Hidrobo, M., Gartaula, H., Van Campenhout, B. and Carrillo, L.(2023). Making complementary agricultural resources, technologies and services more gender responsive. CGIAR GENDER Impact Platform Working Paper #010. Nairobi: CGIAR GENDER Impact Platform. Available at: https://hdl.handle.net/10568/129706.

Kuusaana, E. D. (2022). Customary land governance dynamics and its implications for shea tenure and ecology in selected peri-urban communities in Ghana. *Frontiers in Sustainable Food Systems* 6, https://doi.org/10.3389/fsufs.2022.1033523.

Kristjanson, P. (2020). Closing gender gaps in forest landscape initiatives. *International Forestry Review* 22(S1), 44–54.

Larson, A. M., Solis, D., Duchelle, A. E., Atmadja, S., Resosudarmo, I. A. P., Dokken, T. and Komalasari, M. (2018). Gender lessons for climate initiatives: a comparative study of REDD+ impacts on subjective wellbeing. *World Development* 108, 86–102. https://doi.org/10.1016/j.worlddev.2018.02.027.

Manfre, C., Rubin, D., Allen, A., Allen, A., Summerfield, G., Colverson, K. and Akeredolu, M. (2013). Reducing the gender gap in agricultural extension and advisory services. *Modernizing Extension and Advisory Services*. Discussion Paper. United States Agency for International Development.

Mayoux, L. (2014). GALS catalyst phase 1: community action learning1–15. Hivos. Available at: http://www.galsatscale.net/_documents/GALS0ActionLearningSystem.pdf.

Meinzen-Dick, R., Quisumbing, A., Doss, C. and Theis, S. (2019). Women's land rights as a pathway to poverty reduction: framework and review of available evidence. *Agricultural Systems* 172, 72–82. https://doi.org/10.1016/j.agsy.2017.10.009.

Miller, D. C., Muñoz-Mora, J. C. and Christiaensen, L. (2017). Prevalence, economic contribution, and determinants of trees on farms across Sub-Saharan Africa. *Forest Policy and Economics* 84, 47–61. Available at: https://doi.org/110.1016/j.forpol.2016.12.005.

Mohanraj, P. and Hillenbrand, E. (2015). Transformative tools for monitoring gender behavior change. CARE USA. American Evaluation Association (AEA). Session 1807: Feminist Issues in Evaluation.

Morgan, M., Larson, A. M., Trautman, S., Garner, E., Elias, M. and Meinzen-Dick, R. (2023). Gender transformative approaches to strengthen women's land and resource rights. Bogor: Centre for International Forestry Research (CIFOR) and Nairobi: World Agroforestry (ICRAF) and International Fund for Agricultural Development (IFAD). Available at: https://www.cifor-icraf.org/publications/pdf/project-briefs/GTA-Brief.pdf.

Mulia, R., Nguyen, D. D., Nguyen, M. P., Steward, P., Pham, V. T., Le, H. A., Rosenstock, T. and Simelton, E. (2020). Enhancing Vietnam's nationally determined contribution with mitigation targets for agroforestry: a technical and economic estimate. *Land* 9(12), 528. https://doi.org/10.3390/land9120528.

Müller, J. G., Boubacar, R. and Guimbo, I. D. (2015). The "how" and "why" of including gender and age in ethnobotanical research and community-based resource management. *Ambio* 44(1), 67–78. https://doi.org/10.1007/s13280-014-0517-8.

Mwangi, E., Meinzen-Dick, R. and Sun, Y. (2011). Gender and sustainable forest management in East Africa and Latin America. *Ecology and Society* 16(1).17

Nair, P. R., Kumar, B. M. and Nair, V. D. (2021). *An Introduction to Agroforestry: Four Decades of Scientific Developments*. Berlin: Springer, 666 p.

Nguyen, M. P., Vaast, P., Pagella, T. and Sinclair, F. (2020). Local knowledge about ecosystem services provided by trees in coffee agroforestry practices in Northwest Vietnam. *Land* 9(12), 486. https://doi.org/10.3390/land9120486.

North, D. C. (1991). Institutions. *Journal of Economic Perspectives* 5(1), 97–112.

Pehou, C., Djoudi, H., Vinceti, B. and Elias, M. (2020). Intersecting and dynamic gender rights to néré, a food tree species in Burkina Faso. *Journal of Rural Studies* 76, 230–239. https://doi.org/10.1016/j.jrurstud.2020.02.011.

Quisumbing, A., Meinzen-Dick, R. and Njuki, J. (2019). *Annual Trends and Outlook Report: Gender Equality in Rural Africa: From Commitments to Outcomes*. Washington, DC: International Food Policy Research Institute. https://doi.org/10.2499/9780896293649.

Ragasa, C. (2014). Improving gender responsiveness of agricultural extension. In: Quisumbing, A. R., Meinzen-Dick, R. S., Raney, T. L., Croppenstedt, A., Behrman, J. A. and Peterman, A. (Eds). *Gender in Agriculture: Closing the knowledge gap*. Netherlands: Springer. https://doi.org/10.1007/978-94-017-8616-4_17. 411-430.

Rao, A. and Kelleher, D. (2005). Is there life after gender mainstreaming? *Gender and Development* 13(2), 57–69. Available at: http://www.jstor.org/stable/20053149.

Rigal, C., Wagner, S., Nguyen, M. P., Jassogne, L. and Vaast, P. (2022). ShadeTreeAdvice methodology: guiding tree-species selection using local knowledge. *People and Nature* 4(5), 1233–1248. https://doi.org/10.1002/pan3.10374.

Sarmiento Barletti, J. P., Hewlett, C. and Larson, A. M. (2019). *How Does Context Affect the Outcomes of Multi-stakeholder Forums on Land Use and/or Land-Use Change?: A Realist Synthesis Review of the Scholarly Literature*. Bogor: Center for International Forestry Research (CIFOR). https://doi.org/10.17528/cifor/007284.

Simelton, E., Mulia, R., Rigal, C., Tuan, D. M., Phuong, N. M., North, H. and Xuan, L. H. (2021a). Beyond carbon sequestration–local knowledge about tree functions. Case study from male and female Arabica coffee farmers in Vietnam. Working Paper No. 319. Hanoi: World Agroforestry (ICRAF). https://doi.org/10.5716/WP21025.PDF.

Simelton, E., Mulia, R., Nguyen, T. T., Duong, T. M., Le, H. X., Tran, L. H. and Halbherr, L. (2021b). Women's involvement in coffee agroforestry value-chains: financial training, Village Savings and Loans Associations, and Decision power in Northwest Vietnam. CCAFS Working Paper no. 340. Wageningen: CGIAR Research Program on Climate Change, Agriculture and Food Security (CCAFS). Available at: https://cgspace.cgiar.org/bitstream/handle/10568/111055/WP%20340%20Simelton.pdf.

Soto, E., Mendoza, P. and Aguilar, J. (2022). Manual de buenas prácticas agrícolas para el cultivo del cacao. Lima: Instituto Interamericano de Cooperación para la Agricultura (IICA). Available at: https://repositorio.iica.int/handle/11324/21346.

Sunderlin, W. D., de Sassi, C., Sills, E. O., Duchelle, A. E., Larson, A. M., Resosudarmo, I. A. P., Awono, A., Kweka, D. L. and Huynh, T. B. (2018). Creating an appropriate tenure foundation for REDD+: the record to date and prospects for the future. *World Development* 106, 376–392. https://doi.org/10.1016/j.worlddev.2018.01.010.

Tiendrébéogo, S., Ouedraogo, A., Kabore, R., Zougouri, S., Elias, M., Traoré, A. T., Vinceti, B., Traoré, D. and Yago-Ouattara, E. L. (2020). Enhancing women's rights and lives through gender-equitable restoration in Burkina Faso. *ETFRN NEWS 60: Restoring African Drylands*. ISSN: 1876-5866.

Tseng, T. J., Robinson, B. E., Bellemare, M. F., BenYishay, A., Blackman, A., Boucher, T., Childress, M., Holland, M. B., Kroeger, T., Linkow, B., Diop, M., Naughton, L., Rudel, T., Sanjak, J., Shyamsundar, P., Veit, P., Sunderlin, W., Zhang, W. and Masuda, Y. J. (2021). Influence of land tenure interventions on human well-being and environmental outcomes. *Nature Sustainability* 4(3), 242–251. https://doi.org/10.1038/s41893-020-00648-5.

Villamor, G. B., Akiefnawati, R., van Noordwijk, M., Desrianti, F. and Pradhan, U. (2015). Land use change and shifts in gender roles in central Sumatra, Indonesia. *International Forestry Review* 17(4), 61–75. https://doi.org/10.1505/146554815816002211.

Wilber, K. (1999). *The Collected Works of Ken Wilber, Volume 4*. Boston, MA: Shambhala Press.

Wong, F., Vos, A., Pyburn, R. and Newton, J.(2019). Implementing gender transformative approaches in agriculture. *A Discussion Paper for the European Commission, (May)*. CID: 20.500.12592/wjdzqq.

Zomer, R. J., None Trabucco, A., Coe, R., Place, F., van Noordwijk, M. and Xu, J. (2014). Trees on farms: an update and reanalysis of agroforestry's global extent and socio-ecological characteristics. Working Paper 179. Bogor, Indonesia: World Agroforestry Centre (ICRAF) Southeast Asia Regional Program. https://doi.org/10.5716/WP14064.PDF.

Published by Burleigh Dodds Science Publishing Limited, 2025.

Chapter 23

Women in smallholder pig farming: negotiating inequalities in Uganda's pig value chains

Rosemirta Birungi, Florence Kyazze Birungi, Frank B. Matsiko and Justine Nambi-Kasozi, Makerere University, Uganda; Maria Jones, University of Illinois at Urbana-Champaign, USA; Woldegebrial Zeweld and Dawit Gebregziabher, Mekelle University, Ethiopia

1 Introduction

Women in smallholder rural households play a central role in agriculture, and they are instrumental in upgrading agricultural value chains. In the African context, patterns of gender inequalities in division of labour, access and control of resources, and spending of household income make it difficult for women to participate in and benefit from the development of agricultural value chains (USAID, 2009; FAO, 2018). Evidence suggests that these inequalities are persistent and characterized by limited understanding of women's access to and control of benefits that accrue from participating in value chains (FAO, 2018). The limited understanding greatly explains why upgrading interventions continuously fail to effectively address specific needs of women, and why women continue to be marginalized and excluded from participation, especially in the upgrading of agricultural value chains (Coles & Mitchell, 2011; Kasente, 2012; Campos et al., 2016). Quisumbing et al. (2015) argue that it's because upon upgrading, men own agricultural assets and they capture the majority of benefits, compared to women, unless a deliberate effort is made to change the distribution of benefits which requires proper understanding of the division of labour, access and control of resources and benefits, and factors determining upgrading among women and men involved in the value chains.

http://dx.doi.org/10.19103/AS.2024.0148.34

Pig value chains in Uganda have shown similar patterns of gender inequalities (Mangheni, 2014). Mostly women (about 90%) are involved in doing work of smallholder pig production at the household level, while men mostly do work in smallholder marketing of pigs and pork, and few women in urban communities are involved in pig trading (ILRI, 2013). Despite the observed inequalities in Uganda's pig value chains, women are increasingly involved in pig production work, while men are increasingly involved in pig marketing and pork retailing. Nonetheless, still to date, no studies have been done to explain why women (compared to their male counterparts) in pig value chain continue to mostly do production work with limited participation in marketing.

Failure to fill the information gap would constrain the progress of interventions for value chain development in Uganda and similar countries in the African context. The need is evident in the literature. Studies argue that failure to clarify women's access and control of opportunities and benefits derived from value chain upgrading may lead to directing information and development interventions to unintended beneficiaries (Farnworth et al., 2013; Galiè & Farnworth, 2019) and may lead to widening or deepening the inequalities, especially for women (Achandi et al., 2018).

Limited women's access and control of opportunities and benefits derived from involvement in agricultural value chains is increasingly gaining attention for understanding women and smallholder farming. Upgrading of agricultural value chains is characterized by women increasingly lacking control over benefits accrued from participation and further becoming marginalized, which consequently could make them (women) excluded from involvement in commercial activities of the value chains (Farnworth, 2013; Quisumbing et al., 2015). Studies attribute this problem to gender-based constraints in smallholder farming, such as time poverty, social norms, and money management practices, that routinely limit women's participation in agricultural value chain activities and consequently frustrate their desire to upgrade. Specifically, for pig value chains in Uganda, studies suggest that to avoid the negative effects of gender-based constraints to women's participation, there is need to understand the benefits derived by the different genders (especially women and men) involved, examine who (among men and women) has access to what type of benefits, and what factors determine women's participation in pig value chain (Mangheni, 2014).

2 Case study: pig value chains in Uganda – methods

2.1 Case study method – abduction case study method

Case study method is used to study the pig value chains in Uganda. This method captures the complexity of pig value chains in Masaka district, central Uganda, as a single case which is considered as a functioning unit and is investigated in its natural and contemporary context of Masaka district. The case of pig value chains in Uganda is studied with an interest of generalizing by using the principle of abduction. As an approach to case study research, abduction case study method is used to understand the inequalities experienced by women who are involved in smallholder pig farming in the context of Uganda's pig value chains. Induction case study method is used as a means of reflection on data that is grounded in particular research projects – in this case, the ILRI pig value chain project (ILRI, 2013) and the positioning against different theories such that data can contribute to and develop further the study research questions. This case study induction method aligns with the induction case study method suggested by various scholars such as Ahrens and Chapman (2006) and Dey (2021), arguing that case study researchers may approach the study context with a very loose idea, which would be negotiated as the project's study proceeds.

2.2 Description of case context – the pig value chains in Uganda

In Uganda, pigs are an important source of income for smallholder farmers, especially women (MAAIF, 2016). Smallholder farmers perceive pig farming as a 'live bank' for addressing household basic needs as soon as money is needed. Pigs serve as sources of cash for smallholder farmers in times of need for money to repair houses, pay school fees, rent agricultural land, purchase seed, fertilizers, and other farm inputs (Deka et al., 2007). The International Livestock Research Institute (ILRI) has, for the past decade, supported research and development interventions to improve (upgrade) the pig value chain in Uganda. The ILRI intervention studies were intended to create and increase stakeholders' attention to the concern about smallholder women's participation and the likely negative consequences in pig value chains that were evident among high pig producer communities of Uganda mostly in Masaka, Kamuli, Lira, and Hoima districts (ILRI, 2013). My PhD study supported by ILRI under the supervision of ILRI Gender Specialist (Galie' Alessandra) and the University of Illinois at Urbana-Champaign (USA) was intended to help understand the gender-based determinants of upgrading in pig value chains in Masaka district – a high pig producer community of Central Uganda.

3 Case study: results

The study reveals that various benefits are derived from participation, mostly including income, asset ownership, food and nutrition, social capital, and leisure (Birungi et al., 2020). Women smallholder farmers consider food, nutrition, and leisure as the most important benefits derived from participating in pig value chains, while men consider income (from the sale of pigs and piglets) as the most important benefit derived (Birungi et al., 2020). This study finding showed that men use the income to pay for school fees, raise capital to open/start a side business, acquire land, construct a house, pay house rent, purchase domestic items, acquire other livestock, employ more labour, save resources for future use, raise money for marriage, and pay medical bills. The leisure time for women is used to socialize with colleagues, make new friends, and meet with old friends. Women smallholder farmers derive food and nutrition benefits because they consider that it provides them with food for consumption, a source of alternative food, cooking oil from pork, medicine from pork and oil, and a balanced diet.

The structure of the primary activities of the pig value chain comprises pig farming, trading in live pigs, distribution of pig products, and pork retailing. Table 1 shows the primary activities of the pig value chain and tasks done

Table 1 Primary activities of the smallholder pig value chain in Central Uganda

Primary activities of pig value chain	Tasks done for upgrading in pig value chains
Pig production	• Pig sty maintenance (pig sty construction and repair) • Routine management (feeding, washing/cleaning the sty, monitoring growth, tethering, giving drinking water, ear-tagging, and record keeping) • Breeding (selecting pig breeds, castrating pigs, relieving pig stress, and taking pigs for mating) • Health management (observing sickness, deworming, spraying, treating pigs, and fattening)
Pig trading	• Market search (sourcing buyers, linking with buyers, and price searching) • Pricing (negotiating, determining, prices, and quality assurance) • Selling (delivery and payment receipts)
Pork distribution	• Preparing for slaughter (collecting banana leaves, collecting firewood, cooking/boiling water, looking for slaughter place, verifying pig health status before slaughter, buying polythene bags, hiring a weighing scale, and mobilizing money for slaughter from members). • Slaughtering (slaughter and dressing) • Hygiene maintenance (waste disposal and cleaning the slaughter place) • Transportation/outbound logistics (weighing pork, packaging, and transporting pork)
Pork retailing	• Buying and selling (contacting pork distributors, purchasing pork, weighing pork, displaying the pork in the stalls, selling) • Pork preparation (skinning, dressing) for cooking, culinary services (frying, roasting, serving) • Workplace hygiene maintenance

for each activity required to take the pigs from production to consumption as pork.

Structures differentiate what constitutes men's and women's spheres and often construct opportunities that reproduce divisions, for instance in labour and time allocation, benefit sharing, access and control of resource that are marginalizing (Tobin & Castellanos, 2021). This study reveals that the structure for division of labour in pig value chain activities are skewed with women concentrated in few activities, which include pig sty maintenance, routine management, and pork preparation and culinary services out of a set of 15 tasks for pig value chain that described the process from pig production through pig–pork transformation to pork retailing.

Figure 1 shows the flow of pigs and pig products as well as the gender division of labour in terms of 'who mostly does the activities at each node

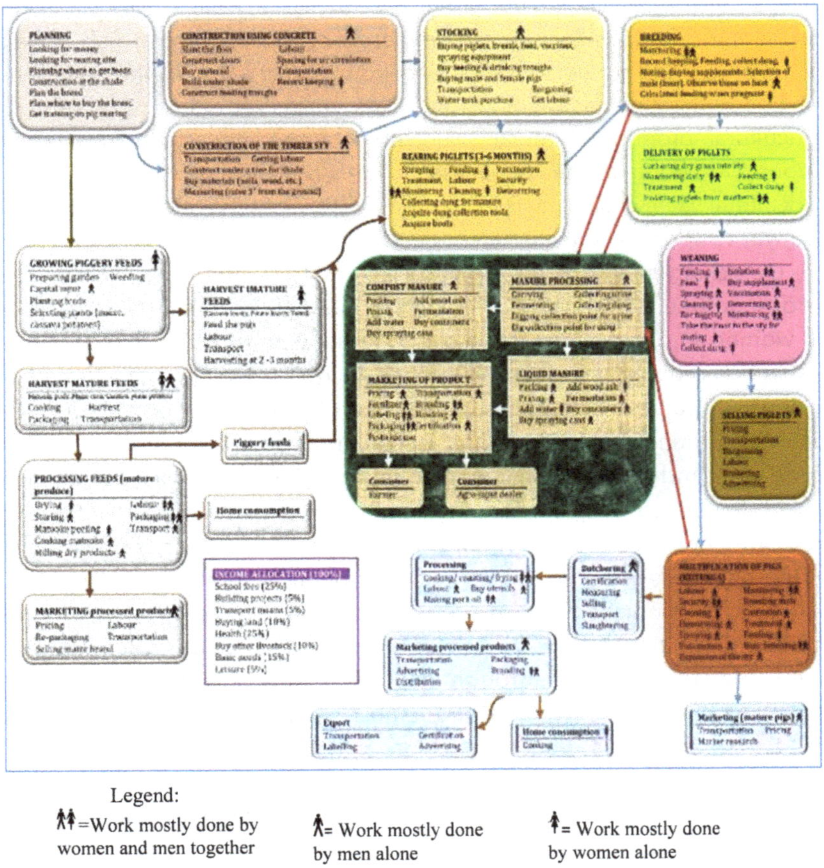

Figure 1 Pig value chain showing the flow of pigs and pig products, and who mostly does value chain activities. *Source:* Authors.

of pig value chain'. This finding entrenched the previous assertion about the structure of intensified smallholder livestock systems that men concentrate on commercial roles such as purchase of animals and feed and sale of livestock products (Dione et al., 2020; Ouma et al., 2018). Hence, the identified skewed division of labour, with women mostly concentrated in traditional roles of care for pigs and pork consumers, while men concentrate in commercial roles of trading pigs and pork would most likely construct inequalities in access and control of economic benefits, which consequently would force women to drop-out of participation in pig value chain.

4 Conclusion and future trends

Future trends suggested by this study are that (pig) value chain development interventions need to further understand and address the centrality of women in transformation (including upgrading) of agricultural value chains. It is evident that as agricultural value chains commercialize, women are at risk of being marginalized and further on, excluded from participating in the commercial (income-generating) activities of the value chain, yet women are central in agricultural production, mostly involved in providing smallholder farming labour. De-risking solutions necessary for sustained women's centrality in agricultural value chains need to be identified and interventions designed as a pathway to addressing the inherent gender inequalities.

De-risking solutions for women's centrality in the value chains would help negotiate pathways for eliminating inequalities that are experienced by women in agricultural value chains. This would involve specific levers that drive actions and decision-making in doing various activities of the agricultural value chain. De-risking solutions to women's inequalities would include creating business value, enhancing asset/resource control and ownership, and strengthening women's functionality, as some of the gender-based levers considered for creating desired transformations through women's centrality in agricultural value chains. These gender-based levers would enable leveraging the gains from uplifting women's participation in commercial value chains. In this regard, the gender-based levers would be applied in ignition across the value chains, as a set of de-risking solutions that could be applicable through local programmes that support women-led farming to address global inequities in agriculture. Specifically, such gender-based levers would focus on igniting women's capabilities for doing sustainable businesses across the value chain as well as igniting men's opportunities for access to markets of smallholder agricultural farm produce.

5 References

Achandi, E. L., et al. (2018). Women's access to agricultural technologies in rice production and processing hubs: a comparative analysis of Ethiopia, Madagascar and Tanzania. *Journal of Rural Studies*, 188–198.

Ahrens, T. & Chapman, C. S. (2006). Doing qualitative field research in management accounting: positioning data to contribute to theory. *Accounting, Organizations and Society*, 299–318.

Birungi, R., et al. (2020). Perceived benefits of upgrading in pig value chains: a gender perspective of smallholder pig farmers in Uganda. *African Journal of Rural Development*, 261–272.

Campos, A. P., et al. (2016). How does the choice of the gender indicator affect the analysis of gender differences in agricultural productivity? Evidence from Uganda. *World Development*, 17–33.

Coles, C. & Mitchell, J. (2011). *Gender and agricultural value chains – a review of current knowledge and practice and their policy implications*. London, UK: Overseas Development Institute.

Deka, R. et al. (2007). Assam's pig sub-sector: Current status, constraints and opportunities. International Livestock Research Institute (ILRI) Project Report, New Delhi, India.

Dey, A. K. (2021). *Case method for digital natives teaching and research: 100 years of case method* (S. Mishra, Ed.) India: Bloomsbury Publishing India.

Dione, M. et al. (2020). Gendered perceptions of biosecurity and the gender division of labour in pig farming in Uganda. AgriGender: *Journal of Gender, Agriculture and Food Security* 5(2): 13–26.

Farnworth, C., et al. (2013). *Transforming Gender Relations in Agriculture in Sub-Saharan Africa*. Stockholm, Sweden: Swedish International Agricultural Network Initiative (SIANI).

FAO. (2018). *The State of Food and Agriculture 2018: Migration, agriculture and rural development*. Rome: Food and Agriculture Organization of United Nations

Galiè, A. & Farnworth, C. (2019). Power through: a new concept in the empowerment discourse. *Global Food Security*, 13–17.

ILRI. (2013). *More pork by and for the poor: Catalysing emerging smallholder pig value chains in Uganda for Food Security and Poverty Reduction*. Kampala, Uganda: International Livestock Research Institute (ILRI).

Kasente, D. (2012). Fair Trade and organic certification in value chains: lessons from a gender analysis from coffee exporting in Uganda. *Gender & Development*, 111–127.

MAAIF. (2016). *Agriculture Sector Strategic Plan (2015/16-2019/20): "Draft"*. Kampala, Uganda: Ministry of Agriculture, Animal Industry and Fisheries.

Mangheni, M. N. (2014). *Review of the Uganda smallholder pig value chain assessment results and suggested potential interventions to improve women's access and control of resources in the pig value chain*. Nairobi: ILRI.

Ouma, E., et al. (2018). African swine fever control and market integration in Ugandan peri-urban smallholder pig value chains: an ex-ante impact assessment of interventions and their interaction. *Preventive Veterinary Medicine*, 29–39.

Quisumbing, A. R., et al. (2015). Gender, assets, and market-oriented agriculture: learning from high-value crop and livestock projects in Africa and Asia. *Agric Hum Values*, 705–725.

Tobin, D. and Castellanos, P. (2021). Gendered differences in decision-making and participation in value chains: lessons from the highlands of Peru. *Development in Practice*, *32*(4), 492–502.

USAID. (2009). *Promoting gender equitable opportunities in agricultural value chains*. Washington, DC: United States Agency for International Development.

Chapter 24

Supporting women smallholders involved in small-scale chicken production

Brigitte Bagnol, Tufts University, USA; Catherine Kaluwa, Jemimah Oduma, Tevin Konde, Angela Opondoh and Diana Muta, University of Nairobi, Kenya; Meghan Stanley, Marieke Rosenbaum and Hellen Amuguni, Tufts University, USA; and Robyn Alders, Tufts University, USA and Australian National University, Australia

1 Introduction

For around two decades, authors have clearly established that women are the main caretakers of small-scale chicken in most countries and that poultry are one of their only sources of income compared to men (Alders, 1996; Kitalyi, 1998; Guèye, 2000, 2005; Alexander et al., 2004; Bagnol, 2009a, 2009b; Abdirahman et al., 2023). In many rural areas around the world, income and productivity generated from small-scale poultry raising are used to support in particular children's education, household health, and family food, and to contribute to numerous feasts and rituals (Dumas et al., 2017; de Bruyn et al., 2015; FAO, 2014; Bagnol, 2001, 2009; Wong et al., 2018). Thus, many

http://dx.doi.org/10.19103/AS.2024.0148.35

programmes aimed at improving livelihoods cantred their interventions on improving small-scale poultry production systems.

Between the years 2000 to 2020, projects aimed at increasing the production of chickens and eggs were developed, focussing on women as the main implementors and beneficiaries of the interventions (Bagnol, 2013, 2009a, 2009b). Even if gender awareness was discussed, it was not the aim of the project to transform power relations between men and women. It was understood that women would have increased bargaining power in the household as the result of increased production and income generated by small-scale chicken production (Bagnol, 2017). Female vaccinators were the ones whose power in society increased the most as a result of their position in the community as 'chicken doctors' (Bagnol, 2012).

As gender approaches in general evolved from awareness to women's empowerment (Kabeer, 1999; Gammage et al., 2016; Agarwal, 1994, 1997, 2000), in recent years, women's empowerment is also being promoted in the poultry sector (Tukahirwa et al., 2022; Abdirahman et al., 2023; Kaluwa et al., 2022; Kyotos et al., 2022; Hillenbrand et al., 2022). Access to information and resources (including financial) alone is not enough to change women's status. A more gender-transformative approach aims to change gender relations and positions of women within their society as well as decision-making power within the household and community. This more holistic approach addressing the barriers to women's full involvement in economic activities and participation in decision-making at the household, community, institutional, and societal levels is required to yield change in women's status in society.

However, such gender-sensitive interventions in the small-scale chicken production sector remain very limited in the absence of specific policies addressing the societal barriers to women's access to and control over resources. In many examples of animal disease outbreaks, for example, as with the highly pathogenic avian influenza (HPAI) pandemic in 2003–2006, gender-blind policies resulted in negative consequences for women (FAO, 2013: 59; Velasco, 2008; Miers, 2008; AED & UNICEF, 2006). Similar to the experience with COVID-19, gender-blind interventions exacerbated existing inequalities (UNFPA, 2020a, 2020b; Wenham et al., 2020; Herten-Crabb and Davies, 2020; Canadian Institutes of Health Research, 2020). These findings highlight the need for gender-transformative approaches that lead to the inclusion of women in the development of animal health policies. However, existing literature is quite scarce on the impact of gender-blind policies. The following case studies examine the differences between gender-blind and gender-aware policies and opportunities for increased women empowerment.

2 Case study 1: Kyeema projects with Newcastle disease vaccination campaigns with female community vaccinators

2.1 Background

The control of Newcastle disease (ND) in small flocks of chickens raised under extensive conditions in rural and remote regions was largely ignored until the development of a thermotolerant ND vaccine that was safe to administer to birds of all ages via eye-drop or drinking water (Spradbrow, 1993–94). Perhaps it also received limited attention because veterinary services in the nineteenth and twentieth centuries were largely dominated by men who focussed on animals (frequently cattle) of importance to men.

On the completion of successful laboratory and field trials with thermotolerant vaccines in Mozambique in the 1990s (Dias et al., 2001), the major challenge was to develop a sustainable system that would facilitate the implementation of vaccination campaigns in rural areas. For ND control programmes to be sustainable, it was essential to address a range of technical, social, cultural, administrative, and economic issues such as community participation, gender-sensitive extension activities, facilitating government policies, training of government, NGO and private sector personnel and farmers, cost recovery, and distribution and marketing networks (Alders, 2001). The gender-oriented methodologies employed by the Kyeema Foundation from 2003 onwards, evolved from this research by the Mozambican National Veterinary Research Institute (*Instituto Nacional de Investigação Veterinária*) with support from the University of Queensland and the Australian Centre for International Agricultural Research (ACIAR, 2023).

2.2 Millions of indigenous chickens, too few animal health agents

Insufficient numbers of public and private sector animal health agents are a problem in many countries (OIE, 2019) meaning that the control of vaccine-preventable animal diseases remains lacking in numerous settings. One solution to this situation was the training of community animal health workers (CAHWs) and community vaccinators. For over four decades, community animal health workers have been trained, frequently by NGOs, to provide basic animal health services to small-scale livestock, especially large and small ruminants in underserved areas. Observations in Mozambique in the late 1990s suggested that once externally funded projects ended, CAHWs had difficulties continuing their services as payment by livestock owners was insufficient. In response, the ND control project decided to focus on supporting government animal health agents to train community vaccinators as a first step.

The idea was that these vaccinators would work part-time as vaccinators while maintaining their own mixed farming enterprise in the hope that this would be a more viable approach (Alders et al., 2002). Another important aspect of the work in Mozambique was that the National Veterinary Services were willing to partner with the Agricultural Extension Services. As the number of agricultural extension agents far outnumbered livestock extension agents, this meant that community vaccinators could be effectively supported in areas where livestock extension agents were scarce. The agricultural extension officers also tended to have stronger relations with women at the community level given their strong involvement with crop production.

2.3 Who manages the chickens and who should be able to vaccinate them?

The ND control activities benefitted from the insights of an anthropologist, with a keen interest in equity and gender, from the early stages (Bagnol, 2001). This resulted in a better understanding of the roles, responsibilities and challenges faced by the various household members (especially women and children) whose decisions directly and/or indirectly impacted the health and productivity of household poultry flocks.

The selection of community vaccinators was done in collaboration with community leaders and representatives. These representatives were asked to select equal numbers of men and women to be trained as vaccinators. In most cases, representatives selected more men than women. However, over time, many communities recognized that female community vaccinators were more likely to remain working in their home areas than their male counterparts and they then commenced selecting more women when replacements were required (Alders et al., 2010).

The three-day community vaccinator training on ND control and poultry husbandry was designed to facilitate women's involvement (Alders et al., 2002). Emphasis was placed on organizing training in short, regular bursts that did not run for the entire day, and was conducted in the village rather than at central training centres. This approach allowed women to participate while also fulfilling multiple roles within their households. The training methodologies took into account the high levels of illiteracy among rural women by using visual aids such as flip charts that the vaccinators could take with them on the successful completion of the course. When new information was introduced that related specifically to women's areas of work, care was taken to ensure that women were given control over this information. The active engagement of female community vaccinators also meant that they communicated effectively with the women in the village who took care of the chickens. One easily recognizable benefit was that when women (rather than only their husbands) knew that the birds were to be vaccinated, they would not open the door to the chicken shelter until the vaccinator had arrived.

2.4 Remuneration of community vaccinators

Ensuring that vaccinators had adequate compensation and motivation to carry out the vaccination campaigns three times per year was a major priority (Bagnol, 2009). As mentioned earlier, women vaccinators have tended to be more willing to continue with their work as they were less likely to leave the village in search of higher paying work. In addition to the fee-for-service payment received by vaccinators during the vaccination campaigns, the increase in their own flock sizes enabled women to diversify their holdings. In both Mozambique and Tanzania, increasing chicken flocks and sales of hens and roosters have allowed women to buy goats, which in many cases were then considered their asset rather than the property of their husband (Villanueva Cabezas, 2020).

2.5 Who implements and manages the Newcastle disease control activities?

Since 2003, the Kyeema Foundation has implemented ND control activities in collaboration with national veterinary services in many countries, supported by a range of donors including the Food and Agriculture Organization of the United Nations, USAID, the Australian Government, and private sector corporate social responsibility funds. In most cases to date, the key implementation partner has been the National Veterinary Services. In recent years, in Ethiopia, training and supervision of community vaccinators was done in collaboration with private sector veterinary vendors, i.e. premises permitted to sell prescribed veterinary medicinal products and allied substances. In the majority of cases, management of these ND control projects has been supported by women project staff. Their commitment to listening, understanding, and responding to learning opportunities in the field have enabled the community-based ND control activities to continue to evolve with a strong emphasis on enabling women at all levels to demonstrate their capabilities and knowledge.

3 Case study 2: women's empowerment along the Newcastle disease vaccine value chain in Kola and Kalama ward, Machakos County, Kenya

This case study is the result of a larger research project (SheVax+) entitled 'Hearing their voices: Action research to support women's agency and empowerment in livestock vaccine distribution, delivery and use in Rwanda, Uganda and Kenya'. It was carried out between 2019 and 2023 and was funded by the International Development Research Centre, under the 'livestock vaccine innovation fund', which was supported by the Bill and Melinda Gates Foundation and Global Affairs Canada, with co-operation between Africa One Health University Network (AFROHUN) and Tufts University in the USA.

The action research took place in Kola and Kalama wards which are located in Machakos town sub-county, Machakos county has a population of 170 606 (Kenya National Bureau of Statistics, 2019) and neighbors Nairobi, the capital city of Kenya. Agricultural crop farming and livestock keeping are common and most smallholder farmers raise indigenous chickens. Machakos country is a major indigenous chicken production area and provides an estimated 862 392 indigenous chickens per year to Nairobi (Okemer, 2019). Although ND is preventable through vaccination, it is not regularly implemented and most chickens die of this disease.

To empower smallholder women to effectively contribute to and benefit from the vaccine supply chain as entrepreneurs, service/product providers, and end users, the gender-related technical, social, cultural, and economic barriers that they encounter, and the opportunities therein need to be addressed. SheVax+ focussed on systemic transformational change and empowerment of women in the livestock vaccine value chain (LVVC) through the engagement of key critical partners in Kenya at national and sub-national levels; analysis of the barriers and opportunities at systems, enterprise and household levels across the LVVC; and identification of opportunities to create an enabling environment for women smallholder poultry farmers that would sustain outcomes after the project life cycle (see Table 1).

In line with the definition of empowerment, three dimensions of power were used. 'Power within' refers to a transformation of individual consciousness, which leads to a form of self-confidence to act. 'Power with' is the power that results from group organization and action to address common concerns. 'Power to' is the power to produce a result or resist change, the power to act.

Table 1 Main interventions carried out at each level of the ecological model

Level	Interventions
Individual	Training on gender issues Training on livestock management
Household	Training of husbands of female participants on gender issues
Community	Training of male gender models Implementation of community dialogues on gender issues
Institutional	Involvement of key stakeholders in a participatory engagement process around gender issues and women's empowerment in the vaccine value chain
Political	Involvement of government stakeholders Development of policy paper

To address the cold chain barrier in Machakos County, 'A women-centered private sector vaccine delivery model' was implemented (*Fig.* 1). The model provides a solar fridge and animal health medicine to Kenya veterinary board-certified female animal health service providers. It emphasizes partnership between the private and public sectors to support women farmers groups, to train and mentor female animal health service providers, and to support the network of women vaccine deliverers. Lastly, it engages the regulatory and policymakers. This model includes some of the systems and regulations that govern vaccine availability, the different LVVC actors and their capacities

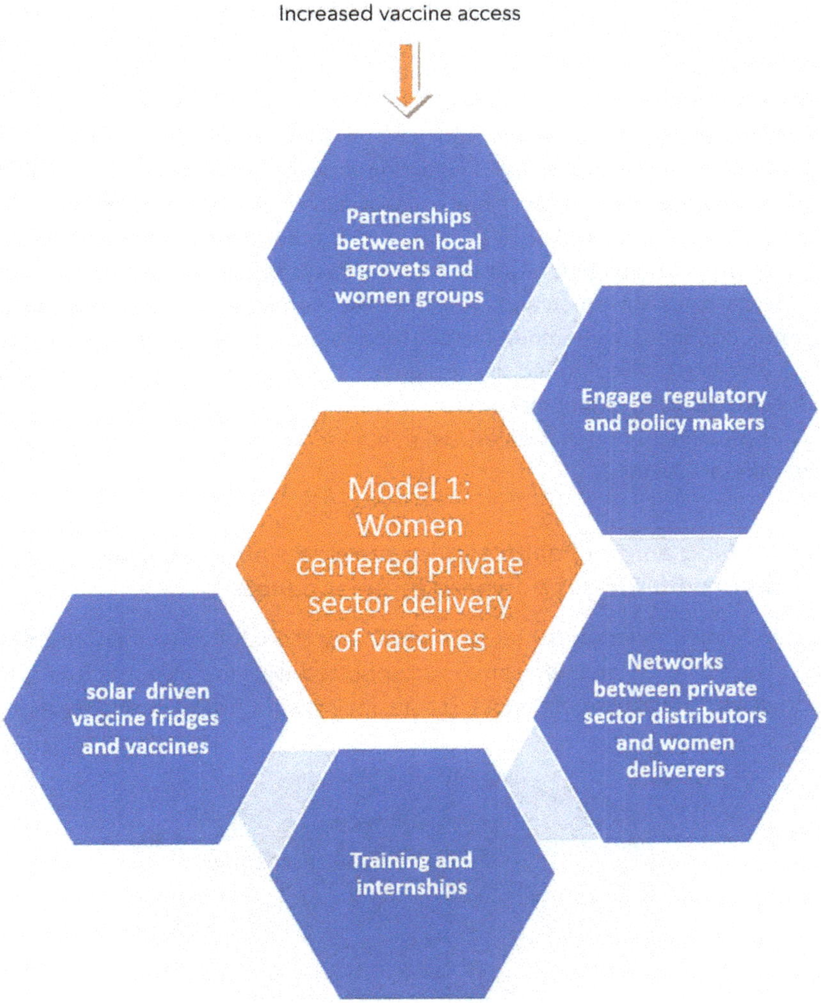

Figure 1 A women-centred private sector delivery model for livestock vaccines in Machakos, Kenya.

and professional knowledge, and the gender dimensions as influenced by attitudes and culture. The model also emphasizes scaling up the agency and empowerment of women smallholder farmers which includes strengthened skills and knowledge, increased economic benefits, and opportunities for women to influence legal and governance structures through united voices and networks that lead to increased vaccine adoption.

Inaccessible vaccines due to the lack of local vaccinators and sellers, unreliable cold storage, lack of knowledge which would increase demand for vaccine services, lack of extension services, social and cultural limitations, and non-gender responsive legal systems and structures are some of the barriers that were identified as impeding women from fully benefitting from livestock vaccines during a baseline assessment (Abdirahman et al., 2023; Kaluwa et al., 2022). These barriers, as well as women's limited decision-making agency in their households and communities, were key reasons for the low adoption of livestock vaccines in Kenya that were addressed by the action research.

SheVax+ aimed to transform the underlying social structures, policies, and norms that perpetuate gender inequalities. Social construction of the feminine and the masculine is based, among other things, on social, cultural, or religious symbols that legitimize the places (privileges – burdens), norms, and social roles of women and men and other non-binary people. This process is interconnected with other factors (economic, ethnic, geographical area, etc.). This understanding led the research team to adopt an ecological model represented in *Fig.* 2 that places the individual in a specific context.

The action research tackled women's empowerment and leadership at four different levels:

3.1 Training animal health service providers and equipping them with fridges powered by solar panels

In this women-centred private sector delivery model, five women and three men were offered an internship and supported with the infrastructure (solar fridges and initial vaccine stock) to start their own agrovet shop. The fridges

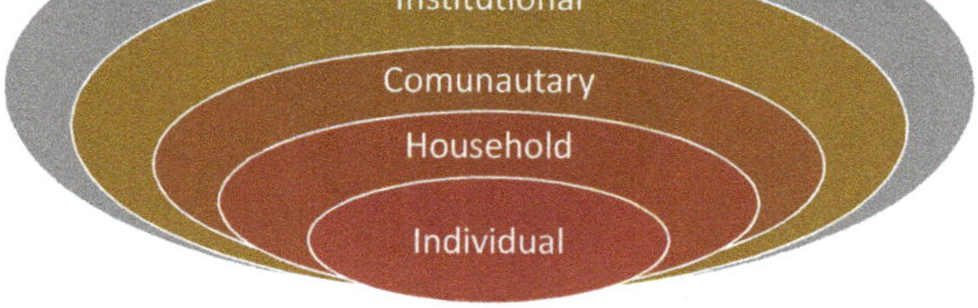

Figure 2 Socio-ecological model of the gender transformative approach

are managed by the animal health service providers (AHSP) but belong to the community. The project established linkages with vaccine distributors to ensure reliable supply of vaccines to the AHSPs. The 2-month internships focussed on how to manage an agrovet shop, purchase and sell products, provide services to customers, maintain the vaccine cold chain and storage, as well as general business management. The internship period was interspersed with didactic lectures and seminars from the project team. As entrepreneurs, the trained AHSP increased vaccine accessibility to smallholder farmers in their communities. They also provided training and raised awareness among smallholder farmers and organized farmer groups to increase vaccine adoption and women empowerment throughout the community.

SheVax+ facilitated training for the AHSPs on animal health, cold chain, and gender transformative approaches to business entrepreneurship. They set up mini-agrovet shops within their communities. In the field, these AHSPs provided two levels of training to the community animal health and vaccine training. The AHSPs utilized training sessions to also undertake vaccination campaigns and expand networks to more farmers. Through community training sessions, and via 'word of mouth' the AHSP network has expanded.

3.2 Raising gender awareness at the community level

Each AHSP trained farmers alongside a male role model. Male role models were trained on gender transformative approaches, and on how to facilitate conversations at the community level (Lemma et al., 2018). They were charged to support the AHSPs, mobilizing farmers especially men within the community to join self-help groups to interrogate and challenge social and cultural norms within households and how they impact their partner's livelihoods.

The SheVax+ team, community development officers and AHSPs have trained 486 women leaders and farmers on gender transformation (challenging social and cultural norms), animal health, vaccines, vaccination, and business entrepreneur skills.

3.3 Government departments and officers and agrovet vaccine distributors

In the first year, SheVax+ identified LVVC key players, their roles and how they impacted women's empowerment and gender equality with a focus on the gender capacities (skills, knowledge, perceptions, attitudes, behaviour) of the actors along the LVVC. In the second year, three critical partner stakeholder engagement meetings were carried out to evaluate progress.

The project has trained critical partners including community development officers and Veterinary Officers from targeted local authorities/

sub-counties. These are supporting the training of smallholder farmers in gender, leadership and other soft skills, and animal husbandry skills and practices. The trained extension workers (community development officers, veterinary officers, and AHSP/community vaccinators) work together in teams to deliver training and form a sustainable network that will likely continue beyond the project period. They have also integrated the gained knowledge and skills into their day-to-day engagement with communities, recognizing the systemic and relational nature of the challenges that communities are facing.

The agrovets are now providing advisory services as well as selling livestock products to farmers. This is a pleasant change. They recognize the need to put up more agrovet shops within communities that will improve sales and extension to farmers.

3.4 Increased access to Newcastle disease vaccines

The model has intensified vaccine adoption through an improved enabling environment, increased vaccine accessibility, and increased demand by women smallholder livestock farmers. SheVax+ focussed on catalysing the engagement of critical partners from the onset, facilitating networking and collaboration between women smallholder farmers and value chain actors with the aim of institutionalizing gender-responsive sustainable change in the LVVC and ensuring that the model is sustainable beyond the project.

The presence of AHSPs at the community level has also improved access to extension, advises, and vaccination. Results from the Kenya endline qualitative survey carried out in January 2023 showed an increase in knowledge and adoption of the vaccination as well as a change in gender awareness at all levels.

3.5 Increased women's empowerment

Baseline survey participants described what empowerment meant from their perspective. It included economic independence: women can negotiate for better buying prices of poultry and goats, increase the size of their flocks, sell for more profit, and female AHSP can own successful businesses such as agrovets. The second aspect of women's understanding of empowerment included knowledge and skills. As a result of the project, women have access to informal training on vaccine handling and administration and to information on animal health and entrepreneurship so they can improve their businesses. They are engaging their partners and family members and sharing information and skills obtained from trainings. The third component of empowerment was opportunities for networking. At the end of the project women's networks and activities along the LVVC has increased. Their capacity to get access to resource and interact with agrovet and livestock government structures has increased allowing them to lobby and request support. The AHSP is also leveraging strong women's networks to mobilize and mass vaccinate poultry at the community level. The fourth aspect of empowerment mentioned was autonomy in decision-making. As a result of the sensitization by the project, women are recognized by their husbands and male community leaders as chicken producers involved in the LVVC and contributing to their household livelihood. They can make decisions and take action to improve their livelihoods and participate and benefit from livestock markets. The male role model, the AHSPs, and the community development officers are holding gender transformative talks every 2 weeks in communities to challenge discriminatory cultural norms and stereotypes.

In addition to qualitative research methods for baseline and endline, the project used an adapted version of the Women Empowerment in Livestock Index (WELI) that included specific elements on vaccine and chickens and to assess the empowerment level of the interventions' beneficiaries (Kaluwa et al., 2022).

By the end of the SheVax+, in September 2023, the project had generated evidence and formulated strategies that helped position women to effectively and efficiently contribute to and benefit from livestock vaccines and enhance women's participation in livestock vaccine distribution, delivery, and use. The intervention tested a gender-focussed model that targeted mainly female AHSPs as vaccinators and livestock medicine sellers and sensitized women, men, and community leaders and stakeholders along the LVVC about gender issues and allowed for the inclusion of women in the VVC as vaccine sellers and as vaccine buyers. In addition, the project sensitized stakeholders along the value chain from the ward to the city capital and involved both the private and public sectors as well as civil society. Finally, the project developed a policy paper to present to stakeholders for the adoption of gender transformative planning by county governments.

4 Case study 3: high pathogenicity avian influenza

In 2003, H5N1 avian influenza began to spread to eight countries in eastern Asia including China, Japan, South Korea, Vietnam, Laos, Cambodia, Thailand, and Indonesia and since then has spread over all continents. A literature review, from a gender perspective, of the impact of the disease among female small-scale chicken producers, focussing specifically on the Asian context, illustrates how gender-blind policies exacerbate the exclusion of women from the poultry value chain.

4.1 Understanding the context: gender dynamic at the household level and women's participation in the planning of policies and high pathogenicity avian influenza responses

Several authors highlighted the fact that women are the main persons caring for small-scale chickens (Velasco et al., 2008; Alders et al., 2014; Dai, 2019b) and played a central role in the rural economy (Velasco et al., 2008). Guèye (2007) warned about the need to conduct research related to responsibilities and ownership within the household to inform to whom compensation should be given when culling of poultry was carried out (McLeod et al., 2007). Indeed, the lack of understanding of the power dynamics at the household level was related to the absence of a gender-aware approach to animal health and HPAI disease outbreaks. Although women owned the chickens, the compensation was often paid to the male's bank account. Velasco et al. (2008) observed that most planners and technicians for animal and human health were men with no understanding of gender analysis and of gender-sensitive planning. These authors identified that the absence of gender-disaggregated data supporting a gender analysis of the impact of avian flu in different contexts, as well as the absence of gender policies in most of the institutions involved in the responses, were barriers to gender awareness and sensitivity. They also noted that the weak participation of women and women-focussed agencies in the planning of avian influenza responses was a hindrance to the inclusion of gender-sensitive approaches.

4.2 Communication and training

Because women were not identified as the main small-scale chicken producers, they were not the main recipients of HPAI information and training, most of which was directed towards men (Velasco et al., 2008). In addition, while women are caring for children, their male partners, the sick, and the elderly, women's role as a caregiver was not taken into consideration in HPAI messaging and communication strategies when dealing with animal-to -human transmission (Velasco et al., 2008; Dai, 2019b).

4.3 Impact of high pathogenicity avian influenza policies

HPAI has a multidimensional impact on people as it includes loss in terms of income, consumption, nutrition, and livelihood (Limon et al., 2009). However, people were affected differently along the value chain. Depending on their specific activities, as producers, sellers, or processors, people were affected by different measures such as market or slaughtering restriction and culling. The death of chickens negatively affected farmers who were market oriented, while wet market closings affected mainly farmers who sold in those spaces. In turn, slaughtering regulation (limitation or ban) impacted the small slaughtering businesses (Miers, 200).

Poultry sale prices fluctuated during the outbreak. The drop in price affected smallholders who reared poultry and negatively impacted both income and consumption. The increase in the price of other sources of animal protein resulted in lower capacity to acquire this type of food and in nutritional deterioration.

HPAI outbreaks disproportionately impacted people who were already vulnerable and those who experienced layers of vulnerabilities (Miers, 2008; Limon et al., 2009). Very poor people, households headed by women, and small farmers depending on poultry for their income were most affected (Limon et al., 2009). Women were affected disproportionally in relation to men as they lost one of their very few sources of income and found it difficult to quickly identify other forms of income (Miers, 2008). As often is the case when there are household income restrictions, children were removed from schools and households reduced their number of meals per day (Miers, 2008).

4.4 Sector restructuring

The HPAI outbreak led, in most Asian countries, to a sector restructuring that affected small-scale actors who were mainly women farmers and traders and transporters. In most cases, the evolution from small-scale chicken production to more intensive chicken rearing systems resulted in men taking over the previously female-dominated poultry production activities (Velasco et al., 2008). Miers' analysis of the situation in Vietnam and Thailand is that small producers, traders, and transporters were expelled from the value chain that became more industrialized (Miers, 2008).

4.5 Recommendations for policies

A series of recommendations were developed by researchers involved in the analysis of the impact of HPAI interventions. Both mass media and interpersonal communication were recognized as important and should be used (AED & UNICEF, 2006). Messages should be specific to different stakeholders. For example, the difference between households that rely on poultry as assets and those that rely on poultry for income should be a primary consideration for the development of all communication strategies and IEC materials targeting small-scale poultry farmers in Cambodia (Hickler, 2007).

Behaviour-change communication efforts should focus mainly on the socioeconomic benefits for people adopting specific recommended measures

(FAO, 2015; Dai, 2019a). In the same vein, Hickler (2007) recommended focussing on risk-perception and not on fear. He suggested connecting messages to local rationales, values, and priorities, especially family well-being and prosperity (as opposed to technical messages) (Hickler, 2007). He also recommended whenever possible to take into consideration pre-existing practices, especially indigenous practices of 'separation' (Hickler, 2007). Moreover, AED and UNICEF (2006) considered it necessary to promote motivations for reporting that resonates with communities' values.

Developing a compensation policy that takes into account ownership of chickens by women and children and different types of producers (small-scale as well as commercial farmers) with a transparent a clear process for obtaining it was considered of utmost importance to encourage stakeholder engagement (AED & UNICEF, 2006).

Finally, researchers strongly recommended taking into consideration the socioeconomic context and that campaigns should be women-friendly, gender, and culturally sensitive including in relation to place and time delivery. Women's mass organizations should be involved in contributing to IEC dissemination according to Velasco et al. (2008). The need to improve the gender balance of animal health workers was also mentioned by Velasco et al. (2008). Considering women as beneficiaries of training for poultry production and care was considered important (Velasco et al., 2008; Dai, 2019b). Overall, a general recommendation lay in the participation of the people affected in the design of the policies, the planning of interventions, and the programmes developed to address the issues identified by them.

5 Discussion of the three case studies

The literature review of the impact of HPAI on smallholder farmers indicates that gender issues were not taken into consideration and women were disproportionally negatively impacted by HPAI in Asia during the 2003–2006 outbreak. The lack of consideration of gender roles and differences in access to and control over resources is referred to as a gender-blind policy. This approach exacerbates existing discrimination and vulnerabilities. Moreover, people suffering from various forms of vulnerabilities are more likely to find themselves in a worse situation after a pandemic event. This was the case in the HPAI outbreak for minorities, marginalized, and impoverished groups. It was also the case during the COVID-19 pandemic. Slowly awareness about gender issues is increasing, but unfortunately, policies and the application of policies where they already exist are too slow.

Projects carried out in Southern African countries by Kyeema Foundation between 2000 and 2012 aimed at training female community vaccinators and carrying out three vaccination campaigns per year with the main aim of reducing chicken mortality, increasing chicken flocks, and improving women's socioeconomic condition. As a result, women's bargaining power within their household and their community increased. They were gender aware in the sense that they were aware of the important role women played in the small-scale production system and they supported women to be trained as chicken farmers, vaccinators, and in project leadership. However, the project was not openly advocating for a gender-transformative approach. No gender training was carried out for policymakers, extension workers, community leaders, community members, or women farmers.

Adopting a gender transformative approach to empower women in the LVVC, the SheVax+ action research was implemented in Uganda, Kenya, and Rwanda by the Africa One Health University Network and Tufts University between 2019 and 2023. It trained female AHSP to deliver vaccine and medicine to rural farmers who were supported by male role models and community gender sensitization sessions to advocate for women's participation in the livestock sector. The project discussed with all stakeholders the underlying stereotypes and barriers that were preventing women from engaging fully in the LVVC and challenged men and women to actively change their perspectives and social roles. Men were invited to assist women with household chores to allow them to engage in poultry activities such as meeting, training, and vaccination. The project aimed at challenging the existing power relation between men and women explicitly. The project developed a policy brief advocating for the following policy recommendation: 'County government departments should adopt gender-transformative planning that addresses specific barriers that women and other marginalized livestock farmers face.' To comply with the aim to promote gender equality and to adopt a gender transformative approach, future projects to support women smallholders involved in poultry farming should follow or develop similar gender-transformative approaches.

If we come back to the three dimensions of power (power within, power to, and power with) the case study on HPAI indicates that none of the notions of power were addressed by the interventions implemented as they were not aiming at giving voice, autonomy, and decision-making to individual or groups. The case study by Kyeema mobilizes mainly the power of the female vaccinator to become visible and relevant actresses in their community giving them to earn a small income and acquire responsibility and respect from their neighbours and to contribute to animal health. The SheVax+ project supported the three forms of power as in addition to establishing the female vaccinators and supporting them to exercise their power to contribute to animal health, they participated in gender training and leadership to raise their inner power, their power within, and their sense of self-worth. By facilitating the networking process between AHSPs, between AHSPs and farmers and between stakeholders, the project strengthened the ability of the females to collaborate with others, to lobby and thus to increase the 'power with' (their capacity and possibilty to work with others).

According to FAO 2017 and the European Institute for Gender Equality (EIGE), women's empowerment encompasses several aspects. First, women's sense of self-worth (power within); second, their right to have and to determine choices (decision-making at individual level); third, their right to have access to opportunities and resources; fourth, their right to have the power to control their own lives, both within and outside the home; and, lastly, their ability to influence the direction of social change to create a more just social and economic order, nationally and internationally (decision-making at political level). Interventions related to women small-holder farmers involved in poultry production should make sure that they contribute to these five aspects of empowerment.

6 Conclusion and future trends in research

Social sciences research contributed to the understanding that the improvement of small-scale chicken production could not be done without the full participation of women. In addition, it is recognized that to include women only as beneficiaries of interventions is not adequate. The full participation of women at all levels of the intervention (designing, planning, implementation, and evaluation) and in all capacities (beneficiaries, decision-makers, experts, leaders, etc.) should be promoted.

Research also contributed to understand that small-scale poultry production is a key element for sustainable development as poultry has a very low negative impact on greenhouse gas emissions.

Quantitative research methods are implementing Women's Empowerment in Agriculture Index (WEIA), Project level WEIA (PRO WEIA), and Women Empowerment in Livestock Index (WELI) to assess the empowerment level of interventions' beneficiaries. These tools must be adapted to local circumstances prior to implementation.

Future trends are as follows:

- Developing qualitative and quantitative indicators and instruments that are efficient, easy to use, and cheap to measure empowerment to grasp the different aspects of empowerment;
- Identifying which of the different levels of intervention is the most efficient to trigger change;
- Analysing the impact of involving women in decision-making in project and in response to animal disease;
- Analysing the impact of working at different levels such as the level of policymaking, veterinarians, and AHSP in terms of women's empowerment;

- Improving the understanding of the full potential of women group for the improvement of small extensive scavenging and extensive scavenging chicken production and for women's empowerment; and
- Studying local conditions that create the risk environments in which HPAI can emerge and spread to inform local policies and interventions.

7 Where to look for further information

On poultry

- Kyeema Foundation - https://kyeemafoundation.org/explore-our-work/resources-2/featuredaffiliatedresearch/#toggle-id-8.
- FAO. 2014. Decision tools for family poultry development. FAO Animal Production and Health Guidelines No. 16. Rome, Italy. http://www.fao.org/docrep/019/i3542e/i3542e.pdf.
- FAO. 2022. Farmer field school for family poultry producers. A practical manual for facilitators. Rome: FAO. https://www.fao.org/3/cc0254en/cc0254en.pdf.

On gender issues

- The European Institute for Gender Equality (EIGE) - https://eige.europa.eu/publications-resources/thesaurus/terms/1246.
- Agrilinks - https://agrilinks.org/activities/feed-future-advancing-womens-empowerment-awe-program.
- International Development Research Centre (IDRC) Livestock Vaccine Innovation Fund - https://idrc-crdi.ca/en/initiative/livestock-vaccine-innovation-fund and https://idrc-crdi.ca/en/research-in-action/empowering-and-engaging-women-livestock-vaccine-value-chains-east-africa.
- Consultative Group on International Agricultural Research (CGIAR) - Livestock and women's livelihoods: a review of the recent evidence - https://cgspace.cgiar.org/bitstream/handle/10568/3017/Discussion_Paper20.pdf.

On measuring empowerment

- Kaluwa, C., Jemimah, O, Abdirahman F. A., Kyotos K. B., Opondoh, A. A., Muchibi, J., Bagnol, B., Rosenbaum M., Onchaga, S., Stanley, M., Amuguni, J. H. (2022) 'Using the Women Empowerment in Livestock Index (WELI) to Examine Linkages Between Women Smallholder Livestock Farmers' Empowerment, and Access to Livestock Vaccines in Machakos County of Kenya: Insights and Critiques.' *Vaccines*. https://www.mdpi.com/2076-393X/10/11/1868/pdf.

- The International Food Policy Research Institute (IFPRI) The Women's Empowerment in Agriculture Index (WEIA) and Project level WEIA (PRO WEIA) - https://www.ifpri.org/publication/womens-empowerment-agriculture-index and https://weai.ifpri.info/versions/pro-weai/.
- International Livestock Research Institute (ILRI) Women Empowerment in Livestock Index (WELI) - https://www.ilri.org/womens-empowerment-livestock-index-weli
- Care. 2021. Costs and benefits of applying a gender transformative approach in agriculture programming. Evaluation of the Ekata model in Burundi. A win-win for gender, agriculture and nutrition. Testing a gender transformative approach from Asia in Africa. https://reliefweb.int/report/burundi/win-win-gender-and-nutrition-testing-gender-transformative-approach-asia-africa.

8 References

Abdirahman, F. A., Wahome, R. G., Kaluwa, C., Oduma, J., Bagnol, B., Rosenbaum, M. and Amuguni, J. H. (2023). Gender barriers and opportunities along the Newcastle disease vaccine value chain in Machakos sub-county. *Tropical and Subtropical Agroecosystems* 26: 1. https://www.revista.ccba.uady.mx/ojs/index.php/TSA/article/view/3915/1944.

ACIAR (2023). Vaccine innovation secures food supply and income source. In: *40 Years of ACIAR*. Canberra, Australia: Australian Centre for International Agricultural Research, pp. 72–73. https://www.aciar.gov.au/sites/default/files/2023-07/ACIAR_40years_book_Accessible_72dpi_1.pdf.

AED & UNICEF (2006). *Participatory Action Research on Avian Flu Communication: Summary Report and Recommendations*. Washington DC: AED.

Agarwal, B. (1994). *A Field of One's Own: Gender and Land Rights in South Asia*. Cambridge: Cambridge University Press.

Agarwal, B. (1997). 'Bargaining' and gender relations: within and beyond the household. *Feminist Economics* 3(1): 1–51.

Agarwal, B. (2000). Conceptualizing environmental collective action: why gender matters. *Cambridge Journal of Economics* 24(3): 283–310.

Alders, R. G. (1996). Facilitating women's participation in village poultry projects: Experiences in Mozambique and Zambia. *Proceedings of the 20th World's Poultry Congress* 3: 441–447.

Alders, R. G. (2001). Sustainable control of Newcastle disease in rural areas. In: Alders, R. G. and Spradbrow, P. B. (Eds), *SADC Planning Workshop on Newcastle Disease Control in Village Chickens. Proceedings of an International Workshop*, Maputo, Mozambique, 6-9 March, 2000. ACIAR Proceedings N° 131, pp. 80–87. https://www.aciar.gov.au/publication/technical-publications/sadc-planning-workshop-newcastle-disease-control-village-chickens.

Alders, R. G., Awuni, J. A., Bagnol, B., Farrell, P. and de Haan, N. (2014). Impact of Avian Influenza on village poultry production globally. *EcoHealth* 11(1): 63–72 https://doi.org/10.1007/s10393-013-0867-x.

Alders, R. G., Bagnol, B. and Young, M. P. (2010). Technically sound and sustainable Newcastle disease control in village chickens: lessons learnt over fifteen years. *World's Poultry Science Journal* 66(3): 433–440. https://doi.org/10.1017/S0043933910000516.

Alders, R., dos Anjos, F., Bagnol, B., Fumo, A., Mata, B. and Young, M. (2002). *Controlling Newcastle Disease in Village Chickens. A Training Manual*. 2nd Edition 2003, ACIAR Monograph N° 86, p. 128, Canberra: ACIAR. http://www.aciar.gov.au/publication/MN086.

Alexander, D. J., Bell, J. G. and Alders, R. G. (2004). Technology review: Newcastle disease with special emphasis on its effect on village chickens. *FAO Animal Production and Health Paper* 161.

Bagnol, B. (2001). The social impact of Newcastle disease control. In: Alders, R. G. and Spradbrow, P. B. (Eds), *SADC Planning Workshop on Newcastle Disease Control. in Village Chickens. Proceedings of the an International Workshop*, Maputo, Mozambique, 6-9 March, 2000. ACIAR Proceedings N° 131, pp. 69–75. https://www.aciar.gov.au/publication/technical-publications/sadc-planning-workshop-newcastle-disease-control-village-chickens.

Bagnol, B. (2009a). Bicycles, boots, t-shirts and percentage over payment of vaccination: what role for community leaders? In: Alders, R. G., Spradbrow, P. B. and Young, M. P. (Eds). *Village Chickens, Poverty Alleviation and Sustainable Control of Newcastle Disease*. Canberra: Australian Centre for International Agricultural Research. ACIAR Proceedings N° 131, pp. 102–107. ISBN 978 1 921531 75 0 (print), ISBN 978 1 921531 76 7 (on line). http://www.aciar.gov.au/publication/PR131.

Bagnol, B. (2009b). Gender Issues in Small-scale Family Poultry Production: experiences with ND and HPAI control. *World's Poultry Science Journal* 65(2): 231–240. (ISSN: 0043-9339).

Bagnol, B. (2009c). Improving village chicken production by employing effective gender sensitive methodologies. In: Alders, R. G., Spradbrow, P. B. and Young, M. P. (Eds), *Village Chickens, Poverty Alleviation and Sustainable Control of Newcastle Disease*. Canberra: Australian Centre for International Agricultural Research. ISBN 978 1 921531 75 0 (print), ISBN 978 1 921531 76 7 (on line). ACIAR Proceedings N° 131, pp. 35–42. http://www.aciar.gov.au/publication/PR131.

Bagnol, B. (2012). *Daktari Wa Kuku. Chickens'doctor. Mission in Tanzania (Sindiga. Phase 2: Regional Newcastle Disease Control*. Maputo: Kyeema Foundation. International Poultry Centre.

Bagnol, B. (2013). *Advocate Gender Issues: A Sustainable Way to Control Newcastle Disease in Village Chickens. Good Practices for Family Poultry Production*. International Network for Family Poultry Development, IFAD, FAO. http://www.fao.org/docrep/018/aq630e/aq630e.pdf.

Bagnol, B. (2017). *Majiri. 4th PRA. Strengthening Food Security Through Family Poultry and Crop Integration in Eastern and Southern Africa*. Maputo: Kyeema Foundation. International Poultry Centre.

Canadian Institutes of Health Research (2020). *Why Sex and Gender Need to Be Considered in COVID-19 Research*. CIHR. Canadian Institutes of Health Research. https://cihr-irsc.gc.ca/e/51939.html.

Dai, P. (2019a). *Gender equality assessment for HPAI risk management along poultry value chain: gaps and opportunities. Myanmar*. Bangkok: Food and Agriculture Organization.

Dai, P. (2019b). *Gender Equality Assessment for HPAI Risk Management Along Poultry Value Chain: Gaps and Opportunities. Cambodia, Myanmar and Lao PDR*. Bangkok: Food and Agriculture Organization.

de Bruyn, J., Wong, J., Bagnol, B., Pengelly, B. and Alders, R. (2015). Family poultry and food and nutrition security. *CABI Reviews* 10(13): 1–9.

Dias, P. T., Alders, R. G., Fringe, R. and Mata, B. V. (2001). Laboratory and field trials with thermostable live Newcastle disease vaccines in Mozambique. In: Alders, R. G. and Spradbrow, P. B. (Eds) *SADC Planning Workshop on Newcastle Disease Control. in Village Chickens. Proceedings of the an International Workshop*, Maputo, Mozambique, 6-9 March, 2000. ACIAR Proceedings No. 103, pp. 91–96. https://www.aciar.gov.au/publication/technical-publications/sadc-planning-workshop-newcastle-disease-control-village-chickens.

Dumas, S. E., Maranga, A., Mbullo, P., Collins, S., Wekesa, P., Onono, M. and Young, S. L. (2017). "Men Are in Front at Eating Time, but not When it Comes to Rearing the Chicken": unpacking the gendered benefits and costs of lLvestock ownership in Kenya. *Food and Nutrition Bulletin* 133(11). https://doi.org/10.1177/0379572117737428.

FAO (2013). Lessons from HPAI – A technical stocktaking of outputs, outcomes, best practices and lessons learned from the fight against highly pathogenic avian influenza in Asia 2005–2011. In: *FAO Animal Production and Health*. Rome: Food and Agriculture Organization of the United Nations.

FAO (2014). Decision tools for family poultry development. In: *FAO Animal Production and Health Guidelines No. 16*. Rome: Food and Agriculture Organization. http://www.fao.org/docrep/019/i3542e/i3542e.pdf.

FAO (2015). *Addressing Avian Influenza A (H7N9) Guidelines for Risk Communication Messaging*. Rome: Food and Agriculture Organization.

FAO (2017). *Strengthening Sector Policies for Better Food Security and Nutrition Results: Gender Equality*. Policy Guidance Note 6. 34pp. Rome: Food and Agriculture Organization.

Gammage, S., Kabeer, N. and van der Meulen Rodgers, Y. (2016). Voice and agency: where are we now? *Feminist Economics* 22(1): 1–29. https://doi.org/10.1080/13545701.2015.1101308.

Guèye, E. F. (2000). The role of family poultry in poverty alleviation, food security and the promotion of gender Equality in rural Africa. *Outlook on Agriculture* 29(2): 129–136.

Guèye, E. F. (2005). Gender aspects in family poultry management systems in developing countries. *World's Poultry Science Journal* 61(1): 39–46.

Guèye, E. F. (2007). Evaluation of the impact of HPAI on family poultry production in Africa. *World's Poultry Science Journal* 63(3): 391–400.

Herten-Crabb, A. and Davies, S. E. (2020). Why WHO needs a feminist economic agenda. *The Lancet* 395(10229): 1018–1020. https://www.thelancet.com/journals/lancet/article/PIIS0140-6736(20)30110-0/fulltext?dgcid=raven_jbs_etoc_email.

Hickler, B. (2007). *Bridging the Gap Between HPAI "Awareness" and Practice in Cambodia*. Cambodia: Food and Agriculture Organization.

Hillenbrand, E., Mohanraj, P., Njuki, J., Ntakobakinvuna, D. and Tasew, S. A. (2022). There is still something missing: comparing a gender-sensitive and gender-transformative approach in Burundi. *Development in Practice*. https://doi.org/10.1080/09614524.2022.2107613.

Kabeer, N. (1999). Resources, agency, achievements: reflections on the measurement of women's empowerment. *Development and Change* 30(3): 435–464.

Kaluwa, C., Oduma, J., Abdirahman, F. A., Kitoga, B. K., Opondoh, A. A., Muchibi, J., Bagnol, B., Rosenbaum, M., Onchaga, S., Stanley, M. and Amuguni, J. H. (2022). 'Using the women empowerment in livestock index (WELI) to examine linkages between women smallholder livestock farmers' empowerment, and access to livestock vaccines in Machakos County of Kenya: insights and critiques. *Vaccines*. https://www.mdpi.com/2076-393X/10/11/1868/pdf.

Kenya National Bureau of Statistics (2019). *Kenya Population and Housing Census Results*. Nairobi: National Bureau of Statistics.

Kitalyi, A. J. (1998). *Village Chicken Production Systems in Rural Africa. Household Foodsecurity and Gender Issues*. FAO Animal Production and Health Paper 142. Rome: Food and Agriculture Organization.

Kyotos, K. B., Oduma, J., Wahome, R. G., Kaluwa, C., Abdirahman, F. A., Opondoh, A., Mbobua, J. N., Muchibi, J., Bagnol, B., Stanley, M., Rosenbaum, M. and Amuguni ,J. H. (2022). Gendered Barriers and Opportunities for Women Smallholder Farmers in the Contagious Caprine Pleuropneumonia Vaccine Value Chain in Kenya. *Animals: An Open Access Journal from MDPI*. https://www.mdpi.com/2076-2615/12/8/1026/pdf.

Lemma, M., Mulema ,A., Kinati, W. and Wieland, B. (2018). *Transforming Gender Relations and Reducing Risk of Zoonotic Diseases among Small Ruminant Farmers in the Highlands of Ethiopia. A Guide for Community cCnversation Facilitators*. Addis Abeba: CGIAR.

Limon, G., Rushton, J., de Haan, N., Schwabenbauer, K. and Ahmed, S. Z. (2009). *Highly Pathogenic Avian Influenza: A Rapid Assessment of Its Socio-economic Impact on Vulnerable Households in Egypt*. Rome: Food and Agriculture Organization.

Mcleod, A., Rushton, J., Riviere-Cinnamond, A., Brandenburg, B., Hinrichs, J. and Loth, L. (2007). Economic issues in vaccination against highly pathogenic avian influenza in developing countries. *Developments in Biologicals* 130: 63–72.

Miers, H. (2008). *Controlling Avian Flu and Protecting People's Livelihoods in the Mekong Region Poverty, Livelihoods and HPAI*. HPAI Research Brief No. 9. London: DFID.

OIE (2019). *The OIE PPP Handbook: Guidelines for Public-private Partnerships in the Veterinary Domain*. Paris, France: World Organisation for Animal Health. https://rr-asia.woah.org/wp-content/uploads/2019/12/oie_ppp_handbook.pdf.

Okemer, B. I. (1919). *'An analysis of Farmers' and Traders Awareness, Perceptions and Effect of Children Value Chain Practices on Newcastle Disease Outbreaks in Kenya*. Nairobi: University of Nairobi. Department of Agricultural Economics.

Spradbrow, P. B. (1993–94). Newcastle disease in village chickens. *Poultry Science Reviews* 5: 57–96.

Tukahirwa, L., Mugisha, A., Kyewalabye, E., Nsibirano, R., Kabahango, P., Kusiimakwe ,D., Mugabi, K., Bikaako, W., Miller, B., Bagnol, B., Yawe, A., Stanley, M. and Amuguni, H. (2023). Women smallholder farmers' engagement in the vaccine chain in Sembabule District, Uganda: barriers and Opportunities. *Development in Practice* 33(4): 416–433. https://doi.org/10.1080/09614524.2022.2105817.

UNFPA (2020a). As pandemic rages, women and girls face intensified risks. https://www.unfpa.org/news/pandemic-rages-women-and-girls-face-intensified-risks.

UNFPA (2020b). Technical brief. Covid-19. A gender lens. Protecting sexual and reproductive health and rights, and promoting gender equality. https://www.unfpa

.org/sites/default/files/resource-pdf/COVID-19_A_Gender_Lens_Guidance_Note.pdf.

Velasco, E., Dieleman, E. and Phuong, T. T. M. (2008). *Avian Influenza External Response Coordination. Study on the Gender Aspects of the Avian Influenza Crisis in Southeast Asia*. Final Report. Herts: European Union & HTSPE.

Villanueva Cabezas, J. P. (2020). *Why Chickens Make a Major Contribution to Women's Empowerment*. Brisbane, Australia: Kyeema Foundation Blog. https://kyeemafoundation.org/womens-empowerment/.

Wenham, C., Smith, J. and Morgan, R. (2020). COVID-19: the gendered impacts of the outbreak. *Lancet* 395: 846–848. https://doi.org/10.1016/S0140-6736(20)30526-2.

Wong, J. T., Bagnol, B., Grieve, H., Jong, J. B., Li, M. and Alders, R. G. (2018). Factors influencing animal-source food consumption in Timor-Leste. *Food Security* 10: 741–762. https://doi.org/10.1007/s12571-018-0804-5.

Index

http://dx.doi.org/7

www.ingramcontent.com/pod-product-compliance
Ingram Content Group UK Ltd.
Pitfield, Milton Keynes, MK11 3LW, UK
UKHW020834131025
8358UKWH00003B/110

9 781801 468053